# Vedic Cosmology

***Brahmāṇḍa Sākṣāt Ṭīkā***
**ब्रह्माण्ड साक्षात्टीका**


*His Holiness*
**Danavir Goswāmī**



Kansas City * Rupanuga Vedic College * Missouri
रूपानुग विद्यापीठ

**ON THE COVER:**

[*Śrī Gaṇeśa* said] Just see these other Vaikuṇṭha companions of the Lord, traveling in this small universe ruled by four-headed Brahmā. And further away those others, moving swiftly in the universe of an eight-headed Brahmā, a world twice as large. And those others in the world of a Brahmā with sixteen heads, a world twice as large again.

*COMMENTARY BY ŚRĪLA SANĀTANA GOSVĀMĪ: To show that these visitors from Vaikuṇṭha were indeed extraordinary, Śrī Gaṇeśa pointed out more of them entering the other side of the universe, a great distance away, on some other mission. And beyond this relatively small universe of four-headed Brahmā, still other Vaikuṇṭha messengers were visiting the world of an eight-headed Brahmā, which was one billion yojanas in diameter, twice as big as this one. They were traveling swiftly (vegataḥ) because that universe is so large. Gaṇeśa further showed the same thing happening in universes still larger.* (Śrī Bṛhad-bhāgavatāmṛta, Part Two; 3:86-87)

Front cover illustration: Tracy Sabin

---

Readers interested in the subject matter are invited to contact the author at:

Danavir Goswami
Rupanuga Vedic College
5201 Paseo
Kansas City, MO 64110
Tel. (816) 924-5619
E-mail: info@rvc.edu
Web: www.rvc.edu

*First Edition*
First printing:2003


Published and bound in India by Gopsons Papers Ltd., Noida 201 301

Dedicated to

My Beloved Spiritual Master

**His Divine Grace A.C. Bhaktivedanta Swami Prabhupāda**
Founder-ācārya of the International Society for Krishna Consciousness

# Contents

**PLATE 1:** ***Devotee***: *Next verse, it describes the different divisions. "Just north of Ilāvṛta-varṣa and going further northward, one after another, are three mountains, namely Nīla Mountain, Śveta Mountain, and Śṛṅgavān Mountain. These mark the borders of the three varṣas, namely Ramyaka, Hiraṇmaya, and Kuru, and separate them from one another. The width of these mountains is..."*
***Prabhupāda:*** *And it was not possible for me to digest. (laughs) Somebody else helped me to... I am a layman. I do not know.*
***Tamāla Kṛṣṇa:*** *How did you write it?*
***Prabhupāda***: *That somebody, Kṛṣṇa, helped me. That He manufactured.*
***Devotee***: *And these mountains, they extend to the beaches. "It is considered, according to the Bhāga..."*
***Prabhupāda***: *When I was writing, I was praying Kṛṣṇa that "I do not actually accommodate all this knowledge. Please help me." Yes. That's all right.* (Room Conversation — June 18, 1977, Vṛndāvana)

Kindly do not overlook Śrīla Prabhupāda's statement, "*That somebody, Kṛṣṇa, helped me. That He manufactured.*" If one is eager to point out Śrīla Prabhupāda's admittance of his disqualification as a cosmologist, then he must be equally enthusiastic to recognize the qualifications of the astronomer who manufactured the Fifth Canto purports— and that is Krishna. [Page 96]

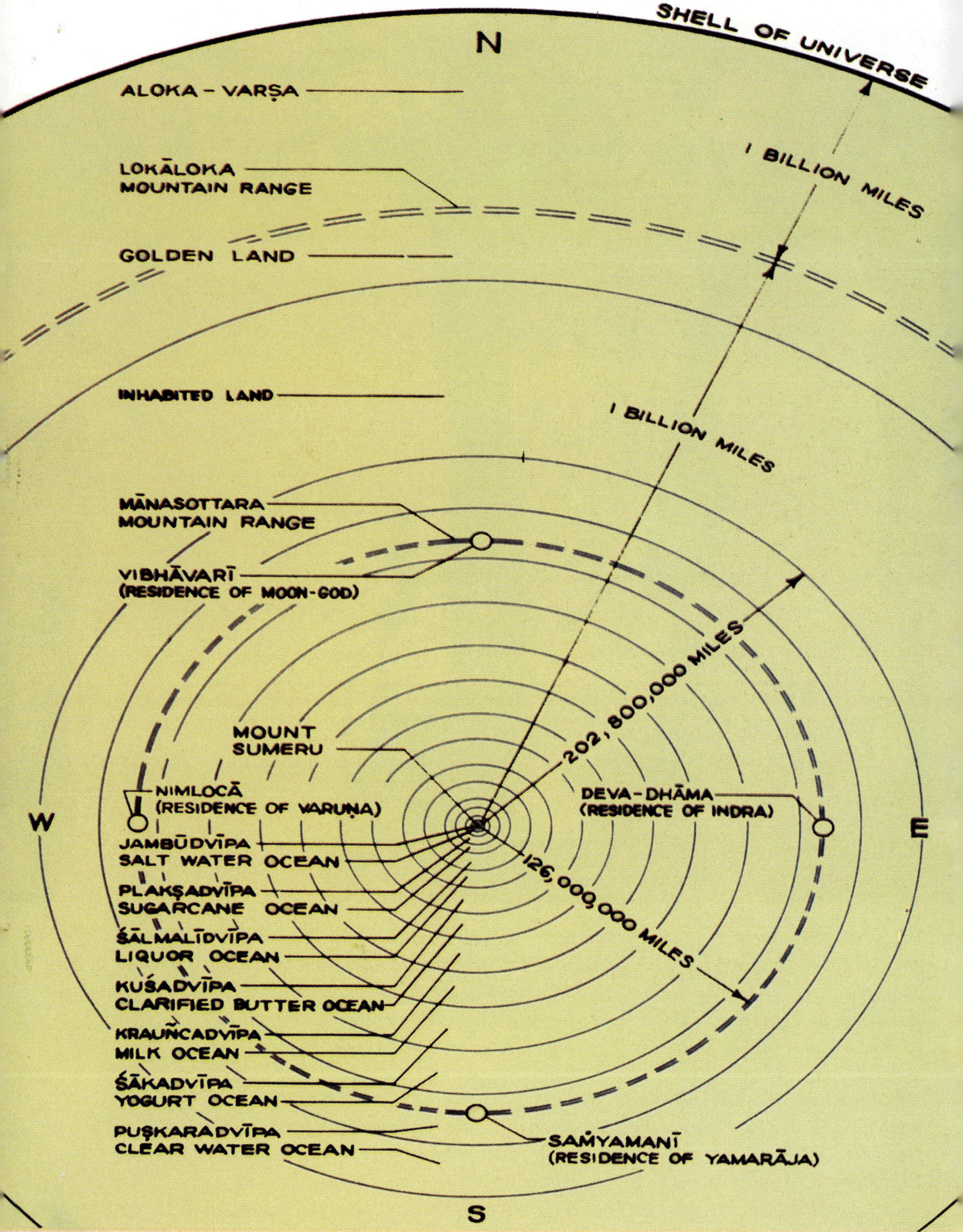


**Plate 2:** The original team of devotees to research the Bhū-maṇḍala under Śrīla Prabhupāda's direct supervision constructed the design of the universe with the Lokāloka Mountain surrounding the Bhū-maṇḍala and rising up like the petals of a lotus flower. They took the Bhāgavatam literally which is the way it was intended thus they envisioned the lotus of Bhū-maṇḍala. [Page 107]

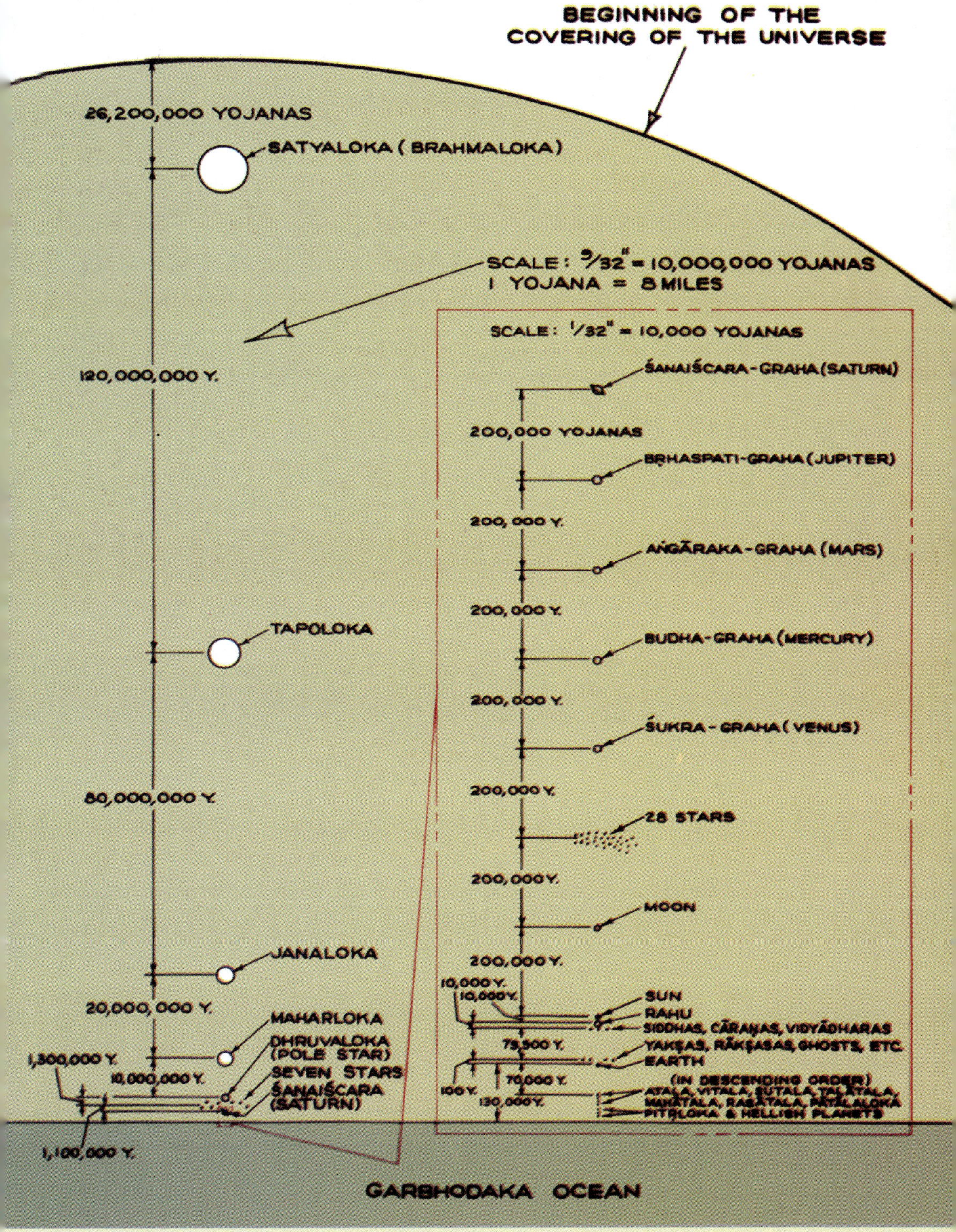


**Plate 3:** This scale drawing shows a basic cross section of the universe in which we live. For simplicity, the planets are represented in a straight line, one above another, although the Bhāgavatam describes that the sun, the moon and other planets are actually revolving around the polestar in their own orbits and at various speeds. [See Figure 2 on Page 135]

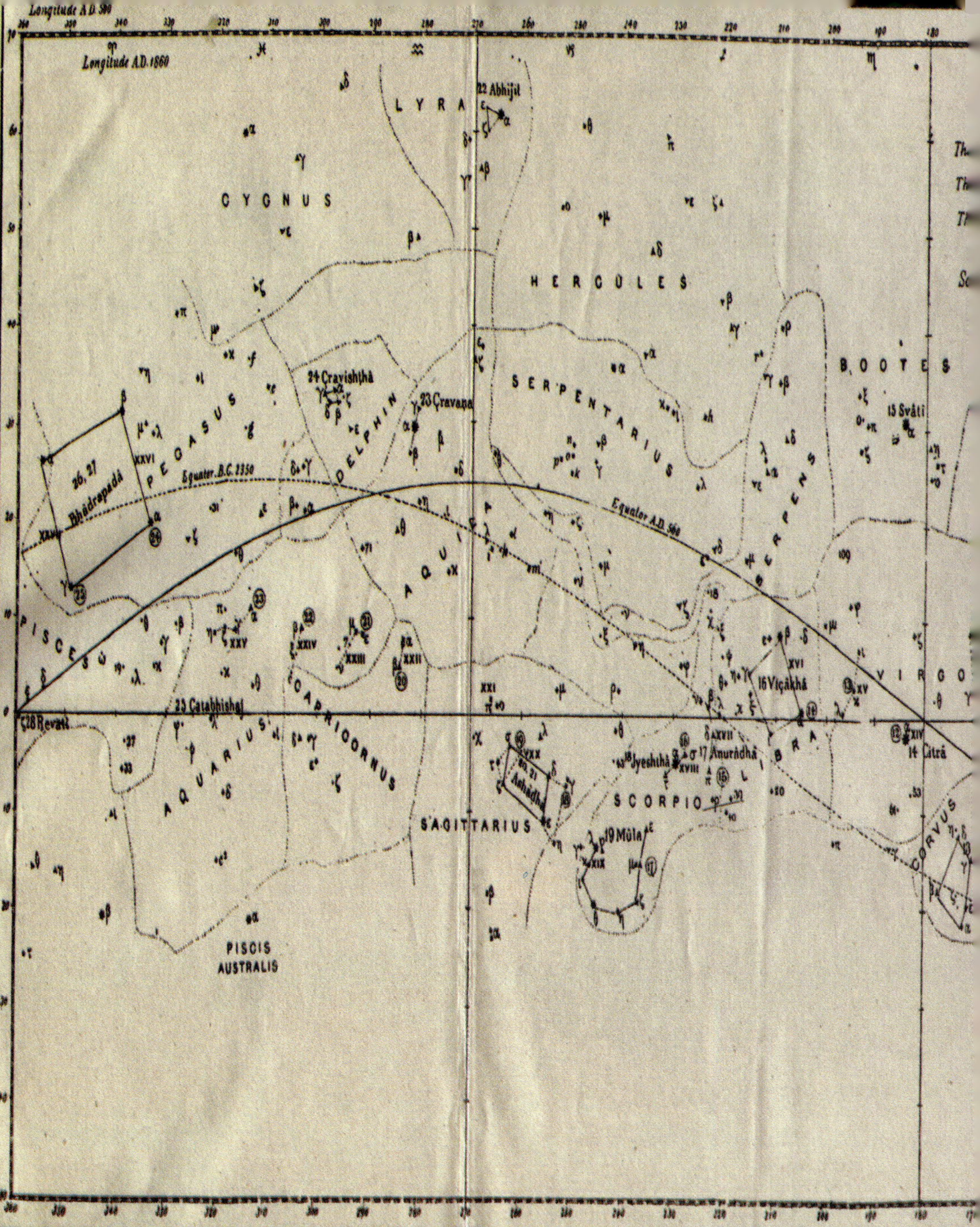


**Plate 4:** In another translation of the Sūrya-siddhānta by Phanindralal Gangooly there is a map upon one map with Hindu, Arab and Chinese names. It can be noted how the cartographer has of different names. In the area named *Andromeda* we find Vedic stars like Asvini, Revati and *Andromeda* which are today considered galaxies lying hundreds of light-years away from us, were theory of galaxies and quasars supposedly occupying distances of thousands and millions of light-accuracy. [Page 68]

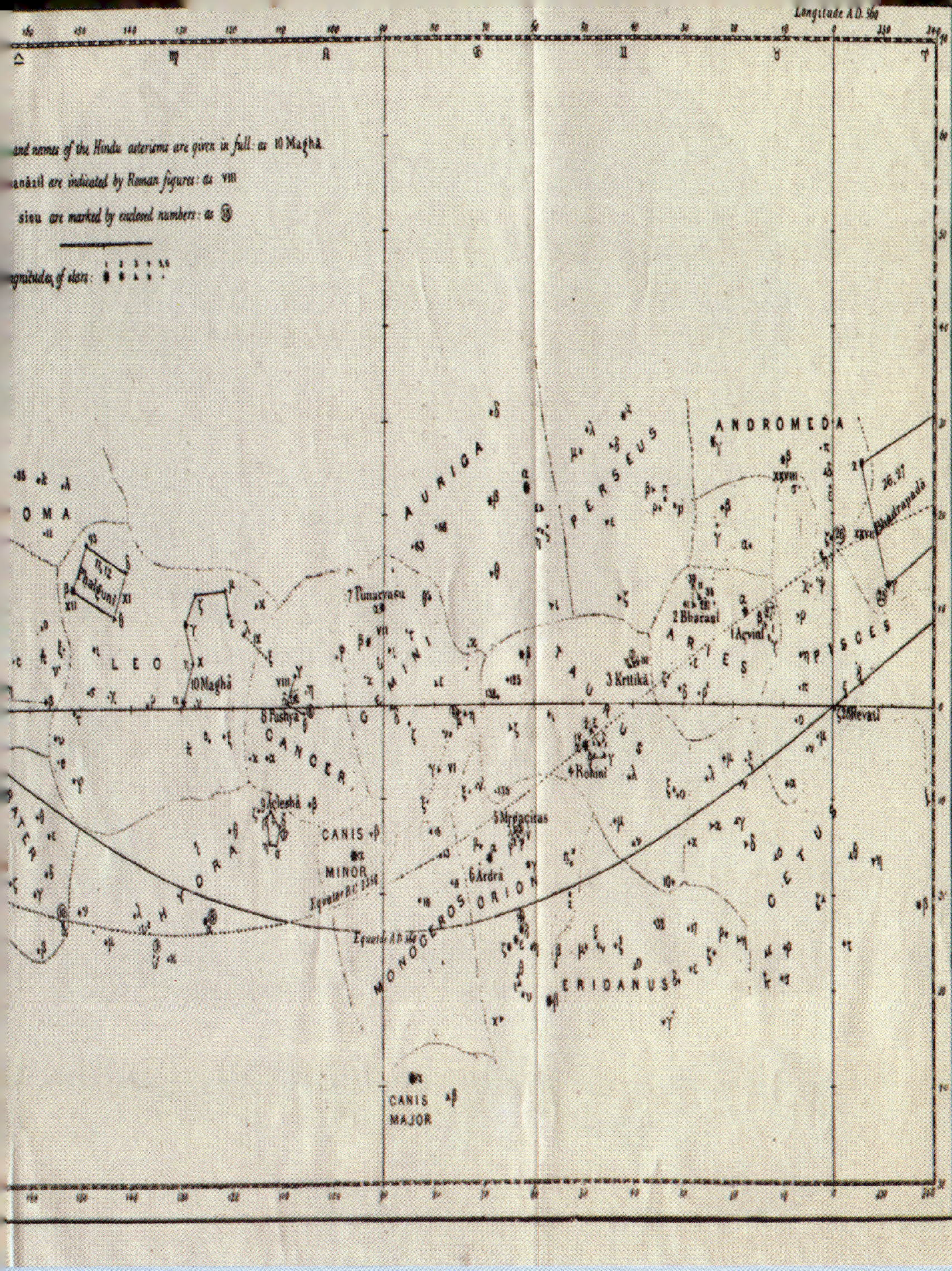


of space and stars containing Arab, Chinese and Hindu conceptions of the sky, all superimposed demonstrated that different cultures describe the same planets, asterisms and constellations by use Bharani within our universe, with no galaxies. Interestingly, on the map printed in 1860 *Orion* and formerly accepted by all astronomers as part of the local sky. It is our contention that the modern years from earth are recent exaggerated fantasies similar to Star Wars, and have no verifiable

**Plate 5:** Śrī Gaṇeśa said; Just see these other Vaikuṇṭha companions of the Lord, traveling in this small universe ruled by four-headed Brahmā. And further away those others, moving swiftly in the universe of an eight-headed Brahmā, a world twice as large [one billion yojanas in diameter]. And those others in the world of a Brahmā with sixteen heads, a world twice as large again. [Page 138]

**Plate 6:** "Kṛṣṇa said [to Brahmā], 'Your particular universe has a diameter of four billion miles; therefore it is the smallest of all the universes. Consequently you have only four heads." [Page 145]

Śrī Maitreya said: "This phenomenal material world is expanded to a diameter of four billion miles, as a combination of eight material elements transformed into sixteen further categories, within and without."

Brahmā said: "My dear son Nārada, know from me that there are seven lower planetary systems out of the total fourteen."

***All these planetary systems are scattered over the complete universe, which occupies an area of two billion times two billion square miles.

Śrī Caitanya Mahāprabhu said: "Śrī Kṛṣṇa smiled and immediately meditated. Unlimited Brahmās arrived instantly. These Brahmās had different numbers of heads. Some had ten heads, some twenty, some a hundred, some a thousand, some ten thousand, some a hundred thousand, some ten million and others a hundred million. No one can count the number of faces they had. There also arrived many Lord Śivas with various heads numbering one hundred thousand and ten million. Many Indras also arrived, and they had hundreds

Śrīla Śukadeva Gosvāmī said: "Learned scholars who are free from mistakes, illusions and propensities to cheat have thus described the planetary systems and their particular symptoms, measurements and locations. With great deliberation, they have established the truth that the distance between Sumeru and the mountain known as Lokāloka is one fourth of the diameter of the universe—or, in other words, 125,000,000 yojanas [1 billion miles]."

# Introduction

The first eight chapters of *Vedic Cosmology* primarily take the form of a debate. On one side are the "Points" representing many of today's ostensibly Vedic-scientific theories. The response comes from our humble self in the form of *"Brahmāṇḍa Sākṣāt Ṭīkā"* or "BST" for short, meaning "Vedic Cosmology taken literally." It should be mentioned that the discussions herein are not merely exercises in theoretical argumentation. Indeed the views expressed in "Points" do weigh heavily at present despite the fact that they are influenced by modern scientific theories which contradict Vedic literature. Thus these crucial issues and their conclusions will determine the degree to which we have understood His Divine Grace A.C. Bhaktivedanta Swami Prabhupāda's instructions on Vedic cosmology and the means of receiving transcendental knowledge.

The "Notes" sections at the end of the earlier chapters provide further evidence via BST as reference information. Throughout the book whenever the terms "Prabhupāda" or "Śrīla Prabhupāda" are seen they refer to His Divine Grace A.C. Bhaktivedanta Swami Prabhupāda, the founder-ācārya of the International Society for Krishna Consciousness (ISKCON). He is also the translator and commentator of Śrīmad-Bhāgavatam (45 volumes), Bhagavad-gītā, Caitanya-caritāmṛta (17 volumes) and dozens of other Vedic translations.

Chapters Nine and Ten provide Śrīla Prabhupāda's unedited words on the subject of building a Vedic Planetarium in Mayapur along with an analysis of the planetarium's correlative constituent— the Bhū-Maṇḍala planetary system. Since this cherished desire of His Divine Grace remains yet unfulfilled, ISKCON devotees worldwide share its legacy as His Holiness Bhaktisvarupa Damodara Swami nicely expresses it:

> *Your concern is very significant because any misconception identified at the later stages of the project would be unpleasant and extremely costly. I do feel there should be a committee to invite and review the proposals for the model of the Vedic Planetarium. You have a very good and valid point and I would like to look into the matter.*

Chapter Eleven comprises all the cosmological topics contained within the Fifth Canto of the Srimad Bhagavatam, both its texts and Śrīla Prabhupāda's purports.

Chapters Twelve, Thirteen and Fourteen comprise the available English translated texts by Western scholars of three famous Jyoti Śāstras namely Sūrya-Siddhānta, Siddhānta-Śiromaṇi, and The Āryabhaṭīya. Although Śrīla Prabhupāda did not recommend that these be consulted, they are nonetheless being used by some of today's Vedically-oriented writings, even antithetically to the direct meaning of the Śrīmad-Bhāgavatam's statements. Thus these nineteenth century texts are included as a reference to prevent their misuse.

*ACKNOWLEDGEMENTS*

Since the beginning of this project seven years ago, Vijnana dasa has been a constant comrade in discussions. Dhira Govinda dasa and his wife Mahalaksmi devi dasi offered valuable suggestions at the outset. Gopiparanadhana Prabhu graciously helped by providing several relevant translations to Sanskrit texts. Similarly, Vamadeva dāsa Brahmacari and Advaita dāsa Brahmacari offered their Bengali translation skills. We thank, as always, the Bhaktivedanta Book Trust International and Bhaktivedanta Book Trust Archives for use of their literary publications and artwork.

However, Arya Siddhanta dasa Brahmacari emerged in the final year as the prime assistant playing a pivotal role in both research and publication. His enthusiasm spread to his brother (Russell), mother (Helen Sabin) and father (Tracy Sabin) who provided graphics advice, contributions and cover artwork respectively. I also wish to thank Umapati Swami for allowing us to use a statement of his. Thanks to Bonnie Parsons for layout help, Bhakta Jesse, Bhaktin Jenny Ledeen for editorial assistance and many others who have not been mentioned. Finally we'd like to thank Radha Jivan dasa for kindly providing a loan for meeting the publishing outlay.

I sincerely hope that *Vedic Cosmology* will assist in supplying a composite cosmological guide to understanding the direct explanations of the Śrīmad-Bhāgavatam as per the sacred desire of Śrīla Prabhupāda.

Danavir Goswami

# 1

# The Universe

## Point #1—"Solar System"

I argue that the Śrīmad Bhāgavatam is giving an excellent description of the solar system, in the spirit of the ancient and medieval systems. But like them, it is not describing the universe of stars and galaxies.

## Brahmāṇḍa Sākṣāt Ṭīkā (BST)#1—

To my mind, use of the expression "the universe of stars and galaxies," propounds a new theory (i.e. "Bhāgavatam's Incomplete Universe Theory") where the universe described in the Śrīmad Bhāgavatam is incomplete. This expression is a polite way of asserting that either the words attributed to Śrīla Śukadeva Goswāmī, Śrīla Vyasadeva, Śrī Caitanya Mahāprabhu and Śrīla Prabhupāda, etc. have been faultily received/understood or that those personalities are Themselves mistaken, or both. Actually the star and galaxy universe is the one mistakenly imagined by modern so-called scientists postulating that the universe is comprised of stars (suns) and galaxies with diameters of hundreds of thousands and millions of light-years. By not rejecting this erroneous theory, we depart from the Vaiṣṇava śāstras and ācāryas. The current astronomical theory is diametrically opposed to the Śrīmad Bhāgavatam's version that our universe is only four billion miles in diameter.

## Point #2—"Bhāgavatam Details"

I propose that the Śrīmad Bhāgavatam was never intended by its author to present a complete description of the material world (what to speak of the spiritual world). There are many topics not touched on by the Bhāgavatam. For example, it does not mention auto mechanics, the geography of North America, the works of Plato, quantum mechanics, DNA, etc., etc. Out of 8,400,000 species, it mentions only a few and does not give a detailed account of zoology or botany. Is this a new theory (i.e. the "Bhāgavatam's Incomplete Account of the Material World Theory"), or is it simply a fact? I would suggest that Vyāsadeva's motive in writing the Bhāgavatam was not to compile a universal encyclopedia.

## BST #2—

Agreed that all details are not given, however whatever details are actually presented in the Śrīmad Bhāgavatam must be taken as true and perfect otherwise the importance of the Bhāgavatam is compromised. For example 8,400,000 species. We cannot conclude this to be a mere estimation.

> *Species of life. This is Vedic knowledge. Exactly giving the number. Exactly. 8,400,000. It doesn't say 9,000,000, or 500,000. No. Exact. Jalajā nava-lakṣāṇi. In the aquatics, there are 900,000 species of life. Now, if you have got your science, now go and see. This is Vedic knowledge. Perfect, without any doubt. Sthāvarā lakṣa-viṁśati. And the plants and trees, they are 2,000,000 species of life.* (Lecture: Śrīmad-Bhāgavatam 2.3.9 — Los Angeles, May 26, 1972)

The Bhāgavatam does not describe in detail every specie within the universe. That is a fact and is not a new theory i.e. the "Bhāgavatam's Incomplete Account of the Material World Theory." However, the "Bhāgavatam's Incomplete Universe Theory" works differently. It does not merely state the obvious but interprets, alters, minimizes or dissents from the Bhāgavatam's statement by presenting a view which differs from what is stated in the Bhāgavatam. For example "Bhāgavatam's Incomplete Universe Theory" might put forward a theory like this:

> "Based on very recent extensive botanical research in Africa and the subsequent discovery of many heretofore unknown species, it must now

> be concluded that the 8,400,000 number of species described in the Vedas can only be attributed to the six other major continents. We are confident that Śrīla Vyāsadeva did not have the advantage of the most modern technology which has aided in the identification of millions, nay billions of newly-discovered species (especially in the germ category.) Most probably he was unaware of all the germs present in Africa because the dark continent might not have been well-travelled or examined during his time period. Therefore we now conclude that there exists not simply 8.4 million species but in fact 20 billion species."

Specifically Bhāgavatam describes the universe as 4 billion miles in diameter and the Bhāgavatam's Incomplete Universe Theory interprets that statement to be incomplete at best or inaccurate at worst, or both. Polite disagreement is still dissent. I find no evidence to indicate that the Bhāgavatam is describing only a portion of the universe in which we dwell.

The job with which Vedic scientists are faced is to utilize their knowledge of the material world to prove to the materially-affected persons that what is stated in the śāstras is perfect and unassailable, even by material calculations.

A devotee should be satisfied with the knowledge and parameters given to him by the spiritual authorities. Lord Caitanya's spiritual master told him not to study Vedanta and He accepted that apparent limitation to great success. There is so much available in the Bhāgavatam's cosmology, more than we can accommodate.

## Point #3—"Celestial Bhāgavatam"

We do not possess the full text of the Bhāgavatam. It is said that on higher planets the Bhāgavatam is much longer than the earthly Bhāgavatam. Clearly, the earthly Bhāgavatam is incomplete in relation to its celestial counterpart.

## BST #3—

We would prefer to say that the Śrīmad Bhāgavatam we have access to on this planet is perfect, transcendental and complete and the Śrīmad Bhāgavatam available to the higher planets is more perfect, transcendental and complete.

## Point #4—"Śrīla Prabhupāda's Dissatisfaction"

Śrīla Prabhupāda himself said in plain words that he was not fully satisfied with his own understanding of the Fifth Canto. Consider the following, from Bhaktivedanta Vedabase:

> ***Prabhupāda:*** *So you can... Why not begin it in my presence?*
> ***Gopāla Kṛṣṇa****: Begin the drawing now so that Prabhupāda can see it before he* leaves.
> ***Indian Astronomer:*** *No, because I am not able to prepare the diagrams and other things. With the help of my friends only I can do. I give the translation to them and all do the work. Simply we can show to them.*
> ***Prabhupāda:*** *I have tried to translate it as far as possible, but I am not satisfied.*
> ***Indian Astronomer:*** *A deep study of the text is absolutely necessary.*
> ***Prabhupāda:*** *Hm?*

This is from a conversation in April 30, 1977, in Bombay. Śrīla Prabhupāda was interviewing an Indian scholar in an effort to find someone who could properly explain the Fifth Canto, and he pointed out his dissatisfaction with his own translation. Unfortunately, the scholar could not help.

So in view of Śrīla Prabhupāda's position, are you sure that you properly understand the Fifth Canto?

## BST #4—

I would not consider Śrīla Prabhupāda's statement "I am not satisfied" to be taken that he was dissatisfied with his own translation. Śrīla Prabhupāda did not wish for the astronomer to adjust his translation of the Fifth Canto nor to retranslate the Fifth Canto. Rather His Divine Grace wanted the astronomer to assist in giving the translation shape by producing a graphic diagram of the universe. Therefore that dissatisfaction would be Śrīla Prabhupāda's longing for an exact working diagram to pattern the planetarium. Consider the following from the same conversation you quoted:

> *Prabhupāda: Ūrdhva-mūlam adhaḥ-śākham. Same thing is explained in the Śrīmad-Bhāgavatam. Now, how it is hanging and where are different*

*situation of the planets and plateaus and hills and..., so as far as the book is concerned, kindly make a diagram so that we shall execute... Immediately we want the diagram how to fix it up so that people can see, "This is the situation." So you make this diagram.* (Conversation with Vedic Astronomer — April 30, 1977, Bombay)

Personally I cannot claim to understand the Fifth Canto completely but I can definitely understand from the Fifth Canto that the diameter of our universe is 4 billion miles since the text says clearly:

*(1st)*

*āṇḍakośo bahir ayaṁ*
*pañcāśat-koṭi-vistṛtaḥ*

*SYNONYMS: vikāraiḥ—by the transformation of the elements; sahitaḥ—along with; yuktaiḥ—being so amalgamated; viśeṣa—manifestations; ādibhiḥ—by them; āvṛtaḥ—covered; āṇḍa-kośaḥ—the universe; bahiḥ—outside; ayam—this; pañcāśat—fifty; koṭi—ten million; vistṛtaḥ—widespread.*

*TRANSLATION: This phenomenal material world is expanded to a diameter of four billion miles,...* (Śrīmad Bhāgavatam 3.11.40 Śrī Maitreya to Vidura)

The statement above made by Śrī Maitreya Muni is straightforward and exact.

*(2nd)*

*etāvāû loka-vinyāso māna-lakṣaṇa-saṁsthābhir*
*vicintitaḥ ka vibhiḥ sa tu*
*pañcāśat-koṭi-gaṇitasya bhū-golasya*
*turīya-bhāgo 'yaṁ lokālokācalaḥ.*

*SYNONYMS: etāvān—this much; loka-vinyāsaḥ—the placing of the different planets; māna—with measurements; lakṣaṇa—symptoms; saṁsthābhiḥ—as well as with their different situations; vicintitaḥ—established by scientific calculations; kavibhiḥ—by learned scholars; saḥ—that; tu—but; pañcāśat-koṭi—500,000,000 yojanas; gaṇitasya—*

*which is measured at; bhū-golasya—of the planetary system known as Bhūgolaka; turīya-bhāgaḥ—one fourth; ayam—this; lokāloka-acalaḥ—the mountain known as Lokāloka.*

*TRANSLATION: Learned scholars who are free from mistakes, illusions and propensities to cheat have thus described the planetary systems and their particular symptoms, measurements and locations. With great deliberation, they have established the truth that the distance between Sumeru and the mountain known as Lokāloka is one fourth of the diameter of the universe—or, in other words, 125,000,000 yojanas [1 billion miles].* (Śrīmad Bhāgavatam 5.20.38; Śrīla Śukadeva Gosvāmī to Mahārāja Parīkṣit)

The diameter of the universe is plainly stated by Śrīla Śukadeva Goswāmī to be *pañcāśat-koṭi (500,000,000 yojanas* or four billion miles.) Since there is agreement between the ācāryas what is the need for speculation on this subject?

## Point #5—"A Radical Position"

To say that everything outside a four-billion-mile diameter shell is imaginary is indeed a radical position. It is practically indefensible within modern society. Unlike Darwin's theory, which has numerous scientific flaws (as admitted recently by many serious scientists), the "Universe Within Four Billion Miles Theory" cannot be supported by scientific evidence or arguments. If you doubt this, please consult with some scientists and see.

## BST #5—

For certain there was no one more radical than Śrīla Prabhupāda when it came to opposing modern scientific theories. He tested them in the laboratory of the Bhāgavatam and then gave his pronouncement. Of course we are not saying that "everything" outside the four-billion-mile diameter shell is imaginary, there are layers, then trillions of other universes, then paravyoma, etc. We do not have a narrow vision of cosmology, it is far vaster that the modern scientists can imagine even using their millions of light year calculations. We do say that the estimates given for the distances of planets within our visible sky by astronomers are inaccurate because they do not tally with the Bhāgavatam. Otherwise, if we do not accept fully the

Bhāgavatam, what does it mean to see through the eyes of the śāstra (śāstra caksusa)?

We should not be intimidated by the scientists.

> *For a common man, both modern science and Vedic wisdom are simply to be accepted because none of the statements either of modern science or of Vedic literature can be verified by him. The alternative for a common man is to believe either of them or both of them. The Vedic way of understanding, however, is more authentic because it has been accepted by the acaryas, who are not only faithful and learned men, but are also liberated souls without any of the flaws of conditioned souls. The modern scientists, however, are conditioned souls liable to so many errors and mistakes; therefore the safe side is to accept the authentic version of Vedic literatures, like Śrīmad-Bhāgavatam, which is accepted unanimously by the great acaryas.* (Śrīmad Bhāgavatam 2.2.26 Purport)

It is not scientific for astronomers to theorize about a universe to which they have such little access. Neither is it wise to assign unverifiable distances to objects seen in the sky through telescopes and insist this is the only feasible explanation. There are so many flaws in that emperical approach where observation and experiment are lacking, what to speak of the total absense of proof.

1) deficient material telescopes
2) our proneness to making mistakes
3) using one's imperfect eyes
4) being subject to illusion
5) using limited brains
6) using restricted intelligence
7) tendency to cheat

> *This vivid description of how the rays of the sun are distributed throughout the different planetary systems of the universe is very scientific. Śukadeva Gosvāmī described these universal affairs to Mahārāja Parīkṣit as he had heard about them from his predecessor. He explained these facts five thousand years ago, but the knowledge existed long, long before because Śukadeva Gosvāmī received it through disciplic succession. Because this knowledge is accepted through the disciplic succession, it is perfect. The history of modern scientific knowledge, on the contrary, does not go back*

> *more than a few hundred years. Therefore, even if modern scientists do not accept the other factual presentations of Śrīmad-Bhāgavatam, how can they deny the perfect astronomical calculations that existed long before they could imagine such things? There is so much information to gather from Śrīmad-Bhāgavatam. Modern scientists, however, have no information of other planetary systems and, indeed, are hardly conversant with the planet on which we are now living.* (Śrīmad Bhāgavatam 5.20.37 Purport)

Attempting to discover universal truths by telescope, mathematics and imaginativeness is defective from the outset. Vaiṣṇavas are not supposed to be concerned with the current theories about the size of the universe presented by the cavalcade of dying so-called scientists.

> *As far as we are concerned, we follow the authority of Vyāsadeva and Śukadeva Gosvāmī, who have described the universal situation according to the Vedic literature. These authorities differ from modern scientists who conclude from their imperfect sensual experience that only this planet is inhabited by living beings whereas the other planets are all vacant or full of dust.* (Śrīmad Bhāgavatam 7.4.17 Purport)

## Point #6 — "Bhū-maṇḍala as Literary Device?!"

If the Bhāgavatam admitted no other reasonable interpretation, then Vaiṣṇavas would, indeed, be forced either to uphold the Four Billion Miles theory or renounce their faith in Vaiṣṇava teachings. However, the Fifth Canto actually demands another interpretation.

Consider Bhū-maṇḍala, which means "earth disk." This structure is described in the Fifth Canto as a flat plate, four billion miles in diameter, marked by various circular features. Is this meant by Śrīla Vyāsadeva to be a physically real object? If we look closely at the text, we see that the answer is no. Consider the following:

"King Parīkṣit inquired from Śukadeva Gosvāmī: My dear lord, you have already affirmed the truth that the supremely powerful sun-god travels around Dhruvaloka with both Dhruvaloka and Mount Sumeru on his right. Yet at the same time the sun-god faces the signs of the zodiac and keeps Sumeru and Dhruvaloka on his left. How can we reasonably accept that the sun-god proceeds with Sumeru and Dhruvaloka on both his left and right simultaneously?"

Śukadeva Goswāmī responds,

> *"When a potter's wheel is moving and small ants located on that big wheel are moving with it, one can see that their motion is different from that of the wheel because they appear sometimes on one part of the wheel and sometimes on another. Similarly, the signs and constellations, with Sumeru and Dhruvaloka on their right, move with the wheel of time, and the antlike sun and other planets move with them. The sun and planets, however, are seen in different signs and constellations at different times. This indicates that their motion is different from that of the zodiac and the wheel of time itself."*

Śukadeva Goswāmī is using the example of ants on a potter's wheel to illustrate the idea of relative motion. However, in the earlier verses alluded to by King Parīkṣit, both the clockwise (with Meru on the right) and the counterclockwise (with Meru on the left) motions of the sun were referring to the same reference frame, namely *Bhū-maṇḍala*. This means that Bhū-maṇḍala is being used for two different purposes. It is the frame of reference for the daily clockwise motion of the sun, and it is the frame of reference for the yearly counterclockwise motion of the sun.

If we carefully study the Fifth Canto, we see that many verses support each of these two uses of Bhū-maṇḍala. Thus Bhū-maṇḍala is not one physical thing. The author of the Bhāgavatam is using it as a literary device to explain at least two different astronomical points. Thus the message of the Fifth Canto lies in the astronomy being explained by Śukadeva Goswāmī. It does not lie in a contradictory and impossible theory of the earth as a disk four-billion miles in diameter.

## BST #6—

There are two types of motion of the sun, namely the gigantic motion in which the sun moves with the constellations around Dhruvaloka and Sumeru and secondly the individual orbit of the sun on his chariot which also moves around Dhruvaloka and Sumeru in the opposite direction. This is *not* a "literary device;" it is a clear explanation of the physical phenomenon which clarifies the apparent contradiction which Parīkṣit Mahārāja raised.

Śrīla Prabhupāda further elaborates on this same subject:

> *We can see that at night, how the whole planetary system is turning around, the pole star being the pivot. Each planet has its orbit fixed but the sun is moving up and down, north and south. It is not that we shall accept the theory that the sun is fixed up and the others are all going around the sun. That is not correct. Even the 7th star is rotating once around the pole star in each 24 hours. The whole universe is just like a big tree, that is a fact. I do not think that the modern astronomers have any such idea that the whole universe is like a big tree. The planets which are full of living entities are one after another, one above the other. The relative positions of the planets is fixed up but the whole thing is turning. The sun is going north and south, it has its own orbit below the moon.* (Letter to: Svarupa Damodara — Auckland 27 April, 1976)

We must not imagine a non-physical literary device which is neither stated nor inferred by the speaker.

The fact is however that according to Vedic literature, "the universe of stars and galaxies" as conceived by modern astronomers is not to be found within our universe. Śrīla Śukadeva Goswāmī is not fantasizing about cosmology. He has accurately described the entire universe with its fourteen planetary systems, from Brahmaloka down to Pātāla and the hellish worlds. The Bhāgavatam provides lucid depiction and precise dimensions of the sun, its orbit, the polestar, the Garbhodaka Ocean, the universal shell with its coverings, everything. There is absolutely no reason to believe that he speaks of only a fragment of the universe.

## Point #7— "Astronomical Contridictions"

> *"The great Śrī Śukadeva Gosvāmī said: My dear King, there is no limit to the expansion of the Supreme Personality of Godhead's material energy. This material world is a transformation of the material qualities [sattva-guṇa, rajo-guṇa and tamo-guṇa], yet no one could possibly explain it perfectly, even in a lifetime as long as that of Brahmā. No one in the material world is perfect, and an imperfect person could not describe this material universe accurately, even after continued speculation. O King, I shall nevertheless try to explain to you the principal regions, such*

> *as Bhū-goloka [Bhūloka], with their names, forms, measurements and various symptoms."*

So a lifetime of Brahmā would not be enough to describe the universe fully. Is it not possible then, that modern astronomers may have discovered some things that Śukadeva Goswāmī left out of his account?

Of course, this is not to say that the astronomers are perfect in everything they say. But even an imperfect person may know something that another, superior person does not discuss.

## BST #7—

If what the scientists postulate is in accord with the Bhāgavatam, then fine, it can be accepted as supporting evidence. If postulations are not in accord with the Bhāgavatam, the postulations must be adjusted, not the Bhāgavatam's explanation.

There are inconsistencies in your presentation. First there is an errant interpretation of the true meaning of the verse by suggesting that there is no need to accept the version of Śrīla Śukadeva Goswāmī about the size of the universe since the subject matter is beyond comprehension. Then in the next breath, you suggest that modern science has now conceived of the universe— albeit with its inferior methods of empiric hypothesis.

> *Prabhupāda: So we are imperfect, just like they are going to study the planets. What they will study from here? This kind of knowledge has no value, speculation.* (Press Interview at Muthilal Rao's House — August 17, 1976, Hydcrabad)

Vedic literature is meant to be accepted as axiomatic, yet you prefer to accept some parts of the Śrīmad Bhāgavatam and reject others. Then with no logical reasoning you accept the modern scientific speculation as self-evidently true. The statements of the Śrīmad Bhāgavatam and the pure devotees like Śrīla Prabhupāda are brilliant and I request you to make your judgments exclusively on Their statements.

The concept of universes is more elaborately explained in the Vedic literature than that presented by modern froggish-speculators. Neither the modern astronomers, nor we, can have direct proof of the universe's planets and distances. All must accept some higher authority. We should not put our faith in deficient methods estimating planetary and cosmic distances. Whether they be mathematical, Red Shift, light wave-bouncing or whatever, the modern so-called scientific methods are faulty and completely unproven. No astronomer has journeyed to another galaxy or star. Why then shall the followers of His Divine Grace A. C. Bhaktivedanta Swami Prabhupāda abandon the perfect and fearless information received via the ripened fruit of all Vedic literature in favor of unsubstantiated guesses?

## Point #8—"Modern Astronomy"

Regarding modern astronomy, you should study it carefully before deriding it. At least you should give a reasonable explanation of how astronomers mistakenly thought that there are heavenly bodies further away than two billion miles. It will also be helpful if you can explain why the earth was mistakenly thought to be a sphere rather than a flat plate. If you preach on these matters in universities, you will be confronted with such questions.

## BST #8—

I have heard from my spiritual master who has concluded that astronomers are worth being derided when they disagree with the Vedic conclusion.

> *Now suppose personally, myself, we are challenging, we are calling them by names, "rascal," but I am not a scientist. I never studied astronomy, astrology, or anything. But why I am telling? What power I have got? But I am challenging on the words of a superior answer. I am confident that the words spoken by Vyāsadeva or Kṛṣṇa, that is perfect. Therefore they are rascals, because their statement do not corroborate. In other words, we can understand. That is our advantage. Bhakti-yogena. If you practice bhakti-yoga, yasya deve parā bhaktir yathā deve tathā gurau, if one is, one has unflinching faith in the Supreme Personality of Godhead and his spiritual master, yathā-deve... Deve means the Supreme Lord, and guru, and guru also, the spiritual master. A person who has got unflinching faith in these two personalities—guru and Kṛṣṇa—then the facility will be that all the revealed scriptures will be manifest automatically, even though he is not, not very learned. The purport of the whole knowledge will be revealed from within, because Kṛṣṇa is within, and the spiritual master is without, so both of them will help. Our students are going so*

> *many places—in the schools, colleges, universities. They're surprised how they're speaking so nicely, because they have got unflinching faith in Kṛṣṇa and the spiritual master. They are not so learned. They have not gone to the college and learned so many things as the materialistic students do, but on this account: bhakti-yogena manasi. Vyāsadeva also did like that. He became so great because he has accepted bhakti-yoga by the order of his spiritual master, Nārada. Before that, he wrote so many books, Purāṇas and others. He was not satisfied. When Nārada said that "You cannot be satisfied by describing so many material things. You try to describe the glories of the Lord..." So after his instruction, bhakti-yogena, by practicing bhakti-yogena, manasī, unto the mind, praṇihite, mind became settled up; intelligence, right intelligence came, and everything became amala. Amala means without any material contamination.* (Room Conversation — New Delhi, December 11, 1971)

If by compromising the literal meaning of the scripture to achieve some left-handed appreciation from scientists we must sacrifice the integrity and the very gist of the teaching, i.e. complete faith in and fidelity to the words of the Bhāgavatam, then I feel we have too much to lose.

## Point #9—"Absurd Literalism"

I propose that the kind of careful reading that I advocated above represents complete faith in and fidelity to the words of the Bhāgavatam. One must consider what the text is really saying, rather than embracing a superficial literalism that reduces the text to an absurdity.

## BST #9—

The text and Śrīla Prabhupāda and Lord Krishna as quoted in the Caitanya-caritāmṛta all say clearly that the visible universe is four billion miles in diameter. That is not absurd nor superficial. Only because of conditioning by modern scientists' theories does one consider that four billion miles is too small to accommodate so many stars and planets. Vaiṣṇavas are not supposed to accept modern scientists as authorities in cosmic matters.

> *Bhakti-prema: So how we will expose it before scientists?*
> *Prabhupāda: We do not require to satisfy the scientists. We have to describe according to our book. That's all. If they can understand, let them understand. Otherwise... It is not our business to satisfy the so-called*

> *scientists. We are giving the real description.* (Room Conversation — June 18, 1977, Vrndavana)

Indications that the Bhāgavatam is incorrect or incomplete are a drastic deviation from parampara and should not be espoused in any media. Let us work together to rectify such misunderstandings regarding Vaiṣṇava cosmological siddhānta as quickly as possible.

Granted, Śrīla Prabhupāda did want to win over scientists— but not by waffling. If the devotees hear any more statements from ISKCON scientists discrediting the Śrīmad Bhāgavatam's version, it will seriously attack their own delicate faith in Vedic scriptures.

## Point #10—"Faith in the Physical"

Should their faith be based on a simplistic physical interpretation of Bhū-maṇḍala when the text of the Fifth Canto indicates otherwise?

## BST #10—

The text of the Fifth Canto does not indicate any other diameter for the universe than four billion miles. To assert that it does is the most ultra extremist interpretation.

Instead of accepting sastric statements at face value, unfortunately you have presented a new theory but at the end, we are left with the same modern view telling us that within this universe there are "many galaxies like the Milky Way distributed over millions of light years of outer space." Accepting the modern scientific notion that there are many suns within our universe is contradictory to the Śrīmad Bhāgavatam and our spiritual master.

The words of the spiritual master and the śāstras are to be taken literally without fanciful so-called scientific interpretation. It is not allowed to pick and choose from Vedic evidence and then manufacture a conclusion in conjunction with modern so-called scientific theories which is contrary to the Śrīmad Bhāgavatam. In order to accept modern astronomy's ideas about the construction of the universe, one must reject the simple and straightforward words which Śrīla Prabhupāda has so many times repeated...only one sun in the universe.

## Point #11—"One Sun"

Modern astronomers say that many other stars are like suns, but even they don't say that they are suns (except in a metaphorical sense). Even in modern English, the word "sun" refers to only one celestial body.

What does it mean to say that an object is "like the sun." If it means that the object gives substantial heat and light to the earth, then scientists will agree that there is only one sun in the universe. Perhaps it means that the object gives off its own heat and light. To be sure, the Purāṇas state that the stars shine with reflected sunlight. However, consider the following quote from the Mahābhārata:

> *"And on this sunlike, divine, wonder-working chariot the wise scion of Kuru flew joyously upward. While becoming invisible to mortals who walk on earth, he saw wondrous airbourne chariots by the thousands. No sun shone there, or moon, or fire, but they shone with a light of their own acquired by their merits. Those lights that are seen as the stars look tiny like oil flames because of the distance, but they are very large. The Pandava saw them bright and beautiful, burning on their own hearths with a fire of their own..."*

So instead of calling them suns, one could call them self-luminous hearths. Scientists could agree with this.

## BST #11—

Modern scientists say explicitly that there are many stars or suns within the sky we see at night. That erroneous theory, presented below, is what I am referring to.

> *Stars are suns. Although a large number of stars are like our nearest star, the Sun, many stars are much smaller, while some are thousands of times larger. Some are thousands of times dimmer than the Sun and others are thousands of times brighter. The Sun appears so bright simply because we happen to be so close to it. The next nearest star, Alpha Centauri, is about 250,000 times further away! It is therefore not surprising that stars appear to us as mere points of light.* (From the internet article "Stars as Suns" given in the Florida State University, Physics Department.)

## Point #12—"Mentioning"

Your view is that if the Bhāgavatam doesn't mention it, then it doesn't exist. But as I have pointed out, there are many things that exist, but are not mentioned in the Bhāgavatam.

## BST #12—

Not at all, my view is that if the Bhāgavatam does mention it, then it does exist.

## Point #13—"Absolutes"

We should both be careful about asserting absolutes, when our statements are inevitably bound by the limitations of our own minds and senses.

## BST #13—

The method of acquiring knowledge which you are proposing is akin to that of the Māyāvāda school, based on "not this, not that" (neti, neti). In essence, it takes great interest in what is not directly spoken in the śāstras or revealed scriptures whereas the Vaiṣṇavas prefer to concern themselves more with the words directly spoken.

Yes, Śrīla Śukadeva Goswāmī has already warned us about thinking we can fully understand the universe. However to have simple faith in the descriptions given by the ācārya is a qualification of a disciple whereas thinking oneself qualified to correct or interpret the ācārya is *over* intelligence and is a negative symptom. Krishna says surrender to Me. If someone says it is not to Krishna to whom we must surrender, but to something beyond Krishna,that is *over* intelligent.

Bhāgavatam says four billion miles—to propose that it is not four billion miles but something beyond four billion miles is *over* intelligent, at least as far as a simple person like me can see. In my opinion, Śrīla Prabhupāda wanted his scientists to establish the model of the universe based wholly on the Bhāgavata version, not to strike a compromise with the modern astronomical theories for gaining popular acceptance.

## Point #14—"The Question"

The question is, what is the Bhāgavatam really saying?

## BST #14—

The Bhāgavatam is saying that the stars we see at night are much closer than we have been told by scientists. They are not galaxies millions of light years away but planets, constellations, space gas, astro dust, etc. which lie within a four billion mile diameter universe. The astronomers are as guilty as the Darwinians at speculating, fantasizing and fudging with their theories. This will be exposed in due course just as the Darwinians are being exposed currently. The Bṛhad-bhāgavatāmṛta corroborates the actual size of the universe in relationship with others of twice and quadruple dimensions.

*paśyeme 'py apare yānti*
*brahmaṇo 'dhikṛte 'lpake*
*brahmāṇḍe catur-āsyasya*
*tathāmī dūrataḥ pare*

*amī cāṣṭa-mukhasyaitad-*
*dvi-guṇe yānti vegataḥ*
*amī tu ṣoḍaśāsyasya*
*brahmāṇḍe dvi-guṇe tataḥ*

*paśya—just see; ime—these; api—also; apare—others; yānti—are going; brahmaṇaḥ—by Brahmā; adhikṛte—presided over; alpake—small; brahmāṇḍe—in the universe; catuḥ-āsyasya—who has four heads; tathā—also; amī—these; dūrataḥ—further away; pare—others; amī—they; ca—and; aṣṭa-mukhasya—of an eight-headed Brahmā; etat—than him; dvi-guṇe—in a universe twice bigger; yānti—are going; vegataḥ—swiftly; amī—these; tu—and; ṣoḍaśa-āsyasya—of a sixteen-headed Brahmā; brahmāṇḍe—in the universe; dvi-guṇe—twice bigger; tataḥ—than that.*

*Just see these other Vaikuṇṭha companions of the Lord, traveling in this small universe ruled by four-headed Brahmā. And further away those others, moving swiftly in the universe of an eight-headed Brahmā, a world*

> *twice as large. And those others in the world of a Brahmā with sixteen heads, a world twice as large again.* (Bṛhad-bhāgavatāmṛta: Part 2; Chapter Three, texts 86-87)

## Point #15—"Not Metaphorically"

Keep in mind that Śukadeva Goswāmī said of the material world that "no one could possibly explain it perfectly, even in a lifetime as long as that of Brahmā."

## BST #15—

The statement "no one could possibly explain it perfectly, even in a lifetime as long as that of Brahmā" does not give license for disregarding or devaluing the precise measurements which Śukadeva Goswāmī provides. Otherwise why would he bother to say anything more than "no one could possibly explain it perfectly, even in a lifetime as long as that of Brahmā?" Obviously he wanted to give some knowledge about the universe as he states later in the same verse you quoted: "O King, I shall nevertheless try to explain to you the principal regions, such as Bhū-goloka [Bhuloka], with their names, forms, measurements and various symptoms."

In the purport to this verse Śrīla Prabhupāda writes:

> *Modern scientists and astronomers try to explain the cosmic situation and the vastness of space, and some of them believe that all the glittering stars are different suns. From Bhagavad-gītā, however, we understand that all these stars (nakṣatras) are like the moon, in that they reflect the sunshine. They are not independent luminaries. Bhūloka is explained to be that portion of outer space through which the heat and light of the sun extend. Therefore it is natural to conclude that this universe extends in space as far as we can see and encompasses the glittering stars. Śrīla Śukadeva Gosvāmī admitted that to give full details of this expansive material universe would be impossible, but nevertheless he wanted to give the King as much knowledge as he had received through the paramparā system. We should conclude that if one cannot comprehend the material expansions of the Supreme Personality of Godhead, one certainly cannot estimate the expansiveness of the spiritual world. The Brahma-saṁhitā (5.33) confirms this:*

*advaitam acyutam anādim ananta-rūpam*
*ādyaṁ purāṇa-puruṣaṁ nava-yauvanaṁ ca*

> *The limits of the expansions of Govinda, the Supreme Personality of Godhead, cannot be estimated by anyone, even a person as perfect as Brahmā, not to speak of tiny scientists whose senses and instruments are all imperfect and who cannot give us information of even this one universe. We should therefore be satisfied with the information obtainable from Vedic sources as spoken by authorities like Śukadeva Gosvāmī.* (Śrīmad Bhāgavatam 5.16.5 Purport)

Kindly keep in mind that it behooves us to accept the statements of Śrīmad Bhāgavatam regarding the size of the universe literally, not metaphorically. There may be issues difficult for us to presently resolve. Yet certain parameters are given and a four-billion mile universe is one of the primary ones.

Śrīla Prabhupāda addresses the topic of "no one could possibly explain it perfectly, even in a lifetime as long as that of Brahma," in the following purport:

> *Therefore, as indicated in Śrīmad-Bhāgavatam by Śukadeva Gosvāmī, we should not try to comprehend the greater mountainous areas of the universe merely by our calculations. Śukadeva Gosvāmī has already stated that such calculations would be very difficult even if one had a duration of life like that of Brahmā. We should simply be satisfied with the statements of authorities like Śukadeva Gosvāmī and appreciate how the entire cosmic manifestation has been made possible by the external energy of the Supreme Personality of Godhead. The measurements given herein, such as 10,000 yojanas or 100,000 yojanas, should be considered correct because they have been given by Śukadeva Gosvāmī. Our experimental knowledge can neither verify nor disprove the statements of Śrīmad-Bhāgavatam. We should simply hear these statements from the authorities. If we can appreciate the extensive energy of the Supreme Personality of Godhead, that will benefit us.* (Śrīmad Bhāgavatam 5.16.11 Purport)

## Point #16—"Far Too Small"

Yes, I know that strictly speaking, the Bhāgavatam sets the diameter of the Brahmāṇḍa to 500 million yojanas, and this equals the diameter of the Bhū-maṇḍala disk, which cuts it in half. At eight miles per yojana, this comes to four billion miles. However, in my opinion, this is far too small for the universe of stars and galaxies, but it does fall in the right range of size for the solar system.

## BST #16—

The premise that the four billion-mile-diameter universe described in the Śrīmad Bhāgavatam in reality defines but a minute part of the universe is not supported anywhere in the Śrīmad Bhāgavatam itself. The Bhāgavatam consistently describes a universe which is divided into three broad planetary divisions or divided into fourteen specific planetary systems, which includes Brahmā's abode, the Pātāla planetary systems, hellish worlds—all together measuring four billion miles in diameter and which is covered by seven layers. These descriptions are not imaginary nor are they exaggerated.

In the beginning of the Second Canto of Śrīmad Bhāgavatam, Śukadeva Goswāmī speaks to Parīkṣit Mahārāja about the components of the universe in which we dwell.

*aṇḍa-kośe śarīre 'smin*
*saptāvaraṇa-saṁyute*
*vairājaḥ puruṣo yo 'sau*
*bhagavān dhāraṇāśrayaḥ*

*TRANSLATION: The gigantic universal form of the Personality of Godhead, within the body of the universal shell, which is covered by sevenfold material elements, is the subject for the virāṭ conception.* (Śrīmad Bhāgavatam 2.1.25)

*pātālam etasya hi pāda-mūlaṁ*
*paṭhanti pārṣṇi-prapade rasātalam*
*mahātalaṁ viśva-sṛjo 'tha gulphau*
*talātalaṁ vai puruṣasya jaṅghe*

*TRANSLATION: Persons who have realized it have studied that the planets known as Pātāla constitute the bottoms of the feet of the universal Lord, and the heels and the toes are the Rasātala planets. The ankles are the Mahātala planets, and His shanks constitute the Talātala planets.* (Śrīmad Bhāgavatam 2.1.26)

*dve jānunī sutalaṁ viśva-mūrter*
*ūru-dvayaṁ vitalaṁ cātalaṁ ca*
*mahītalaṁ taj-jaghanaṁ mahīpate*
*nabhastalaṁ nābhi-saro gṛṇanti*

*TRANSLATION: The knees of the universal form are the planetary system of the name Sutala, and the two thighs are the Vitala and Atala planetary systems. The hips are Mahītala, and outer space is the depression of His navel.* (Śrīmad Bhāgavatam 2.1.27)

*uraḥ-sthalaṁ jyotir-anīkam asya*
*grīvā mahar vadanaṁ vai jano 'sya*
*tapo varāṭīṁ vidur ādi-puṁsaḥ*
*satyaṁ tu śīrṣāṇi sahasra-śīrṣṇaḥ*

*TRANSLATION: The chest of the Original Personality of the gigantic form is the luminary planetary system, His neck is the Mahar planets, His mouth is the Janas planets, and His forehead is the Tapas planetary system. The topmost planetary system, known as Satyaloka, is the head of He who has one thousand heads.* (Śrīmad Bhāgavatam 2.1.28)

*dyaur akṣiṇī cakṣur abhūt pataṅgaḥ*
*pakṣmāṇi viṣṇor ahanī ubhe ca*
*tad-bhrū-vijṛmbhaḥ parameṣṭhi-dhiṣṇyam*
*āpo 'sya tālū rasa eva jihvā*

*TRANSLATION: The sphere of outer space constitutes His eyepits, and the eyeball is the sun as the power of seeing. His eyelids are both the day and night, and in the movements of His eyebrows, the Brahmā and similar supreme personalities reside. His palate is the director of water, Varuṇa, and the juice or essence of everything is His tongue.* (Śrīmad Bhāgavatam 2.1.30)

*vaiśvānaraṁ yāti vihāyasā gataḥ*
*suṣumṇayā brahma-pathena śociṣā*
*vidhūta-kalko 'tha harer udastāt*
*prayāti cakraṁ nṛpa śaiśumāram*

*TRANSLATION: O King, when such a mystic passes over the Milky Way by the illuminating Suṣumṇā to reach the highest planet, Brahmaloka, he goes first to Vaiśvānara, the planet of the deity of fire, wherein he becomes completely cleansed of all contaminations, and thereafter he still goes higher, to the circle of Śiśumāra, to relate with Lord Hari, the Personality of Godhead.*

*PURPORT: The polar star of the universe and the circle thereof is called the Śiśumāra circle, and therein the local residential planet of the Personality of Godhead (Kṣīrodakaśāyī Viṣṇu) is situated. Before reaching there, the mystic passes over the Milky Way to reach Brahmaloka, and while going there he first reaches Vaiśvānara-loka, where the demigod controls fire. On Vaiśvānara-loka the yogī becomes completely cleansed of all dirty sins acquired while in contact with the material world. The Milky Way in the sky is indicated herein as the way leading to Brahmaloka, the highest planet of the universe.* (Śrīmad Bhāgavatam 2.2.24)

*tad viśva-nābhiṁ tv ativartya viṣṇor*
*aṇīyasā virajenātmanaikaḥ*
*namaskṛtaṁ brahma-vidām upaiti*
*kalpāyuṣo yad vibudhā ramante*

*TRANSLATION: This Śiśumāra is the pivot for the turning of the complete universe, and it is called the navel of Viṣṇu [Garbhodakaśāyī Viṣṇu]. The yogī alone goes beyond this circle of Śiśumāra and attains the planet [Maharloka] where purified saints like Bhṛgu enjoy a duration of life of 4,300,000,000 solar years. This planet is worshipable even for the saints who are transcendentally situated.* (Śrīmad Bhāgavatam 2.2.25)

*atho anantasya mukhānalena*
*dandahyamānaṁ sa nirīkṣya viśvam*
*niryāti siddheśvara-yuṣṭa-dhiṣṇyaṁ*
*yad dvai-parārdhyaṁ tad u pārameṣṭhyam*

*TRANSLATION: At the time of the final devastation of the complete universe [the end of the duration of Brahmā's life], a flame of fire emanates from the mouth of Ananta [from the bottom of the universe]. The yogī sees all the planets of the universe burning to ashes, and thus he leaves for Satyaloka by airplanes used by the great purified souls. The duration of life in Satyaloka is calculated to be 15,480,000,000,000 years.*

*PURPORT: In the creation of the Lord there are many wonderful things we can see with our own eyes every day and night, but we are unable to reach them equipped by modern materialistic science. We should not, therefore, depend on the fragmentary authority of materialistic science for knowing things beyond the range of scientific purview.* (Śrīmad Bhāgavatam 2.2.26)

*ete sṛtī te nṛpa veda-gīte*
*tvayābhipṛṣṭe ca sanātane ca*
*ye vai purā brahmaṇa āha tuṣṭa*
*ārādhito bhagavān vāsudevaḥ*

*TRANSLATION: Your Majesty Mahārāja Parīkṣit, know that all that I have described in reply to your proper inquiry is just according to the version of the Vedas, and it is eternal truth. This was described personally by Lord Kṛṣṇa unto Brahmā, with whom the Lord was satisfied upon being properly worshiped.* (Śrīmad Bhāgavatam 2.2.32)

Then later in the Second Canto Śukadeva quotes a discussion between Brahmā and Nārada which confirms the same universal set-up. Brahmā and Śukadeva, two mahajanas, describe the entire universe with all upper, middle and lower planets, space, coverings, Milky Way, Sisumara, Polestar, etc. The details of both of these discussions are said by Śukadeva Goswāmī to have been directly spoken by Lord Krishna to Brahmā.

Nārada asked Brahmā:

*My dear father, please describe factually the symptoms of this manifest world. What is its background? How is it created? How is it conserved? And under whose control is all this being done?* (Śrīmad Bhāgavatam 2.5.2)

Brahmā replies:

> *My dear son Nārada, know from me that there are seven lower planetary systems out of the total fourteen. The first planetary system, known as Atala, is situated on the waist; the second, Vitala, is situated on the thighs; the third, Sutala, on the knees; the fourth, Talātala, on the shanks; the fifth, Mahātala, on the ankles; the sixth, Rasātala, on the upper portion of the feet; and the seventh, Pātāla, on the soles of the feet. Thus the virāṭ form of the Lord is full of all planetary systems.* (Śrīmad Bhāgavatam 2.5.40-41)

Śrīla Prabhupāda emphasizes accepting the authority of this information without interpretation as follows:

> PURPORT: *The statements of Śrīmad-Bhāgavatam are accepted by great ācāryas like Śrīdhara Svāmī, Rāmānujācārya and Vallabhācārya. Lord Śrī Caitanya Mahāprabhu specifically accepts Śrīmad-Bhāgavatam as the spotless Vedic authority, and as such no sane man can ignore the statements of Śrīmad-Bhāgavatam when it is spoken by the self-realized soul Śrīla Śukadeva Gosvāmī, who follows in the footsteps of his great father, Śrīla Vyāsadeva, the compiler of all Vedic literatures. (*Śrīmad Bhāgavatam *2.2.26)*

And Śrīla Prabhupāda further corroborates the same understanding which Brahmā, Śukadeva and Nārada are holding, i.e. that the universe which is being described is one in which the fourteen planetary systems are housed within a four billion mile diameter.

> *PURPORT: Modern enterprisers (the astronauts who travel in space) may take information from Śrīmad-Bhāgavatam that in space there are fourteen divisions of planetary systems. The situation is calculated from the earthly planetary system, which is called Bhūrloka. Above Bhūrloka is Bhuvarloka, and the topmost planetary system is called Satyaloka. These are the upper seven lokas, or planetary systems. And similarly, there are seven lower planetary systems, known as Atala, Vitala, Sutala, Talātala, Mahātala, Rasātala and Pātāla lokas. All these planetary systems are scattered over the complete universe, which occupies an area of two billion times two billion square miles. The modern astronauts can travel only a few thousand miles away from the earth, and therefore their attempt to travel in the sky is something like child's play on the shore of an expansive ocean. The moon is situated in the third status of the upper*

> *planetary system, and in the Fifth Canto of Śrīmad-Bhāgavatam we shall be able to know the distant situation of the various planets scattered over the vast material sky. (Śrīmad Bhāgavatam 2.5.40-41)*

Later the same information is substantiated in the Fifth Canto.

ISKCON scientists should be working with full faith in the words of the Śrīmad Bhāgavatam and Śrīla Prabhupāda; 1) describing the defects in the modern emperic methods, 2) demonstrating the plausibility and infallibility of Bhāgavata cosmology.

> *They cannot go even to the moon planet, what to speak of others. Therefore it is very big for us. When we speak of universe we cannot think of. They simply calculate, the scientists, by light year, and this year, that year, the speed. (Śrīmad-Bhāgavatam 2.1.1-5 — Melbourne, June 26, 1974)*

In summary, I have difficulty with the new theory which you propose for the following reasons:

1) It assigns drastically anomalous meanings to the Śrīmad Bhāgavatam's description of the universe.
2) It causes confusion in regards to terms such as Indra (god of the heavenly planets), Mount Meru, Lord Brahmā, the sun god, the subterranean abodes of the asuras, the abodes of Agni, the polestar (Dhruvaloka—,) etc., because according to the new theory, such nomenclature has no place in the Śrīmad Bhāgavatam's description of the universe which the new theory says is only the local solar system. Brahmā doesn't live within the local solar system. How does one reconcile these matters?
3) It agrees with the modern scientific theory that our visible sky comprises galaxies, etc. occupying distances of hundreds of millions of light-years away. The modern scientific theory has been totally denounced by our spiritual master Śrīla Prabhupāda.
4) It turns the entire discourse of Śukadeva Goswāmī to Mahārāja Parīkṣit upside-down and makes a mockery of the Śrīmad Bhāgavatam.
5) It asks people to throw away the direct explanations given in the Śrīmad Bhāgavatam and replace them with crude guesswork and wishful thinking.

# NOTES

## *Note #1: Universal Size

From Vedic literature we find that our universe, compared to other universes, is relatively small. The Brahmā in charge of this universe has only four heads, whereas Brahmās managing larger universes have proportionately more heads.

> *Śrī Caitanya Mahāprabhu said: "Śrī Kṛṣṇa smiled and immediately meditated. Unlimited Brahmās arrived instantly. These Brahmās had different numbers of heads. Some had ten heads, some twenty, some a hundred, some a thousand, some ten thousand, some a hundred thousand, some ten million and others a hundred million. No one can count the number of faces they had. There also arrived many Lord Śivas with various heads numbering one hundred thousand and ten million. Many Indras also arrived, and they had hundreds of thousands of eyes all over their bodies. 'Kṛṣṇa said [to Brahmā of this universe], 'Your particular universe has a diameter of four billion miles; therefore it is the smallest of all the universes. Consequently you have only four heads.'" (*Caitanya-caritāmṛta Madhya *21.66, 67, 68, 84)*

Śrīmad Bhāgavatam (3.11.40) confirms this distance of four billion miles.

*vikāraiḥ sahito yuktair*
*viśeṣādibhir āvṛtaḥ*
*āṇḍakośo bahir ayaṁ*
*pañcāśat-koṭi-vistṛtaḥ*

> *"This phenomenal material world is expanded to a diameter of four billion miles, as a combination of eight material elements transformed into sixteen further categories, within and without, ...."*

Half of the universe is filled with the water of the Garbhodaka Ocean and the other half is filled with planets floating in space. (Caitanya-caritāmṛta, Ādi 5.96).This huge egglike universe (Brahmāṇḍa) is covered by seven layers of material elements; earth, water, fire, air, etc. The first layer is ten times the diameter of the universe, i.e. forty billion miles. The next is ten times that, namely 400 billion miles. Each layer is successively ten times greater than the previous one.

*kṣity-ādibhir eṣa kilāvṛtaḥ*
*saptabhir daśa-guṇottarair aṇḍa-kośaḥ*

*"Every universe is covered by seven layers—earth, water, fire, air, sky, the total energy and false ego—each ten times greater than the previous one." (Śrīmad-Bhāgavatam 6.16.37)*

Altogether, the universe of floating planets and its immense coverings has a radius of approximately 46.2 quadrillion miles, or a diameter of approximately 92.4 quadrillion miles. (See Figure 1)

To calculate the circumference of the universe we multiply the diameter times pi (3.14) which equals approximately 290 quadrillion miles or 36.3 quadrillion yojanas (a yojana equals eight miles). This figure brings us into the realm of calculations presented by Śrīla Bhaktisiddhānta Sarasvatī in the Siddhānta-śiromaṇi.

*"Śrīla Bhaktisiddhānta Sarasvatī Ṭhākura, one of the greatest astrologers of his time, gives information from Siddhānta-śiromaṇi that this universe measures 18,712,069,200,000,000 x 8 miles. This is the circumference of this universe. According to some, this is only half the circumference."* (Caitanya-caritāmṛta *Madhya 21.84 Purport)*

If 18,712,069,200,000,000 yojanas is half the circumference, then the full circumference is 37,424,138,400,000,000 yojanas, quadrillion figure. Śrīla Prabhupāda always supported the Śrīmad Bhāgavatam statement that the part of the universe that contains planets and the Garbhodaka ocean, which is the relatively hollow part within the coverings, is four billion miles in diameter.

Prabhupāda: The whole universe, diameter, is pañcaśat-koṭi-yojana... fifty crore yojanas, pañcaśat. So one yojana equal to eight miles, one crore equal to ten million.
Parāmahaṁsa: Four billion miles is the diameter.
Prabhupāda: Is the diameter.
Parāmahaṁsa: You gave that in The Teachings of Lord Caitanya also.
Prabhupāda: Yes. And the sun is in the middle.
Parāmahaṁsa: So two billion miles from the edge of the universe.
Prabhupāda: Yes. (Morning walk: May 12, 1975)

## FIGURE 1—COVERINGS CALCULATION

2,080,000,000
41,600,000,000
416,000,000,000
4,160,000,000,000
41,600,000,000,000
416,000,000,000,000
4,160,000,000,000,000
41,600,000,000,000,000
_______________________
46,209,680,760,000,000 miles

| | |
|---|---|
| (4.16 billion mi.) | = radius of universe |
| (40 billion mi.) | = 1st layer |
| (400 billion mi.) | = 2nd layer |
| (4 trillion mi.) | = 3rd layer |
| (40 trillion mi.) | = 4th layer |
| (400 trillion mi.) | = 5th layer |
| (4 quadrillion mi.) | = 6th layer |
| (40 quadrillion mi.) | = 7th layer |
| (46.2 quadrillion mi.) | = radius of the universe & coverings |

circumference = 2 pi (r) radius

radius X 2 = 92,426,419,360,000,000 miles

92,426,419,360,000,000 X pi or 3.14= 290,218,956,790,400,000 miles (circumference)

Divided by 8= 36,277,369,598,800,000 yojanas

Compare to 37,424,138,400,000,000 yojanas (figure given in *Siddhānta-śiromaṇi*)

36,277,369,598,800,000 yojanas / 37,424,138,400,000,000 yojanas= 96.93575096120316% rounded off to 97%

Sometimes it is reasoned that Śrīla Prabhupāda suggested a different, far greater, size for our universe. Such ideas are based on statements such as the following, found in the Śrīmad Bhāgavatam (1.9.29, purport):

> *Even attempting to reach the highest planet will take millions of years at a speed of millions of miles per hour.*

There are many statements like this. The point Śrīla Prabhupāda is making is a philosophical one—that the material creation is beyond our calculations, what to speak of the spiritual world! In the above statement as well as those cited below, Śrīla Prabhupāda is not giving exact astronomical distances of the universe. Rather he is establishing an essential philosophical understanding. Consider the following statements:

> *The conditioned souls may try to understand the Supreme Personality of Godhead for many billions of years through their mental speculative processes, traveling at the speed of the mind or the wind, but still the Absolute Truth will remain inconceivable to them because a materialistic person cannot measure the length and breadth of the Supreme Personality of Godhead's unlimited existence.* (Śrīmad-Bhāgavatam 6.4.23 Purport)

> *Just like if you start your jet plane from here to see where is the original cause, panthās tu koṭi-śata-vatsara, and go on for many, many millions of years with the speed of mind,* panthās tu koṭi-śata-vatsara-sampra... Koṭi-śata-vatsara... *Millions and millions of years... So with the mental speed, airplane, if you start today and go for many millions of years, still it will be not possible to find out wherefrom the explosion is coming.* (Lecture:, April 2, 1975, Māyāpur)

> *Space is unlimited, and the Brahmā-saṁhitā suggests: Suppose one travels by spacecraft for millions of years at the velocity of the wind or even the speed of mind. Everyone knows that the mind is so swift that in even one ten-thousandth of a second it can take us millions of miies. If we have seen something millions of miles away, the mind can go there immediately. But even if we can travel at that speed on a spacecraft manufactured by* muni-puṅgavānām, *the greatest scientists and most thoughtful men, will that be perfection? No. The Brahmā-saṁhitā says,* so 'py asti yat-prapada-sīmny avicintya-tattve*: still this creation will remain inconceivable to our understanding.* (Teachings of Queen Kuntī: 14)

The precise distances for the size of the universe (an inner diameter of four billion miles for the space-and-planet-filled portion) are given in the Fifth Canto of the Śrīmad Bhāgavatam. Śrīla Prabhupāda adamantly confirms their accuracy in his purports. It is not justifiable to suggest, therefore, that Śrīla Prabhupāda is inferring that our universe is something other than four billion miles in diameter.

Yet if we're determined to build a hypothesis on the statement, "Even attempting to reach the highest planet will take millions of years at a speed of millions of miles per hour," then we cannot rule out the possibility of material planets beyond our universe. There are many universes, each with countless planets. To travel from one brahmāṇḍa to another (and there are innumerable brahmāṇḍas) would require billions of light-years. In this way, Śrīla Prabhupāda's statement is easily verified. Vedic cosmology involves vast distances of incalculable light-years. However, according to Vedic authority, such distances are not within our visible universe.

In conclusion we humbly present that the circumference of our Brahmāṇḍa, including the seven coverings, according to the Śrīmad Bhāgavatam, is approximately thirty-six quadrillion yojanas. That is a fact. It is also correct that the *Siddhānta-śiromaṇi* proposes essentially the same distance for the circumference. Śrīla Jiva Goswāmī has explained in his Sandarbha that of all scriptures, Śrīmad Bhāgavatam is the most exalted authority. Even if there appears to be a contradiction between śāstras, the version of the Śrīmad Bhāgavatam is to be accepted above all else.

## *Note #2: Universal Tree

Pusṭa Kṛṣṇa: Is the earth also spinning?
Prabhupāda: No.
Pusṭa Kṛṣṇa: Or is it simply the sun's movements that causes the day and night, everything?
Prabhupāda: No planet is fixed except the sun. All are fixed up. But the whole thing is moving. That is Bhāgavatam. And that you can see at night.
Gurudāsa: What'd he say?
Pañca-draviḍa: No planet is fixed. The earth and sun, they're all moving.
Gurudāsa: The sun is fixed.
Tamāla Kṛṣṇa: Yeah, but they're all moving.
Hari-śauri: But they're all moving.

Prabhupāda: It is like a tree, just like this. This is moving, and the sun is moving, and all other planets, they are fixed.
Hari-śauri: How is it we see the moon coming in every day, then?
Prabhupāda: Hm?
Hari-śauri: How is it that we see the moon moving?
Prabhupāda: Moon, that is..., of course, I do not remember, but (laughs) the whole planetary system is moving.
Tamāla Kṛṣṇa: And the sun is fixed, but the whole, it's also moving.
Prabhupāda: Yes. Sun is also moving.
Puṣṭa Kṛṣṇa: Everything is revolving around that polestar?
Prabhupāda: Hm.
Tamāla Kṛṣṇa: Yes.
Jayapatākā: Everything is moving around the sun, and the sun is moving around the polestar.
Puṣṭa Kṛṣṇa: No. Everything is not moving around the sun.
Hari-śauri: Like that tree, if the tree revolves by itself, like that, then the sun(?) is going around the whole thing.
Prabhupāda: This is also another intelligent.
Puṣṭa Kṛṣṇa: Yes.
Hari-śauri: The tree is just moving like that, and the different branches, they remain in the same position in relationship to the center of the tree.
Puṣṭa Kṛṣṇa: But the sun has it's own...
Hari-śauri: But the sun has an orbit around the whole thing...
Puṣṭa Kṛṣṇa: It's own course.
Hari-śauri: So the whole universe moves around. Every twenty-four hours it does one rotation. And then the sun is also round all that.
Puṣṭa Kṛṣṇa: Every twenty-four hours?
Prabhupāda: The moon.
Gurudāsa: Acintya-bhedābheda-tattva.
Prabhupāda: Eh?
Gurudāsa: It seems to be one and different simultaneously.
Prabhupāda: No, no, not like that point. (laughs)
Satsvarūpa: Śrīla Prabhupāda, are the planets shaped liked balls or more like plates? Because it's, it's hard to understand, 'cause they're called dvīpas, "islands." Their roundness is the roundness of a plate or like a ball?
Prabhupāda: Which one?
Satsvarūpa: The earth planet?
Prabhupāda: If it is like a tree, then these things can be as dvīpa, island.
Tamāla Kṛṣṇa: Wow. You know...
Prabhupāda: Eh?

Tamāla Kṛṣṇa: The scientists are getting smashed to bits by your statements, Śrīla Prabhupāda. This destroys their whole theory. Orbs, round spheres. I think that this Māyāpura building, we must build a big planetarium in it.
Prabhupāda: Yes. That, that I am going to do, Vedic planetarium.
Tamāla Kṛṣṇa: Oh, boy. You're going to bring a lot of.... A lot of scientists will come here just to dispute this.
Prabhupāda: Yes.
Puṣṭa Kṛṣṇa: Wonderful attraction.
Prabhupāda: World people will come to see the way the planetary systems...
Tamāla Kṛṣṇa: We should advertise it very widely that this is the actual, factual explanation of the universe.
Prabhupāda: This will be automatically advertised. As soon as the temple is finished, people will come like anything.
Hari-śauri: They're going to need somebody to...
Prabhupāda: The thing is, on principle, we shall only go against them.
Haṁsadūta: Contrary.
Prabhupāda: On principle. Whatever they say "Yes," we say "No." (laughter)
Pañca-draviḍa: What is the explanation of an eclipse?
Prabhupāda: Eh?
Pañca-draviḍa: Eclipse.
Prabhupāda: Eclipse are.... So that Rāhu planet comes to attack the sun and the moon. And when it comes, it becomes dark.
Pañca-draviḍa: How does it attack?
Prabhupāda: How do I attack you?
Pañca-draviḍa: When we see part of the moon disappear, what is actually happening?
Prabhupāda: Happening, that attack.... Suppose if I come to attack you and if I am in front, then you cannot see.
Tamāla Kṛṣṇa: Yeah. Like now I can't see.
Prabhupāda: Eh?
Tamāla Kṛṣṇa: Yeah. Just like Prabhupāda's.... I'm standing behind him, and when he attacks you, I can't see you.
Nanda-kumāra: By the scientist's calculation of what's happening with the revolving, they can predict an eclipse.
Prabhupāda: Hm?
Nanda-kumāra: They can say in advance when an eclipse will come.
Prabhupāda: Well we can also predict.
Pañca-draviḍa: :Yeah. They predicted Lord Caitanya's appearance.
Balavanta: And five thousand years ago, in Kurukṣetra, they predicted when Kṛṣṇa went...

Pañca-draviḍa: So apparently they can predict Rāhu's attack also.
Tamāla Kṛṣṇa: That's all right.
Gurudāsa: Actually, even the scientists say there is astrological reasons behind Rāhu the planet.
Pañca-draviḍa: So they're predicting five thousand years before the scientists. This is the explanation. (Morning Walk — March 18, 1976, Māyāpura)

# 2

# Jyoti Shastras

## Brahmāṇḍa Sākṣāt Ṭikā (BST) —"Validity"

We are sure that the Jyoti Śāstras have validity since Śrīla Bhaktisiddhānta Prabhupāda translated Sūrya-siddhānta. Unfortunately at present his Bengali translation is still unavailable and the only works done in English have been done by English and American scientists. These however cannot be accepted as authentic since the persons translating lack the spiritual qualifications.

> *No mundane scholar can translate or reveal the true import of the Vedic mantras (hymns).* (Śrīmad-Bhāgavatam 2.4.22 Purport)

> *avaiṣṇava-mukhodgīrṇaṁ*
> *pūtaṁ hari-kathāmṛtam*
> *śravaṇaṁ naiva kartavyaṁ*
> *sarpocchiṣṭaṁ yathā payaḥ*

> *The words or songs of a person not fixed in Vaiṣṇava behavior, not strictly following the rules and regulations and chanting the Hare Kṛṣṇa mantra should not be accepted by pure devotees.* (Śrīmad-Bhāgavatam 6.16.33 Purport)

Śrīla Prabhupāda expressed interest in the Jyoti Śāstra translations by Śrīla Bhaktisiddhānta Sarasvatī Ṭhākura yet he did not do so for works done by Western scholars. We have included these works in chapters 12-14 at the rear of this book simply to avail the reader of the translations which are quoted often these days.

> *Svarūpa Dāmodara: Actually we started looking some on this astronomy thing. About this Sūrya-siddhānta. Sadāpūta is especially interested*

*because he's a mathematician. So normally astronomy is subject of mathematics. So he found this copy in Princeton just recently, last month or so. There we found that the idea is very similar to modern science in the Sūrya-siddhānta. But we heard that Śrīla Bhaktisiddhānta Prabhupāda translated Sūrya-siddhānta, but we are not getting a copy of it. So he asked me to look in Calcutta, and some astrologers.*
*Prabhupāda: He had some disciples, some astronomer. His name can be had from Calcutta.*
*Pradyumna: Mohinī-mohana?*
*Prabhupāda: Ah! Mohinī Śāstrī. Mohinī Śāstrī. He was his disciple.*
*Pradyumna: I went to try... He's dead now.*
*Prabhupāda: He is also. He's passed away.*
*Pradyumna: And I saw his son. I saw his son. I asked if he had any of Śrīla Prabhupāda's books on jyotir, but he did not... He is practicing jyotir, but not...*
*Prabhupāda: Mohinī Śāstrī's son?*
*Pradyumna: His son. Sāmba...*
*Prabhupāda: My Guru Mahārāja was a great astronomer.*
*Svarūpa Dāmodara: I think that would be our proper research, to look into it.*
*Prabhupāda: From Sūrya-siddhānta he got the "Bhaktisiddhānta." He got this title Sūrya-siddhānta. So when he became Vaiṣṇava... A Vaiṣṇava he was as a Bhaktisiddhānta.*
*Svarūpa Dāmodara: We saw some translations in English, Sūrya-siddhānta. That was in Library of Congress in Washington. But the translation there was very... Was not good. It was all wrong interpretations.*
*Prabhupāda: Who translated?*
*Svarūpa Dāmodara: Some English authors, outsiders.*
*Prabhupāda: He was also astronomer?*
*Svarūpa Dāmodara: I think he was from Germany. He knew something about Sanskrit, but translation was just like to criticize those Indian, old Indian astronomers. Something like, very..., not even using even respectful words, but offensive sometimes, so we didn't bother to go through those books.*
*Pradyumna: There's copy of Sūrya-siddhānta by Bhaktisiddhānta in London at India Office Library. They have a copy in Bengali.*
*Svarūpa Dāmodara: In Bengali?*
*Pradyumna: Yes, they have original, these. But they do not let... They have a funny... You cannot copy the whole book with a xerox there. You only can see it and copy by hand. They won't let you do it. But there's a copy there, Samadhi Press.*

*Prabhupāda: It must be very old paper.*
*Pradyumna: Yes, very old.*
*Prabhupāda: To copy in hand, how long it take? Very big book?*
*Pradyumna: No. Not so big. And they also have that paṇḍita who used to live with Bhaktivinoda? They had that tall Sarasvatī, that Sararati Satuspati... Bhaktisiddhānta... His name was... His books are also there, two, three books. They have published jointly from Saraswati Bhavan. Bhaktivinoda Ṭhākura...*(Room Conversation with Svarūpa Dāmodara January 30, 1977, Bhubaneshwar)

## BST—"Premise"

In his Anubhāṣya commentary on this verse of Caitanya-caritāmṛta, Śrīla Bhaktisiddhānta Sarasvatī quotes from Sūrya-siddhānta 12.90, "The circumference of the sphere of the Brahmandee in which the sun's rays spread is 18,712,080,864,000,000 yojanas". Then he quotes Siddhānta-śiromaṇi, Goladhyaya Bhuvana-kosa: "Some astronomers have asserted the circumference of the circle of heaven to be 18,712,069,200,000,000 yojanas in length. Some say that this is the length of the zone binding the two hemispheres of the Brahmāṇḍa. Some Pauranikas say that this is the length of the circumference of the Lokāloka Parvata [adrsya-drsyaka-girim] (Sūrya-siddhānta).

The key to understanding these verses is to measure the circumference around the outside of the universe which includes its coverings (vesṭanam-the gird.) In this way the 37.4 quadrillion yojanas correctly describes the circumference of the entire brahmāṇḍa at its gird, including inhabited and uninhabited portions. Similarly, 18.7 quadrillion yojanas describes the circumference of Lokāloka Mountain (again including the coverings) as seen from outside the universe.

## Point #1—"Lokāloka"

According to the Bhāgavatam, Lokāloka mountain has a circumference equal to half the circumference of the Brahmāṇḍa. It is situated halfway between the center and the outer edge of the Brahmāṇḍa. The shells start after one passes the outer edge. How, then, can you include the coverings in Lokāloka mountain?

## BST #1—

Looking at the universe from the outside which includes the coverings, the circumference is 37.4 quadrillion yojanas. This figure is derived from the four billion mile diameter and the seven coverings each ten times greater than the previous one as stated in the Bhāgavatam. Therefore the Bhāgavatam statement coincides with the circumference figure cited by Śrīla Bhaktisiddhānta Sarasvatī and Sūrya-siddhānta.

Regarding the circumference of Lokāloka Mountain; it is half of the circumference of the universe therefore it is 18.7 quadrillion yojanas. If again one looks at the universe from the outside then it is easy to picture the circular Lokāloka Mountain (a circle half the size of the inside visible portion of the universe) with its circumference measured at the outside of the universe just as the universe's circumference is measured from the outside.

In discussing this subject with senior devotees/scientists no one has had any problems conceptualizing the logic of measuring from the outside.

## Point #2—"Shells"

The verses from Sūrya-siddhānta and Siddhānta-śiromaṇi plainly refer to the Brahmāṇḍa and to Lokāloka mountain. There is no justification for bringing in the shells of the universe, which are not even mentioned in Siddhānta-śiromaṇi or the Sūrya-siddhānta. Note also that the 18 quadrillion mile figure in these texts is defined as the distance traversed by the moon in its orbit in one kalpa. It has nothing to do with the shells.

## BST #2—

Neither do the Siddhānta-śiromaṇi or the Sūrya-siddhānta mention that the circumference given as 37.4 quadrillion yojanas refers to the inside of the universe. That is perhaps how the ambiguity has arisen. It makes perfect sense however for someone describing the circumference of a spherical object to measure it from the outside. Due, possibly, to our being inside the universe we have misconstrued the circumference to refer to the inside.

The distance of the moon's orbit is coincidentally equal to the distance of the circumference of the brahmāṇḍa. It does not negate the fact that the circumference of the brahmāṇḍa is a measurable, real distance.

## Point #3—"Sūrya-siddhānta Contradicts Bhāgavatam?"

The 18 quadrillion mile circumference of the Brahmāṇḍa or Lokāloka mountain clearly contradicts the version given in the Bhāgavatam. It is therefore interesting that Śrīla Bhaktisiddhānta cited it.

## BST #3—

Quite to the contrary, the Śrīmad Bhāgavatam, the Sūrya-siddhānta and the Siddhānta-śiromaṇi are in complete agreement. Had Sūrya-siddhānta meant that the 18 quadrillion yojanas circumference referred to the inside of the brahmāṇḍa there would have been great need for explanation since it contradicts every explicit numerical distance given. Eighteen quadrillion yojanas is not possible for a universe whose diameter is four billion miles. The Bhāgavatam is not incorrect, the new theory is incorrect.

Corroboration between Sūrya-siddhānta and Śrīmad Bhāgavatam is clear: both describe the Brahmā egg as comprising the entire universe. The sungod in the Sūrya-siddhānta describes the exact same universe as does Śukadeva Goswāmī in the Śrīmad Bhāgavatam. Both describe in detail the chariot of the sungod with which he traverses his path.

Even a translator of Sūrya-siddhānta comments,

> *"This is the form of the proportion from which is derived the rule as stated in the text, only the latter designates the product of the multiplication of the moon's orbit by her number of revolutions as the orbit of the ether (akasa), or the circumference of the Brahma-egg, within which the whole creation, as above taught, is enclosed. This is the same thing with attributing to the outermost shell of the universe one complete revolution in an AEon (kalpa), of 4,3200,000,000 years."* (Burgess, Ebenezer, trans., 1860. *The Sūrya-siddhānta*, Ganguli, P., ed., Delhi: Motilal Banarsidass (reprinted in 1989). [page 295]

This "the outermost shell of the universe" is referring to the circumference which is being described by the 18 quadrillion yojanas figure. This is also consistent with the actual distance around the Brahmā egg from the outside. None of the Vaiṣṇava commentators, including Śrīla Bhaktisiddhānta

Sarasvatī Goswāmī and Śrīla Prabhupāda, have indicated or imagined a massive contradiction between the Śrīmad Bhāgavatam and the Sūrya-siddhānta and Siddhānta-śiromaṇi.

There is no contradiction and thus no necessity to bring in modern astronomical speculation about distances between planets within the universe. It appears however that Śrīla Prabhupāda was aware that the 37.4 quadrillion figure referred to the outside otherwise he would have indicated that there was a vast dichotomy between the Jyoti śāstras, Śrīla Bhaktisiddhānta Sarasvatī Ṭhākura and the Śrīmad Bhāgavatam itself.

Please consider this fact. The same Śrīla Bhaktisiddhānta Sarasvatī Ṭhākura while commenting on Śrī Caitanya Bhāgavata in his Gauḍīya Bhāṣya writes:

> "*While describing the prabhava and vaibhava features of the Lord in his Laghu-bhagavatamrta (Purva 62), Srila Rupa Goswami writes: Lord Ananta, the plenary portion of Sri Baladeva who holds the universes on His hoods, resides in Patalaloka.*" (Śrī Caitanya Bhāgavata, Ch. 1 verse 16; Purport)

Must we speculate that the Pātālaloka mentioned here is imaginary in order to keep the hope that the Bhāgavatam's version of the universe is erroneous? Must we conjecture that the Lord Ananta described here is not real?

Later in the same purport, Śrīla Bhaktisiddhānta Sarasvatī Ṭhākura states:

> *The word prakanda-sarira in this verse is explained in the Caitanya-caritamrta (5.119) as follows: "The universe, which measures five hundred million yojanas in diameter, rest on one of His hoods like a mustard seed."*

Are we to believe that the person whom you claim is contradicting Śrīmad Bhāgavatam (i.e. *Śrīla Bhaktisiddhānta Sarasvatī Ṭhākura)* in his earlier statement is now agreeing with the Śrīmad Bhāgavatam's size for the universe, namely five hundred million yojanas? Has he contradicted himself? No, the contradictions lie not with him nor with any Vaiṣṇava śāstra nor with any Vaiṣṇava ācārya, rather they lie with those who lack faith in the Śrīmad Bhāgavatam and the ācāryas.

## Point #4—"Moon Motion"

I am sorry to say that your interpretation of the 18 quadrillion yojana circumference simply does not agree with the texts of the Sūrya-siddhānta and the Siddhānta-śiromaṇi. The Sūrya-siddhānta clearly bases this circumference on the motion of the moon, and there is nothing coincidental about this.

## BST #4—

Circumference means by definition, the distance around an object. Not discounting the accuracy of the Sūrya-siddhānta's method of obtaining a figure for the circumference of the brahmāṇḍa, still that figure must agree with the actual distance around the brahmāṇḍa. In this case the moon's motion does in fact concur with the actual distance around the brahmāṇḍa which is measured by the standard 2 pi r formula.

## Point #5—"Inside Circumference?"

Both the Sūrya-siddhānta and the Siddhānta-śiromaṇi say clearly that the 18 quadrillion yojana circumference "refers to the inside of the universe," since they both define it in terms of the motion of sunlight within the universe. However, neither text refers to the shells, and neither text says anything about 37.4 quadrillion yojanas.

## BST #5—

Negative. Neither in the Sūrya-siddhānta nor the Siddhānta-śiromaṇi does one find such a statement that the 18 quadrillion yojana circumference refers to the "inside of the universe." Such a statement would be preposterous since it would contradict the infallible statements of the Bhāgavata Purāṇa, the Bṛhad-bhāgavatāmṛta, Caitanya-caritāmṛta, as well as such exalted ācāryas as Brahmā, Nārada, Vyāsa, Maitreya, Śukadeva Goswāmī, Viśvanātha Cakravartī Ṭhākura, Śrīla Jiva Goswāmī, Śrīla Sanatana Goswāmī, Śrīla Prabhupāda and many, many others including Lord Krishna Himself.

How could Bhaskara, being such an expert mathematician, mistakenly propose a circumference of 18 quadrillion yojanas for a four billion mile diameter universe? The large figure does however fit the circumference of the universe measuring from the outside including its shells.

Furthermore the 37.4 quadrillion yojana is indeed stated in Caitanya-caritāmṛta:

> *"According to the Siddhānta-śiromaṇi this universe measures 18,712,069,200,000,000 x 8 miles., This is the circumference of this universe. According to some, this is only half the circumference."*

[This is a quote from Bhaktisiddhānta Sarasvatī and Śrīla Prabhupāda: 2 X 18.7 quadrillion equals 37.4 quadrillion yojanas]

## Point #6—"Calculating by Revolutions"

Please consider the excerpts from the two texts that are included below.

I. According to the Sūrya-siddhānta:

> *In a great yuga (of 4,320,000 years):*
>
> *1.30. There are 57,753,336 revolutions of the moon and 2,296,832 revolutions of the planet Mars.*
>
> *12.81. Multiply the number of the said revolutions of the moon in a Kalpa by the Moon's orbit (to be declared in Sloka 85th): the product is equal to the orbit of heaven (or the circumference of the middle of the Brahmanda): to this orbit the rays of the Sun reach.*
>
> *12.85. The orbit of the Moon is 324,000 (Yojanas) and that of the Sighrocheha of Mercury, beyond the Moon is 1,043,209.*

There are 1,000 great yugas in a kalpa. Therefore the required calculation is:

324,000 x 1000 x 57,753,336 = 18,712,080,864,000,000.

This is the actual calculation of the 18 quadrillion yojana circumference. The role of the moon in this calculation is no coincidence. All other calculations are simply speculation, including the calculation leading to

37.4 quadrillion yojanas. This latter figure does not agree at all closely with 18 quadrillion, whereas the above calculation based on the moon gives the exact 18 quadrillion yojana figure.

> *12.90. The circumference of the sphere of the Brahmandee in which the Sun's rays spread is 18,712,080,864,000,000 Yojanas.*

Note the two references to the sun's rays in 12.81 and 12.90. The sun's rays reach to the "circumference of the middle of the Brahmanda," and this circumference is exactly 18,712,080,864,000,000 yojanas, according to the Sūrya-siddhānta. Note that the ray's of the sun cannot be expected to penetrate the shells covering the outside of the Brahmanda, beginning with the shell of earth. Therefore, this cannot refer to the outermost shell. Indeed, the "phrase "middle of the Brahmanda" refers to the Lokāloka mountain, which is the limit of the sun's rays, according to the Bhāgavatam.

## BST #6—

The term *vestanam* refers to the gird or binding around the universe. When a circle is bound it is bound from the outside. Similarly in common terminology when one refers to the circumference of a sphere it refers to the outside of the sphere. The circumference of the inside would only be applicable if the speaker were to specifically indicate that he wanted to measure the inside of a sphere. Please note the following word meanings and translation to the verse from Sūrya-Siddhānta.

> *yojanaih—measured in yojanas; tat—that; brahma-anda—of the universe; kataha—like a kettle; samputa—of the container; tate—along the edge; kecit—some; jaguh—declare; vestanam—the gird;*
>
> *Some declare that this [18,712, 080,864,000,000 yojanas] indicates the gird of the edge of the kettle-like vessel of the universal egg.*

## Point #7—"Siddhānta-śiromaṇi"

Please consider the following—according to Siddhānta-śiromaṇi:

> *3.67. Some astronomers have asserted the circumference of the circle of heaven to be 18,712,069,200,000,000 Yojanas in length. Some say that*

*this is the length of the zone which binds the two hemispheres of the Brahmanda. Some Pauranikas say that this is the length of the circumference of the Lokaloka Parvata.*

*3.68. Those, however, who have had a most perfect mastery of the clear doctrine of the sphere, have declared that this is the length of the circumference bounding the limits to which the darkness dispelling rays of the sun extend.*

Here the 18 quadrillion yojana circumference is said clearly to bound the limits to which the sun's rays spread. In other words, they can spread this far and no farther. If the shell of earth begins at a radius of 250,000,000 yojanas, and the sun's rays cannot penetrate it, then it makes no sense to say that a circumference of 18 quadrillion yojanas bounds "the limits to which the darkness dispelling rays of the sun extend." Indeed, according to the Bhāgavatam the sun's rays are blocked by the Lokāloka mountain at a radius of 125,000,000 yojanas.

## BST #7—

As mentioned before, it is a question of vantage point. By viewing from the outside, the universal globe also bounds the rays of the sun.

Incidentally, if you say with such certainty that the radius of the Lokāloka Mountain is correctly 125,000,000 yojanas, which is a statement taken from the Śrīmad Bhāgavatam, how can you then argue that the Śrīmad Bhāgavatam is incorrect when it asserts that the diameter of the universe is *pañcāśat-koṭi-vistṛtaḥ* (five hundred thousand yojanas?)

It is inconsistent to select some verses from the śāstra to accept and reject others. On what basis do you accept the radius of Lokāloka Mountain as a fourth of that diameter of the universe when you also reject the diameter of the universe— both given in the Śrīmad Bhāgavatam? There must be a standard authority and that standard authority is Śrīmad Bhāgavatam.

*We cannot simply accept that part of the scripture we like, and reject what we don't like, and still expect to get the result. For example, a hen lays eggs with its back part and eats with its beak. A farmer may consider, "The front part of the hen is very expensive because I have to feed it. Better to cut it off." But if the head is missing there will be no eggs anymore, because the body is dead. Similarly, if we reject the difficult part of the scriptures and obey the part we like, such an interpretation will not help us. We have to accept all the injunctions of the scripture as they are given, not only those that suit us.* (Science of Self Realization Ch.4)

I see two major problems with your interpreting subsidiary texts, i.e. Sūrya-siddhānta and Siddhānta-śiromaṇi.

Firstly, although these texts are valid, they cannot be used to contradict the most important of all scriptures, Śrīmad Bhāgavatam. In other words, Śrīmad Bhāgavatam states clearly that the universe is four billion miles in diameter and it is not proper to use Sūrya-siddhānta and Siddhānta-śiromaṇi as evidence against Śrīmad Bhāgavatam.

Secondly, one should not interpret statements from Sūrya-siddhānta and Siddhānta-śiromaṇi about the circumference of the universe to mean the inside of the universal shells when there is no evidence to substantiate it.

The Vedic scriptures and ācāryas consistently use diameters calculated in terms of yojanas as the means of measuring the sizes of universes. Please consider the following quotation from Caitanya-caritāmṛta:

*kona brahmāṇḍa śata-koṭi, kona lakṣa-koṭi*
*kona niyuta-koṭi, kona koṭi-koṭi*

*SYNONYMS: kona brahmāṇḍa—some universes; śata-koṭi—one billion yojanas; kona—some; lakṣa-koṭi—one trillion yojanas; kona—some; niyuta-koṭi—ten trillion; kona—some; koṭi-koṭi—one hundred trillion.*

*TRANSLATION: "'Some of the universes are one billion yojanas in diameter, some one trillion, some ten trillion and some one hundred trillion yojanas. Thus they are almost unlimited in area.*

*PURPORT: A yojana equals eight miles.* (Caitanya-caritāmṛta, Madhya 21.85)

*brahmāṇḍānurūpa brahmāra śarīra-vadana*
*ei-rūpe pāli āmi brahmāṇḍera gaṇa*

> *SYNONYMS: brahmāṇḍa-anurūpa—according to the size of a universe; brahmāra—of Lord Brahmā; śarīra-vadana—heads on the body; ei-rūpe—in this way; pāli āmi—I maintain; brahmāṇḍera gaṇa—all the innumerable groups of universes.*
>
> *TRANSLATION: "'According to the size of the universe, there are so many heads on the body of Brahmā. In this way I maintain innumerable universes [brahmāṇḍas].* (Caitanya-caritāmṛta, Madhya 21.86)

Here Lord Caitanya is narrating a pastime of Lord Krishna in Dvārakā. Lord Krishna makes it crystal clear that our universe measures only 4 billion miles in diameter because our Brahmā only has four heads whereas other universes measure up to eight hundred trillion miles in diameter (which is 200,000 times the diameter of our universe). This is reasonable because just a few verses earlier Lord Krishna described that the size of the universe is proportionate to the number of heads of the Brahmā managing it. The Brahmās who visited Dvārakā had up to one hundred million heads.

> *"These Brahmās had different numbers of heads. Some had ten heads, some twenty, some a hundred, some a thousand, some ten thousand, some a hundred thousand, some ten million and others a hundred million. No one can count the number of faces they had."* (Caitanya-caritāmṛta; Madhya 21.68)

## Point #8—"Sun's Size"

Verse 4.1 of the Sūrya-siddhānta gives the diameters of the sun and the moon as 6,500 and 480 yojanas, respectively. Given 5 miles per yojana, the resulting lunar diameter of 5X480=2,400 miles is about 11% higher than the modern value. The corresponding earth-moon distance of about 258,000 miles is high by 8.3%. However, the sun's diameter comes to 5X6,500 = 32,500 miles, which is far too small.

It is easy to see why the diameter of the moon should be reasonably accurate. The dimensions of the moon and its orbit were well known in ancient times. For example, the lunar diameter given by Ptolemy in his Planetary Hypotheses falls within about 7% of the modern value, if we convert his earth-diameters into miles using the modern diameter of the earth (Swerdlow, 1968).

It is also easy to see why the diameter for the sun is too small. Ancient astronomers tended to greatly underestimate the earth-sun distance, and the table shows that this also happened in the Sūrya-siddhānta. The angular diameter of the sun is easily seen to be about the same as that of the moon—about ½ degree. This angular diameter, combined with a small earth-sun distance, leads inevitably to a small estimate for the diameter of the sun. Ptolemy's solar diameter figure is similar to the Sūrya-siddhānta's.

**BST #8—**

From this statement it appears that in your eyes, even Sūrya-siddhānta is discredited when it goes against modern astronomy. Although no distance to the sun from the earth can be proven, you hold modern science as correct above the Jyoti Śāstras and above the Śrīmad Bhāgavatam. Regrettably to say that it is clear that you are biased in favor of modern astronomy and thus are not qualified to comment on either the Jyoti Śāstras or the Bhāgavata cosmology.

# NOTES

## *Note #1—"Śrīla Bhaktisiddhānta Sarasvatī Ṭhākura (1874-1937)"

At the time of initiation, Gaura Kisora Das Babaji didn't change Sarasvatī Ṭhākura's name. Earlier in his life he had written a book called "Siddhānta Candrika"— a book on astrology. He received the name "Siddhānta Sarasvatī" from a society of astrologers, in recognition of his contribution in the field of astrology; later adding the prefix "Bhakti" himself when he took sannyasa. (*Bhaktisiddhānta* by Bhakti Vikasa Swami P. 18)

## *Note #2— "Śrīla Bhaktisiddhānta's Commentary"

At present, the only available authorized translation from Sanskrit of any Jyoti Śāstras verses are those included within his commentary on the Caitanya-caritāmṛta by Śrīla Bhaktisiddhānta Sarasvatī Goswāmī Mahārāja. These verses were translated into Bengali from the Sanskrit by Śrīla Bhaktisiddhānta Sarasvatī Ṭhākura. Sections A, B, D and F below are translated into English from the Bengali by Bengali-speaking ISKCON devotees Vamadeva dāsa Brahmacari and Advaita dāsa Brahmacari. Translation into English of the Sanskrit verses in sections C and E are by Gopīparāṇadhana dāsa Adhikārī.

*(A: Bengali)*

*brahmāṇḍa śata-koṭi yojana dharile*
*tadardha pañcāśat koṭi yojana haya*

*brahmāṇḍa—the universe; śata—hundred; koṭi—ten millions; yojana—lengths of eight miles; dharile—if we consider; tadardha—half of that; pañcāśat—fifty; koṭi—ten millions; yojana—lengths of eight miles; haya—becomes.*

*If we consider the universe to be 100 crore yojana (eight billion miles), then half of it becomes fifty crore yojana (four billion miles.)*[from Bengali]*

*(B: Bengali)*

*manu likhiyāchena:*
*svayam ca eva ātmano dhyānāt tad aṇḍam dvidhā*
*sūrya-siddhānte 12 āḍhyāye 90 śloke*

*manu—Manu; likhiyāchena—has written; svayam—personally; ca—also; eva—certainly; ātmanaḥ—within Himself; dhyānāt—by meditation; tad—that; aṇḍam—the universal globe; dvidhā—divided into two parts; sūrya-siddhānte—Sūrya-siddhānta; 12 āḍhyāye—chapter twelve; 90 śloke—text ninety.*

*Translation: Manu has written: "by meditating on this (in Manu's opinion), he divided the universe into two." (Sūrya-siddhānta; chapter twelve, text ninety)**

*(C: Sanskrit)*

*kha-vyoma-kha-traya-kha-sāgara-ṣaṭka-nāga-*
*vyomāṣṭa-śūnya-yama-rūpa-nagāṣṭa-candrāḥ*
*brahmāṇḍa-samputa-paribhramaṇaṁ samantād*
*abhyantare dina-karasya kara-prasāraḥ*

*kha—zero; vyoma—zero; kha-traya—three zeroes; kha—zero; sāgara—four; ṣaṭka—six; nāga—seven; vyoma—zero; aṣṭa—eight; śūnya—zero;*

*yama—two; rūpa—one; naga—seven; aṣṭa—eight; candrāḥ—and one (Reading these digits backwards, as is the standard encoding, gives the number 18,712,080,764,000,000.); brāhmaṇḍa—of the universal egg; sampuṭa—of the container; paribhramaṇam—the circumference; samantāt—in all directions; abhyantare—within; dina-karasya—of the sun; kara—of the rays; prasāraḥ—the extent.*

*The circumference of the universe is 18,712,080,764,000,000 yojanas. Within this area extend the rays of the sun.*

*(D: Bengali)*

*siddhānta-śiromaṇi graha gaṇite madhyam adhikāre*
*kakṣā prakrame tathā golādhyāye bhuvana-kośa 67 śloke*

*siddhānta-śiromaṇi—Siddhānta-śiromaṇi; graha—planets; gaṇite—to calculate; madhyam—middle; adhikāre—portion; kakṣā—orbit; prakrame—orbits; tathā—as well as; gola—round; ādhyāye—chapter; bhuvana-kośa—of all the worlds; 67 śloke—text 67.*

*Translation: In the same book Siddhānta-śiromaṇi, there is a section called "planetary mathematics" and within that there is another small section or middle portion called "orbits" and "universal globes" and within these is another small section called "bhuvana-kośa" or planetary systems. Within that section the 67th sloka is as follows: [This corresponds to Chapter 3 verse 67] [from Bengali]*

*(E: Sanskrit)*

*koṭi-ghair nakha-nanda-ṣaṭ-kha-nakha-*
*bhū-bhūbhṛd-bhujāṅgendubhir*
*jyotiḥ-śāstra-vido vadanti*
*nabhasaḥ kakṣām imāṁ yojanaiḥ*

*tad brahmāṇḍa-kaṭāha-sampuṭa-*
*taṭe kecij jagur veṣṭanaṁ*
*kecit procur adṛśya-dṛśyaka-*
*giriṁ paurāṇikāṁ sūrayaḥ*

*koṭi-ghaiḥ—times ten million; nakha—two; nanda—nine; ṣaṭ-kha —six and zero; nakha—two; bhū—one; bhū-bhṛt—seven; bhujaṅga—eight; indubhiḥ—and one (These digits, read backwards, give the number*

*187,120,692, which should then be multiplied by ten million.); jyotiḥ-śāstra—astronomical science; vidaḥ—those who know; vadanti—say; nabhasaḥ—of the heavens; kakṣām—the circumference; imām—this; yojanaiḥ—measured in yojanas; tat—that; brahma-aṇḍa—of the universe; kaṭāha—like a kettle; sampuṭa—of the container; taṭe—along the edge; kecit—some; jaguḥ—declare; veṣṭanam—the gird; kecit—others; procuḥ—say; adṛśya-dṛśyaka—separating the visible and invisible worlds; girim—the mountain; paurāṇikam—described in the Purāṇas; sūrayaḥ—learned authorities.*

*Persons expert in astronomical science say that the circumference of the heavens measures 18,712,069,200,000,000 yojanas. Some declare that this indicates the gird of the edge of the kettle-like vessel of the universal egg, while other learned authorities say that it is the [distance from the center of the universe to] mountain, described in the Purāṇas, which separates the visible and invisible worlds.[From Sanskrit]*

Note by Gopīparāṇadhana dāsa Adhikārī regarding the verse "E"above:

"I am presuming that there was a mistake in printing the Sanskrit, because otherwise the number does not match the one Śrīla Bhaktisiddhānta Sarasvatī specifies or approximately equal the number in the previous verse. The change I made in the text of the verse, in the first line, was to replace ṣaṭka ("six") with ṣaṭ-kha ("six and zero").

*(F: Bengali)*

*18,712,069,200,000,000 yojana*
*kha-kakṣā uhāke keha keha*
*brahmāṇḍa kaṭāha dvayer milan sthaler parimāṇ*

*18,712,069,200,000,000—eighteen quadrillion, seven hundred and twelve trillion, sixty-nine billion, two hundred million; yojana—lengths of eight miles; kha—ether, space; kakṣām—orbits; uhāke—to that; keha—someone; keha—again someone; brahmāṇḍa—universe; kaṭāha-dvayer—halves of a shell; milan—conjunction; sthale—place; beṣṭan—circumference; parimāṇ—measurement; balen—says;*

*According to the Siddhānta-śiromaṇi, this universe measures 18,712,069,200,000,000 x 8 miles. This is the circumference of this* *universe. According to some, this is only half the circumference. *[from Bengali]*

Note: The verse that *Śrīla Bhaktisiddhānta Sarasvatī* is commenting upon is from Caitany-caritāmṛta:

*kṛṣṇa kahe, "ei brahmāṇḍa pañcāśat koṭi yojana*
*ati kṣudra, tāte tomāra cāri vadana*

*SYNONYMS: kṛṣṇa kahe—Kṛṣṇa said; ei brahmāṇḍa—this universe; pañcāśat koṭi yojana—four billion miles; ati kṣudra—very small; tāte—therefore; tomāra—your; cāri vadana—four* faces.

*TRANSLATION: "Kṛṣṇa said, 'Your particular universe has a diameter of four billion miles; therefore it is the smallest of all the universes. Consequently you have only four heads.* (Caitanya-caritāmṛta, Madhya 21.84)

Śrīla Prabhupāda's purport:

*Śrīla Bhaktisiddhānta Sarasvatī Ṭhākura, one of the greatest astrologers of his time, gives information from the Siddhānta-śiromaṇi that this universe measures 18,712,069,200,000,000 x 8 miles. This is the circumference of this universe. According to some, this is only half the circumference.* (Caitanya-caritāmṛta, Madhya 21.84)

Nothing within Śrīla Bhaktisiddhānta's Bengali commentary to Caitanya-caritāmṛta on this verse indicates that the large figure for the circumference refers to inside the universe, rather it supports the conclusion that the circumference refers to outside the coverings.

## *Note #3— "Bhaskaracharya II (1114-1185)"

Bhaskaracharya otherwise known as Bhaskara is probably the most well known mathematician of ancient Indian today. Bhaskara was born in 1114 A.D. according to a statement he recorded in one of his own works. He was from Bijjada Bida near the Sahyadri mountains. Bijjada Bida is thought to be present day Bijapur in Mysore state. Bhaskara wrote his famous Siddhānta Siromani in the year 1150 A.D. It is divided into four parts; Lilavati (arithmetic), Bijaganita (algebra), Goladhyaya (celestial globe), and Grahaganita (mathematics of the planets). Much of Bhaskara's work in the Lilavati and Bijaganita was derived from earlier mathematicians; hence it is not surprising that Bhaskara is best in dealing with indeterminate analysis. In connection with the Pell equation, $x^2=1+61y^2$, nearly solved by Brahmagupta, Bhaskara gave a method (Chakravala process) for solving the equation.

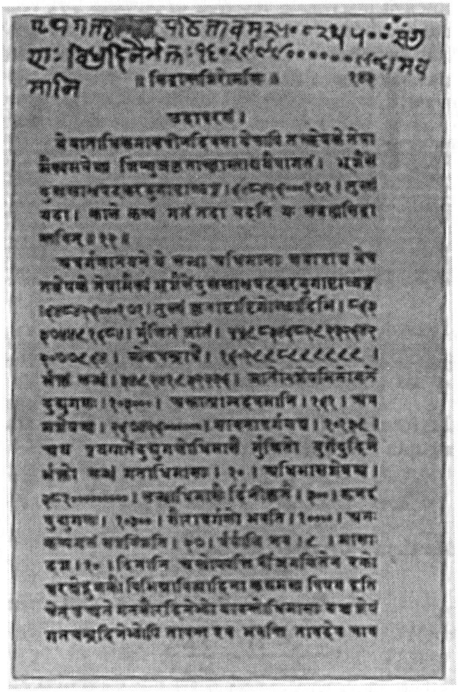

Picture of Goladhyaya.

Bhaskara was somewhat of a poet as were many Indian mathematicians at this time.

Here is couple of verses found in Lilavati that demonstrates his poetic inclination.

बाले मरालकुलमूल दलानिसप्त
तीरे विलास भरमंथरगाण्यपश्यम्
कुर्वचकेलि कलहं कलहंसयुग्मम्
शेषं जले वदमरालकुल प्रमाणम्

***O girl! out of a group of swans, 7/2 times the square root of the number are playing on the shore of a tank. The two remaining ones are playing with amorous fight, in the water. What is the total number of swans? (from http://www.math.sfu.ca/histmath/India/12thCenturyAD/Siddhanta_Siroman.html)***

*In the Goladhyaya chapter of his Siddhānta Śiromaṇi, Bhaskara gave value for the ratio of the circumfrence to the diameter.* ***When the diameter of a circle is multiplied by 3927 and divided by 1250, the quotient is the near circumfrence.*** *He calculated the area and volume of the sphere as follows:* ***In a circle one quarter of the diameter multiplied by 4 is the net all round of the ball. The content or surface of the sphere multiplied by the diameter and divided by 6 is the precise solid or cubic content within the sphere.*** *(See Bibliotheca Indica. Trans. of the Goladhyaya of the Siddhānta-śiromaṇi III., 21-44).*

## *Note #4— "Circumference of the Universe"

Using the formula given by Bhaskara:

The diameter of the inside of the universal sphere is 4,000,000,000 (four billion miles) multiplied by 3927 and divided by 1250 = 12,566,400,000 miles is the near circumference of the inside of the universal sphere.

This is corroborated by the standard Western method for calculating the circumference is 2 pi X radius:

2 X 3.14159 (pi approximately) = 6.28318
6.28318 X 2,000,000,000 (two billion miles) = 12,566,360,000,000 billion miles

## *Note #5— "Jyoti Śāstras Agree with the Purāṇas"

The Jyoti Śāstras agree with the Purāṇas, Bhagavad-gītā, Caitanya-caritāmṛta and all other authorized Vedic scriptures in regard to the universal principles. This is confirmed by Śrīla Prabhupāda himself as follows:

> *These calculations are given in the authentic astronomy book known as the Sūrya-siddhānta. A Bengali translation of this book was compiled by the great professor of astronomy and mathematics Bimal Prasād Datta, later known as Bhaktisiddhānta Sarasvatī Gosvāmī, who was our merciful spiritual master. He was honored with the title Siddhānta Sarasvatī for translating the Sūrya-siddhānta, and the title Gosvāmī Mahārāja was added when he accepted sannyāsa, the renounced order of life. (Caitanya-caritāmṛta Ādi 3.8 Purport)*

Sūrya-siddhānta agrees exactly with the Śrīmad Bhāgavatam about the duration of a manvantara...

स एवं स्वान्तरं निन्ये युगानामेकसप्ततिम्
वासुदेवप्रसङ्गेन परिभूतगतित्रय"

*sa evaṁ svāntaraṁ ninye*
*yugānām eka-saptatim*
*vāsudeva-prasaṅgena*
*paribhūta-gati-trayaḥ*

> *SYNONYMS: saḥ—he (Svāyambhuva Manu); evam—thus; sva-antaram—his own period; ninye—passed; yugānām—of the cycles of four ages; eka-saptatim—seventy-one; vāsudeva—with Vāsudeva; prasaṅgena—by topics connected; paribhūta—transcended; gati-trayaḥ—the three destinations.*

TRANSLATION: He passed his time, which lasted seventy-one cycles of the four ages [71 x 4,320,000 years], always thinking of Vāsudeva and always engaged in matters regarding Vāsudeva. Thus he transcended the three destinations. (Śrīmad Bhāgavatam 3.22.36)

the period of Brahmā's lifespan…

त्रिलोक्या युगसाहस्रं बहिराब्रह्मणो दिनम्
तावत्येव निशा तात यन् निमीलति विश्वसृक्

*tri-lokyā yuga-sāhasraṁ*
*bahir ābrahmaṇo dinam*
*tāvaty eva niśā tāta*
*yan nimīlati viśva-sṛk*

*SYNONYMS: tri-lokyāḥ—of the three worlds; yuga—the four yugas; sāhasram—one thousand; bahiḥ—outside of; ābrahmaṇaḥ—up to Brahmaloka; dinam—is a day; tāvatī—a similar (period); eva—certainly; niśā—is night; tāta—O dear one; yat—because; nimīlati—goes to sleep; viśva-sṛk—Brahmā.*

*TRANSLATION: Outside of the three planetary systems [Svarga, Martya and Pātāla], the four yugas multiplied by one thousand comprise one day on the planet of Brahmā. A similar period comprises a night of Brahmā, in which the creator of the universe goes to sleep.* (Śrīmad Bhāgavatam 3.11.22)

The description of the current seventh Manu's reign…

श्रीशुक उवाच मनुर्विवस्वत" पुत्र"
श्राद्धदेव इति श्रुत" सप्तमो वर्तमानो यस्

*śrī-śuka uvāca*
*manur vivasvataḥ putraḥ*
*śrāddhadeva iti śrutaḥ*
*saptamo vartamāno yas*

*SYNONYMS: śrī-śukaḥ uvāca—Śrī Śukadeva Gosvāmī said; manuḥ—Manu; vivasvataḥ—of the sun-god; putraḥ—son; śrāddhadevaḥ—as*

*Śrāddhadeva; iti—thus; śrutaḥ—known, celebrated; saptamaḥ—seventh; vartamānaḥ—at the present moment; yaḥ—he who; tat—his; apatyāni—children; me—from me; śṛṇu—just hear.*

*TRANSLATION: Śukadeva Gosvāmī said: The present Manu, who is named Śrāddhadeva, is the son of Vivasvān, the predominating deity on the sun planet. Śrāddhadeva is the seventh Manu.* (Śrīmad Bhāgavatam 8.13.1)

That Brahmā is now living in the second half of his life…

यद्दर्धमायुषस्तस्य परार्धमभिधीयते
पूर्व” परार्धोऽपक्रान्तो ह्यपरोऽद्य प्रवर्तते

*yad ardham āyuṣas tasya*
*parārdham abhidhīyate*
*pūrvaḥ parārdho 'pakrānto*
*hy aparo 'dya pravartate*

*SYNONYMS: yat—that which; ardham—half; āyuṣaḥ—of the duration of life; tasya—his; parārdham—a parārdha; abhidhīyate—is called; pūrvaḥ—the former; para-ardhaḥ—half of the duration of life; apakrāntaḥ—having passed; hi—certainly; aparaḥ—the latter; adya—in this millennium; pravartate—shall begin.*

*TRANSLATION: The one hundred years of Brahmā's life are divided into two parts, the first half and the second half. The first half of the duration of Brahmā's life is already over, and the second half is now current.* (Śrīmad Bhāgavatam 3.11.34)

The number of manus within a day of Brahmā…

*ekāttara catur-yuge eka manv-antara*
*caudda manv-antara brahmāra divasa bhitara*

*SYNONYMS: ekāttara—seventy-one; catuḥ-yuge—in cycles of four ages; eka—one; manu-antara—period of a Manu; caudda—fourteen; manu-antara—periods of Manu; brahmāra—of Lord Brahmā; divasa—a day; bhitara—within.*

*TRANSLATION: Seventy-one divya-yugas constitute one manv-antara. There are fourteen manv-antaras in one day of Brahmā. (Caitanya - caritāmṛta Ādi 3.8)*

the duration of the celestial days of the devas…

*satya, tretā, dvāpara, kali, cāri-yuga jāni*
*sei cāri-yuge divya eka-yuga māni*

*SYNONYMS: satya—Satya; tretā—Tretā; dvāpara—Dvāpara; kali—Kali; cāri-yuga—four ages; jāni—we know; sei—these; cāri-yuge—in the four ages; divya—divine; eka-yuga—one age; māni—we consider.*

*TRANSLATION: We know that there are four ages [yugas], namely Satya, Tretā, Dvāpara and Kali. These four together constitute one divya-yuga. (Caitanya-caritāmṛta; Ādi 3.8)*

That the present millennium is the 28th… (see Sūrya-siddhānta chapter 1.15-23)

*'vaivasvata'-nāma ei saptama manv-antara*
*sātāiśa catur-yuga tāhāra antara*

*SYNONYMS: vaivasvata-nāma—named Vaivasvata; ei—this; saptama—seventh; manu-antara—period of Manu; sātāiśa—twenty-seven; catuḥ-yuga—cycles of four ages; tāhāra—of that; antara—period.*

*TRANSLATION: The present Manu, who is the seventh, is called Vaivasvata [the son of Vivasvān]. Twenty-seven divya-yugas [27 x 4,320,000 solar years] of his age have now passed.* (Caitanya-caritāmṛta Ādi 3.9)

Sūrya-siddhānta (Chapter 2 1-5) describes how planets are drawn in orbits by cords of air. This is precisely corroborated within the Śrīmad Bhāgavatam…

यथा मेढीस्तम्भ आक्रमणपशव" संयोजितास्त्रिभिस्त्रिभि" सवनैर्यथास्थानं
मण्डलानि चरन्त्येवं भगणा ग्रहादय एतस्मिन् अन्तर्बहिर्योगेन कालचक्र
आयोजिता ध्रुवमेवावलम्ब्य वायुनोदीर्यमाणा आकल्पान्तं परिचङ् क्रमन्ति
नभसि यथा मेघा" श्येनादयो वायुवशा" कर्मसारथय" परिवर्तन्ते एवं
ज्योतिर्गणा" प्रकृतिपुरुषसंयोगानुगृहीता" कर्मनिर्मितगतयो भुवि न पतन्ति.

*yathā meḍhīstambha ākramaṇa-paśavaḥ saṁyojitās tribhis tribhiḥ savanair yathā-sthānaṁ maṇḍalāni caranty evaṁ bhagaṇā grahādaya etasminn antar-bahir-yogena kāla-cakra āyojitā dhruvam evāvalambya vāyunodīryamāṇā ākalpāntaṁ paricaṅ kramanti nabhasi yathā meghāḥ śyenādayo vāyu-vaśāḥ karma-sārathayaḥ parivartante evaṁ jyotirgaṇāḥ prakṛti-puruṣa-saṁyogānugṛhītāḥ karma-nirmita-gatayo bhuvi na patanti.*

*SYNONYMS: yathā—exactly like; meḍhīstambhe—to the pivot post; ākramaṇa-paśavaḥ—bulls for threshing rice; saṁyojitāḥ—being yoked; tribhiḥ tribhiḥ—by three; savanaiḥ—movements; yathā-sthānam—in their proper places; maṇḍalāni—orbits; caranti—traverse; evam—in the same way; bha-gaṇāḥ—the luminaries, like the sun, the moon, Venus, Mercury, Mars and Jupiter; graha-ādayaḥ—the different planets; etasmin—in this; antaḥ-bahiḥ-yogena—by connection with the inner or outer circles; kāla-cakre—in the wheel of eternal time; āyojitāḥ—fixed; dhruvam—Dhruvaloka; eva—certainly; avalambya—taking support of; vāyunā—by the wind; udīryamāṇāḥ—being propelled; ā-kalpa-antam—until the end of the creation; paricaṅ kramanti—revolve all around; nabhasi—in the sky; yathā—exactly like; meghāḥ—heavy clouds; śyena-ādayaḥ—birds such as the big eagle; vāyu-vaśāḥ—controlled by the air; karma-sārathayaḥ—whose chariot drivers are the results of their own past activities; parivartante—move around; evam—in this way; jyotiḥ-gaṇāḥ—the luminaries, the planets and stars in the sky; prakṛti—of material nature; puruṣa—and of the Supreme Personality, Kṛṣṇa; saṁyoga-anugṛhītāḥ—supported by the combined efforts; karma-nirmita—caused by their own past fruitive activities; gatayaḥ—whose movements; bhuvi—on the ground; na—not; patanti—fall down.*

*TRANSLATION: When bulls are yoked together and tied to a central post to thresh rice, they tread around that pivot without deviating from their proper positions—one bull being closest to the post, another in the middle, and a third on the outside. Similarly, all the planets and all the hundreds and thousands of stars revolve around the polestar, the planet of Mahārāja Dhruva, in their respective orbits, some higher and some lower. Fastened by the Supreme Personality of Godhead to the machine of material nature according to the results of their fruitive acts, they are driven around the polestar by the wind and will continue to be so until the end of creation. These planets float in the air within the vast sky, just as clouds with hundreds of tons of water float in the air or as the great śyena eagles, due to the results of past activities, fly high in the sky and have no chance of falling to the ground.* (Śrīmad Bhāgavatam 5.23.3)

## *Note #6— "Odd Bedfellows"

I did a calculation and found that the outside circumference of the brahmāṇḍa with its coverings equals approximately 18,712,069,200,000,000 x 8 miles doubled which reconciles the extraordinary numbers mentioned in the Jyoti Śāstras for the circumference of the universe.

However there is a book where a scientist commenting on the Bhāgavatam's cosmology stated that the four billion miles declared in the Śrīmad Bhāgavatam as the diameter of the universe only refers to our solar system and not the universe, etc. The author attempts to justify this on the basis of the following quotation:

> *Śrīla Bhaktisiddhānta Sarasvatī Ṭhākura, one of the greatest astrologers of his time, gives information from the Siddhānta-śiromaṇi that this universe measures 18,712,069,200,000,000 X 8 miles. This is the circumference of this universe. According to some, this is only half the circumference.* (Caitanya-caritāmṛta, Madhya 21.84)

An elaborate theory has been built postulating that our solar system alone measures 4 billion miles and that Śrīla Prabhupāda and Śukadeva Goswāmī were not really referring to the whole universe which contained suns, galaxies and quasars billions of light years away. The theory maintains that Śrīla Bhaktisiddhānta Sarasvatī's statement (above) refers to the inside circumference of the universe. This conclusion is induced from reading the Sūrya-siddhānta translated and commented upon by Pandit Bapu Deva Sastri and Lancelot Wilkinson. I acquired that book and on page 126 Bhaskara, the actual author of the book, is speaking, it reads:

> *66. The earth and its mountains, others and also the orbits of the constellations and planets and the Lokas which, it is said, are arranged one above the other, are all included in what has been denominated the Brahmanda (universe).*
>
> *67. Some astronmers have asserted the circumference of the circle of Heaven to be 18,712,069,200,000,000 Yojanas in length. Some say that this is the length of the zone which binds the two hemispheres of the*

*Brahmanda. Some Pauranikas say that this is the length of the circumference of the Lokaloka Parvata.*

*68. Those, however, who have had a most perfect mastery of the clear doctrine of the sphere, have declared that this is the length of that circumference bounding the limits, to which the darkness dispelling rays of the Sun extend.*

*69. But let this be the length of the circumference of the Brahmanda or not: [of that I have no sure knowledge] but it is my opinion that each planet traverses a distance corresponding to this number of Yojanas in the course of a Kalpa or day of Brahma and that it has been called the Kahkasha by the ancients. (end)*

These passages do not indicate that the circumference is to be measured of the inside of the universe. The author of the solar system theory suggests furthermore that perhaps modern astronomers are viewing galaxies existing within the coverings of the universe. Such a attempt to synthesize modern astronomical theory with the Vedic cosmology is not objectionable provided there really is correlation. However, to say that Śrīmad Bhāgavatam and Śrīla Prabhupāda are mistaken or in polite terms, "only referring to the solar system" is simply an unjust concoction.

In another translation of the Sūrya-siddhānta by Phanindralal Gangooly there is a map of of space and stars containing Arab, Chinese and Hindu conceptions of the sky, all superimposed upon one map with Hindu, Arab and Chinese names. It can be noted how the cartographer has demonstrated that different cultures describe the same planets, asterisms and constellations by use of different names. In the area named *Andromeda* we find Vedic stars like Asvini, Revati and Bharani within our universe, with no galaxies. Interestingly, on the map printed in 1860 *Orion* and *Andromeda* which are today considered galaxies lying hundreds of light-years away from us, were formerly accepted by all astronomers as part of the local sky. It is our contention that the modern theory of galaxies and quasars supposedly occupying distances of thousands and millions of light-years from earth are recent exaggerated fantasies similar to Star Wars, and have no verifiable accuracy.

# 3

# Solar System

## Point #1—"Words"

I see two problems in understanding Vedic cosmology: First is the use very flowery language which is admitted in the Vedic literature itself and second is the difficulty in translating the Sanskrit language.

For example: What is a universe, what is a solar system? How do you to use these words? After all, they are just words, English words. And now we figure out mathematically, that the rings of the Bhū-maṇḍala corresponds with the measurements of the scientific orbits of the luminary planets. So that could mean; what somebody has translated as universe could be actually what science labels as solar system. And the solar system is actually a unit in it self, a small universe, if you consider the vast empty space outside the solar system.

## Brahmāṇḍa Sākṣāt Ṭikā (BST)#1—

Perhaps you have overlooked the fact that planetary distances are discussed at length in the Śrīmad Bhāgavatam which was written over 5,000 years ago in the Sanskrit language.

> *This vivid description of how the rays of the sun are distributed throughout the different planetary systems of the universe is very scientific. Śukadeva Gosvāmī described these universal affairs to Mahārāja Parīkṣit as he had heard about them from his predecessor. He explained these facts five*

*thousand years ago, but the knowledge existed long, long before because Śukadeva Gosvāmī received it through disciplic succession. Because this knowledge is accepted through the disciplic succession, it is perfect. The history of modern scientific knowledge, on the contrary, does not go back more than a few hundred years. Therefore, even if modern scientists do not accept the other factual presentations of Śrīmad-Bhāgavatam, how can they deny the perfect astronomical calculations that existed long before they could imagine such things? There is so much information to gather from Śrīmad-Bhāgavatam. Modern scientists, however, have no information of other planetary systems and, indeed, are hardly conversant with the planet on which we are now living.* (Śrīmad Bhāgavatam 5.20.37)

Using the 5.20.2 with Jambūdvīpa having a breadth of 100,000 yojanas:

| | | |
|---|---|---|
| I. | 1. Jambūdvīpa | 100,000 |
| | 2. Salt Water Ocean | 100,000 |
| II. | 3. Plakṣadvīpa | 200,000 |
| | 4. Sugarcane Ocean | 200,000 |
| III. | 5. Śālmalidvīpa | 400,000 |
| | 6. Liquor Ocean | 400,000 |
| IV. | 7. Kuśadvīpa | 800,000 |
| | 8.Clarified Butter Ocean | 800,000 |
| V. | 9. Krauñcadvīpa | 1,600,000 |
| | 10. Milk Ocean | 1,600,000 |
| VI. | 11. Śākadvīpa | 3,200,000 |
| | 12. Yogurt Ocean | 3,200,000 |
| VII. | 13. Puṣkaradvīpa | 6,400,000 |
| | 14. Clear Water Ocean | 6,400,000 |

| | |
|---|---|
| Total 7 islands & 7 oceans | 25,400,000 yojanas (203,200,000 miles) |
| | + |
| Loka-varṣa (inhabited) | 15,800,000 yojanas (126,400,000 miles) |
| | + |
| Aloka-varṣa (uninhabited) | 125,000,000 yojanas (1,000,000,000 |
| miles) | + |
| Golden land | 83,800,000 yojanas (670,400,000 miles) |
| | 250,000,000 yojanas (2,000,000,000 miles) |

Based on the quote from Viśvanātha Cakravartī Ṭhākura.

The descriptions given in the Śrīmad Bhāgavatam regarding the size of the universe are unmistakably referring to the entire universe (*aṇḍa, aṇḍa-kośa*, or *brahmāṇḍa*) and not to just the local solar system. The Vedic conception of a universe includes the seven coverings of earth, water, fire, air, etc. which enclose the smaller inner circle where the planets revolve. Inside the coverings there is a Garbhodaka Ocean filling half the universe, one sun, one moon and millions of planets comprising fourteen planetary systems from Pātālaloka to Satyaloka. The Śrīmad Bhāgavatam and its speakers make no errors. The Sanskrit terms *aṇḍa* or *aṇḍa-kośaḥ* both refer directly to the egg-shaped universe. Kindly consider the following examples.

Mahārāja Parīkṣit asks Śukadeva Goswāmī about details of the universe employing the Sanskrit word *aṇḍa-kośa. (Śrīmad Bhāgavatam 2.8.16)*

*pramāṇam aṇḍa-kośasya bāhyābhyantara-bhedataḥ*

*Also, please describe the inner and outer space of the universe by specific divisions*

In reply Śukadeva Goswāmī cites a discussion where Maitreya Muni speaks to Vidura. Maitreya Muni also uses the same word *aṇḍa-kośa* as he describes the universe. Note that he refers to the entire universe which expands to four billion miles.

*āṇḍakośo bahir ayaṁ pañcāśat-koṭi-vistṛtaḥ*

*This phenomenal material world is expanded to a diameter of four billion miles.* (Śrīmad Bhāgavatam 3.11.40)

*daśottarādhikair yatra*
*praviṣṭaḥ paramāṇuvat*
*lakṣyate 'ntar-gatāś cānye*
*koṭiśo hy aṇḍa-rāśayaḥ*

*The layers or elements covering the universes are each ten times thicker than the one before, and all the universes clustered together appear like atoms in a huge combination.* (Śrīmad Bhāgavatam 3.11.41)

*so 'śayiṣṭābdhi-salile*
*āṇḍakośo nirātmakaḥ*
*sāgraṁ vai varṣa-sāhasram*

*For over one thousand years the shiny egg lay on the waters of the Causal Ocean in the lifeless state.* (Śrīmad Bhāgavatam 3.20.15)

Then Śukadeva cites a discussion where Lord Kapila gives an extremely detailed description of the universe to His mother Devahuti. Here also Lord Kapila uses the word *aṇḍaṁ* to indicate the universal egg.

*etad aṇḍaṁ viśeṣākhyaṁ*
*krama-vṛddhair daśottaraiḥ*
*toyādibhiḥ parivṛtaṁ*
*pradhānenāvṛtair bahiḥ*
*yatra loka-vitāno 'yaṁ*
*rūpaṁ bhagavato hareḥ*

*This universal egg, or the universe in the shape of an egg, is called the manifestation of material energy. Its layers of water, air, fire, sky, ego and mahat-tattva increase in thickness one after another. Each layer is ten times bigger than the previous one, and the final outside layer is covered by pradhāna. Within this egg is the universal form of Lord Hari, of whose body the fourteen planetary systems are parts.* (Śrīmad Bhāgavatam 3.26.52; Lord Kapila to Devahuti)

Then Śukadeva Goswāmī personally explains how Lord Vamanadeva pierced the covering of the universe (*aṇḍa-kaṭāha*) which consists of seven layers—earth, water, fire, etc. (Śrīmad Bhāgavatam 5.17.1)

Then Śrīla Prabhupāda in his purport Śrīmad Bhāgavatam 5.20.38 quotes from the Jyotir Veda a long Sanskrit description from Śrīla Viśvanātha Cakravartī Ṭhākura. Here, the Ṭhākura speaks of the Bhūsphere wherein the "whole universe" is discussed as *āṇḍa-golaka.*

Then Śukadeva Goswāmī explicitly describes the distance between the sun and the inner circumference of the universe employing the term *aṇḍa.*

*aṇḍa-madhya-gataḥ sūryo*
*dyāv-ābhūmyor yad antaram*
*sūryāṇḍa-golayor madhye*
*koṭyaḥ syuḥ pañca-viṁśatiḥ*

> *The sun is situated [vertically] in the middle of the universe, in the area between Bhūrloka and Bhuvarloka, which is called antarikṣa, outer space. The distance between the sun and the circumference of the universe is twenty-five koṭi yojanas [two billion miles]. The sun is situated [vertically] in the middle of the universe, in the area between Bhūrloka and Bhuvarloka, which is called antarikṣa, outer space. The distance between the sun and the circumference of the universe is twenty-five koṭi yojanas [two billion miles].* (Śrīmad Bhāgavatam 5.20.43 Śukadeva Goswāmī to Mahārāja Parīkṣit)

Then at the very end of the Fifth Canto, Śrīla Śukadeva Goswāmī summarizes by stating that the universe which he has been describing throughout the Śrīmad Bhāgavatam which is the same universe described by the other authorities he cited is the entire universe which is divided into fourteen parts. He further explains that this is the version of all the Purāṇas; *āṇḍa-kośo yaś caturdaśadhā purāṇeṣu vikalpita upagīyate.*

This version is also corroborated by many other authorities which are cited elsewhere in the Śrīmad Bhāgavatam such as Brahmā to Nārada, Nārada to Yudhiṣṭhira, and Citraketu to Lord Anantadeva. No where are terms such as *āṇḍa-kośa*, *āṇḍa* or *brahmāṇḍa* used to indicate merely a solar system as indicated by modern astronomers. They only refer to the entire universe with its fourteen planetary systems ranging from Satyaloka down to Pātālaloka. There is no room for speculating that the descriptions of the universe given in the Fifth Canto or anywhere else within the Bhāgavatam refer only to the modern concept of the solar system or to merely a portion of the universe.

## Point #2—"Mathematics"

There is an other language called mathematics, which is a universal language requiring no translation. Mathematical evidence is for me "hard" evidence and thus the strongest points proving that the ancient Indians had higher knowledge which had an extraterrestrial source i.e. Paramatma, Channeling, Demigods, etc.

## BST #2—

Mathematics may be a viable tool, but if we use it to speculate it has no value. Mathematics is a logical system which tells us absolutely nothing about the "real world." Physics is the interpretive bridge between the abstract mathematical world and the "real world." Unfortunately, many scientists don't realize that their interpretations are based on certain assumptions about the world, assumptions which are often held as beliefs. One of those assumptions is that there is a direct correlation between mathematics and reality.

A science background is impressive and can be a valuable instrument in the Lord's hands to counteract the injustices being propounded by ignorant persons in the name of science and truth. To do so however, if I may say, requires the scientist to become really pure and faithful in the words of the Vedas and the ācāryas. With such education this may not be easy since it is difficult to break free from attachments, especially ones which have been held as proven, hard knowledge and scientific evidence. Often mathematics today is put forward as being an almost unassailable, pure, totally accurate body of science which ultimately answers all questions. I agree that mathematics may be a viable tool however success depends on the qualifications of the person using the tool and eventually mathematics have their limits.

The mathematics to which we have access is very limited. The material mathematicians and scientists are unable to understand most physical phenomena, including subtle elements such as mind, intelligence, ego; what to speak of the spirit soul or the supreme spirit soul. These things are not material and thus not accessible by material mathematics. Even for Lord Brahmā to understand how to construct the universe required that he be granted special spiritual empowerment from the Supreme Lord Krishna. How will one explain intelligence or consciousness in terms of mathematics? If material intelligence created mathematics then what generated intelligence, etc.?

## Point #3—"Brahmāṇḍa"

The accurate translation of the word Brahmāṇḍa is the English word solar system, according to my opinion and the opinion of others. And as there are many Brahmāṇḍas in different sizes there are many different solar system in different sizes each with one sun. This opinion is supported mathematically.

## BST #3—

How can you define a Brahmāṇḍa— lit. "Brahmā's egg" (Brahmā + aṇḍa,) without Brahmā? The modern scientists' idea of our solar system includes several planets and a sun but no Brahmaloka. Claiming the scientists' solar system to be a Brahmāṇḍa without including Brahmā is word-jugglery to absurdity. Then after saying Brahmāṇḍa means solar system, i.e. one sun, will you turn around and use the modern scientist's definition of a solar system as being only a tiny part of the universe of "galaxies and stars." Finally, using that logic one ends up with a fictitious so-called Bhāgavatam solar system-universe comprising only the scientist's little solar system without Brahmaloka and Pātālaloka—which are rejected since modern science considers such things as mythological. Then we have supposedly made a synthesis between modern science and the Vedas but actually we have accepted the scientist's theory of the universe. Back in the well Dr. Frog.

## Point #4—"Stars"

Please tell me Mahārāja, what are these stars out there if there is only one sun in the Brahmāṇḍa and if we can not see beyond the coverings?

## BST #4—

The Sanskrit word used most often for stars or constellation of stars is *nakṣatraṁ*. Lord Krishna states in the Bhagavad-gītā that "among stars I am the moon." Śrīla Prabhupāda nicely explains in his purport to Śrīmad Bhāgavatam 5.21.11 as follows:

> *In Bhagavad-gītā (10.21) Kṛṣṇa says, nakṣatrāṇām ahaṁ śaśī: "Of stars I am the moon." This indicates that the moon is similar to the other stars. The Vedic literature informs us that within this universe there is one sun, which is moving. The Western theory that all the luminaries in the sky are different suns is not confirmed in the Vedic literature. Nor can we assume that these luminaries are the suns of other universes, for each universe is covered by various layers of material elements, and therefore although the universes are clustered together, we cannot see from one universe to another. In other words, whatever we see is within this one universe. In*

*each universe there is one Lord Brahmā, and there are other demigods on other planets, but there is only one sun.*

## Point #5—"Solar System Theory"

Strictly speaking, the Bhāgavatam sets the diameter of the Brahmāṇḍa to 500 million yojanas, and this equals the diameter of the Bhū-maṇḍala disk, which cuts it in half. At a traditional length of about eight miles per yojana, this comes to four billion miles. This is far too small for the universe of stars and galaxies, but it does fall in the right range of size for the solar system. (For example the orbit of Uranus has a radius of about 1.8 billion miles.)

## BST #5—

The Sanskrit term *aṇḍa* or *aṇḍa-kośaḥ* both refer directly to the egg-shaped universe not to a minute portion of the universe which you term "solar system." Please consider:

अण्डकोशे शरीरेऽस्मिन् सप्तावरणसंयुते
वैराज" पुरुषो योऽसौ भगवान् धारणाश्रय"

*aṇḍa-kośe śarīre 'smin*
*saptāvaraṇa-saṁyute*
*vairājaḥ puruṣo yo 'sau*
*bhagavān dhāraṇāśrayaḥ*

*SYNONYMS: aṇḍa-kośe—within the universal shell; śarīre—in the body of; asmin—this; sapta—sevenfold; āvaraṇa—coverings; saṁyute—having so done; vairājaḥ—the gigantic universal; puruṣaḥ—form of the Lord; yaḥ—that; asau—He; bhagavān—the Personality of Godhead; dhāraṇā—conception; āśrayaḥ—object of.*

*TRANSLATION: The gigantic universal form of the Personality of Godhead, within the body of the universal shell, which is covered by sevenfold material elements, is the subject for the virāṭ conception.* (Śrīmad Bhāgavatam 2.1.25 Śukadeva Goswāmī to Mahārāja Parīkṣit)

Note that in the verse above Śukadeva Goswāmī clearly outlines the parameters of his discourse, i.e. the entire universe which is covered by seven layers.

वर्षपूगसहस्रान्ते तदण्डमुदके शयम्
कालकर्मस्वभावस्थो जीवोऽजीवमजीवयत्

*varṣa-pūga-sahasrānte*
*tad aṇḍam udake śayam*
*kāla-karma-svabhāva-stho*
*jīvo 'jīvam ajīvayat*

*SYNONYMS: varṣa-pūga—many years; sahasra-ante—of thousands of years; tat—that; aṇḍam—the universal globe; udake—in the causal water; śayam—being drowned; kāla—eternal time; karma—action; svabhāva-sthaḥ—according to the modes of nature; jīvaḥ—the Lord of the living beings; ajīvam—nonanimated; ajīvayat—caused to be animated.*

*TRANSLATION: Thus all the universes remained thousands of eons within the water [the Causal Ocean], and the Lord of living beings, entering in each of them, caused them to be fully animated.* (Śrīmad Bhāgavatam 2.5.34 Brahmā to Nārada)

Note in the verse above that Lord Brahmā, the architect of the universe, uses the same word *aṇḍam* to refer to the exact same object described by Śukadeva Goswāmī, namely "the universal globe."

प्रमाणमण्डकोशस्य बाह्याभ्यन्तरभेदत"
महता चानुचरितं वर्णाश्रमविनिश्चय"

*pramāṇam aṇḍa-kośasya*
*bāhyābhyantara-bhedataḥ*
*mahatāṁ cānucaritaṁ*
*varṇāśrama-viniścayaḥ*

*SYNONYMS: pramāṇam—extent and measurement; aṇḍa-kośasya—of the universe; bāhya—outer space; abhyantara—inner space; bhedataḥ—by division of; mahatām—of the great souls; ca—also; anucaritam—character and activities; varṇa—castes; āśrama—orders of life; viniścayaḥ—specifically describe.*

*TRANSLATION: Also, please describe the inner and outer space of the universe by specific divisions, as well as the character and activities of the great souls, and also the characteristics of the different classifications of the castes and orders of social life.* (Śrīmad Bhāgavatam 2.8.16 Mahārāja Parīkṣit to Śukadeva Goswāmī)

It is most interesting to observe that in the verse above, Parīkṣit Mahārāja requests detailed measurements (*pramāṇam)* about the universe (*aṇḍa-kośasya.*) Śukadeva Goswāmī replies exactly to the point citing the discussion between two exalted scientists Maitreya to Vidura.

विकारै" सहितो युक्तैर् विशेषादिभिरावृत"
आण्डकोशो बहिरयं पञ्चाशत्कोटिविस्तृत"

*vikāraiḥ sahito yuktair*
*viśeṣādibhir āvṛtaḥ*
*āṇḍakośo bahir ayaṁ*
*pañcāśat-koṭi-vistṛtaḥ*

*SYNONYMS: vikāraiḥ—by the transformation of the elements; sahitaḥ—along with; yuktaiḥ—being so amalgamated; viśeṣa—manifestations; ādibhiḥ—by them; āvṛtaḥ—covered; āṇḍa-kośaḥ—the universe; bahiḥ—outside; ayam—this; pañcāśat—fifty; koṭi—ten million; vistṛtaḥ—widespread.*

*TRANSLATION: This phenomenal material world is expanded to a diameter of four billion miles, as a combination of eight material elements transformed into sixteen further categories, within and without, as follows.* (Śrīmad Bhāgavatam 3.11.40)

In his purport, Śrīla Prabhupāda explicitly instructs that the entire universe extends diametrically to four billion miles."

*PURPORT: As explained before, the entire material world is a display of sixteen diversities and eight material elements. The analytical studies of the material world are the subject matter of Sāṅkhya philosophy. The first sixteen diversities are the eleven senses and five sense objects, and the eight elements are the gross and subtle matter, namely earth, water, fire, air, sky, mind, intelligence and ego. All these combined together are distributed throughout the entire universe, which extends diametrically to four billion miles. Besides this universe of our experience, there are innumerable other universes. Some of them are bigger than the present one, and all of them are clustered together under similar material elements as described below.*

If one speaks in learned circles it is expected that he will quote from authoritative Vedic sources to support his statements. I can provide numerous more scriptural references supporting the fact that the Śrīmad Bhāgavatam describes the entire universe.

You have given no such supportive evidence to substantiate your claim that the Bhāgavatam only speaks of the solar system. Rather you have casually dismissed the entire the mature fruit of the desire tree of Vedic literatures *(nigama-kalpa-taror galitaṁ phalaṁ)* in favor of spurious modern astronomical data.

# 4
# Modern Science & the Vedas

## Point #1—"Mysterious Hypothesis"

We should also mention another hypothesis based on the traditions surrounding the Bhāgavatam itself. This is that the text is a product of divine inspiration, and therefore it may encode levels of meaning not understood even by the author himself. According to this hypothesis, the Bhāgavatam may contain information that was unknown to human society in the period of its composition, but which can be discerned in the text now on the basis of our more advanced scientific knowledge. In this case, a simple model of the ecliptic, which could have been based on Greek astronomy, is supplemented by information reflecting advanced knowledge of planetary distances."

## Brahmāṇḍa Sākṣāt Ṭīkā (BST)#1—

Greek astronomy, however laudable, cannot be considered superior to the wisdom of Śrīla Śukadeva Goswāmī. Nor can we accept a mysterious hypothesis where Śukadeva Goswāmī is supposed to be unaware of what he is talking about. Please consider the following verse and purport:

*sa saṁvṛtas tatra mahān mahīyasāṁ*
*brahmarṣi-rājarṣi-devarṣi-saṅghaiḥ*
*vyarocatālaṁ bhagavān yathendur*
*graharkṣa-tārā-nikaraiḥ parītaḥ*

*Śukadeva Gosvāmī was then surrounded by saintly sages and demigods just as the moon is surrounded by stars, planets and other heavenly bodies. His presence was gorgeous, and he was respected by all.*

*PURPORT: In the great assembly of saintly personalities, there was Vyāsadeva the brahmarṣi, Nārada the devarṣi, Paraśurāma the great ruler of the kṣatriya kings, etc. Some of them were powerful incarnations of the Lord. Śukadeva Gosvāmī was not known as brahmarṣi, rājarṣi or devarṣi, nor was he an incarnation like Nārada, Vyāsa or Paraśurāma. And yet he excelled them in respects paid. This means that the devotee of the Lord is more honored in the world than the Lord Himself. One should therefore never minimize the importance of a devotee like Śukadeva Gosvāmī.* (Śrīmad Bhāgavatam 1.19.31)

Śrīla Śukadeva Goswāmī is the speaker of Śrīmad Bhāgavatam which is said to be the ripened fruit of all Vedic literatures.

*nigama-kalpa-taror galitaṁ phalaṁ*
*śuka-mukhād amṛta-drava-saṁyutam*
*pibata bhāgavataṁ rasam ālayaṁ*
*muhur aho rasikā bhuvi bhāvukāḥ*

*O expert and thoughtful men, relish Śrīmad-Bhāgavatam, the mature fruit of the desire tree of Vedic literatures. It emanated from the lips of Śrī Śukadeva Gosvāmī. Therefore this fruit has become even more tasteful, although its nectarean juice was already relishable for all, including liberated souls.* (Śrīmad Bhāgavatam 1.1.3)

Śrīla Śukadeva Gosvāmī delivers unadulterated perfect knowledge.

*Such a stage is attained by knowledge and renunciation, as we have already explained (Śrīmad Bhāgavatam 1.2.12), and perfect knowledge, as delivered by Śrīla Śukadeva Gosvāmī, results in*

> *the attainment of the transcendental service of the Lord.* (Śrīmad Bhāgavatam 1.18.16)

Modern so-called science or astronomy is by no means more advanced than that given in the Śrīmad Bhāgavatam any more than a glow worm can shed light on the sun for clearer vision. That Śukadeva Goswāmī did not understand the cosmology of the Bhāgavatam and that he and Śrīla Prabhupāda are ignorant of what they are being used for and only now with the help of modern science and novel speculation are we able to discern the true import—is unsupported by Vedic śāstra.

According to King Parīkṣit, Śrīla Śukadeva Goswāmī knew everything: "My dear lord, O great devotee sage, you are omniscient." (Śrīmad Bhāgavatam 5.13.26) Had he not been so, why would such personalities as Nārada and Vyāsa waste their valuable time listening to him speak?

> *Apart from our modern distinguished estimations of where the planetary systems are located, we can understand that the sky and its various planets were studied long, long before Śrīmad-Bhāgavatam was compiled. Śukadeva Gosvāmī explained the location of the planets, and this indicates that the information was known long, long before Śukadeva Gosvāmī related it to Mahārāja Parīkṣit. The location of the various planetary systems was not unknown to the sages who flourished in the Vedic age.* (Śrīmad Bhāgavatam 5.16.1 Purport)

## Point #2—"Intellectual Honesty"

You are worried that the teachings of the Vedas will be betrayed and I do understand your concern. This is actually the big mistake of Christianity, since they adjusted the scriptures according to political circumstances. In this context I would like to cite an example of the early Christian Church. In the sixth century, in the Synode of Constantinople (543 A.D.) and in the Council of Constantinople (553 A. D.) as the "intellectuals" of that time, due to political motives aiming to please Justinian I, the Byzantine ruler, dropped the philosophical principles of karma and reincarnation and introduced the dogma of the eternal hell. They ordered the excommunication of Origenes of Alexandria, one of the most influential of the early Christians, and banned his related teachings about karma and reincarnation. This crime on humanity was then signed by Pope Vigilus in the same year.

Once more in the course of history the intellect had sold itself to the baser drives of mankind, and with this, the whole world fell, for the next two thousand years, into a darkness of ignorance whose terrible effects can even be seen today. If this betrayal had not been carried out against the principles of life, against the meaning of truth, against intellectual honesty and against the allegiance to laws and methods of the intellect, today our planet would certainly look different, in any case, better. This example shows clearly why the service to the truth is one of the highest commandments for the brahmanas. It also shows how the intellectuals can change into intensely diabolical behavior by not following this commandment.

## BST #2—

What you described happened in Christianity is what we do when we speculate on the version of Śrīmad Bhāgavatam.

## Point #3—"Interpreting"

Well, I see a different between *interpreting* scriptures and *changing* them. This is not the same thing.

## BST #3—

Misinterpreting them means changing them. As soon as we propound something contrary to Śrīla Prabhupāda's conclusions in the name of mathematics, Aryabhatta, common sense, or anything else, we have failed as true followers and effective disciples of our disciplic succession. I perceive a desire among some of our scientists to become imitation Aryabhattas by presenting their own theories— never mind whether the theories are in agreement with Śrīla Prabhupāda and Śrīmad Bhāgavatam. This is the disease of the *munis* who always differ with other *munis* attempting to establish the superiority of their intelligence. Because of scientific prejudice we doubt the Vedic version. Very dangerous.

## Point #4—"Planetary Orbits"

I THINK WE DON'T EVEN KNOW what the Vedic version of cosmology exactly is, its much too complicated for our brain. But as I said, we can see that the dimensions of planetary orbits and the diameters of the planets itself were known to the Indians for many thousands of years and we here in the West only know of them for the past 100 or 200 years and that's something.

## BST #4—

To some extent it may be said that understanding the cosmos is extremely difficult for the tiny, limited jivas. However we are indeed able to accept what we hear from authorized sources. That much we can know and that is eons more than the modern so-called scientists know about the universe. Śrīmad Bhāgavatam is the top source of all Vedic literature. If we don't have faith in that then we are bewildered to float in the changing astro-mentalsphere—i.e. sometimes accepting this theory, sometimes rejecting it, speculating, etc.

Planetary orbits were described in Śrīmad Bhāgavatam. Why do you suggest that it was necessary to wait until other Indian astronomers provided the mathematical calculations? Such biases against the Vedic literature are to be expected among non-devotees but not amongst devotees. Śrīla Prabhupāda put his faith in the literal Bhāgavatam:

> *So why shall we not believe just that statement of planetary system? Planetary system is already there, but they[the speakers of the Śrīmad Bhāgavatam]are foretelling what in future, it will happen. That is my conviction. Therefore I do not believe anyone except Bhāgavata, Bhagavad-gītā. That is my science. They speculate. I don't believe it. Why shall I...? And in the beginning Vyāsadeva said, kim anyaiḥ śāstraiḥ: "Take only Śrīmad-Bhāgavatam as the book of knowledge. Bas. You need not read any other." Nigama-kalpa-taror galitaṁ phalam idam"This is essence of all Vedic knowledge."* (Morning Conversation — April 23, 1977, Bombay)

## Point #5—"Building Bridges"

If we can not build a bridge between science and the Vedas, between scriptures and common sense, then I guess the best is, in the moment, we sit back and wait until some angel comes from the sky and explains all it in a better way. And until then, there is no need for any temple of Vedic Planetarium, for any exhibitions, universities or to write any papers or even to discuss this topics or to preach. We simply sit back and wait and chant—bhajananandis ki jaya!

## BST #5—

Our task is to present what is explained in the Bhāgavatam. We have no obligation to pander to the so-called common sense of the ordinary people or scientists. If they can understand and accept the Vedic version, very good. If not, we will not fudge the Vedic version to pacify them, although we can try our best to present the facts expertly without compromising. The planetarium must subscribe totally to the Bhāgavatam otherwise it is ludicrous to make an attempt. We can succeed in preparing it provided we are faithful to the spiritual authorities.

## Point #6—"Appeal"

By presenting the Bhāgavata Purāṇa in a non-literal manner, we are able to appeal to the modern man who is accustomed to view things through astronomy and mathematics. Scholars gain new respect for the Vedas by learning that modern science is fully corroborated therein. Please note some comments from scholars about our non-literal interpretations.

> *It is also becoming clear that the Purāṇas are not to be taken literally and once the code underlying their vocabulary is understood, they are revealed as books of the most subtle scientific and spiritual knowledge.*

> *The Bhāgavatam includes some modern understandings of the science of astronomy, not just mythology...*

> *The interpretation of this mythology brings it into focus and ties it to what we have learned in the field of modern astronomy.*

## BST #6—

Yes, if one interprets the Śrīmad Bhāgavatam in such a way to concur with modern astronomy, then naturally those dedicated to modern astronomy will applaud. It is not at all surprising. Yet the monumental error of this interpretation and applause concludes that the Bhāgavatam is viewed in a non-literal perspective—the same prejudice which has flawed Western scholar's vision of the Vedas for hundreds of years remains. Mundane scholars traditionally view Lord Krishna as a mythological character who spoke a charming, yet irrelevant treatise called Bhagavad-gītā.

There would be no harm in harmonizing the Vedas with modern science if that was indeed the case. It seems that you believe in modern science (which is not surprising, most people do) and you are presenting the Vedic version in a way that agrees with modern science. But much of modern science is completely wrong. What we have is the Bhāgavatam clearly enunciating a four-billion mile diameter visible universe with one sun while modern astronomy claims a visible universe comprised of millions of light-years, and thousands of suns. The ill-fated attempt to merge them has brought about astronomical speculation.

## Point #7—"Faith"

If a person develops faith in the authority of the Vedas by seeing that they contain truth in different departments of knowledge like astronomy, then it is easier for such a person to understand the philosophy and practice of Krishna consciousness, which are described in the Vedas.

People today find it very interesting that the *Bhāgavata* understanding of the solar system encodes planetary distance information and sophisticated understandings of the structure and the relative motion of various planets within the universe. The new interpretation whereby the Bhāgavatam describes the local solar system presents understandings with great empiric accuracy, compared to modern science, close to a one-to-one match, without the use of technology. In contrast, the early Greek and medieval calculations were very far from matching our modern scientific calculations.

## BST #7—

Although members of the Vaiṣṇava community and the modern scientific community may, for the moment, rejoice in the proposed synthesis of Vedic and modern scientific cosmology, the ramifications of such a theory are staggering. The Vaiṣṇava community will have endorsed a drastic deviation from parampara i.e. that the Bhāgavatam presents an incorrect or incomplete account of universal cosmology. Or even worse, ISKCON may pay millions of dollars to perpetuate such a bogus theory in its Māyāpur planetarium.

Such "preaching" may attract some people to read the Vedic literature and it may help some unconvinced devotees to have more faith in the descriptions given in Śrīmad Bhāgavatam, but it ultimately undermines the words of the ācāryas and introduces new, speculated interpretations. And what happens when the modern theories change? Evidence from mundane scientific research can sometimes support the Vedic conclusions although it can only take them so far. Eventually we have to surrender to Krishna, accepting the words of guru, sadhu and śāstra as authoritative. It is a huge leap to try to force Uranus's estimated orbit to become the Bhū-maṇḍala planetary system described in the Fifth Canto of Śrīmad Bhāgavatam. Although there is no support from śāstra, one may gather support from some scientists because it supports their theory.

## Point #8—"Hubble Helps"

I maintain that the Fifth Canto has several levels of valid, non-literal interpretation that are built into the text itself. One valid interpretation treats Bhū-maṇḍala as a planisphere map of the earth globe. Another treats Bhū-maṇḍala as a highly accurate model of the solar system. Since Bhū-maṇḍala is being used in the text to represent different things, we are not required to take the Fifth Canto as a single, complete, and literal description of the universe. I also point out that the Bhāgavatam does not describe the vast array of stars, galaxies, quasars, etc. that modern astronomers observe with instruments such as the Hubble telescope, etc.

## BST #8—

Every beginner to Krishna consciousness knows ***not*** to put value in empirical evidence derived from faulty sense perception nor limited mental speculation founded upon impaired man-made instruments such as Hubble telescopes. If all statements found in the Bhāgavatam are judged and condemned according to modern science, we are giving modern science a position superior to the Śrīmad Bhāgavatam.

> *"Devotees are prepared to smash the scientists when it comes to evolution and that is laudable, but why should they become timid when it comes to cosmology? We keep trying to bend and twist Śrīmad Bhāgavatam to fit the modern theories of the universe. We say things like, 'Four billion*

*miles in diameter refers to miles of the demigods' and so on. But why not establish Hubble as wrong and Kṛṣṇa as right? What is Hubble anyway, besides a big mirror attached to a computer? Why not establish that the moon and stars are of the same nature? that our sun illumines the whole universe?*

*"I know it is a daunting task, but someone has to be up to it. The benefit of being a scientist is not that you can lick the big guys' boots and say, 'You are really right, and we will bend the scriptures to fit your theories.' The benefit is that you can smash them in their own language."*(From a letter by His Holiness Umapati Swami; November 25, 2002 to an ISKCON scientist)

This topic has been discussed elaborately by Śrīla Prabhupāda. All of modern astronomical galactic distances are based on theories; i.e. the "big bang," the red shift, Hubble's constant, etc. The big bang is obviously speculation. Some scientists say that the red shift attributed to different brightness's of stars is not due to the Doppler Theory which means their distances from earth cannot be calculated that way. Hubble's constant keeps changing. So how can thoughtful persons put their trust in such flawed ideas based on modern science and calculations acquired through astronomical estimates and data extrapolated from telescopes, etc.? Aside from errors in modern science which are evident, there are those that our spiritual master instructed us about.

Please note Śrīla Prabhupāda's reply to a devotee scientist and original member of the Bhaktivedanta Institute:

*Regarding details of the universe, be satisfied by reading only Bhagavatam. What is the use of reading other books—you are not going there. Some portion of the earth is flat. When you stand in any place you see flat, so for us to some extent it appears flat, but it is round.*

*Simply keep faith that whatever we describe, that is a fact. After all, we are an insignificant creature in the universe, so whether you take the modern scientists or Sukadeva Goswami, it is inconceivable. It is best to keep faith in Sukadeva Goswami, because actually our only business is to go back to home, back to Godhead.* (Letter to: Madhava — Vrindaban 3 October, 1976)

## Point #9—"Hard Core"

I think I can see your point. Your point is correct according to the traditions. But this is not totally satisfying for me, then the problem is, that out there are millions and millions of people who are suffering like anything and those are part and parcels of God as well. And the question is: How do we build the bridge to them?

I have the following proposal: Around the hard core, which is not compatible with the masses of non devotees, we create a soft circle which is a separate institution and the bridge to the outer world. The hard core; the scriptures, the archives, ISKCON, the modern disciplic succession and so on, we keep as it is and (one of) its function is, that nothing gets changed. They are the guardians of the scriptures.... But you are right, they are not compatible with the masses of non devotees, modern science and so on.

So the soft circle is very similar to the community in the beginning of ISKCON, the Tompkins Square Park Community. At that time a lot of the philosophy was not revealed and for example the 5th Canto was not yet translated. You could join just for fun and bring any of your musical instruments. You could join that community without being confronted with hundreds of rules and regulations, dress codes, behaviour patterns and so on.

This soft circle is for all those million people out there who don't want and cannot accept these hundreds of rules and regulations, dress codes, behaviour patterns and those far out scriptures. And as well for those who don't think they are on the absolute platform, for example those who are into modern science and common sense.

This soft circle is based on monotheistic interfaith, like Kṛṣṇa, Kristos and Christos and its task is to present the similarities of the monotheistic world religions and monotheistic yoga paths. It will present the basic philosophy of guna, karma and reincarnation and the way out of it by chanting the holy names of God. It is basically a kitchen religion with vegetarian cooking classes and a lot of music.

To this outer circle there should be no public deity worship, since that does not fit in to the idea of monotheistic interfaith (for example Islam and Christianity is turned away) and it calls for all those rules and regulations, dress codes, behaviour patterns. For those people who want to be involved in these activities, they can move into the hard core. And I think a lot of sincere people will choose that way.

The outer circle will avoid the differences of the monotheistic world religions and monotheistic yoga paths and simply concentrate on the similarities. It will show to the world that these differences are archaic and obsolete. The outer circle is a counter weight to religious fanatics and terrorists. It is a counterweight to the coming war of the cultures; Christianity against Islam against Jews and so on.

The outer circle is based on Bhagavad-gītā 18.66 "Abandon all varieties of religion and just surrender unto Me. I shall deliver you from all sinful reactions. Do not fear." This concept will make the outer circle very popular. It is like a antidote for the sick social body of our days human society. It is the right medicine for our time, place and circumstances, but of course it needs a functional hard core in order to exist. It is like a plant who climbs along a pole to give its fruits.

## BST #9—

My concern is that ISKCON faithfully understand and perpetuate the teachings of the Śrīmad Bhāgavatam, etc. as presented by His Divine Grace A. C. Bhaktivedanta Swami Prabhupāda. As of late I have noticed an increasing tendency for ISKCON scientists to make compromises with modern mundane science theories of cosmology at the cost, it appears to me, of slighting the sacred Vedic texts and their sages. Why compromise with ignorance?

> *Prabhupāda: That is, but we have taken that they are talking all nonsense. And whatever we present, they'll take it as mythology.*
> *Rūpānuga: But like Mādhava Prabhu pointed out, we have to make some explanation. We should explain because they...*
> *Prabhupāda: That you do, but the position is like this. Our conclusion is that whatever they are saying, that is imperfect. That is not possible. And that is a fact. Therefore they change, after fifty years, they change. Because it's speculation. Therefore we say totally they are wrong, and they will take totally "From mythology," like that, this is going on.* (Room Conversation — July 5, 1976, Washington, D.C.)

As soon as someone says that all references found in the Śrīmad Bhāgavatam and other authentic Vedic literatures, as well as all statements made by Śrīla Prabhupāda describing the size of our universe as being four billion miles in diameter are fallacious— he has drastically deviated from parampara. Or if in more polite terms, one says that four billion miles can only describe our solar system and not the entire universe, he is no longer qualified to espouse Vaiṣṇava siddhānta in any Vaiṣṇava media.

> *Śrīla Prabhupāda: Yes. That is not possible for any human being. A human being has imperfect senses. So how can he teach perfect knowledge? Suppose you see the sun as a disc. You have no means to approach the sun. If you say that we can see the sun by telescope and this and that, they are also made by you, and you are imperfect. So how can your machine be perfect? Therefore, your knowledge of the sun is imperfect. So don't teach about the sun unless you have perfect knowledge. That is cheating.*
> *Bob: But what about to teach that it is supposed that the sun is 93,000,000 miles away?*
> *Śrīla Prabhupāda: As soon as you say "it is supposed," it is not scientific.*
> *Bob: But I think that almost all science, then, is not scientific.*
> *Śrīla Prabhupāda: That is the point!*
> *Bob: All science is based on, you know, suppositions of this or that.*
> *Śrīla Prabhupāda: Yes. They are teaching imperfectly.* (Perfect Questions Perfect Answers 1: Kṛṣṇa, the All-Attractive)

## Point #10—"Too"

Apart from this, the Purāṇas indicate that the stars are nearby and that they shine by the reflected light of the sun. In this sense, it could be said that the Purāṇas are too heliocentric.

## BST #10—

Pardon me, the Purāṇas are not *too* anything! Have we forgotten that the revealed scriptures are to be taken as axiomatic truth. What makes us think that we are qualified to pass judgment on the perfect śāstras?

स्वत"प्रमाण वेदप्रमाणशिरोमणि लक्षणा करिले स्वत"प्रमाणताहानि

*svataḥ-pramāṇa veda——pramāṇa-śiromaṇi*
*lakṣaṇā karile svataḥ-pramāṇatā-hāni*

*SYNONYMS: svataḥ-pramāṇa—self-evident; veda—the Vedic literatures; pramāṇa—evidence; śiromaṇi—topmost; lakṣaṇā—interpretation; karile—doing; svataḥ-pramāṇatā—self-evidence; hāni—lost.*

*TRANSLATION: The self-evident Vedic literatures are the highest evidence of all, but if these literatures are interpreted, their self-evident nature is lost.*

*PURPORT: We quote Vedic evidence to support our statements, but if we interpret it according to our own judgment, the authority of the Vedic literature is rendered imperfect or useless. In other words, by interpreting the Vedic version one minimizes the value of Vedic evidence. When one quotes from Vedic literature, it is understood that the quotations are authoritative. How can one bring the authority under his own control? That is a case of principiis obsta.* (Caitanya-caritāmṛta; Ādi 7.133)

The brilliance of a scientist may be compared to the expertise of an experience cook. The cook mixes appropriate spices in precise quantities, heating them just to the proper temperature in pure ghee, then he combines spices with vegetables and produces a delightful dish. If however at some point, the cook gets the terrible notion to add motor oil to the composition, the entire preparation is ruined. Similarly, adding empirical mental speculation to the pure Vedic literature is the half a cup of motor oil which ruins one's understanding.

*Modern scientific calculations are subject to one change after another, and therefore they are uncertain. We have to accept the calculations of the Vedic literature. These Vedic calculations are steady; the astronomical calculations made long ago and recorded in the Vedic literature are correct even now. Whether the Vedic calculations or modern ones are better may remain a mystery for others, but as far as we are concerned, we accept the Vedic calculations to be correct.*
(Śrīmad Bhāgavatam 5.22.9)

## Point #11—"Good Lighting"

I think that whatever your personal convictions may be, you should have mercy on the people in universities, who will have enough trouble accepting Kṛṣṇa Consciousness without having to swallow a 4-billion mile universe. There are alternative interpretations of the Fifth Canto which are well supported by the text, and which put the Bhāgavatam in a good light from a scientific standpoint. I would recommend emphasizing these points rather than taking your approach.

## BST #11—

I feel pity for one who does not accept the distances as explained in the Śrīmad Bhāgavatam by universal authorities such as Lord Brahmā, Lord Krishna, Śukadeva Goswāmī, etc. yet welcomes those of modern astronomers. Modern astronomy includes the theory of universal distances in terms of millions of light-years which are nothing but assumptions according to the Red Shift Theory based on the conjecture of the Big Bang Theory.

> *Prabhupāda: "Big bang." One scientist says, "First of all there was explosion." But nobody will question, "How the explosion took place?" Because they are śūdras. They have no brain. The rascal scientist says. "There was explosion," and he accepts. That's all. This is going on.* (Lecture Śrīmad-Bhāgavatam 1.15.40 — Los Angeles, December 18, 1973)

> *Prabhupāda: Big bang. "And all of a sudden explosion," this is all rascal proposal. Therefore the people have become so rascal, guided by these rascals, "All of a sudden, by chance," and they accept them as scientist... This is the drawback of the present civilization. Because they are śūdras, like animals, they have got no brain to answer that "How all of a sudden there can be explosion." They have no brain even to ask. Rather, they are giving Nobel Prize. This rascal is speaking like a rascal, still, he should get Nobel Prize. That is the defect. The so-called scientists, begin, all beginning is like that. "There were chemicals, these chemicals." Now wherefrom these chemicals came? Who placed these chemicals? They do not ask. Because they are fools, the other fools bluff them and they accept it. But we are not going to accept. We shall inquire. And that is human intelligence.* (Morning Walk — December 18, 1973, Los Angeles)

Hubble's Law describes that after the Big Bang, the evolution of the expanding Universe occurs. It can be most simply stated as

$$v = H \times D \text{ (or } H=v/D\text{)}$$

where v is the velocity of recession of the receding object, D is its distance from the point of observation, and H is Hubble's constant.

Heaven forbid if one dismisses Bhāgavata cosmological information with the wave of a wand and in the next instant expresses unshakable faith in Hubble's telescope and the accompanying theory of planetary distances based on the ever-shifting Red Shift theory.

> *The best present calculation of H gives an age within a factor of 2 of that implied by stellar evolution and cooling-rate dating—not a bad match by current standards. Underlying whatever theory may be applied is an assumption that all the objects have identical velocity histories, which is not necessarily the case. Nor is it even certain whether the velocity of recession is: (1) less than the "escape velocity" (which means the universe is oscillating); (2) greater than it (which means it is "evaporating"); or (3) equal to the escape velocity (which means it is in a static state, sometimes described as "heat death")* (Arthur Ross, freelance science writer describing Yale Scientists' Project Quest)

Furthermore many respected scientists do not accept the Red Shift Theory. Halton Arp served on the staff of the Hale Observatory at Mt. Palomar and is currently doing research at the Max Planck Institute near Munich, West Germany, maintains that red shifts are not due to the Doppler effect, rather, it has something to do with the actual physical state of the object. (Vedic Cosmography And Astronomy 7.B. Anomalous Red Shifts: The Observations of Halton Arp)

> *What is your material knowledge? What do you know? You are trying to go to the moon planet. That is also you committing mistake. And what about the millions of universes?*
> *Therefore our knowledge is very, very limited, poor. Poor fund of knowledge, and still we are proud. Sapari phora-phorayate.(?) Just like a small fish, you have seen in the pond. Little water, say one feet—(makes sound) "phor, phor, phor, phor." But the big fishes, they are in the deep in the water, big fish. So we take knowledge from the big fish, not from the small fish, "phor, phor, phor, phor." So phora-phorayate(?), this kind of knowledge will not satisfy us. We take from the big fish, one who knows. Then who knows? Kṛṣṇa knows. And one who knows from Kṛṣṇa, he knows. Ācāryavān puruṣo veda. One who is ācāryavān, who has taken shelter of ācārya, he knows. Why? Because ācārya receives knowledge from Kṛṣṇa and he distributes knowledge to his disciple. That is perfect knowledge. Ācāryavān puruṣaḥ. Tad-vijñānārthaṁ sa gurum evābhigacchet. These are the instruction. (*Śrīla Prabhupāda's Lecture Śrīmad-Bhāgavatam 3.28.21 — Nairobi, November 1, 1975)

> *Now we have to accept knowledge from Vedas, Vedic knowledge, not this rascal's knowledge. Rascal knowledge is that "Yes," as soon as he comes to the imperfect point, "yes, we are trying." You are trying. What is this trying? Trying means that your knowledge is imperfect. And another rascal will come, he will say, "Now here is the perfect." And ten years after, another rascal will come," No, this is not perfect. This is perfect." This is going on. This is called scientific advancement. This is... Advancement means... But we don't change our Vedic knowledge. We do not say, "Now, Kṛṣṇa, five thousand years ago, said like this. Now we are advanced. We change this line." So change means imperfect knowledge. No change means that is perfect knowledge. So we follow that no change policy. No, not that because I think I have become now advanced, I change this to that. That mean I am not advanced. My knowledge is imperfect. Therefore I am changing.* (Śrīla Prabhupāda's Lecture on Śrīmad Bhāgavatam 1.3.15 — Los Angeles, September 20, 1972)

## Point #12—"Prabhupāda Didn't Understand"

Be that as it may, the fact remains that Śrīla Prabhupāda himself admitted that he was not an expert astronomer.

## BST #12—

Yes, that is true, Śrīla Prabhupāda called himself a layman. However, he was admittedly helped in presenting the Fifth Canto by the greatest astronomer. Please read on.

> Yaśodānandana: *It is mentioned that this Sumeru Parvata, the mountain's height is the same as the width of Jambūdvīpa. So this mountain comes up to here, the same distance as this, 100,000 yojanas. So it's perfectly... This is the same length on this side and also like this. It's made like a big, a big cone on top. It describes, "Of that mountain, Sumeru Parvata, 16,000 yojanas, or 128,000 miles, are within, under." So Sumeru is like this, and it also goes under the Jambūdvīpa planetary system. And therefore the mountain's head above the earth, above here, there is 84,000 yojanas, 672,000 miles above the level. And the mountain's width, the mountain on top, is considered to be 32,000 yojanas, or 256,000 miles. And in the bottom it's 16,000 yojanas. Scientists cannot conceive of this. Their estimation of a mountain is that it must be bigger in the bottom and end*

*up smaller at the top, but Mount Meru is twice as big in the top than it is in the bottom. They cannot understand. Next verse, it describes the different divisions. "Just north of Ilāvṛta-varṣa and going further northward, one after another, are three mountains, namely Nīla Mountain, Śveta Mountain, and Śṛṅgavān Mountain. These mark the borders of the three varṣas, namely Ramyaka, Hiraṇmaya, and Kuru, and separate them from one another. The width of these mountains is..."*

Prabhupāda: And it was not possible for me to digest. (laughs) Somebody else helped me to... I am a layman. I do not know.

Tamāla Kṛṣṇa: How did you write it?

Prabhupāda: Th*at somebody, Kṛṣṇa, helped me. That He manufactured.*

*Yaśodānandana: And these mountains, they extend to the beaches. "It is considered, according to the Bhāga..."*

*Prabhupāda: When I was writing, I was praying Kṛṣṇa that "I do not actually accommodate all this knowledge. Please help me." Yes. That's all right.* (Room Conversation — June 18, 1977, Vṛndāvana)

Kindly do not overlook Śrīla Prabhupāda's statement, " *That somebody, Kṛṣṇa, helped me. That He manufactured.*" If one is eager to point out Śrīla Prabhupāda's admittance of his disqualification as a cosmologist, then he must be equally enthusiastic to recognize the qualifications of the astronomer who manufactured the Fifth Canto purports— and that is Krishna.

## Point #13—"Surrendering to Science"

The *Śrīmad-Bhāgavatam*'s description of the universe is very different from modern science's description. Although the *Bhāgavatam*'s presentation of Earth as the disk-shaped Bhū-maṇḍala may look unrealistic, careful study shows that the *Bhāgavatam* uses Bhū-maṇḍala to represent at least four reasonable and consistent models: (1) a polar-projection map of the Earth globe, (2) a map of the solar system, (3) a topographical map of south-central Asia, and (4) a map of the celestial realm of the demigods.

## BST #13—

Although alluring to some, within non-literal interpretations is the incumbent proposition that the Śrīmad Bhāgavatam's description of the universe is, in reality, nothing more than what modern science calls our solar system. Consequently, through surrender to modern science, one abandons the true statements of the Bhāgavatam. Śrīla Prabhupada's followers must be dedicated to preserve the authentic conclusions of Vedic śāstra not dedicated to proving Śrīmad Bhāgavatam's statements concerning the universe are incorrect or incomplete through use of word jugglery.

> *Prabhupāda: The whole universe, diameter, is pañcaśat-koṭi-yojana... fifty crore yojanas, pañcaśat.So one yojana equal to eight miles, one crore equal to ten million.*
> *Paramahaṁsa: Four billion miles is the diameter.*
> *Prabhupāda: Is the diameter.*
> *Paramahaṁsa: You gave that in The Teachings of Lord Caitanya also.*
> *Prabhupāda: Yes. And the sun is in the middle.*
> *Paramahaṁsa: So two billion miles from the edge of the universe.*
> *Prabhupāda: Yes.* (Morning walk: May 12, 1975)

Śrīla Prabhupāda's books are perfect—one must approach them in the right mood. It is transcendental knowledge coming from Krishna, the Supreme Personality of Godhead to us through disciplic succession. It is the perfect way of acquiring knowledge. There is no need whatsoever to consult modern science for knowledge of the universe. For example, modern science contends that there are many suns in our sky but this is not supported by Vedic evidence.

> *By the force of the pregnancy of Diti, the light of the sun and moon was impaired in all the planets, and the demigods of various planets, being disturbed by that force, asked the creator of the universe, Brahmā, "What is this expansion of darkness in all directions?"*
>
> *PURPORT: It appears from this verse of Śrīmad-Bhāgavatam that the sun is the source of light for all the planets in the universe. The modern scientific theory which states that there are many suns in each universe is not supported by this verse. It is understood that in each universe there is only one sun, which supplies light to all the planets.* (Śrīmad Bhāgavatam 3.15.2 and Purport)

The Bhāgavatam description is completely different from what is currently accepted by science, yet it is correct, although it is difficult for us to fathom. We are biased by the Western ideas. Recently a devotee speaking with scientists presented the Vedic view that the earth is stationary and the sun moves relative to the earth, not the earth around the sun. Several physicists including an expert on relativity, Jonathan Scott, concurred that such a thing had already been postulated. He said that the relative effect of a rotating earth or a rotating universe is the same and couldn't be distinguished by observational data. Scott admitted that since the relative motions are the same, the earth may well be stationary and the universe rotating around it as the Vedic scriptures describe. The whole universe (Śiśumāra-cakra) rotates 360 degrees every 24 hours causing the rising and setting of the sun, moon, planets, stars, etc. Also the sun moves around the earth and one rotation takes a full year. It is not the earth rotating at all.

## Point #14—"Polar Projection Map Or?"

Consider Bhū-maṇḍala, which means "earth disk." This structure is described in the Fifth Canto as a flat plate, four billion miles in diameter, marked by various circular features. Is this meant by Śrīla Vyāsadeva to be a physically real object? If we look closely at the text, we see that the answer is no. Consider the following:

"King Parīkṣit inquired from Śukadeva Gosvāmī: My dear lord, you have already affirmed the truth that the supremely powerful sun-god travels around Dhruvaloka with both Dhruvaloka and Mount Sumeru on his right. Yet at the same time the sun-god faces the signs of the zodiac and keeps Sumeru and Dhruvaloka on his left. How can we reasonably accept that the sun-god proceeds with Sumeru and Dhruvaloka on both his left and right simultaneously?"

Śukadeva Goswāmī responds,

> *"When a potter's wheel is moving and small ants located on that big wheel are moving with it, one can see that their motion is different from that of the wheel because they appear sometimes on one part of the wheel and sometimes on another. Similarly, the signs and constellations, with Sumeru and Dhruvaloka on their right, move with the wheel of time, and the antlike sun and other planets move with them. The sun and planets, however, are seen in different signs and constellations at different times. This indicates that their motion is different from that of the zodiac and the wheel of time itself."*

Śukadeva Goswāmī is using the example of ants on a potter's wheel to illustrate the idea of relative motion. However, in the earlier verses alluded to by King Parīkṣit, both the clockwise (with Meru on the right) and the counterclockwise (with Meru on the left) motions of the sun were referred to the same reference frame, namely Bhū-maṇḍala. This means that Bhū-maṇḍala is being used for two different purposes. It is the frame of reference for the daily clockwise motion of the sun, and it is the frame of reference for the yearly counterclockwise motion of the sun.

If we carefully study the Fifth Canto, we see that many verses support each of these two uses of Bhū-maṇḍala. Thus Bhū-maṇḍala is not one physical thing. The author of the Bhāgavatam is using it as a literary device to explain at least two different astronomical points. Thus the message of the Fifth Canto lies in the astronomy being explained by Śukadeva Goswāmī. It does not lie in a contradictory and impossible theory of the earth as a disk four-billion miles in diameter.

Are we flat earthers? A literal, physical interpretation of Bhū-maṇḍala would require this. Yet Śrīla Prabhupāda himself said that the earth is a globe. Indeed, if one carefully follows the words of the text (which is the true meaning of literal reading), then we see that Bhū-maṇḍala is not meant as a physical object. In fact, one valid interpretation of Bhū-maṇḍala is that it is a polar projection map of the earth.

## BST #14—

Part of the problem may be your opening statement; "Consider Bhū-maṇḍala, which means "earth disk." We would be better to use the definition of the Śrīmad Bhāgavatam and Śrīla Prabhupāda for the Bhū-maṇḍala (Śrīmad Bhāgavatam 5.16.1)

> भूमण्डल bhū-maṇḍala—of the planetary system known as Bhū-maṇḍala;

You say that Bhū-maṇḍala is not to be taken as a physical object because Śukadeva Goswāmī explains two motions with reference to the sun and Bhū-maṇḍala namely the daily and yearly. But Bhū-maṇḍala is a real physical object relative to the rotation of the constellations and the sun's movement. Please consider the following statement by Śrīla Prabhupāda.

> *From the Srimad-Bhagavatam we learn that the universe is like a tree with the roots being upwards. The pole star which is situated within the constellation is the root. The universe is pivoting around the pole star. That is one movement. The second movement is that the sun is revolving*

> *around the universe, or as if it were going around the tree. That is the second movement. The sun is not fixed. If this is proved, then their whole scientific theory collapses. According to them, the stars are all suns. The stars may have the same composition as the sun but they are not suns. The Brahma-samhita says that the sun is the king with unlimited light and temperature. The Srimad-Bhagavatam sublime literature describes all these things and it was written five thousand years ago. And they say that five thousand years ago there was no civilization. That Srimad-Bhagavatam has this information proves that Indian civilization is the oldest civilization.* (Letter to: Svarupa Damodara — Bombay 21 November, 1975)

In the letter above written to the director of the Bhaktivedanta Institute, ISKCON's scientific preaching arm, Śrīla Prabhupāda further demonstrates how the daily and yearly sun-earth functions are simultaneously executed. The sun moves with the constellations around the pole star producing the year and the sun's up and down motions produce the days.

At first you cite Śrīmad Bhāgavatam with reference to the daily and yearly sun-earth movements as if you seriously respect its words and then unceremoniously denounce Śrīmad Bhāgavatam as presenting a fantasy which is not reality. I see no logic in this suddenly jumping to the conclusion that Bhū-maṇḍala is not a physical reality simply because of the description of the sun's daily and yearly motions. Certainly Śrīla Prabhupāda did not subscribe to such whimsy. Similarly, kindly note in the following conversation how Śrīla Prabhupāda explains that the sun travels at sixteen thousand miles per second. It is this *dakṣiṇāyana-mārga* and *uttarāyaṇa-mārga* movement which accounts for day and night.

> *Tamāla Kṛṣṇa: All the books from Rādhā-Dāmodara temple, those original texts, they're all there. So he's going to consult your Guru Mahārāja's book on... The astronomy book? What is the name of that book?*
> *Prabhupāda: Sūrya-siddhānta.*
> *Tamāla Kṛṣṇa: We are going to consult some other texts to get the full picture.*
> *Prabhupāda: And what is from Bhāgavata? The sun movement?*
> *Tamāla Kṛṣṇa: It's just like there's a couple of things that we want to get very clear. Like it's described that one of the axles...*

*Prabhupāda: Uttarāyaṇa, dakṣiṇāyana. This Sumeru Mountain... So six months, northern side; six months, southern side.*
*Tamāla Kṛṣṇa: Yes. Some things are quite clear. Then there's a description, though, that there's an axle, it's described, a tie made of wind going from the chariot to Dhruvaloka. So things like that we want to get a..., further information on. Just so that it can be somehow demonstrated. Our... To understand these things for the purpose of making an exhibit requires a very clear picture. So that's the only reason we're looking to other books. But only bona fide books. The ācāryas' commentaries, like that. And then it has to agree with the Bhāgavatam. If in any way there's a discrepancy, we choose the Bhāgavatam as the authority.*
*Prabhupāda: On the whole, the sun is not fixed up.*
*Tamāla Kṛṣṇa: No, not at all. It's moving. As a matter of fact, it describes sometimes it moves in one way with Meru at its right side, and then sometimes it moves the other way with Meru at its left side.*
*Prabhupāda: That is dakṣiṇāyana-mārga, dakṣiṇāyanam, uttarāyaṇa.*
*Tamāla Kṛṣṇa: It's very much moving and at very high speed.*
*Prabhupāda: Yes. I calculated sixteen thousand miles per second, so far I remember.*
*Tamāla Kṛṣṇa: Yes. I think that's right.*
*Prabhupāda: And it is Sūryaloka. It has population.*
*Tamāla Kṛṣṇa: All of the personalities there have very dazzling bodies.*
*Prabhupāda: Fiery body.* (Room Conversation With Son (Vrindavan De) July 5, 1977, Vṛndāvana)

# 5

# Cosmography

## Point #1—"Yogurt"

Let us discuss the Yogurt Ring-Ocean in the Bhū-maṇḍala. Please tell me Mahārāja, do you actually believe there is some yogurt floating around there in outer space? I mean, it is a fact that one speaker was criticized by devotees when he said in a lecture that, "This is not actually yogurt as you buy it in the supermarket."

Since we have the mathematical proof that the Bhū-maṇḍala disk is the ecliptic plane of our solar system, lets consider (for a short moment) the Bhū-maṇḍala is made of elements like for example yogurt and so on. That would mean that half of the stars in the night sky are not visible, since light doesn't travel trough yogurt that well. So for me, the dimension of that yogurt ocean is very interesting, but that term "yogurt" is in itself flowery language as described in Bhagavad-gītā. In other words, could you please tell me what "dadhi-manda—emulsified yogurt" exactly means ... cosmologically ?

## Brahmāṇḍa Sākṣāt Ṭīkā (BST)#1—

The Bhāgavatam speaks of "*dadhi-maṇḍa*—emulsified yogurt." These are not flowery words. Flowery words means descriptions of enjoyments on the higher planets.

*"Men of small knowledge are very much attached to the flowery words of the Vedas, which recommend various fruitive activities for elevation to heavenly planets, resultant good birth, power, and so forth. Being desirous of sense gratification and opulent life, they say that there is nothing more than this."*

In another verse, of the Fifth Canto of Śrīmad Bhāgavatam the words "*dadhi-maṇḍa-udena*" are described as "by an ocean containing water resembling churned yogurt." Those different oceans are mentioned throughout the Bhāgavatam in different cantos; 5th, 7th, 9th, etc. These statements are to be taken literally, not metaphorically. This same question was presented to Śrīla Prabhupāda.

*Tamāla Kṛṣṇa: Just like as we're talking, we're a little... Sometimes it's difficult to understand the Bhāgavatam.*
*Prabhupāda: But I take it simply—that there is ocean, and it was churned. So there is no difficulty.*
*Tamāla Kṛṣṇa: No. Mostly we have understood. Only in one place we are a little...*
*Prabhupāda: But nobody has seen that ocean. And nobody can believe that ocean can be churned. Would you believe it? Because it is. And the Vāsuki was taken as rope.*
*Bhakti-prema: And this Mandara mountain, fifty thousand miles high, was taken there, carried by.*
*Prabhupāda: And it was born by tortoise incarnation.*
*Bhakti-prema: This is combined with description.*
*Prabhupāda: Acintyāḥ khalu ye bhāvā na tāṁs tarkeṇa yojayet. What is beyond your conception, don't try to. So that is Vedic civilization. They were satisfied with information received from the Vedas.*
*Bhakti-prema: The first and last thing we have to prove logically that this is...*
*Prabhupāda: No, no. Logically you cannot. Acintya. Logic comes when it is conceivable; but it is inconceivable. Where is your logic?*
*Bhakti-prema: But first to prove them right understanding we have to bring them.*
*Prabhupāda: It is very difficult. Acintya.*
*Yaśodā-nandana: If such a thing as going from one place to the other on earth, this is cintya, this is conceivable.*

> *Prabhupāda: Argument is there when it is conceivable. It is inconceivable. Kṛṣṇa lifted Govardhana. Seven years old, six years old. How it is conceivable? But devotees, they are crying, "Oh, Giri-vara-dhārī." And the Māyāvādīs they say kalpanāyā. The Akhandananda was saying. And Śukadeva Gosvāmī is narrating something kalpanāyā and wasting his time? This is acintya for the Māyāvādīs. They say kalpanāyā. These Akhandananda and other Māyāvādīs, they explain Bhāgavata-kalpanāyā.* (Discussion about Bhū-maṇḍala — July 5, 1977, Vṛndāvana)

Let us take careful note that it is the Māyāvādīs who say that Bhāgavatam is imagination (*kalpanā*) or that Śukadeva Goswāmī is imagining the dimensions of the universe. Those who say that Śukadeva Goswāmī is imagining that the universe is four billion miles in diameter are influenced by the Māyāvāda tendency.

## Point #2—"Flat Earthers?"

The Bhāgavatam describes the universe literally as a flat disk (Bhū-maṇḍala) situated in the middle of a sphere. The sun is described as orbiting in a circular path situated parallel to this disk and very close to it. In modern astronomy, the ecliptic is the plane of the sun's apparent orbit around the earth, as seen by an earthly observer. By examining the Bhāgavatam's description of the sun's orbit, we can see that Bhū-maṇḍala must closely correspond to the ecliptic if we take it as a literal disk.

In the Bhāgavatam, the sun's orbit is described in terms of a metaphorical solar chariot. The Bhāgavatam verse 5.21.13 gives a description of this chariot, where it is said that the chariot wheel is composed of the parts of the year (Samvatsara in Sanskrit):

> *The chariot of the sun-god has only one wheel, which is known as Saṁvatsara. The twelve months are calculated to be its twelve spokes, the six seasons are the sections of its rim, and the three cātur-māsya periods are its three-sectioned hub. One side of the axle carrying the wheel rests upon the summit of Mount Sumeru, and the other rests upon Mānasottara Mountain. Affixed to the outer end of the axle, the wheel continuously rotates on Mānasottara Mountain like the wheel of an oil-pressing machine.*

## BST #2—

Some confusion may arise from the use of the terms "earth," "earth planet," "earth globe," "earthly planetary system," "the world" etc. Śrīla Prabhupāda interchanges these words and their corresponding meanings freely.

1) The "earth, et al" can mean Bhū-maṇḍala , i.e. the massive earthly planetary system spanning the diameter of the universe.
2) "Earth" can also refer to Jambūdvīpa, which is the middle of the seven dvīpas or islands or planetary systems within Bhū-maṇḍala.
3) Jambūdvīpa is comprised of nine varṣas, khandas, dvīpas or valleys of which Bhārata-varṣa is one. "Earth" can therefore also indicate Bhārata-varṣa which formerly was known as Ajanābha-varṣa before the rule of Bharata Mahārāja.

The Bhāgavatam describes Bhū-maṇḍala as resembling a lotus flower. "Bhū-maṇḍala" is translated by Śrīla Prabhupāda as the earthly planetary system. Maṇḍala means planetary system.

यो वायं द्वीप" कुवलयकमलकोशाभ्यन्तरकोशो नियुतयोजनविशाल" समवर्तुलो
यथा पुष्करपत्रम.

*yo vāyaṁ dvīpaḥ kuvalaya-kamala-kośābhyantara-kośo niyuta-yojana-viśālaḥ samavartulo yathā puṣkara-patram.*

*SYNONYMS: yaḥ—which; vā—either; ayam—this; dvīpaḥ—island; kuvalaya—the Bhūloka; kamala-kośa—of the whorl of a lotus flower; abhyantara—inner; kośaḥ—whorl; niyuta-yojana-viśālaḥ—one million yojanas (eight million miles) wide; samavartulaḥ—equally round, or having a length and breadth of the same measurement; yathā—like; puṣkara-patram—a lotus leaf.*

*TRANSLATION: The planetary system known as Bhū-maṇḍala resembles a lotus flower, and its seven islands resemble the whorl of that flower. The length and breadth of the island known as Jambūdvīpa, which is situated in the middle of the whorl, are one million yojanas [eight million miles]. Jambūdvīpa is round like the leaf of a lotus flower.* (Śrīmad Bhāgavatam 5.16.6)

The original team of devotees to research the Bhū-maṇḍala under Śrīla Prabhupāda's direct supervision constructed the design of the universe with the Lokāloka Mountain surrounding the Bhū-maṇḍala and rising up like the petals of a lotus flower. They took the Bhāgavatam literally which is the way it was intended thus they envisioned the lotus of Bhū-maṇḍala. See below:

> *Yaśodā-nandana: So the whole planetary system of Bhū-maṇḍala is like a big lotus flower with very high, high petals. It is very wonderful.*
> *Tamāla Kṛṣṇa: Which is that Lokāloka?*
> *Yaśodā-nandana: This is the Lokāloka mountain.*
> *Tamāla Kṛṣṇa: Just see how much it is, Śrīla Prabhupāda. It's like the whole Bhū-maṇḍala is protected by a huge mountain. And only in the very center are the inhabited lands of Bhū-maṇḍala.*
> *Prabhupāda: The sun is above them?*
> *Tamāla Kṛṣṇa: Not above Lokāloka. Inside, within.*
> *Yaśodā-nandana: If this would be on the floor, that means Mount Meru would be 84,000 yojanas. Then, above Mount Meru, 16,000 yojanas above, is the sun. But then, so that the rays of the would not penetrate in that land all around, this great mountain, Lokāloka, extends all the way up to Svarga. So it is like a big, big cup, in which the middle of the cup, or the bottom, there is all of these planets, all of this Bhū-maṇḍala.* (Bhū-maṇḍala Diagram Discussion — July 2, 1977, Vṛndāvana)

## Point #3—"Not Flat Earth—Petal Earth"

Your interpretation leaves us with a flat earth, which we know Śrīla Prabhupāda, the Vedas, and common sense reject.

## BST #3—

By charging those who accept the literal statements of the Śrīmad Bhāgavatam as being "flat earthers" one suggests that they believe that the planet on which we live is flat. Śrīla Prabhupāda agrees that the earth (Bhū) is round (gola) yet he addresses this topic from the broader perspective of the greater earthly planetary system.

> *In Vedic literature it is bhū-gola, jagad-aṇḍa. These words are there. We can see also it is round, jagad-aṇḍa. The universe is round. And Goloka. Or Bhū-gola. Bhū-gola, the earth is round. So in the Vedic literatures...*

> *Therefore their knowledge is also imperfect because they do not refer to the Vedic literatures. It is already there. Bhū-gola. Bhū means the earth; gola means round. It is already there. And the geography's called, according to Sanskrit, it is called Bhū-gola. Long, long ago, before Galileo.* (Morning Walk—December 9, 1973, Los Angeles)

> *Jagad-aṇḍa means... Aṇḍa, it is round. Aṇḍa means egg shape. We don't say it is flat. Śāstra never says, "Flat." Then you correct, "No, no, it is not flat. It is round." This is śāstra. Long, long years, in the beginning, it was called round, aṇḍa. Goloka, go, Goloka, Bhurloka. Goloka means round. Goloka Vṛndāvana. Everything is round. So jagad-aṇḍa. Aṇḍa, aṇḍa means egg shape. Jagad-aṇḍa-koti. The each and every universe is egg shape; it is covered.* (Śrīmad-Bhāgavatam 3.28.21—Nairobi, November 1, 1975)

So we followers of Śrīla Prabhupāda also say that the earth is round. Still if one insists that we discuss this globe that the world calls earth, Śrīla Prabhupāda refers to the lotus shape of Bhū-maṇḍala as round and this part upon which we live, this earth is one petal of that lotus.

> *Lotus is also round. So within one lotus petal this round or that round, there is. So in one lotus petal you are conditioned. You cannot go. There are so many petals. You are conditioned with one petal. You do not know what is going on outside. That we have to hear from authority. Unless you are obstinate, you have to accept if you are convinced. There are so many millions of stars and moons that we cannot go. Yes. That fact we have learned from Bhāgavatam. That is you are unable to, but the fact is that you are conditioned. You cannot go beyond that conditioned. So we are also conditioned. But as far as possible we take description from Bhāgavata, try to. That is our... Suppose here is India, here is Los Angeles. You start from India, Los Angeles..., or India, you'll come to Los Angeles. And again return to India. Similarly you start from this again going. No, we don't say flat. That is the explanation. Petal.*

But if one still disagrees with Śrīla Prabhupāda's version, one is prejudiced. Accepting the Śrīmad Bhāgavatam literally does not mean one thinks the earth is flat. Bhāgavatam says Bhū-maṇḍala is lotus-like (kamalasya) and I agree with the Bhāgavatam yet you say Bhāgavatam says it is flat.

*Prabhupāda: You are Western. You are prejudiced. That is reason. If I can understand why you...*
*Tamāla Kṛṣṇa: I can understand blindly, but I don't want to do that.*
*Prabhupāda: I do not understand blindly.*
Tamāla Kṛṣṇa: No.
*Prabhupāda: Then.*
*Tamāla Kṛṣṇa: But I don't want to...*
*Prabhupāda: Still we differ. That means one of us is prejudiced.*
*Tamāla Kṛṣṇa: Well it's not Your Divine Grace. (laughter) I mean I'm sorry I have to take this thankless task to ask all these questions.*
*Prabhupāda: You are also prejudiced. It is written in the śāstra. In that sense we are also prejudiced.*
*Tamāla Kṛṣṇa: Yeah, but we're correctly prejudiced.*
*Prabhupāda: Just like I take Kṛṣṇa's word. Bas, fact. You can say that you are prejudiced, you see. This is the book.*
*Tamāla Kṛṣṇa: That is a nice explanation, the petals. Prabhupāda: Everyone is prejudiced. But who is rightly prejudiced who is wrongly prejudiced. Yes. Now who is correct? Who will decide? I accuse you, you accuse me. But who is correct? Who will decide? That is śāstra. There is no question of argument. That is called pratiṣṭhā. You will never come to any conclusion by arguing. I think you are wrong, you think I am wrong. Somebody must decide. Judge. And that is śāstra. They say, we are saying so many things. In the court there are two parties. I say something, another party says. The judge is there. He has now decision. Yes. That the judge gives. Judges according to lawbook. Tad-vijñānārthaṁ sa gurum evābhigacchet. So we have to approach the right person who can give the judgment.* (Discussion about Bhū-maṇḍala — July 5, 1977, Vṛndāvana)

## Point #4—"All-inclusive Attempt"

The presentation of four non-literal interpretations of the Fifth Canto is the most coherent and consistent view I have heard yet. It is the best attempt I have seen at squaring ALL (not just some) of Śrīla Prabhupāda's comments and the rest of the Vedic presentation found in the various Purāṇas etc.

## BST #4—

If we accept the Śrīmad Bhāgavatam's direct statements, we have a complete cosmological understanding. If we accept these new theories we are left without scriptural answers to even basic questions. Then we are taken back to the position where we are totally at the mercy of modern science and our own speculation to answer questions about the cosmos. Is this what we desire or is this what Śrīla Prabhupāda taught us? Based on the new theories how would we answer these questions?

1. What is the size of our universe?
2. How far is Satyaloka from earth?
3. Why did the divine inspiration not allow Śukadeva Goswāmī to understand what he was talking about but it did allow us with the help of modern science to understand it?
4. Why did Śukadeva Goswāmī say one thing when he actually meant another?
5. Why were the entire fourteen planetary systems elaborately described in the Śrīmad Bhāgavatam as part of the four billion mile universe?

## Point #5—"Foolish Fundamentalists"

I am feeling rather frustrated by your refusal to concede any point, no matter what arguments are given to back it up. Also, I am rather dismayed by your harsh language, your extreme statements, and your evident conviction that your views are the absolute truth.

Our respective positions can be summed up as follows:

I maintain that the Fifth Canto has several levels of valid, non-literal interpretation that are built into the text itself. One valid interpretation treats Bhū-maṇḍala as a planisphere map of the earth globe. Another treats Bhū-maṇḍala as a highly accurate model of the solar system. Since Bhū-maṇḍala is being used in the text to represent different things, we are not required to take the Fifth Canto as a single, complete, and literal description of the universe. I also point out that the Bhāgavatam does not describe the vast array of stars, galaxies, quasars, etc. that modern astronomers observe with instruments such as the Hubble telescope, etc. Thus I maintain that the Bhāgavatam gives a highly sophisticated and scientific description of the earth and the solar system, and that it leaves out any discussion of galaxies, quasars, and most stars. It was the prerogative of the author of the Bhāgavatam to mention or omit topics according to his desire.

Your view, if taken literally (which is certainly what you intend) maintains that the earth is flat, since Bhū-maṇḍala is certainly described literally as a flat plate. However, Śrīla Prabhupāda himself said that the earth is a globe, even though the Fifth Canto literally says otherwise.

By putting us into the flat earth camp, and the tiny universe camp, you are playing into the hands of the mundane scholars. Scholars at universities have generally (but not always) viewed Puranic cosmology literally up until now. They say that it should be taken literally as primitive, pre-scientific mythology. If we go with you, scholars will take for granted that the text should be taken literally, and they will deplore the power of religious fundamentalism to make intelligent people embrace outmoded medieval ideas.

So the line up will be: (a) devotees back the tiny universe, and (b) scholars marvel at the foolishness of fundamentalists. It is one thing to criticize the glaring errors of Darwinism and another to claim that all stars and galaxies are within 2 billion miles (what to speak of the flat earth).

## BST #5—

How can we, as representatives of the Brahmā-Madhva-Gauḍīya sampradaya, support a theory that does not accept our brahmāṇḍa as being four billion miles inside diameter? It is unthinkable. Yet your new theory adamantly demands that we do so. No 4 billion mile universe. Why not? Because we cannot realize how it can be so and/or because it does not agree with modern science's view.

The evidence is so overwhelmingly complete yet you obstinately resist accepting the straightforward statements of the śāstras and the spiritual master. How can one misconstrue verses such as these?

विकरै?सहितो युक्तैर् विसेसादिभिरवृह्
अन्दकोसो बहिरयम् पन्चसत्कोतिविस्त्रह्

*vikāraiḥ sahito yuktair*
*viśeṣādibhir āvṛtaḥ*
*āṇḍakośo bahir ayaṁ*
*pañcāśat-koṭi-vistṛtaḥ*

*SYNONYMS: vikāraiḥ—by the transformation of the elements; sahitaḥ—along with; yuktaiḥ—being so amalgamated; viśeṣa—manifestations; ādibhiḥ—by them; āvṛtaḥ—covered; āṇḍa-kośaḥ—the universe; bahiḥ—outside; ayam—this; pañcāśat—fifty; koṭi—ten million; vistṛtaḥ—widespread.*

*TRANSLATION: This phenomenal material world is expanded to a diameter of four billion miles, as a combination of eight material elements transformed into sixteen further categories, within and without, as follows.* (Śrīmad Bhāgavatam 3.11.40)

The exact words पञ्चाशत्कोटि *pañcāśat-koṭi* are used in both verses from the Bhāgavatam and the exact translation is 50 kotis. A koti equals 10,000,000 (ten million) so the number becomes 500,000,000 or five hundred million yojanas. A yojana equals eight miles so the total diameter of the entire universe is 4,000,000,000 (four billion) miles. Why one would insist upon concocting different interpretations is the real mystery of the universe.

*एतावाल्लोकविन्यासो मानलक्षणसंस्थाभिर्विचिन्तित" कविभि"*
*स तु पञ्चाशत्कोटिगणितस्य भूगोलस्य तुरीयभागोऽयं लोकालोकाचल".*

*etāvāl loka-vinyāso māna-lakṣaṇa-saṁsthābhir vicintitaḥ kavibhiḥ*
*sa tu pañcāśat-koṭi-gaṇitasya bhū-golasya turīya-bhāgo 'yaṁ*
*lokālokācalaḥ.*

*SYNONYMS: etāvān—this much; loka-vinyāsaḥ—the placing of the different planets; māna—with measurements; lakṣaṇa—symptoms; saṁsthābhiḥ—as well as with their different situations; vicintitaḥ—established by scientific calculations; kavibhiḥ—by learned scholars; saḥ—that; tu—but; pañcāśat-koṭi—500,000,000 yojanas; gaṇitasya—which is measured at; bhū-golasya—of the planetary system known as Bhūgolaka; turīya-bhāgaḥ—one fourth; ayam—this; lokāloka-acalaḥ—the mountain known as Lokāloka.*

*TRANSLATION: Learned scholars who are free from mistakes, illusions and propensities to cheat have thus described the planetary systems and their particular symptoms, measurements and locations. With great deliberation, they have established the truth that the distance between Sumeru and the mountain known as Lokāloka is one fourth of the diameter of the universe—or, in other words, 125,000,000 yojanas [1 billion miles].*

*PURPORT: Śrīla Viśvanātha Cakravartī Ṭhākura has given accurate astronomical information about the location of Lokāloka Mountain, the movements of the sun globe and the distance between the sun and the circumference of the universe. However, the technical terms used in the astronomical calculations given by the Jyotir Veda are difficult to translate into English. Therefore to satisfy the reader, we may include the exact Sanskrit statement given by Śrīla Viśvanātha Cakravartī Ṭhākura, which records exact calculations regarding universal affairs.* (Śrīmad Bhāgavatam 5.20.38 and Purport)

# NOTES

## *Note #1—"Long Sanskrit Quote"

In the Fifth Canto of Śrīmad Bhāgavatam there are a few, long Sanskrit prose texts with no translations placed within purports by Śrīla Prabhupāda. The longest of such texts is found in 5.20.38 as follows:

*PURPORT: Śrīla Viśvanātha Cakravartī Ṭhākura has given accurate astronomical information about the location of Lokāloka Mountain, the movements of the sun globe and the distance between the sun and the circumference of the universe. However, the technical terms used in the astronomical calculations given by the Jyotir Veda are difficult to translate into English. Therefore to satisfy the reader, we may include the exact Sanskrit statement given by Śrīla Viśvanātha Cakravartī Ṭhākura, which records exact calculations regarding universal affairs.*

*sa tu lokālokas tu bhū-golakasya bhū-sambandhāṇḍa-golakasyety arthaḥ; sūryasy eva bhuvo 'py aṇḍa-golakayor madhya-vartitvāt kha-golam iva bhū-golam api pañcāśat-koṭi-yojana-pramāṇaṁ tasya turīya-bhāgaḥ sārdha-dvādaśa-koṭi-yojana-vistārocchrāya ity arthaḥ bhūs tu catus-triṁśal-lakṣonapañcāśat-koṭi-pramāṇā jñeyā; yathā meru-madhyān mānasottara-madhya-paryantaṁ sārdha-sapta-pañcāśal-lakṣottara-koṭi-yojana-pramāṇam; mānasottara-madhyāt svādūdaka-samudra-paryantaṁ ṣaṇ-ṇavati-lakṣa-yojana-pramāṇaṁ tataḥ kāñcanī-bhūmiḥ sārdha-sapta-pañcāśal-lakṣottara-koṭi-yojana-pramāṇā evam ekato meru-lokālokayor antarālam ekādaśa-śal-lakṣādhika-catuṣ-koṭi-*

*parimitam anyato 'pi tathatyeto lokālokāl loka-paryantaṁ sthānaṁ dvāviṁśati-lakṣottarāṣṭa-koṭi-parimitaṁ lokālokād bahir apy ekataḥ etāvad eva anyato 'py etāvad eva yad vakṣyate, yo 'ntar-vistāra etena hy aloka-parimāṇaṁ ca vyākhyātaṁ yad-bahir lokālokācalād ity ekato lokālokaḥ sārdha-dvādaśa-koṭi-yojana-parimāṇaḥ anyato 'pi sa tathety evaṁ catus-triṁśal-lakṣonapañcāśat-koṭi-pramāṇā bhūḥ sābdhi-dvīpa-parvatā jñeyā; ata evāṇḍa-golakāt sarvato dikṣu sapta-daśa-lakṣa-yojanāvakāśe vartamāne sati pṛthivyāḥ śeṣa-nāgena dhāraṇaṁ dig-gajaiś ca niścalī-karaṇaṁ sārthakaṁ bhaved anyathā tu vyākhyāntare pañcāśat-koṭi-pramāṇatvād aṇḍa-golaka-lagnatve tat tat sarvam akiñcit-karaṁ syāt cākṣuṣe manvantare cākasmāt majjanaṁ śrī-varāha-devenotthāpanaṁ ca durghaṭaṁ syād ity adikaṁ vivecanīyam.* (Śrīmad Bhāgavatam 5.20.38)

Many a reader has been intrigued by this lengthy uncharted text. Now, by the kindness and expertise of His Grace Gopīparāṇadhana Prabhu who has provided Sanskrit translation, we can understand the meaning of Śrīla Viśvanātha Cakravartī Ṭhākura.

*sa tu lokālokas tu bhū-golakasya bhū-sambandhāṇḍa-golakasyety arthaḥ;*

*And that Lokāloka [is the one-fourth part] of the Bhū sphere, which means the sphere of the whole universe, which is connected with Bhū.*

*sūryasya eva bhuvo 'py aṇḍa-golakayor madhya-vartitvāt kha-golam iva bhū-golam api pañcāśat-koṭi-yojana-pramāṇaṁ tasya turīya-bhāgaḥ sārdha-dvādaśa-koṭi-yojana-vistārocchrāya ity arthaḥ bhūs tu catus-triṁśal-lakṣonapañcāśat-koṭi-pramāṇā jñeyā;*

*Since Bhū is present, just like the sun, in the middle zone of the universal egg, thus the sphere of Bhū, like the sphere of the heavens, measures fifty crore yojanas, the one-quarter part of which is, in other words, a width extending 12½ crore yojanas. More properly, however, Bhū should properly be understood to measure 34 lakhs less than 50 crore.*

*yathā meru-madhyān mānasottara-madhya-paryantaṁ sārdha-sapta-pañcāśal-lakṣottara-koṭi-yojana-pramāṇam; mānasottara-madhyāt svādūdaka-samudra-paryantaṁ ṣaṇ-ṇavati-lakṣa-yojana-pramāṇaṁ tataḥ kāñcanī-bhūmiḥ sārdha-sapta-pañcāśal-lakṣottara-koṭi-yojana-pramāṇā evam ekato meru-lokālokayor antarālam ekādaśa-lakṣādhika-*

*catuṣ-koṭi-parimitam anyato 'pi tathatyeto lokālokāl loka-paryantaṁ sthānaṁ dvāviṁśati-lakṣottarāṣṭa-koṭi-parimitaṁ lokālokād bahir apy ekataḥ etāvad eva anyato 'py etāvad eva yad vakṣyate, yo 'ntar-vistāra etena hy aloka-parimāṇaṁ ca vyākhyātaṁ yad-bahir lokālokācalād ity ekato lokālokaḥ sārdha-dvādaśa-koṭi-yojana-parimāṇaḥ anyato 'pi sa tathety evaṁ catus-triṁśal-lakṣonapañcāśat-koṭi-pramāṇā bhūḥ sābdhi-dvīpa-parvatā jñeyā;*

*This is reasoned as follows: From the middle of Meru to the middle of Mānasottara is an extent of one crore, 57½ lakh yojanas; from the middle of Mānasottara to the ocean of sweet water is an extent of 96 lakh yojanas; from there to the Golden Land is an extent of one crore, 57½ lakh yojanas; also the intermediate distance on one side of Meru to Lokāloka is 4 crore, 11 lakh yojanas, and the same on the other side—consequently the space extending from Lokāloka to the loka is 8 crore, 22 lakhs. When it is said that beyond Lokāloka as well, there is the same distance on one side and also on the other, this is properly explained by saying that the measurement of the aloka region is equal to the extent of the inner part. The same distance extending outward from Lokāloka Mountain as 12½ crore yojanas on one side, is the same also on the other side. Thus we should understand that Bhū along with its oceans, islands, and mountains extends 50 crore minus 34 lakhs.*

*ata evāṇḍa-golakāt sarvato dikṣu sapta-daśa-lakṣa-yojanāvakāśe vartamāne sati pṛthivyāḥ śeṣa-nāgena dhāraṇaṁ dig-gajaiś ca niścalī-karaṇaṁ sārthakaṁ bhaved anyathā tu vyākhyāntare pañcāśat-koṭi-pramāṇatvād aṇḍa-golaka-lagnatve tat tat sarvam akiñcit-karaṁ syāt cākṣuṣe manvantare cākasmāt majjanaṁ śrī-varāha-devenotthāpanaṁ ca durghaṭaṁ syād ity-adikaṁ vivecanīyam.*

*Therefore, since there is an empty space present of 17 lakh yojanas on all sides, it is feasible that the earth is held by Śeṣa Nāga and kept steady by the elephants of the directions. Otherwise, if we gave a differing explanation, making Bhū extend 50 crores, it would be stuck to the shell of the universe and these ideas [of Śeṣa Nāga and the elephants keeping the earth in place] would be meaningless. It would also not have been feasible in the Cākṣūṣa-manvantara for the earth to suddenly become submerged and then be lifted by Śrī Varāhadeva. We should take these and similar facts into consideration.*

## *Note #2—"Niyuta"

In the Śrīmad Bhāgavatam, 5.16.5 it says that the width of Jambūdvīpa is one million yojanas (eight million miles.)

*yo vāyaṁ dvīpaḥ kuvalaya-kamala-kośābhyantara-kośo niyuta-yojana-viśālaḥ samavartulo yathā puṣkara-patram.*

*SYNONYMS: yaḥ—which; vā—either; ayam—this; dvīpaḥ—island; kuvalaya—the Bhūloka; kamala-kośa—of the whorl of a lotus flower; abhyantara—inner; kośaḥ—whorl; niyuta-yojana-viśālaḥ—one million yojanas (eight million miles) wide; samavartulaḥ—equally round, or having a length and breadth of the same measurement; yathā—like; puṣkara-patram—a lotus leaf.*

*TRANSLATION: The planetary system known as Bhū-maṇḍala resembles a lotus flower, and its seven islands resemble the whorl of that flower. The length and breadth of the island known as Jambūdvīpa, which is situated in the middle of the whorl, are one million yojanas [eight million miles]. Jambūdvīpa is round like the leaf of a lotus flower. (Śrīmad Bhāgavatam 5.16.5)*

Two verses later, the distance is given as 100,000 yojanas or 800,000 miles for the width of Jambūdvīpa.

*eṣāṁ madhye ilāvṛtaṁ nāmābhyantara-varṣaṁ yasya nābhyām avasthitaḥ sarvataḥ sauvarṇaḥ kula-giri-rājo merur dvīpāyāma-samunnāhaḥ karṇikā-bhūtaḥ kuvalaya-kamalasya mūrdhani dvā-triṁśat sahasra-yojana-vitato mūle ṣoḍaśa-sahasraṁ tāvat āntar-bhūmyāṁ praviṣṭaḥ.*

*SYNONYMS: eṣām—all these divisions of Jambūdvīpa; madhye—among; ilāvṛtam nāma—named Ilāvṛta-varṣa; abhyantara-varṣam—the inner division; yasya—of which; nābhyām—in the navel; avasthitaḥ—situated; sarvataḥ—entirely; sauvarṇaḥ—made of gold; kula-giri-rājaḥ—the most famous among famous mountains; meruḥ—Mount Meru; dvīpa-āyāma-samunnāhaḥ—whose height is the same measurement as the width of Jambūdvīpa; karṇikā-bhūtaḥ—existing as the pericarp; kuvalaya—of this planetary system; kamalasya—like a lotus flower; mūrdhani—on the top; dvā-triṁśat—thirty-two; sahasra—thousand; reyojana—yojanas*

> *(eight miles each; vitataḥ—expanded; mūle—at the base; ṣoḍaśa-sahasram—sixteen thousand yojanas; tāvat—so much; āntaḥ-bhūmyām—within the earth; praviṣṭaḥ—entered.*
>
> *TRANSLATION: Amidst these divisions, or varṣas, is the varṣa named Ilāvṛta, which is situated in the middle of the whorl of the lotus. Within Ilāvṛta-varṣa is Sumeru Mountain, which is made of gold. Sumeru Mountain is like the pericarp of the lotuslike Bhū-maṇḍala planetary system. The mountain's height is the same as the width of Jambūdvīpa—or, in other words, 100,000 yojanas [800,000 miles]. Of that, 16,000 yojanas [128,000 miles] are within the earth, and therefore the mountain's height above the earth is 84,000 yojanas [672,000 miles]. The mountain's width is 32,000 yojanas [256,000 miles] at its summit and 16,000 yojanas at its base.* (Śrīmad Bhāgavatam 5.16.8)

The difference is due to the usage of the Sanskrit word *niyuta*. *Niyuta* is defined as "a very high number, generally a million" in the Monier-Williams Sanskrit Dictionary. In verse five *niyuta* it taken as "one million" and in the seventh verse *niyuta* is taken as "a very high number." The distance of 100,000 yojanas or 800,000 miles for the width of Jambūdvīpa is confirmed not only in the seventh verse of the sixteenth chapter of the Fifth Canto of Śrīmad Bhāgavatam but also in other places.

## *Note #3—"Brahmapurī"

Śrīmad Bhāgavatam 5.16.29 gives an account for the size of Brahmapurī, the "summer resort residence" of Lord Brahmā which is situated upon Mount Sumeru.

> *meror mūrdhani bhagavata ātma-yoner madhyata upakḷptāṁ purīm ayuta-yojana-sāhasrīṁ sama-caturasrāṁ śātakaumbhīṁ vadanti.*
>
> *SYNONYMS: meroḥ—of Sumeru Mountain; mūrdhani—on the head; bhagavataḥ—of the most powerful being; ātma-yoneḥ—of Lord Brahmā; madhyataḥ—in the middle; upakḷptām—situated; purīm—the great township; ayuta-yojana—ten thousand yojanas; sāhasrīm—one thousand; sama-caturasrām—of the same length on all sides; śāta-kaumbhīm—made entirely of gold; vadanti—the great learned sages say.*

*TRANSLATION: In the middle of the summit of Meru is the township of Lord Brahmā. Each of its four sides is calculated to extend for ten million yojanas [eighty million miles]. It is made entirely of gold, and therefore learned scholars and sages call it Śātakaumbhī.*

The Sanskrit words used are *ayuta-yojana-sāhasrīṁ*. *Ayuta* translates as ten thousand, *yojana* is the standard measurement for distances used in the Vedas which corresponds to eight miles and *sāhasrīṁ* means "one thousand" or "any very large number" (according to Monier-Williams Sanskrit Dictionary). Therefore an alternative meaning for this verse would be that each of the four sides of Brahmapurī is calculated to extend to the very large number of ten thousand yojanas.

This version appears to be accepted also by Śrīla Prabhupāda who wrote in the Chapter Summary to Śrīmad Bhāgavatam's sixteenth chapter of the Fifth Canto as follows:

*"On the summit of Mount Sumeru is Brahmapurī, the residence of Lord Brahmā. Each of its four sides is 10,000 yojanas long."*

## *Note #4—"Mountain Lengths"

A question may arise in regards to Śrīmad Bhāgavatam verse 5.16.8. The translation reads:

*Going from south to north, the length of each mountain is one tenth that of the previous mountain, but the height of them all is the same.*

However in drawing the Jambūdvīpa diagram as described, we have noticed that the intended meaning could read:

Going from south to north, the length of each mountain decreases by one tenth from the previous mountain, but the height of them all is the same.

The relevant Sanskrit is:

पूर्वस्मादुत्तर उत्तरो दशांशाधिकांशेन दैर्घ्य एव ह्रसन्ति.

*pūrvasmād uttara uttaro daśāṁśādhikāṁśena dairghya eva hrasanti* pūrvasmāt—than the former; pūrvasmāt—than the former; uttaraḥ—further north; uttaraḥ—further north; daśa-aṁśa-adhika-aṁśena—by one tenth of the former; dairghyaḥ—in length; eva—indeed; hrasanti—become shorter.

## *Note #5—"Bhārata-varṣa's Former Name"

Many times Śrīla Prabhupāda referred to this planet as Bhārata-varṣa which was former known as Ilāvṛta-varṣa. He also refered to Bhārata-varṣa as formerly known Ajanābha which is corroborated in the Śrīmad Bhāgavatam 5.7.3 as follows:

अजनाभं नामैतद्वर्षं भारतमिति यत आरभ्य व्यपदिशन्ति

*ajanābhaṁ nāmaitad varṣaṁ bhāratam iti yata ārabhya vyapadiśanti*

*SYNONYMS: ajanābham—Ajanābha; nāma—by the name; etat—this; varṣam—island; bhāratam—Bhārata; iti—thus; yataḥ—from whom; ārabhya—beginning; vyapadiśanti—they celebrate.*

*TRANSLATION: Formerly this planet was known as Ajanābha-varṣa, but since Mahārāja Bharata's reign, it has become known as Bhārata-varṣa.*

*PURPORT: This planet was formerly known as Ajanābha because of the reign of King Nābhi. After Bharata Mahārāja ruled the planet, it became celebrated as Bhārata-varṣa.*

The latter of the two usages (i.e. *ajanābha-varṣa*)is especially pertinent since is directly stated in the śāstra as well.

## *Note #6—"Arjuna Conquers Harivarsa"

*Arjuna was becoming successful in his conquests. He first overpowered the king of the Kulindas, then the Anastas, then the Kalkuttas. He defeated the many kings on the island continent of Sakala, then encountered King Bhagadatta, a powerful warrior. The battle with Bhagadatta lasted eight days, but finally when the king found himself overpowered, he agreed to accept Yudhisthira's rule.*

*Arjuna continued toward the north. He reached the mountains and vanquished the many barbarian tribes. Arriving in Kashmir, he subdued the region's kshatriyas, then passed over the Himalayas. To the north of the mountains lay the land inhabited by the Kimpurushas, semi-celestial beings born of Yaksha women. Incensed at the approach of a human army, they fought a fierce battle, but finally Arjuna subdued them. The Pandava then conquered the land of the Guhakas, another tribe of celestial fighters, followed by a region inhabited by Gandharvas.*

*Finally Arjuna arrived in Harivarsha, the land where the Northern Kurus dwelt. At the border, a number of powerful, large-bodied guards stopped him. They said, "O Arjuna, this land cannot be entered by humans. If you try you will perish along with your army. Indeed, even if you were able to enter, you would not see anything, because human eyes cannot see this land or its residents. Go back. There is nothing to be conquered here. Your conquests are already sufficient. We are pleased with you and will happily offer you a gift. What would you like.'"'*

*Arjuna bowed respectfully to the divine beings and said, "I desire Yudhisthira's imperial dignity. If you accept him as the emperor of this wide earth, then please give something as a tribute."*

*The Northern Kurus immediately offered Arjuna a large number of celestial clothes and ornaments, saying, "We know you and your brothers to be great servants of the Supreme Lord. Go now with our blessings. May you always gain victory."*

*Taking all the wealth he had won, Arjuna returned to Indraprastha and reported his success to Yudhisthira. (Mahābhārata: The Rajasuya Sacrifice p. 171)*

## *Note #7—"Arjuna Goes to Heaven"

*Yudhisthira then informed Arjuna of Vyasadeva's instructions. He told him that he should leave as soon as possible for the Himalayas and, by meditating on the mantras he was now going to repeat, he should seek the gods' audience in order to receive their weapons. "Allow me to initiate you today, O virtuous one, and go at once to propitiate Indra. Being pleased with you I am sure Indra will give you his weapons and, by his order, so will the other gods."*

*Arjuna fully controlled his mind and senses. Then, with due rites, Yudhisthira bestowed the Pratismriti mantras on him. When he was finished Yudhisthira stood up, with tears pricking his eyes as he thought of Arjuna's separation. "Now go, dear brother."*

*Arjuna smiled at Draupadi, circumambulated his brothers and Dhaumya, and then began running swiftly along the path, frightening creatures along the way with his speed. By chanting the mantras Yudhisthira had given him, he felt himself travelling over mountains and forests at the speed of mind. By the end of the first day he arrived at the great Mandara mountain. Arjuna stopped and looked around. The mountain was beautiful with its bluish stone rising up into the clouds. It was covered with blossoming trees, their many-colored flowers creating rich tableaus and their scent captivating his mind. The sound of peacocks, cranes and cuckoos filled the air, and he could see Siddhas and Charanas sporting on the mountain slopes. Arjuna decided to climb the mountain and begin his austerities there.*

*Upon reaching a plateau high on the mountain, he suddenly heard a voice resounding in the sky. "Stop!" Arjuna looked around and saw an ascetic sitting at the foot of a tree. The tawny-colored brahmin had a brilliant aura. His lean body was covered in deerskin and his matted locks hung down to his shoulders. The ascetic said, "O child, who are you? You appear to be a kshatriya. Do you not know that this is the abode of peaceful brahmins who are free from anger? Even to have been able to reach this spot indicates that you already attained a high state of purity. Now perfect your life. You have no need of weapons here. Throw them away."*

*Arjuna made it clear that he had no plans to renounce his weapons. "I need my weapons to serve my elder brother Yudhisthira, who has been wrongfully deprived of his kingdom. O brahmin, I will not cast them aside."*

*Seeing Arjuna's dedication to duty, the ascetic replied, "O slayer of foes, I am pleased with you. Know me to be Shakra, king of the gods. Ask me for a boon."*

*Arjuna was overjoyed. Here was Indra!…*

*[After some time] Indra disappeared, and Arjuna was left alone. He decided to remain where he was and begin his austerities and worship of Shiva. He took off his armor and set his weapons down by the same tree where Indra had sat. The tree was situated next to a flowing river, and Arjuna bathed in its clear waters, then sat down to meditate. As he did so, conchshells were heard in the sky and showers of flowers fell from the heavens. Arjuna was pleased by the auspicious sign, and he sat with half-closed eyes, controlled his breathing, and began to intone prayers to Shiva.*

*During the first month of his meditations, Arjuna ate only fruit on every third day. During the second month, he ate fruit only on every sixth day, and during the third month he ate fruit once a fortnight. During the fourth month he lived only on air, standing on the tips of his toes with his arms upraised. He bathed three times a day and kept his mind and senses under perfect control. As a result of his austerities, Arjuna began to glow like the sun.*

*In the heavens the celestial rishis grew anxious. They approached Shiva and said, "Arjuna has become greatly effulgent due to his austerities on the breast of the Himalayas. The earth is becoming heated by his asceticism and is sending forth smoke. O chief of the gods, you should stop him before he upsets the universal order by the power of his penance."*

*Shiva replied, "You need not feel anxiety on account of Phalguni. He does not wish to attain heaven or prosperity. I know his purpose and will satisfy him today."*

*The rishis bowed before Shiva and returned to their own abodes. Shiva then assumed the form of a tall, powerful, golden-hued hunter, and descended to the place where Arjuna was meditating. Uma, his consort, accompanied him in a similar costume, and many of his goblin followers, who assumed various forms and wore bright garments, followed them. Many of the female goblins assumed the forms of beautiful women and also descended. As Shiva appeared on the mountainside with his followers, the mountain seemed ablaze with beauty. Strangely, however, all nature became silent. Even the springs and waterfalls ceased their sounds.*

*When Shiva was gone, Arjuna stood for a few moments gazing at the sky, awed by what had just transpired. He had seen the god of gods. The unknowable deity had touched him with his own hand. Arjuna now considered his enemies already vanquished and his ends accomplished. He possessed Shiva's irresistible weapon. Although he could not discharge the Pashupata at human foes, its very possession made him feel invincible. While Arjuna stood absorbed in thought, Varuna appeared before him accompanied by the rivers personified and many Nagas, Siddhas and other lesser gods. Kuvera was also in attendance, his body resembling pure gold and seated upon a splendid chariot. Kuvera was accompanied by countless Yakshas. They stood before Arjuna, illuminating everything with their bodily effulgence. Arjuna also saw in the sky the god of justice, Yamaraja, approaching on his chariot with mace in hand and flanked by the personified forms of Death and Time.*

*On a nearby mountain summit, Arjuna saw Indra and his queen Sachi seated on the back of the celestial elephant Airavata and appearing like the rising sun. A white umbrella was being held over his head and he was surrounded by Gandharvas and rishis, who were eulogizing him with Vedic hymns.*

*Arjuna chanted numerous prayers in praise of the gods and offered them fruits and water. As he stood gazing in amazement, Yamaraja spoke, his voice as deep as autumnal rain clouds. "Behold, O Arjuna, that all the world's protectors, the Lokapalas, have come here. We shall bestow upon you divine eyes. O sinless one, you were formerly the rishi Nara. At Brahma's behest, you have taken your birth as a mortal. You shall vanquish in battle the powerful Bhishma and the many other fierce warriors headed by Drona. The enemies of the gods, Daityas and Danavas, have taken their birth in the world of men. These shall all be slain in a great battle in which you will play the main role. Your fame on earth shall be eternal."*

*Yamaraja offered his personal weapon, a celestial mace, to Arjuna. The Pandava received the weapon along with the mantras for hurling and withdrawing it. After that, Varuna gave Arjuna his inescapable Nagapasha nooses. Arjuna also received Kuvera's weapon, the Antardhana, which could baffle and render unconscious even mighty Asuras. Then Indra spoke: "O subduer of enemies, you shall perform a great work for the gods. Now you must ascend to heaven. There I will give you my own weapon and other irresistible astras belonging to the celestials. My chariot, driven by my servant Matali, will soon carry you there."*

*The Lokapalas then vanished, leaving Arjuna alone. As he looked up, he saw what appeared to be a second sun gradually increasing in brilliance in the sky. The whole region vibrated as the radiance increased. Arjuna realized that this was Indra's chariot descending to earth. The chariot was huge, and it was drawn by ten thousand celestial horses of greenish and golden hue. Matali, in golden armor, was holding the reins. As the chariot approached, Arjuna could see upon it innumerable swords and maces, as well as all kinds of missiles. Axes, darts, lances, and spears of every kind were placed all around the chariot. Celestial thunderbolts and brilliant lightning bolts glowed from its sides. Huge-bodied, fierce Nagas with fiery mouths were seated on the chariot, along with great silver canons furnished with wheels and capable of sending celestial missiles a vast distance.*

*Arjuna saw Indra's dark blue flagstaff, the Vaijayanta, standing in the middle of the chariot as straight as a bamboo and as tall as a great palm tree. Then the chariot halted and Matali stood before Arjuna. Bowing low, he said, "O fortunate one, I have been instructed to take you to Shakra. Your father, encircled by the celestials, rishis, Gandharvas and Apsaras, wishes to see you in heaven. Shiva has also ordered me. Ascend this chariot and come to the heavenly regions."*

*Arjuna again bathed for purification, then offered prayers to Mount Mandara: "O king of mountains, you are the refuge of pious sages seeking heaven. You have sheltered me and I have lived here happily. I offer you my thanks and bid you farewell." With that, Matali led Arjuna to a flight of crystal steps leading up to the chariot.*

*Matali urged the steeds which drew the chariot with the speed of the mind high into the sky. As they entered the celestial regions, Arjuna saw thousands of wondrous chariots carrying shining beings of great beauty. He saw kings and warriors who had been slain in battle and had attained effulgent celestial forms. As they moved along the paths of the gods, Arjuna saw Siddhas, Charanas, rishis, Guhyakas and Apsaras. The region glowed with a brilliance born of the acsetic merits of such beings. Amazed, Arjuna inquired into their identities. The charioteer replied, "These are pious persons, O son of Kunti, stationed in their respective spheres. You have seen their shining abodes from the earth as stars in the firmament."*

*The chariot passed through the gate of Indra's planet, where Arjuna saw Airavata, the great white elephant with four tusks. The elephant resembled Mount Kailash with its four summits. Passing through divine regions meant for pious men who had performed great sacrifices and asceticism, they arrived at last in Amaravati, Indra's capital city.*
*Arjuna gazed with wide open eyes at the astonishing city. Flowers from all the seasons bloomed simultaneously, and there were groves of sacred trees. Fragrant breezes moved among their golden leaves, creating cascades of brilliant light. Countless mansions and palaces stretched into the distance. Arjuna saw the famous Nandana gardens, in which the gods sported with Apsaras.*

*"Those who turn their backs in battle can never see this place," Matali said, seeing Arjuna's amazement. "Neither can those who are wicked-minded, who have not performed sacrifice, who have not abstained from liquor and meat, who have not bathed in holy rivers nor given charity to brahmins ever reach this abode."*

*As they proceeded through the city, Arjuna beheld celestial cars traveling by the occupants' wills. Gandharvas and Apsaras eulogized Arjuna as he passed. He heard the sound of celestial music and of thousands of conchshells and drums. Arjuna entered Indra's magnificent assembly hall and met with the Sadhyas, Maruts, Vishvadevas, and twin Ashvinis. He saw the Adityas, Vasus, Rudras, and the many royal saints headed by King Dilipa. Narada Rishi was also present along with the Gandharva leader, Tumbaru.*

*After offering his respects to the assembled personalities, Arjuna approached his father, the king of the gods. Indra was seated beneath his white umbrella and was being fanned by golden-handled and perfumed chamaras. Many pure brahmins praised him with hymns from the Rig and Yajur Vedas, and the Gandharvas and Charanas played celestial instruments of all descriptions for his pleasure.*

*Arjuna prostrated himself before Indra, who got down from his throne and embraced him. The thousand-eyed god took Arjuna by the hand and led him back up the flight of golden steps to his bejeweled throne. Indra then had his son sit next to him on the throne, where he shone like a second Indra. With his perfumed hand, Indra affectionately stroked Arjuna's head. Out of love, Indra gently patted and rubbed his arms, which were like gold columns. The deity gazed without satiation at his son's face. Seated together on the throne, Arjuna and Indra appeared like the sun and the moon together in a clear sky.*

*Indra gave Arjuna the celestial sight to perceive Amaravati's divine beauty. The city was unimaginably opulent. The Pandava looked around at the numerous effulgent rishis who sat reciting sacred texts from the Vedas. Groups of Gandharvas headed by Tumbaru sang hymns in voices which entirely captivated the mind. As they sang, Apsaras performed exquisite dances full of gestures and sidelong glances. They shook their breasts and moved their broad hips.*

*The celestials offered Arjuna the sacred arghya and washed his feet. At Indra's command, Arjuna was then escorted to Indra's palace. Arjuna took up residence there and Indra instructed him in how to use the celestial weapons. (Mahābhārata: p25)*

## *Note #8—"Badarik Ashram"

*When they reached the Himalayas, Lomasha warned the party to proceed with caution. "There is Mount Meru and the peaks of Kailash, Gandhamadana, Trishringa and Makaragiri," he said, pointing ahead to the splendid mountain rising into the clouds. "Here reside thousands of invisible celestial beings. The Kimpurushas, Yakshas, Kinnaras, Suparnas, Nagas and Rakshasas, all as swift as the wind and as powerful as a thousand mighty elephants, move about on these mountain ranges. Sinful men cannot gain access here, for they will be destroyted by the celestials."*

*As Lomasha prayed to the gods for protection, Yudhisthira said to Bhima, "Carefully guard Draupadi, dear brother. We shall now proceed into these sacred mountains. By virtue of our asceticism and the pious merits we have earned through sacrifice, we may all be allowed to approach this holy region."*

*Lomasha had told the brothers that they would meet Arjuna in the mountains as he descended from heaven onto the summit of the*

*Gandhamadana mountain. They decided to go there and await Arjuna's arrival. The path to the mountain peak, however, was fraught with danger. Bhima then placed Draupadi on his shoulders, and the party, with only a few brahmins, made their way into the Himalayan range along the craggy mountain paths.*

*The terrain was rugged and the climb arduous. At last they came to the Alaknananda river, said to descend to earth from the heavens. They worshipped the holy river and bathed in its crystal clear waters. The brothers looked around at the brilliant scenery that surrounded them on the high mountain plateau. It seemed as if they had arrived on a heavenly planet. Blossoming trees of every color gave off celestial fragrances. The ground was carpeted with soft bluish grasses and expanses of wild flowers. Transparent lakes filled with blue lotuses and crowded with swans and charavarkas lay amid groves of fruit trees. The musical sounds of cukoos, peacocks and countless other birds filled the air. Lomasha told the Pandavas that Indra came daily to this region to perform his rituals and prayers.*

*The brothers saw in the distance what appeared to be a number of massive white mountains, but Lomasha told them it was the bones of Naraka, a mighty Asura whom Vishnu had slain in a past age. The rishi related the story as they traveled onward toward Gandhamadana. As they approached the mountain, the path became increasingly difficult. Leaving behind the heavenly terrain, they moved slowly along snow-bound and rocky paths.*

*Bhima embraced his son. "O invincible hero, O child, we are proceeding to the peak of Mount Gandhamadana where we shall again see our brother Arjuna. Your mother is worn out with fatigue and cannot continue. Nor shall we find it easy ourselves to follow the tortuous mountain paths, covered as they are by ice and snow. You should therefore carry us though the skies."*

*Ghatotkacha bowed. "With pleasure. I have at my command hundreds of sky-ranging Rakshasas. " Ghatotkacha gently placed Draupadi on his shoulders. He looked upon her as he would look upon his own mother, Hidimbi. Other huge-bodied Rakshasas suddenly appeared and picked up the travelers. Lomasha, however rose unaided into the sky by his own mystic power, appearing like a second sun.*

*The Rakshasas flew swiftly through the heavens along the Siddha's path. As the party traveled they saw beneath them the beautiful lands inhabited by Vidyadharas, Kinnaras, Kimpurushas, Gandharvas, and other divine beings. They saw great forests intersected by rivers and filled with elephants, bears, apes, rurus, surabhi cows and buffaloes. They passed over the country of the Uttara-Kurus. Finally, they all arrived at the Gandhamadana mountain. The Rakshasa took them high onto the mountain to Badarika Ashram, the ancient hermitage of Nara and Narayana Rishis.*

*The Pandavas gazed around in wonder. Even though they were high in the mountains the entire region was adorned with celestial trees in full blossom. They saw the great Badari tree spread out like a vast umbrella of lush green foliage, from which the Ashram had derived its name. The tree exuded freshness and calm, and it invited shelter beneath its boughs. Its branches were loaded with ripened fruits. Colorful birds, intoxicated with the honey falling from the fruits, flew in and out of the tree, filling the air with music.*

*The Pandavas approached the Badari tree with folded palms. It was surrounded by soft bluish-green grass. Countless thatched cottages, the dwellings of numerous rishis, stood around the tree. The whole region was lit by its own effulgence, requiring neither sunlight nor moonlight, and it was free from excessive heat and cold. The brothers felt both their physical pains an their mental anxieties dissipating. Even their hunger and thirst abated as their minds filled with a deep peace.*

*As they looked about, they noticed the many sacrificial altars tended by ascetics, and they heard brahmins chanting in melodious voices. Nearby, the cool and clear waters of the Ganges flowed on its way to Bharata, after having descended from heaven.*

*Some of the foremost rishis living in the hermitage came forward to receive the travelers. By their mystic vision they already knew who their guests were, and after accepting their obeisances, they honored them with offerings of pure water, fruits and flowers…*

*Bhima faced the wind and began to climb the mountain. He traveled swiftly, resembling a furious elephant crashing through the forest.*

*The region through which Bhima passed became increasingly beautiful. Not only were the trees in full bloom, but the rocks were black stone inlaid with gems. It seemed to Bhima as if the Goddess Earth had raised her arm adorned with sparkling ornaments in the form of this mountain. Bhima felt a cool breeze fanning his body, and his energy increased. On all sides he saw numerous Gandharvas and Siddhas sporting on the mountain slopes with their consorts, the beautiful celestial maidens turning their heads sideways to look at the Pandava as he rushed past.*

*Bhima scoured the mountainside as he ran, looking everywhere for the lotuses, but he did not find the flowers. He continued higher into the mountain, roaring with exultation at his own strength and power. Hearing the tremendous noise, animals at a distance became afraid. The mountain tigers left their dens and ran about. Some of them tried to attack Bhima, but he slapped them aside as he ran. He lifted elephants and tossed them at other elephants, clearing his path.*

*Bhima then came to a plantain forest. He entered it with force and broke some of the tall trees by hitting against them with his arms and thighs. He was still shouting, and his roars now mixed with the terrified cries of the forest animals. The tumult he created carried for miles.*

*Suddenly, Bhima noticed aquatic birds rising into the air not far off, frightened by the noise. Realizing there must be water nearby, and that lotuses grow on water, he made his way toward the birds. Soon, he saw a delightful lake. The Pandava went down to the water which was adorned by innumerable lotuses and lilies. He dived in and swam about like a maddened elephant. Refreshed and enlivened, he rose from the water and let out a terrific blast from his conch shell. Examining the lotuses, however, he realized that they were not the same as the one Draupadi desired. He set off up the mountain again, following a path leading up from the lake.*

*[Hanuman said:] Do you not know that you cannot proceed further? This path leads to heaven and men cannot access it. Only those who are successful in ascetic practices can go to the celestial regions. Therefore, give up your quest and turn back. Or rest here awhile first and partake of the sweet fruits and cool water. O foremost of men, do not try to force your way past and thus die for nothing."*

*Intrigued, Bhima responded politely. "Who are you, O respectable one? Why are you in the form of a monkey? I myself belong to the royal order.*

*Bhima explained that he first had to find the source of the thousand-petalled lotuses and Hanuman showed him the way to the forest where they grew. "There is the path which leads to the Saugandhika forest, and there you will see Kuvera's gardens, which are guarded by the Yakshas and Rakshasas. In a great lake lie the flowers which you seek for your wife."*

*Soon Bhima's mind again drifted to his surroundings. He was moving swiftly again, but the beauty of the woodlands, groves, orchards, lakes and rivers was not lost on him. Still the cool breeze carried that captivating fragrance from the blossoming trees. Herds of wild elephants roved about like masses of clouds, while buffaloes, bears, leopards and deer moved here and there.*

*Bhima pressed on. Just after noon, he at last arrived in the Saugandhika region. There he saw the lake filled with fresh golden lotuses, exactly like the one that had blown to Draupadi's feet. Swans swam upon the lake, and other water birds mingled with them, all making delightful sounds. The lake seemed to be fed by mountain springs that fell into it in cascades that sparkled in the sun. A canopy of green and golden trees, which swayed gently in the breeze, provided shade along the sandy lakeshore. Heaps of precious stones lay here and there. Along with the thousand-petalled lotuses, other charming flowers of a dark blue hue grew on stalks made of vaidurya gems. Bhima's mind was stolen by their beauty.*

*But the thousands of Yaksha and Rakshasa guards Kuvera had deployed to protect his lake saw Bhima arrive. [After having been defeated by Bhima] The Krodhavasas ran back in terror to Kuvera and told him what had happened. The god only smiled and said, "Let Bhima take as many lotuses as he likes for Draupadi. I already knew he would be coming." (Mahābhārata: TO BADARIK ASHRAM p.269-273)*

## *Note #9—"Kuvera's Abode"

*Seeing Yudhisthira there, the Yaksha guards folded their palms and bowed their heads in humility. The Pandava apologized for his brother's acts and consoled them with gentle words. Looking around at the beauty of the region, Yudhisthira desired to remain there for some time with his brothers. He asked the guards to request Kuvera's permission for them to stay. Hearing that Pandu's sons had come there, Kuvera was pleased and he gave his assent. Thus Yudhisthira and his party remained for some time on the high slopes of Gandhamadana, enjoying the heavenly atmosphere.*

*After a few days, Yudhisthira desired to see Kuvera. Sitting by the lakeside one morning, he told Bhima. "O brother, we have seen all the holy places with Lomasha and the other brahmins. Let us now see that most holy of places, Kuvera's abode, always frequented by Siddhas and rishis. Can you think of some way by which we can enter the heavenly region?"*

*But as Yudhisthira finished speaking, a celestial voice resounded in the sky: "O King, you will not be able to go to the inaccessible abode of Vaishravana by this way. Go back to Badarika Ashram. From that place you will go to the hermitage of Vrishaparva, and then to the hermitage of Arstishena. From there you will see Kuvera's abode."*
*As the divine voice spoke, a cool, fragrant breeze blew down a shower of flowers. Dhaumya said, "This order of the gods cannot be ignored. We should leave immediately."*

*It was almost five years since Arjuna had departed. Lomasha told the Pandavas that they could expect him to return shortly. He thus instructed them to go to a higher part of the mountain to meet him. Taking up their few possessions, and with Dhaumya, Lomasha and a handful of other brahmins accompanying them, the Pandavas climbed to a high plateau on Gandhamadana. As the celestial voice had predicted, they came to the ashrams of the royal sage Vrishaparva and the great rishi, Arshtisena. Remaining with them for some time, the Pandavas received spiritual instructions from them.*

*Arshtisena told them that it would not be possible for them to go further up the mountain. "Above here lies the path to the higher planets. No man can go there unless he is entirely pure and sinless. Along that path you will find the gods' sporting ground. Even from here we can hear the sound of their kettledrums, tabors and vinas."*

*Arshtisena told the brothers that he had been to the mountain's summit and seen the gods for himself. He'd also seen the Gandharvas,*

*Vidhyadaras, Kimpurushas and other heavenly beings consorting with the Apsaras. The great lord of wealth Kuvera could often be seen there being worshipped by the best of the Yakshas. "At that time, this entire region resounds with the melodic chanting of the Samaveda. Daityas, Danavas and Siddhas alike join in the worship of that illustrious Lokapala, Kuvera."*

*The rishi suggested that the brothers should reside peacefully at his ashram until Arjuna returned. Looking especially at Bhima he said, "Do not be restless here. Rather, live at ease. Be peaceful. The time will come soon enough for you to display your strength. Without doubt you will crush your enemies and govern the earth."*

*Surveying the region, the Pandavas felt as if they had already gone to the heavenly planets. Every kind of tree was present, all of them blossoming and bending with the weight of ambrosial fruits. Sweetly singing birds played in the lush foliage while peacocks, with their tails outspread, appeared like the trees' crowns. The lakes teemed with waterfowl and were adorned with lotuses, and the air rang with the hum of maddened bees who drank the nectar falling from the lotus cups. The brothers were fanned by a balmy breeze that carried celestial fragrances and fine drops of cooling spray. There were countless varieties of animals, all of them peaceful and harmless.*

*Bhima found a narrow passage by which he could scale the mountain, which was inaccessible to ordinary men. Grasping hold of creepers and crevices in the rocks, he climbed swiftly. As he reached the summit he came to a vast plateau where he beheld Kuvera's abode. It was adorned with glowing golden and crystal buildings of celestial design. The heavenly mansions were surrounded by high golden walls encrusted with gems of every color.*

*Bhima leaned on the end of his bow and gazed in wonder at the city. Ramparts and towers stood all around it and it was graced by huge gates and rows of tall flagstaffs, their colorful pennants fluttering in the wind. In the gardens outside the city lay heaps of gems, and along its walls garlands were hung. The trees were beyond human imagination, and around them Bhima saw many beautiful Apsaras dancing to the strains of heavenly music. (Mahābhārata P. 283)*

# 6

# The Verticle Map

## Point #1—"The Visible Sky"

You maintain that the Bhāgavatam should be taken literally, and that it states that all objects seen in the sky are within a 4-billion mile diameter sphere. This means that they are all within 2 billion miles of where we stand, at the center. You insist that this must be accepted by all devotees as parampara. You declare that any statement to the contrary is rank heresy, and that our preaching to scientists must present this point.

## Brahmāṇḍa Sākṣāt Ṭīkā (BST)#1—

An astrophysicist to whom I spoke said that there is a great deal of interpretation involved in the calculations of the vast distance scales given in modern astronomy and cosmology. The modern estimates are based largely on measurement of parallax, which uses the assumption that the Earth is moving around the Sun with a fixed background of stars, but if that is not the case, then the large distances are unnecessary. Additionally, our spiritual master, Śrīla Prabhupāda, the great commentator on Śrīmad Bhāgavatam, strongly emphasizes the fact that all the planetary systems are within the two billion mile radius— which unfortunately you deplore. We cannot ignore the following statement.

> *Modern enterprisers (the astronauts who travel in space) may take information from Śrīmad-Bhāgavatam that in space there are fourteen*

> *divisions of planetary systems. The situation is calculated from the earthly planetary system, which is called Bhūrloka. Above Bhūrloka is Bhuvarloka, and the topmost planetary system is called Satyaloka. These are the upper seven lokas, or planetary systems. And similarly, there are seven lower planetary systems, known as Atala, Vitala, Sutala, Talātala, Mahātala, Rasātala and Pātāla lokas. All these planetary systems are scattered over the complete universe, which occupies an area of two billion times two billion square miles.* (Śrīmad Bhāgavatam 2.5.40-41 Purport)

Consider the following chart compiled directly from the Śrīmad Bhāgavatam (SB) and purports by Śrīla Prabhupāda wherein he quotes Śrīla Viśvanātha Cakravartī Ṭhākura (VCT).

## Figure 2—UNIVERSE VERTICLE DISTANCE CHART

(According to Śrīmad Bhāgavatam and Śrīla Viśvanātha Cakravartī Ṭhākura)

+ **Universal Coverings** (260,000,000 yojanas or 2,080,000,000 miles above the sun [VCT & Visnu Purāṇa)
+ **Satyaloka** (233,800,000 yojanas, or 1,870,400,000 miles above sun[VCT)
+ **Tapoloka** (113,800,000 yojanas or 910,400,000 miles above the sun {VCT])
+ **Janaloka** (33,800,000 yojanas or 270,400,000 miles above the sun [VCT])
+ **Maharloka** (13,800,000 yojanas or 110,400,000 miles above the sun [VCT])
+ **Dhruvaloka** (3,700,000 yojanas or 29,600,000 miles above the sun[SB])
+ **Sapta Rsi-7 Sages** (2,400,000 yojanas or 19,200,000 miles above the su[SB])
+ **Saturn** (1,300,000 yojanas or 10,400,000 miles above the sun[SB])
+ **Jupiter** (1,100,000 yojanas or 8,800,000 miles above the sun[SB])
+ **Mars** (900,000 yojanas or 7,200,000 miles above the sun[SB])
+ **Mercury** (700,000 yojanas or 5,600,000 miles above the sun[SB])
+ **Venus** (500,000 yojanas or 4,000,000 miles above the sun[SB])
+ **Twenty-eight stars** (300,000 yojanas or 2,400,000 miles above the sun[SB])
+ **Moon** (100,000 yojanas or 800,000 miles above sun[SB])

SUN (at middle of the universe; size:10,000 yojanas or 80,000 miles in diameter)

- **Rāhu** (10,000 yojanas or 80,000 miles below sun[SB])
- **Siddhaloka, Caranaloka & Vidyadhara Loka** (20,000 yojanas or 160,000 miles below sun[SB])
- **Antariksa; Yaksas, Raksasas, Ghosts** (99,900 yojanas or 799,200 miles below sun[SB])
- **Earth** (100,000 yojanas or 800,000 miles below sun[SB])
- **Atala** (110,000 yojanas or 880,000 miles below sun[SB])
- **Vitala** (120,000 yojanas or 960,000 miles below sun [SB])
- **Sutala** (130,000 yojanas or 1,040,000 miles below sun [SB])
- **Talatala** (140,000 yojanas or 1,120,000 miles below sun [SB])
- **Mahatala** (150,000 yojanas or 1,200,000 miles below sun [SB])
- **Rasatala** (160,000 yojanas or 1,280,000 miles below sun [SB])
- **Pātāla** (170,000 yojanas or 1,360,000 miles below sun [SB])
- **Pitriloka and Hellish Planets** (Slightly above Garbhodaka Ocean [SB])
-**Śeṣa Nāga is lying on the Garbhodaka Ocean** (200,000 yojanas or 1,600,000 miles below sun [VCT])
- **Garbhodaka Ocean** (249,800,000 yojanas or 1,998,400,000 miles deep [VCT])

The entire fourteen planetary systems are included here, the Garbhodaka Ocean is included, and the beginning of the coverings of the universe. Since Satyaloka is the highest and most distant planetary system described in Vedic literature how then can we conclude that Śukadeva Goswāmī or Śrīla Prabhupāda or any ācārya is intimating that the four billion mile universe is too small to accommodate the planets (graha), planetary systems (loka), the stars (nakṣatra), the luminaries (tārāṇām), etc.?

## Point #2—" The Yojana & *Śrī Bṛhad-bhāgavatāmṛta*"

We observe that the rate of passage of time is much slower on Satyaloka than it is on the earth. We suggest that there might also be a comparable transformation of space in the region of Satyaloka. Thus, while a yogī traveling to Satyaloka may experience that he is crossing 2 billion miles, from our point of view he might be covering a much greater distance. We therefore suggest that when the Vedic literature speaks of a distance of 2 billion miles to the shell of the universe, it is referring to this distance as it would be perceived by the demigods, yogīs, and ṛṣis who can actually make this trip.

## BST #2—

We do not find any evidence of measuring universal distances in variant yojanas within the śāstras, indeed we find that the yojana (eight miles) is the standard unit of measurement used throughout the Vedic śāstras which describe the entire universe from top to bottom and even beyond. Please note the following verse from ***Śrī Bṛhad-bhāgavatāmṛta,*** Chapter Two, Text 225 by Śrīla Sanātana Goswāmī translated by His Grace Gopīparāṇadhana dāsa Adhikārī.

ब्रह्माण्डात्कोटिपञ्चाशद् योजनप्रमिताद्बहि"
यथोत्तरं दशगुणान्य् अष्टावावरणानि हि

*brahmāṇḍāt koṭi-pañcāśad-*
*yojana-pramitād bahiḥ*
*yathottaraṁ daśa-guṇāny*
*aṣṭāv āvaraṇāni hi*

*brahma-aṇḍāt—of the egg of the universe; koṭi-pañcāśat—five hundred million; yojana—yojanas; pramitāt—whose extent; bahiḥ—outside;*

*yathā-uttaram—one after the other; daśa-guṇāni—ten times larger; aṣṭau—eight; āvaraṇāni—covering shells; hi—indeed.*

*Outside the egg of the universe, which extends for five hundred million yojanas, are eight covering shells, each ten times thicker than the one before.*

*COMMENTARY: The eight shells covering the universe consist of earth, water, fire, air, ether, ego, mahat, and pradhāna. These layers form the stock of subtle causal ingredients from which all things in material existence are produced. Some philosophers consider the covering of earth to be part of the internal structure of the universe and so recognize only seven shells. That idea, however, conflicts with the measurement for the diameter of the universe as five hundred million yojanas, and it also leaves the transformations of earth within the universe with no identifiable primary source.*

We see once again that our universe is *koṭi-pañcāśat yojana* (five hundred million *yojanas* or four billion miles). Furthermore it will be helpful for us to consider the following verse from the same book.

पश्येमेऽप्यपरे यान्ति ब्रह्मणोऽधिकृतेऽल्पके
ब्रह्माण्डे चतुरास्यस्य तथामी दूरत" परे

अमी चाष्टमुखस्यैतद् द्विगुणे यान्ति वेगत"
अमी तु षोडशास्यस्य ब्रह्माण्डे द्विगुणे तत"

*paśyeme 'py apare yānti*
*brahmaṇo 'dhikṛte 'lpake*
*brahmāṇḍe catur-āsyasya*
*tathāmī dūrataḥ pare*

*amī cāṣṭa-mukhasyaitad-*
*dvi-guṇe yānti vegataḥ*
*amī tu ṣoḍaśāsyasya*
*brahmāṇḍe dvi-guṇe tataḥ*

*paśya—just see; ime—these; api—also; apare—others; yānti—are going; brahmaṇaḥ—by Brahmā; adhikṛte—presided over; alpake—small; brahmāṇḍe—in the universe; catuḥ-āsyasya—who has four heads; tathā—*

*also; amī—these; dūrataḥ—further away; pare—others; amī—they; ca—and; aṣṭa-mukhasya—of an eight-headed Brahmā; etat—than him; dvi-guṇe—in a universe twice bigger; yānti—are going; vegataḥ—swiftly; amī—these; tu—and; ṣoḍaśa-āsyasya—of a sixteen-headed Brahmā; brahmāṇḍe—in the universe; dvi-guṇe—twice bigger; tataḥ—than that.*

*Just see these other Vaikuṇṭha companions of the Lord, traveling in this small universe ruled by four-headed Brahmā. And further away those others, moving swiftly in the universe of an eight-headed Brahmā, a world twice as large. And those others in the world of a Brahmā with sixteen heads, a world twice as large again.*

*Commentary by Śrīla Sanātana Gosvāmī: To show that these visitors from Vaikuṇṭha were indeed extraordinary, Śrī Gaṇeśa pointed out more of them entering the other side of the universe, a great distance away, on some other mission. And beyond this relatively small universe of four-headed Brahmā, still other Vaikuṇṭha messengers were visiting the world of an eight-headed Brahmā, which was one billion yojanas in diameter, twice as big as this one. They were traveling swiftly (vegataḥ) because that universe is so large. Gaṇeśa further showed the same thing happening in universes still larger.* (Śrī Bṛhad-bhāgavatāmṛta, Part Two; 3:86-87)

## *Special Note—Gopīparāṇadhana dāsa Adhikārī's Note

In discussing the above two verse quoted from *Śrī Bṛhad-bhāgavatāmṛta*, His Grace Gopīparāṇadhana dāsa Adhikārī wrote the following note:

*The geometric formula for calculating the circumference of a circle from its radius is 2-pi-r, or twice the radius (in other words, the diameter) times the number pi. Given a diameter of 500,000,000 yojanas, the circumference should be 1,570,796,326 yojanas. This is about ten million times smaller than the number given in the two verses from the Caitanya-caritāmṛta purport, but we can explain this by assuming that the smaller measurement is only of the inside of the universe, not counting the covering shells of elements. Each consecutive layers is ten times wider than the one before it.*

## Point #3—"Light-years"

Caitanya-caritāmṛta Madya Lila 21.84 states that the diameter of this universe is 4 billion miles. This yields a circumference of approximately 12.566 billion miles. Yet in the purport Śrīla Prabhupāda cites information from Śrīla Bhaktisiddhānta Sarasvatī indicating that the circumference of the universe is 18,712,069,200,000,000 x 8, or 149,696,553.6 billion miles. If we are to take this figure seriously, then we must accept that there exist different scales of distance that can be applied to the universe. In Chapter 1 we calculated on the basis of this figure (plus some considerations involving the length of the yojana) that the radius of the universe must be about 5,077 light-years. This would mean that the diameter of the fiery shell (marking the ultimate limit for the travel of material light) must be 4,442 X 5,077, or some 22.5 million, light-years, a respectable distance even by modern cosmological standards.

## BST #3—

No, since the 18 quadrillion figure is the circumference of the outer globe, the light years is only 5,077 using the 8-mile yojana. Also the fiery shell is not to be included here because it is not visible. We cannot see beyond the first shell.

## Point #4—"Space"

In Śrīmad Bhāgavatam 3.26.52 purport Śrīla Prabhupāda states, "The space within the hollow of the universe cannot be measured by any human scientist or anyone else." This also suggests that something unexpected must happen to space (as well as time) as one approaches the universal shell, for it hardly seems impossible to measure a distance of 2 billion miles in ordinary space. That such a transformation of space and time should occur is in agreement with the basic character of the universal coverings themselves. As one passes from covering to covering, the nature of the material manifestation is progressively transformed, until finally one emerges into a purely spiritual realm (Śrīmad Bhāgavatam 2.2.28p). Thus, it would not be surprising if transformations of the material energy and its laws of operation were to occur as one approached the first universal shell.

## BST #4—

There are standard increments of measurement within the universe both in time and space although the perceptions of those may be different. For example a man being punished in hell is actually travelling 792,000 miles within a very short time but to him it seems like an intolerably long amount of time. In time and its counterpart space, the standard measurements are calculated from the atom and are given in the Śrīmad Bhāgavatam. It is significant also to notice that Śrīla Prabhupāda links time, space and planetary orbits throughout the entire universe in his purport.

स काल” परमाणुर्वै यो भुङ्क्ते परमाणुताम्
सतोऽविशेषभुग्यस्तु स काल”चपरमो महान्

*sa kālaḥ paramāṇur vai*
*yo bhuṅkte paramāṇutām*
*sato 'viśeṣa-bhug yas tu*
*sa kālaḥ paramo mahān*

*SYNONYMS saḥ—that; kālaḥ—eternal time; parama-aṇuḥ—atomic; vai—certainly; yaḥ—which; bhuṅkte—passes through; parama-aṇutām—the space of an atom; sataḥ—of the entire aggregate; aviśeṣa-bhuk—passing through the nondual exhibition; yaḥ tu—which; saḥ—that; kālaḥ—time; paramaḥ—the supreme; mahān—the great.*

*TRANSLATION: Atomic time is measured according to its covering a particular atomic space. That time which covers the unmanifest aggregate of atoms is called the great time.*

*PURPORT: Time and space are two correlative terms. Time is measured in terms of its covering a certain space of atoms. Standard time is calculated in terms of the movement of the sun. The time covered by the sun in passing over an atom is calculated as atomic time. The greatest time of all covers the entire existence of the nondual manifestation. All the planets rotate and cover space, and space is calculated in terms of atoms. Each planet has its particular orbit for rotating, in which it moves without deviation, and similarly the sun has its orbit. The complete calculation of the time of creation, maintenance and dissolution, measured in terms of the circulation of the total planetary systems until the end of creation, is known as the supreme kāla.*

अणुर्द्वौ परमाणू स्यात् त्रसरेणुस्त्रय” स्मृत”
जालार्करश्म्यवगत” खमेवानुपतन्न् अगात्

*aṇur dvau paramāṇū syāt*
*trasareṇus trayaḥ smṛtaḥ*
*jālārka-raśmy-avagataḥ*
*kham evānupatann agāt*

*SYNONYMS: aṇuḥ—double atom; dvau—two; parama-aṇu—atoms; syāt—become; trasareṇuḥ—hexatom; trayaḥ—three; smṛtaḥ—considered; jāla-arka—of sunshine through the holes of a window screen; raśmi—by the rays; avagataḥ—can be known; kham eva—towards the sky; anupatan agāt—going up.*

*TRANSLATION: The division of gross time is calculated as follows: two atoms make one double atom, and three double atoms make one hexatom. This hexatom is visible in the sunshine which enters through the holes of a window screen. One can clearly see that the hexatom goes up towards the sky. (Śrīmad Bhāgavatam 3.11.5)*

The yojana is used in this reference on the hellish planets, it is used to measure distances throughout the entire universe. All the ācāryas are using the same term of yojanas. Consider the following:

योजनानां सहस्राणि नवतिं नव चाध्वन”
त्रिभिर्मुहूर्तैर्द्वाभ्यां वा नीत” प्राप्नोति यातना”

*yojanānāṁ sahasrāṇi*
*navatiṁ nava cādhvanaḥ*
*tribhir muhūrtair dvābhyāṁ vā*
*nītaḥ prāpnoti yātanāḥ*

*SYNONYMS: yojanānām—of yojanas; sahasrāṇi—thousands; navatim—ninety; nava—nine; ca—and; adhvanaḥ—from a distance; tribhiḥ—three; muhūrtaiḥ—within moments; dvābhyām—two; vā—or; nītaḥ—brought; prāpnoti—he receives; yātanāḥ—punishments.*

*TRANSLATION: Thus he has to pass ninety-nine thousand yojanas within two or three moments, and then he is at once engaged in the torturous punishment which he is destined to suffer.*

*PURPORT: One yojana is calculated to be eight miles, and he has to pass along a road which is therefore as much as 792,000 miles. Such a long distance is passed over within a few moments only. The subtle body is covered by the constables so that the living entity can pass such a long distance quickly and at the same time tolerate the suffering. This covering, although material, is of such fine elements that material scientists cannot discover what the coverings are made of. To pass 792,000 miles within a few moments seems wonderful to the modern space travelers. They have so far traveled at a speed of 18,000 miles per hour, but here we see that a criminal passes 792,000 miles within a few seconds only, although the process is not spiritual but material.* (Śrīmad Bhāgavatam 3.30.24)

In time; muhurta is defined as 48 minutes, but is also used to indicate a moment in time. Like we say, "just wait a moment or just wait a few moments." Here moment can mean moments, minutes, or even hours.

*The four yugas are calculated in terms of the heavenly calendars and accordingly are twelve thousand years in terms of the heavenly planets. This is called a divya-yuga, and one thousand divya-yugas make one day of Brahmā. The creation during the day of Brahmā is called kalpa, and the creation of Brahmā is called vikalpa. When vikalpas are made possible by the breathing of Mahā-Viṣṇu, this is called a mahā-kalpa. There are regular and systematic cycles of these mahā-kalpas, vikalpas and kalpas.* (Śrīmad Bhāgavatam 2.10.46)

*The age of Brahmā is calculated in terms of divya years, which are distinct from the solar years of human beings. The divya years are calculated in Bhagavad-gītā (8.17): sahasra-yuga-paryantam ahar yad brahmaṇo viduḥ. Brahmā's one day is equal to one thousand times the aggregate of the four yugas (calculated to be 4,300,000 years).* (Śrīmad Bhāgavatam 3.8.22 Purport)

In the example above, we see that time is standard throughout the universe yet it is perceived differently by those situated in different places.

## Point #5—"Spatial Perspectives"

As for as the figures given in the Śrīmad Bhāgavatam, as much as there are different time dimensions among the fourteen planetary systems, there certainly are "space dimensions" too. Certain contradictions, can simultaneously be right also, because the laws of math and physics are multifarious in different areas of universe.

Apart from these, the information from the Jyotisha books (again two kinds, astronomical and predictive astrological) cannot be tallied one to one with the Purāṇas, because they are calculations based on the views from the angle of earthly planet to the sky. Basic figures are taken from smrtis and Purāṇas, but a lot of earthly angle of view for the purpose of calculating the influences of planets on the native is taken. It is like the difference one can see between the supply engineer's, electrical engineer's and the construction engineer's blue print plan of the same building.

Also when Brahmā speaks to Manu, or Maitreya speaks to the prajapathis, or Maitreya speaking with Vidura or Uddhava the distances and sizes presented will be from totally different perspectives, because of the speaker and listener's orientation in the universe. Even from the same vakta (speaker)the vertical descriptions and the horizontal descriptions will have to differ, especially in a multi dimensional area like Jambūdvīpa etc.... One may have to make separate drawings of these and compute them within the mind?!

Just like we have different elevation plans front , side ,etc, and the floor plan and utilities plan of a constructed building even within same three dimensions. In this way a minimum of eighteen models have to be made to get an earthly idea of the structure of universe.

## BST #5—

When we speak of time within the finite fourteen planetary systems there is a ratio, i.e. one day of Brahmā equals 4 billion three hundred million years on earth; one day on the heavenly planets equals six months on earth, etc. The total time within the universe is also given exactly in time calculated from the atom. Śrī Maitreya Muni made his statement about the size of the universe as four billion miles immediately after explaining time within the universe.

In terms of universal space we have a concurrence between Lord Kṛṣṇa, Lord Caitanya, Maitreya, Vyāsadeva, Śukadeva Goswāmī and Śrīla Prabhupāda who all agree that the diameter of the universe is *pañcāśat-koṭi-*

*yojana* (fifty million yojanas or four billion miles). The misunderstanding arises because Śrīla Bhaktisiddhānta Thakura quotes the Siddhānta-śiromaṇi as saying the universe's circumference is either approximately 150 quadrillion or 300 quadrillion miles in circumference. If this figure is referring to the circumference of the outside of the brahmāṇḍa (which includes all the coverings) then it is consistent with the Bhāgavatam's stated diameter of the universe. If, as some devotee scientists hold, it refers to the inside of the visible universe then one must speculate, as you have done, to reconcile the immense difference in distances.

Consider the following:

अन्दमध्यगत?सुर्यो द्यवभुम्योर्यदन्तरम्<br>
सुर्यन्दगोल्योर्मध्ये कोत्य?स्यु?पन्चविम्सतिह्

*aṇḍa-madhya-gataḥ sūryo*<br>
*dyāv-ābhūmyor yad antaram*<br>
*sūryāṇḍa-golayor madhye*<br>
*koṭyaḥ syuḥ pañca-viṁśatiḥ*

*SYNONYMS: aṇḍa-madhya-gataḥ—situated in the center of the universe; sūryaḥ—the sun globe; dyāv-ābhūmyoḥ—the two planetary systems Bhūrloka and Bhuvarloka; yat—which; antaram—in between; sūrya—of the sun; aṇḍa-golayoḥ—and the globe of the universe; madhye—in the middle; koṭyaḥ—groups of ten million; syuḥ—are; pañca-viṁśatiḥ—twenty-five.*

*TRANSLATION: The sun is situated [vertically] in the middle of the universe, in the area between Bhūrloka and Bhuvarloka, which is called antarikṣa, outer space. The distance between the sun and the circumference of the universe is twenty-five koṭi yojanas [two billion miles].*

*PURPORT: The word koṭi means ten million, and a yojana is eight miles. The diameter of the universe is fifty koṭi yojanas (four billion miles). Therefore, since the sun is in the middle of the universe, the distance between the sun and the edge of the universe is calculated to be twenty-five koṭi yojanas (two billion miles).* (Śrīmad Bhāgavatam 5.20.43 and Purport)

Here we see that the yojana is used equally to describe regions within the universe, from the middle to the edge. Therefore the hypothesis that a "celestial yojana" is different from an "earthly yojana" is not supported by śāstra. In actuality, time is also standard in that a *parārdha* measures a particular amount of time. That period of time will be perceived differently by Brahmā (as his fifty years) or by earthlings (as 155 trillion years).

*kṛṣṇa kahe, "ei brahmāṇḍa pañcāśat koṭi yojana*
*ati kṣudra, tāte tomāra cāri vadana*

*SYNONYMS: kṛṣṇa kahe—Kṛṣṇa said; ei brahmāṇḍa—this universe; pañcāśat koṭi yojana—four billion miles; ati kṣudra—very small; tāte—therefore; tomāra—your; cāri vadana—four faces.*

*TRANSLATION: "Kṛṣṇa said, 'Your particular universe has a diameter of four billion miles; therefore it is the smallest of all the universes. Consequently you have only four heads.* (Caitanya-caritāmṛta; Madhya 21.84)

## Point #6—"Layers"

Regarding the layers, I think they are either subtle invisible energies, or they are flowery language, or they are something in another dimension or all of those things together or even more … Who knows? Could you please tell me where Śrīla Prabhupāda mentions that we can not see through those layers?

## BST #6—

Yes.

*In Bhagavad-gītā (10.21) Kṛṣṇa says, nakṣatrāṇām ahaṁ śaśī: "Of stars I am the moon." This indicates that the moon is similar to the other stars. The Vedic literature informs us that within this universe there is one sun, which is moving. The Western theory that all the luminaries in the sky are different suns is not confirmed in the Vedic literature. Nor can we assume that these luminaries are the suns of other universes, for each universe is covered by various layers of material elements, and therefore although the universes are clustered together, we cannot see from one universe to another. In other words, whatever we see is within this one universe. In each universe there is one Lord Brahmā, and there are other demigods on other planets, but there is only one sun.* (Śrīmad Bhāgavatam 5.21.11 Purport)

*If we look as far as we can see — up to the sky — our vision is still confined within only one universe...* (Easy Journey to Other Planets 2: Varieties of Planetary Systems)

*We can see only one universe, and in one universe there are innumerable planets.* ( Śrī Caitanya-caritāmṛta, Madhya-lila 20.146-151 — New York, December 3, 1966)

*You see this universe, as far as you can see up to the sky. That is only one universe.* (Śrīla Prabhupāda: Engagement Lecture — Buffalo, April 23, 1969)

*Man (2): The astronomers also accept that there are innumerable universes and they say that in each one they have millions and millions of self-luminous stars, and in this one also, but you say that in this one there is only the sun.*
*Prabhupāda: But do you think that I have to accept the astronomers blindly?*
*Man (2): Well, they say...*
*Prabhupāda: No. Why? Why? If there are innumerable suns, why they are not present at night? Why you are suffering from want of one sun?*
*Man (2): Because we are much further away from the others than our particular one.*
*Prabhupāda: No. At least, we are not able to bring them into service. We are depending only on one sun in this universe. You may say there are innumerable suns, but why the astronomers or the scientists could not take advantage of another sun at night? Why it is darkness?*
*Man (2): Because we're much closer to this one than the others. We're much closer to this one.*
*Prabhupāda: Yes. That is... What is that?*
*Madhudviṣa: He says that the earth is much more closer to this sun than it is to other suns.*
*Prabhupāda: Anyway, we do not take. But we, according to our Vedic information, this is only one universe, which is within our vision, this sky, the dome. That is one universe. The other universes are outside this*

*universe. That is the Vedic information.* (Bhagavad-gītā 2.9 — Auckland, February 21, 1973)

*You can see the border of the universe, the round sky.* (Śrīmad-Bhāgavatam 5.5.1-8 — Stockholm, September 8, 1973)

*You have no estimation of this sky. Not only this, this sky; even in the material world, this is one universe. What sky we see, that is of one universe. But there are innumerable universes. That is material world. And beyond that, there is spiritual sky.* (Lecture: Śrīmad-Bhāgavatam 2.1.1 — Paris, June 9, 1974)

*Krishna dāsa: Śrīla Prabhupāda? Using modern observatories or large telescopes, scientists have photographed other brahmandas. At least to my impression they appear to be other solar systems. Like they have a sun and planets around them. Does that mean the material covering between the different brahmandas is invisible?*
*Prabhupada: Yes.*
*Krishna dāsa: Is that correct?*
*Tamala Krsna: But one of the coverings is earth.*
*Prabhupada: This is the covering of the universe. Covering you as far as you can see, that is covering.*
*Tamala Krsna: Aren't there coverings of the 8 elements?*
*Prabhupada: Yes, that is above.*
*Tamala Krsna: So he's saying that they have seen beyond that.*
*Prabhupada: How they can see?*
*Krishna dāsa: That was my question.*
*Tamala Krsna: Prabhupada says how can they see?*
*Prabhupada: No, they take all these stars as sun. But that is not fact. Sun is one. In each universe there is one sun in the middle. In this universe in the middle, from the circumference, 200,000,000,000's, no, two billion.* (Morning Walk — July 24, 1975, Los Angeles)

# NOTES

## *Note #1—"Polestar"

नान्यैरधिष्ठितं भद्र यद्भ्राजिष्णु ध्रुवक्षिति
यत्र ग्रहर्क्षताराणां ज्योतिषां चक्रमाहितम्
मेढ्यां गोचक्रवत्स्थास्नु परस्तात्कल्पवासिनाम्
धर्मोऽग्नि" कश्यप" शुक्रो मुनयो ये वनौकस"
चरन्ति दक्षिणीकृत्या भ्रमन्तो यत्सतारका"

*nānyair adhiṣṭhitaṁ bhadra*
*yad bhrājiṣṇu dhruva-kṣiti*
*yatra graharkṣa-tārāṇāṁ*
*jyotiṣāṁ cakram āhitam*
*meḍhyāṁ go-cakravat sthāsnu*
*parastāt kalpa-vāsinām*
*dharmo 'gniḥ kaśyapaḥ śukro*
*munayo ye vanaukasaḥ*
*caranti dakṣiṇī-kṛtya*
*bhramanto yat satārakāḥ*

*SYNONYMS: na—never; anyaiḥ—by others; adhiṣṭhitam—was ruled; bhadra—My good boy; yat—which; bhrājiṣṇu—brightly glowing; dhruva-kṣiti—the land known as Dhruvaloka; yatra—where; graha—planets; ṛkṣa—constellations; tārāṇām—and stars; jyotiṣām—by luminaries; cakram—encirclement; āhitam—is done; meḍhyām—around a central pole; go—of bulls; cakra—a multitude; vat—like; sthāsnu—stationary; parastāt—beyond; kalpa—a day of Brahmā (millennium); vāsinām—those who live; dharmaḥ—Dharma; agniḥ—Agni; kaśyapaḥ—Kaśyapa; śukraḥ—Śukra; munayaḥ—great sages; ye—all of them who; vana-okasaḥ—living in the forest; caranti—move; dakṣiṇī-kṛtya—keeping it to their right; bhramantaḥ—circumambulating; yat—which planet; satārakāḥ—with all the stars.*

*TRANSLATION: The Supreme Personality of Godhead continued: My dear Dhruva, I shall award you the glowing planet known as the polestar, which will continue to exist even after the dissolution at the end of the millennium. No one has ever ruled this planet, which is surrounded by all the solar systems, planets and stars. All the luminaries in the sky circumambulate this planet, just as bulls tread around a central pole for the purpose of crushing grains. Keeping the polestar to their right, all the stars inhabited by the great sages like Dharma, Agni, Kaśyapa and Śukra circumambulate this planet, which continues to exist even after the dissolution of all others. (Śrīmad Bhāgavatam 4.9.20, Śrīmad Bhāgavatam 4.9.21, Śrīmad Bhāgavatam 4.9.20-21)*

*bha-gaṇāḥ—the luminaries, like the sun, the moon, Venus, Mercury, Mars and Jupiter; graha-ādayaḥ—the different planets; etasmin—in this; antaḥ-bahiḥ-yogena—by connection with the inner or outer circles; kāla-cakre—in the wheel of eternal time; āyojitāḥ—fixed; dhruvam—Dhruvaloka; eva—certainly; avalambya—taking support of; vāyunā—by the wind; udīryamāṇāḥ—being propelled; ā-kalpa-antam—until the end of the creation; paricaṅ kramanti—revolve all around; nabhasi—in the sky; yathā—exactly like; meghāḥ—heavy clouds; śyena-ādayaḥ—birds such as the big eagle; vāyu-vaśāḥ—controlled by the air; karma-sārathayaḥ—whose chariot drivers are the results of their own past activities; parivartante—move around; evam—in this way; jyotiḥ-gaṇāḥ—the luminaries, the planets and stars in the sky; prakṛti—of material nature; puruṣa—and of the Supreme Personality, Kṛṣṇa; saṁyoga-anugṛhītāḥ—supported by the combined efforts; karma-nirmita—caused by their own past fruitive activities; gatayaḥ—whose movements; bhuvi—on the ground; na—not; patanti—fall down.*

*TRANSLATION: When bulls are yoked together and tied to a central post to thresh rice, they tread around that pivot without deviating from their proper positions—one bull being closest to the post, another in the middle, and a third on the outside. Similarly, all the planets and all the hundreds and thousands of stars revolve around the polestar, the planet of Mahārāja Dhruva, in their respective orbits, some higher and some lower. Fastened by the Supreme Personality of Godhead to the machine of material nature according to the results of their fruitive acts, they are driven around the polestar by the wind and will continue to be so until the end of creation. These planets float in the air within the vast sky, just as clouds with hundreds of tons of water float in the air or as the great śyena eagles, due to the results of past activities, fly high in the sky and have no chance of falling to the ground.* (Śrīmad Bhāgavatam 5.23.3)

## Comment:

That the planets are driven in their orbits by the wind (vayuna) is also confirmed in the Sūrya-siddhānta (7.73)translation; "The circle of asterisms, bound at the two poles, impelled by the provector (pravaha) winds, revolves eternally: attached to that are the orbits of the planets, in their order."

## *Note #2—"Planetary Distances"

The Śrīmad Bhāgavatam elaborately describes the exact distances of the planets in the universe. Given the precision and corroboration between the Bhāgavatam and the Jyoti śāstras in calculating orbits and eclipses, the following figures should be taken as literal.

> *Summarizing the entire description of the planetary systems of the universe, Śrīla Viśvanātha Cakravartī Ṭhākura says that one who is able to meditate upon this arrangement as the virāṭ-rūpa, or viśva-rūpa, the external body of the Supreme Personality of Godhead, and worship Him three times a day by meditation will always be free from all sinful reactions. Viśvanātha Cakravartī Ṭhākura estimates that Dhruvaloka, the polestar, is 3,800,000 yojanas above the sun. Above Dhruvaloka by 10,000,000 yojanas is Maharloka, above Maharloka by 20,000,000 yojanas is Janaloka, above Janaloka by 80,000,000 yojanas is Tapoloka, and above Tapoloka by 120,000,000 yojanas is Satyaloka. Thus the distance from the sun to Satyaloka is 233,800,000 yojanas, or 1,870,400,000 miles. The Vaikuṇṭha planets begin 26,200,000 yojanas (209,600,000 miles) above Satyaloka. Thus the Viṣṇu Purāṇa describes that the covering of the universe is 260,000,000 yojanas (2,080,000,000 miles) away from the sun. The distance from the sun to the earth is 100,000 yojanas, and below the earth by 70,000 yojanas are the seven lower planetary systems called Atala, Vitala, Sutala, Talātala, Mahātala, Rasātala and Pātāla. Below these lower planets by 30,000 yojanas, Śeṣa Nāga is lying on the Garbhodaka Ocean. That ocean is 249,800,000 yojanas deep. Thus the total diameter of the universe is approximately 500,000,000 yojanas, or 4,000,000,000 miles.* (Śrīmad Bhāgavatam 5.23.9)

> *This chapter describes the planet Rāhu, which is 10,000 yojanas (80,000 miles) below the sun, and it also describes Atala and the other lower*

*planetary systems. Rāhu is situated below the sun and moon. It is between these two planets and the earth. When Rāhu conceals the sun and moon, eclipses occur, either total or partial, depending on whether Rāhu moves in a straight or curving way. Below Rāhu by another 1,000,000 yojanas are the planets of the Siddhas, Cāraṇas and Vidyādharas, and below these are planets such as Yakṣaloka and Rakṣaloka. Below these planets is the earth, and 70,000 yojanas below the earth are the lower planetary systems—Atala, Vitala, Sutala, Talātala, Mahātala, Rasātala and Pātāla.* (Śrīmad Bhāgavatam 5.24 Chapter Summary)

*upavarṇitaṁ bhūmer yathā-sanniveśāvasthānam avaner apy adhastāt sapta bhū-vivarā ekaikaśo yojanāyutāntareṇāyāma-vistāreṇopakḷptā atalaṁ vitalaṁ sutalaṁ talātalaṁ mahātalaṁ rasātalaṁ pātālam iti.*

*SYNONYMS: upavarṇitam—stated previously; bhūmeḥ—of the planet earth; yathā-sanniveśa-avasthānam—according to the arrangement of the different places; avaneḥ—the earth; api—certainly; adhastāt—beneath; sapta—seven; bhū-vivarāḥ—other planets; eka-ekaśaḥ—in succession, up to the outer limit of the universe; yojana-ayuta-antareṇa—with an interval of ten thousand yojanas (eighty thousand miles); āyāma-vistāreṇa—by width and length; upakḷptāḥ—situated; atalam—named Atala; vitalam—Vitala; sutalam—Sutala; talātalam—Talātala; mahātalam—Mahātala; rasātalam—Rasātala; pātālam—Pātāla; iti—thus.*

*TRANSLATION: My dear King, beneath this earth are seven other planets, known as Atala, Vitala, Sutala, Talātala, Mahātala, Rasātala and Pātāla. I have already explained the situation of the planetary systems of earth. The width and length of the seven lower planetary systems are calculated to be exactly the same as those of earth.* (Śrīmad Bhāgavatam 5.24.8)

## *Note #3—Lord Siva

*Within the universal form is an open space of five hundred million yojanas. That open space is surrounded by seven coverings, each one ten times greater than the one before it.* (Satvata Tantra: Lord Siva speaking Pātāla 1; verse 33)

## *Note #4—Kṛṣṇadāsa Kaviraja Goswāmī

*brahmāṇḍa-pramāṇa pañcāśat-koṭi-yojana*
*āyāma, vistāra, dui haya eka sama*

*SYNONYMS: brahmāṇḍa-pramāṇa—measurement of the universe; pañcāśat—fifty; koṭi—ten millions; yojana—lengths of eight miles; āyāma—length; vistāra—breadth; dui—both of them; haya—are; eka sama—one and the same.*

*TRANSLATION: The universe measures five hundred million yojanas. Its length and breadth are one and the same.* (Caitanya-caritāmṛta: Ādi 5.97)

# 7

# Bhagavat Cosmology

## Brahmāṇḍa Sākṣāt Ṭīkā (BST) "Premise"—

We have been taught in Krishna consciousness that one should abandon the speculative habit and instead accept the śāstras as they are. We have been warned repeatedly not to speculate about the scriptures, especially if our speculation contradicts the direct meaning stated in the scripture. We must not propose speculative ideas based on non-literal interpretations going far beyond our designated portfolio. We were not asked to re-size the universe to fit modern sciences' theories. The standard Vaiṣṇava method of understanding Śrīmad Bhāgavatam is literal. There is no possibility of offering "non-literal" interpretations which oppose the literal statements of the Śrīmad Bhāgavatam.

Here are a few statements from Śrīla Prabhupāda and other ācāryas explaining the importance of taking Śrīmad Bhāgavatam (including the Fifth Canto) literally.

> *We have translated the full text—first we present the original Sanskrit text, its transliteration, the English equivalent for each word, then the translation, and then a purport, or explanation of the verse. In this way, there are eighteen thousand verses in Śrīmad-Bhāgavatam. We are translating everything literally. You can see. Each and every verse is being done like that for the whole Bhāgavata Purāṇa. The opinion of the ācāryas, the great saintly sages who are the preachers of the Bhāgavata philosophy, is nigama-kalpa-taror galitaṁ phalam: this is the ripened*

> *fruit of the Vedic desire tree (Śrīmad-Bhāgavatam 1.1.3). It is accepted by all the Indian scholars, and Lord Caitanya especially preached this Bhāgavatam.* (Science of Self Realization: 6)

In this statement Śrīla Prabhupāda includes all 18,000 verses of the Śrīmad Bhāgavatam. He says that the whole Śrīmad Bhāgavatam is being presented literally and the Fifth Canto is also part of the Śrīmad Bhāgavatam. The Fifth Canto is not meant to be removed and then operated upon by mental speculation. The Fifth Canto is also spoken by Śrīla Śukadeva Goswāmī and recorded by Śrīla Vyāsadeva, the same spiritual authorities who teach us brahma jñāna, vairāgya and the ten primary subject matters including sarga (creation of the universe), visarga (sub creation), sthānam (planetary systems), poṣaṇam (protection by the Lord), ūtayaḥ (the creative impetus), manvantara (the change of Manus), īśa-anukathāḥ (the science of God), nirodhaḥ (returning home back to Godhead), muktiḥ (liberation), and āśrayaḥ (the summum bonum.)

The third division described in the Śrīmad Bhāgavatam is sthānam (planetary systems) which is presented in the Fifth Canto.

> *Śrīla Madhvācārya has quoted the following statement from the Brahmāṇḍa Purāṇa: "One should have complete faith in transcendental literature such as Śrīmad-Bhāgavatam and other literature that directly glorifies the Supreme Personality of Godhead.* (Śrīmad Bhāgavatam 11.3.26 Purport)

If while studying Śrīmad Bhāgavatam one finds a section which is difficult to understand, he should accept it whole-heartedly and pray to the Lord and the spiritual master for help. Then he should try to increase his devotion. It is not that he must make up a temporary theory to adjust the difficult section.

> *"The Vedic injunctions are self-authorized, and if some mundane creature adjusts the interpretations of the Vedas, he defies their authority. It is foolish to think of oneself as more intelligent than Śrīla Vyāsadeva. He has already expressed himself in his sūtras, and there is no need of help from personalities of lesser importance. His work, the Vedānta-sūtra, is as dazzling as the midday sun, and when someone tries to give his own interpretations on the self-effulgent sunlike Vedānta-sūtra, he attempts to cover this sun with the cloud of his imagination. The Vedas and Purāṇas*

> *are one and the same in purpose."* (Lord Caitanya to Sarvabhauma Bhattacarya from Śrīmad Bhāgavatam Introduction)

Lord Caitanya compares the *Vedānta-sūtra* to the sun. Similarly, the ripened fruit of all Vedic literature, Śrīmad Bhāgavatam, is the brilliant essence of the sun (purāṇa-arkaḥ). As *Vedānta-sūtra* is to be accepted as it is, so the Śrīmad Bhāgavatam must be accepted. To deny any portion of the great scripture by presenting one's own interpretation is an insult to the author.

> ***Ms. Nixon:*** *No, I agree with you one hundred percent.... One thing I'm very curious about: can the Vedas be taken symbolically as well as literally?*
> ***Śrīla Prabhupāda:*** *No. They must be taken as they are, not symbolically. That is why we are presenting the Bhagavad-gītā As It Is.* (Science of Self Realization; 1)

As the entrance examination, *Bhagavad-gītā* must be delivered and received as it is, so the graduate study, namely Śrīmad Bhāgavatam, must be treated the same way. It is not that when one reaches the graduate level he gives up direct aural reception and begins speculating.

> *David Lawrence: Where, is it the demoness Pūtanā takes Kṛṣṇa and Kṛṣṇa sucks her breast?*
> *Śyāmasundara: Should it be taken literally or allegorically?*
> *Prabhupāda: No, literally, literally.*
> *David Lawrence: Yeah, literally as a physical fact.*
> *Prabhupāda: Oh yes, oh yes.*
> *David Lawrence: Now, having said that then, if one goes on logically...*
> *Prabhupāda: You'll find so many demons like Pūtanā even at the present moment...*
> *David Lawrence: One or two of the stories there's a reference to Kṛṣṇa and the unmarried gopīs, saying that He treated them like dolls yet they were well pleased with Him...Is that to be taken again literally or is there, as you say, a figurative meaning there in terms of faith?*
> *Prabhupāda: No, that is literally. Because gopīs were just dancing according to Kṛṣṇa's desire. They are so devotees that whatever Kṛṣṇa desires, they are prepared to do.* (Morning Walk — August 30, 1973, London)

It is interesting to note that Śrīla Prabhupāda upheld the eternal Vedic law of literal comprehension in relationship with the pastime of Putana the witch. David Lawrence had just read Śrīla Prabhupāda's *Krishna Book* where the Putana witch is told to have been killed by baby Krishna and then expanded into her ghastly gigantic body—eight miles long. The Bhāgavatam Fifth Canto's cosmology is not any more unbelievable than an eight-mile long witch yet Śrīla Prabhupāda did not pander to a reader's unfamiliarity with the subject matter.

> *Sadāpūta: I was wondering, perhaps these periods that they have here correspond with some of these floods, like this is between two Manus and there is a flood of the whole world, and that might pile up a huge layer of mud and rock. And I was thinking that might correspond with some of these layers here because they are filled with remains of fish that look as though they were buried, things like that.*
> *Prabhupāda: That is also imagination. Again you are bringing imagination, speculation.*
> *Svarūpa Dāmodara: So actually it's very subtle and critical.*
> *Prabhupāda: As soon as you bring your imaginations, they'll bring their own imagination. "Perhaps," "It may be," they say*
> *Rūpānuga: Our imagination is as good as your imagination.*
> *Prabhupāda: We are to take the Vedic injunction axiomatic. (Room Conversation — July 5, 1976, Washington, D.C.)*

The conversation above is profoundly applicable to the present day. Śrīla Prabhupāda was holding a meeting with his Bhaktivedanta Institute (BI) scientists discussing how to present Vedic conclusions in opposition to modern science. His disciple-scientists wanted to amalgamate the Vedic evidence with the modern scientific views when His Divine Grace abruptly reminded them not to speculate.

Śrīla Prabhupāda's BI scientists had taken the liberty of mixing so-called scientific evidence such as geological remains with evidence from the Śrīmad Bhāgavatam, but that was speculation. As abandoning the literal statements of Śrīmad Bhāgavatam and replacing them with non-literal interpretations, we flagrantly disregard all of the directions from the ācāryas as pointed out above.

In the same conversation he said,

> *"Our conclusion is that whatever they are saying, that is imperfect. That is not possible. And that is a fact. Therefore they change, after fifty years, they change. Because it's speculation. Therefore we say totally they are wrong, and they will take totally "From mythology," like that, this is going on."*

His advice to the ISKCON scientists was to present arguments based on the literal understanding of the Vedic śāstra without trying to adjust them to tally with modern science. Śrīla Prabhupāda's concluding advice to his scientists at Washington D.C. is succinctly stated below:

> *You present your own as it is Bhāgavatam, try to explain them to your best capacity scientifically. If you bring this, bring that... All of them are imperfect. So what is the use of wasting time with something which is imperfect?*

## Point #1—" Non-Literal Conception vs. Literal Conception"

For centuries, Puranic cosmology has seemed incomprehensible to most observers, and this has encouraged many people either to summarily reject it or to accept it literally with unquestioning faith. This has certainly been true in recent times. If we go back to the medieval period, we likewise find that astronomers such as Bhaskaracarya and Paramesvara dismiss Puranic cosmology as imaginary, even though they use Puranic ideas and terms in their astronomical writing. At the same time, traditional commentators, such as Sridhara Swami, discuss the cosmology of the Bhāgavatam from the standpoint of direct literal acceptance.

## BST #1—

The basis for the belief that Bhaskaracarya and Paramesvara dismissed Puranic cosmology is the commentaries on his work written by biased Western scholars or paid Indian pundits. All these Western scholars, committed to the Western view of astronomy and religion, i.e. Christianity, carry contaminations. It was part of the British colonial strategy to deprecate all authentic Vedic literature as mythological in order to convert the natives residing in India into Christians.

## Point #2—"Bhāgavatam's Violations?"

If we take it literally, the cosmology of the Bhāgavatam differs radically from modern astronomy, and it also suffers from internal contradictions and violation of common sense.

## BST #2—

What appears as "internal contradictions" is due to our inability to comprehend the vast scope of Vedic wisdom. Is it not childish for us, no matter how supposedly competent in modern science and mathematics we are, to claim a superior view to the Bhāgavata Purāṇa? We should not arrogantly critique the Vedic perfect knowledge lest we commit the grave offense of *śruti-śāstra-nindanam* blaspheming Vedic literature. Please consider Śrīla Prabhupāda's advice on apparent contradictions.

> *Generally speaking, every word in scripture has a literal meaning, but people cannot understand it properly because they do not hear from the proper person. They interpret instead. There is no need to interpret the words of God. Sometimes the words of God cannot be understood by an ordinary person; therefore we may require the transparent medium of the guru. Since the guru is fully cognizant of the words spoken by God, we are advised to receive the words of the scriptures through the guru. There is no ambiguity in the words of God, but due to our imperfect knowledge, we sometimes cannot understand. Not understanding, we try to interpret, but because we are imperfect, our interpretations are also imperfect. The conclusion is that the words of God, the scriptures, should be understood from a person who has realized God.* (Śrīla Prabhupāda discussing Origen)

Within an Aesops Fable there was a lake. So one small lamb was drinking water on the other side, and on this side stood a tiger. The tiger challenged to the lamb that, "Why you are muddying the water." So the lamb said, "Sir, I am here, a long distance away from you, I am not muddying the water." So anyway, the tiger picked up some quarrel with the lamb and killed him. So the idea was to kill him, but the tiger found out some fault in the lamb to justify his meal. The English proverb, "give the dog a bad name and hang it" is also appropriate. First we give the Vedic literature the bad name of "violating common sense," and containing "internal contradictions," then we attempt to destroy it with the weapon of modern science. Kindly read an excerpt of the following conversation regarding the attempt to argue with Śrīmad Bhāgavatam:

> *Harikeśa: Śrīla Prabhupāda, I was wondering if it was more important to understand the universal make up or simply to accept what Śrīmad-Bhāgavatam says.*
> *Prabhupāda: That means you don't accept Śrīmad-Bhāgavatam?*
> *Harikeśa: No. But some of the things in Śrīmad-Bhāgavatam are difficult to understand.*
> *Prabhupāda: Rascal, fool. Therefore it is difficult for you.*
> *Harikeśa: Yes.*
> *Prabhupāda: You should not... You understand or not understand, you have to accept it.*
> *Harikeśa: That is why we are asking so many questions about the moon and the planets.*
> *Prabhupāda: Yes, so where is the doubt? It is a fact that they did not go to the moon. That's a fact.*
> *Harikeśa: I'm just trying to understand for preaching purposes.*
> *Prabhupāda: Why do you try to understand?*
> *Harikeśa: Because I have a hard time explaining about the make up of the universe as far as...*
> *Prabhupāda: Then don't understand them also, you don't understand here also. Then what you will do?*
> *Harikeśa: Well, the difficulty I was having about this mountain, this Meru. It sounds... It's very difficult to explain that to someone.*
> *Prabhupāda: No, no, that is difficult, but everything is difficult for you. Because you want to see. You have not seen their arrangement; neither you have seen our arrangement. So but your proposition is you don't believe what you don't see. You have not seen neither of them, so you have to remain silent.*
> *Harikeśa: So the best position is just remain silent until realization comes?*
> *Prabhupāda: No. You should... Śukadeva Gosvāmī says. You have to believe that. It is an idea. If you believe or not believe, it doesn't matter. But you can get an idea about the planetary situation. That's all. Neither you can go there; neither you can see it. An idea is given, that is all. But there is no argument.* (Morning Walk — June 10, 1975, Honolulu)

## Point #3—"Sun's Orbit"

However, these very contradictions point the way to a different understanding of Bhāgavata cosmology in which it emerges as a deep and scientifically sophisticated system of thought.

Study of the contradictions shows that they are caused by overlapping self-consistent interpretations that use the same textual elements to expound different ideas. For example, the Bhāgavatam states that the sun god orbits Mount Meru with the mountain to his left and to his right and this is questioned in the text itself by King Parīkṣit. It turns out, however, that the orbiting with Meru on the right refers to the daily motion of the sun relative to the earth, and it is part of the interpretation of Bhū-maṇḍala as a planisphere map of the earth. The orbiting with Meru on the left refers to the sun's yearly motion relative to the stars, and it is part of the interpretation of Bhū-maṇḍala as a geocentric map of the solar system.

## BST #3—

I find this to be a most unethical argument. Let us not sentence Bhāgavata Purāṇa to a "Life in Prison"—guilty of breaching common sense. That prison would be our own limited material intelligence with its prison cell made of four dingy non-literal interpretive walls.

> *Prabhupāda: No, accept or not accept, the whatever is description there, in Bhāgavatam, we are accepted. Just like when we speak of the hell, everyone is little anxious, "Where are the hell?" So that is described. In the down this universe there are lokas.*
> *Harikeśa: I become very fascinated when I read these descriptions in Śrīmad-Bhāgavatam about the creation and the universal...*
> *Prabhupāda: Yes, yes. That's gist idea of the universal position. That is sufficient.*
> *Harikeśa: So, in other words, Śukadeva Gosvāmī is always bringing Mahārāja Parīkṣit to the platform of "Why bother with all of this? Simply perform devotional service"?*
> *Prabhupāda: Huh? No, no. He's giving full knowledge of the universal affairs. He also says at the end that "I have described whatever I have heard." That's all.*

*Harikeśa: So we should simply describe it without being concerned that the scientific mind may make sense out of it or not?*
*Prabhupāda: What is this nonsense scientific? That is... We reject immediately. What is scientific? A tiny brain, what is their science? Phene bare dhake nate ute. A snake catcher... There is a kind of snake which has no poison. So he cannot catch even that non-poisonous snake, and he's trying to catch one cobra. So these scientists, what is their value? What they have done anything contribution to the world for the benefit of the human society?* (Morning Walk — June 10, 1975, Honolulu)

The fact that literal Bhāgavatam cosmology differs radically from modern astronomy does not disqualify it, rather that difference is the emblem of its perfection. Common sense is a relative consideration. Where is the common sense that a poor, tiny jiva soul inside a defective human body stranded upon a minute portion of the planet earth, straining his six-inch neck to view the sky with its millions of stars declares that Brahmā, Lord Kṛṣṇa and Śukadeva Goswāmī were mistaken about cosmology.

Please don't become upset if I say that apparently you have failed to comprehend a basic principle of studying Vedic literature, i.e. that the Śrīmad Bhāgavatam must be taken literally. Proposing non-literal interpretations for the Bhāgavatam is not the correct way of understanding the knowledge. The Bhāgavatam must be accepted literally. The Bhāgavatam's description of the size of the universe is well supported through identical evidence related by several recognized ācāryas and these conclusions are corroborated in other authentic śāstras such as Bṛhad-bhāgavatāmṛta, Śrī Caitanya-caritāmṛta, etc. Consider this quotation from Śrīmad Bhāgavatam itself saying that the Śrīmad Bhāgavatam describes the cosmology of the entire universe including earth, sky, heavenly planets and lower planetary systems.

द्वीपवर्षसमुद्राणां गिरिनद्युपवर्णनम्
ज्योतिश्चक्रस्य संस्थानं पातालनरकस्थिति"

*dvīpa-varṣa-samudrāṇāṁ*
*giri-nady-upavarṇanam*
*jyotiś-cakrasya saṁsthānaṁ*
*pātāla-naraka-sthitiḥ*

*SYNONYMS: dvīpa-varṣa-samudrāṇām—of the continents, great islands and oceans; giri-nadī—of the mountains and rivers; upavarṇanam—the detailed description; jyotiḥ-cakrasya—of the celestial sphere; saṁsthānam—the arrangement; pātāla—of the subterranean regions; naraka—and of hell; sthitiḥ—the situation.*

*TRANSLATION: The Bhāgavatam gives an elaborate description of the earth's continents, regions, oceans, mountains and rivers. Also described are the arrangement of the celestial sphere and the conditions found in the subterranean regions and in hell.* (Śrīmad Bhāgavatam 12.12.16)

We are not at liberty to correct the Bhāgavatam. If one is unable to fully accept the statements of Śrīmad Bhāgavatam, he should silently and humbly go back to studying the Bhagavad-gītā and confirm whether he is performing his basic practices of sadhana bhakti according to the instructions of the spiritual master. Kindly consider me your friend when I say that one who accepts some portions of the Śrīmad Bhāgavatam and rejects other portions falls into the pit of *nāma-aparādha* called *śruti-śāstra-nindanam* or vilifying Vedic literature.

Let us understand the Śrīmad Bhāgavatam as Śrīla Prabhupāda taught:

*So one planet to another, or one planet to the Sumeru hills, there is long, millions of miles, made of gold. They are gold. Similarly copper. Similarly, there are different oceans. Now, we may not believe, but there is no question of believing. You cannot say, "There is no such thing." But we can say because we get the information from the śāstra. You have no evidence, so you cannot say "No." You can say "Maybe," but we don't say "Maybe." There is because we get the information from the śāstra.* (Lecture on Śrīmad-Bhāgavatam 6.1.33 — San Francisco, July 18, 1975)

Let us not become opportunistic by giving interpretations of the Śrīmad Bhāgavatam's Fifth Canto cosmology which are light years away from the presentation of Śrīla Prabhupāda. It is very good that ISKCON scientists should try to explain Bhāgavata cosmology but they must not succumb to the speculative and cheating mentality. If someone desires to espouse one's own imaginative ideas, let them do so without taking advantage of the Śrīmad Bhāgavatam.

*Indian man: It must be understood literally, you say.*

*Prabhupāda: Yes, why not? Suppose you have got some philosophy. So you can explain your philosophy differently. Why should you take Bhagavad-gītā and explain your philosophy? Is it honesty?*
*Indian man: All the ācāryas have been doing it.*
*Prabhupāda: No ācāryas are doing it. All lower-class men. No ācāryas do it. Rāmānujācārya, Madhvācārya, Nimbārka, these are ācāryas. Śaṅkarācārya, Caitanya, they never did it. Outsiders, who did not care for the authority of the ācārya, they did it. Otherwise, we are the ācārya sampradāya. They'll never do that. Ācāryavān puruṣo veda. Ācāryopāsanam. That is bona fide process of knowledge. Ācāryopāsanam. Amānitvam adambhitvam. Ācāryopāsanam. This is the process of knowledge. Evaṁ paramparā-prāptam imaṁ rājarṣayo viduḥ. This is the ācārya. Ācārya will never interpret things like that. You see Rāmānujācārya's comments on Bhagavad-gītā. Nothing changed. But in every śloka he has given evidence from the Vedas, from the Upaniṣads. Ācārya will never change.* (Interview with Professors O'Connell, Motilal and Shivaram — June 18, 1976, Toronto)

# 8

# Copernican Revolution

From the historical point of view, when India was colonized by the British years ago, some Westerners, motivated by a desire to decrease the faith of the indigenous people in their own scriptures and a hope to convert them to Christianity, sought to discredit the statements of India's scriptures. The cosmological descriptions in the Purāṇas, with their oceans of milk and yogurt and mountains many miles high, appeared to be a weak point that could be used to dismantle faith in the entire worldview, and so they were ruthlessly ridiculed.

In this chapter, we shall review some of the events and the methodology used. It is interesting to observe that the means used by the British for discrediting the Purāṇas (especially the Bhāgavata Purāṇa or Śrīmad Bhāgavatam) was to prepare English translations and commentaries for some of the prominent Jyoti Śāstras such as Sūrya-siddhānta and Siddhānta-śiromaṇi. Those are the translations which are presented in the later chapters of this book. Here, we shall take a look at the Jyoti śāstra translators/ commentators for the works included in this compilation and examine their comments. It is obvious that the translator/commentators aimed at establishing Western astronomical concepts in place of the Vedic cosmology through slanting interpretations of the Jyoti Śāstras. They were so extraordinarily successful that to the present day their Western views permeate the minds of the so-called educated persons interested in and "authorities" upon Vedic cosmology. Our motive for presenting this information is not merely to cast blame, rather it is to trace the misconceptions which exist today and threaten to compromise the faith and direct understanding of the Vaiṣṇava community, especially within ISKCON.

For example Capt. James Herbert, Deputy Surveyor General of British India, a scientist of high repute, was appointed to the position of "royal" astronomer at the Lucknow observatory. In 1836 from there "A Brief Account of the Solar System" was published in which one finds the heliocentric theory set forth in simple terms, the planets described, the probability of extra-terrestrial life, comets, etc., all aptly couched in a biblical idiom that would have been more likely acceptable in parts of India where the rulers were Muslim nawabs. In the book's concluding rationale for the study of astronomy we find a dominant theme in colonial discourse about science: 1) the idea that colonialism helped create the conditions for European progress in science; and 2) the idea that progress in science is itself a justification for European colonialism. Consider the following extract from "A Brief Account of the Solar System" in English; with the Translation into Hindustani; Arranged as Reading Lessons for the Use of Schools (Calcutta, 1836: Baptist Mission Press), pp. 105-108

*The Benefit of Astronomy to Man*

1. *The study of astronomy enlarges the mind as much as faith in astrology enfeebles it. Astronomy leads the mind up to God and fills it with sublime conceptions of His power and wisdom. On a due acquaintance with astronomy depends the perfection of navigation, geography, chronology, commerce, ...*

2. *By the learned and useful calculations of astronomers, the surface of our globe has been measured with scientific accuracy. The distances of kingdoms, capes, continents, and cities have been laid down in miles and furlongs, and above all by charts and maps. The great ocean is now everywhere intersected by the lines of science and has become a well known highway for our fleets and navies.*

3. *By knowledge of astronomy the mariner is enabled with his compass to guide his ship through the pathless ocean without an object to direct him except the moon and stars by night.*

4. *Thus the various productions of India as cotton, indigo, sugar, silk, spices, saltpeter, ivory, precious stones, et cetera are carried to Europe.*

5. *And India receives in exchange the treasures of Europe: cloth, lead, metal, telescopes, watches, mathematical instruments, and above all, the wisdom of the best books teaching science and virtue, for knowledge of which the European nations are famed.*

6. *But for astronomy, the pathless ocean would be a barrier between the nations of the world, and the distant inhabitants of the earth never would have met. Duly instructed in this science, the adventurous mariner launches his ship into the deep, laden with all the commerce, varieties, luxuries, animals, and inventions of foreign lands.*

7. *And boldly spreads his sails to the breeze and guided only by his compass, by the sun and the moon and stars of heaven he weathers many a mid-night storm. A solitary wanderer in the unfathomable deep until at length he reaches the far-distant port.*

In short, science puts a benign face on colonial exploitation: in return for its raw materials, India not only gets processed goods from Europe, but also its science. Even though India loses its culture in the bargain.[1]

## Copernican Revolution in India

Focusing on what might be called the Indian version of the Copernican revolution, let us consider the work of a handful of Indian astronomers who participated in the transformation of cosmological knowledge. This involves some scholars active at the beginning of the 19th century, at a time when Sanskrit came into direct contact with the growing imperial presence of the British. The aggressive writings on Hinduism by Christian missionaries and civil servants of the East India Company who were sympathetic to Christianity had been going on for centuries, but became especially prominent in the 1830s and 1840s when the paradigm shift to Western astronomy or Copernican heliocentrism was finally affected.

Lancelot Wilkinson (who figures prominently in the Copernican revolution of India) had been appointed the British Political Agent to the court of Bhopal sometime before 1829, and continued there until his death in 1841. The Agent's residence was located in the nearby town of Sihore, and attached to the residence was a school, which Wilkinson made into a Sanskrit school with a special emphasis on the study of the astronomical Siddhāntas. This became known as the 'Sabha," or intellectual circle of Lancelot Wilkinson.

Wilkinson was interested in the Siddhāntas, whose serious study, he felt, had largely vanished by the beginning of the 19th Century. In addition to studying Sanskrit and Jyoti Śāstra himself, he promoted the study, edition and publication of various Indian astronomical texts. Wilkinson believed that the best way to introduce the modern Copernican system of astronomy to learned Indians, especially to the whole class of Indian astronomer/astrologers, was through the medium of Sanskrit, and in particular through the instrumentality of the Siddhāntic model of the cosmos. Wilkinson reasoned, that since the Jyoti Śāstras made use of many of the necessary principles of geometry, trigonometry, and arithmetic, it would be a short step to move from the Siddhāntic to Copernican scientific models, which could be presented in a way that would not alarm the whole class of Jyotiṣa paṇḍits.

One of the most talented of Wilkinson's sabha was a Brahmin from Ahmadnagar district, known as Bapu Deva Sastri. Wilkinson and Sastri combined to translate and comment upon the Sūrya-siddhānta and Siddhānta-śiromaṇi. They presented both an English version as well as a version in Hindi since the texts so particularly fulfilled one of the ideals of the Asiatick Society, namely to revive ancient Indian learning as a vehicle for advancing European visions of scientific knowledge.

Wilkinson and Sastri endorsed the English heliocentric model as the correct one whereby the earth is a globe with the continents located where they are found by modern sailors and explorers, who are seen as the pioneers of a modern, observation-based scientific geography. In general there was a certain awe by the people of India toward the English for their observational instruments and exploratory zeal. Wilkinson's sabha considered the Siddhāntas a basic and reliable source for arguments about such things as the reason for the days getting longer and shorter, and so on. The Purāṇas are cited occasionally when they accord with the modern scientific view, but they generally come in for harsh treatment. The Purāṇic geography of the earth with its seven concentric oceans and continents is explicitly rejected. [2]

During that time period, an independent (rajput) princely state called Kotah (situated in today's Rajasthan) naturally felt sensitive about its territorial integrity at a time when the British were encroaching on all sides. In 1832 when a British survey party from Delhi entered Kotah's dominions and attempted to map the region, the Kotah *râjrâna* (king), Râm Singh, suspected that this was a prelude to annexation. After all, mapping (surveying the realm) was the most symbolic act of empire in the colonial era. To the *râj râna* it implied a claim on his territory by the British. Kotah was located not far from Jaipur, which in the 18th century had been India's leading center for the study of astronomy. The British resident in Kotah (in effect, an ambassador), Lancelot Wilkinson was instructed by the British government in Calcutta to allay the fears of the *râjrâna*. This he did by invoking the authority of an ancient text of Indian astronomy (the *Siddhântaśiromaṇi* of Bhâskara II) on the importance of knowing the exact coordinates (latitude and longitude) of the king's residence (for astrology, this was essential). Wilkinson's tactic worked, far better, in fact, than anticipated, because it gave to the court at Kotah and thus India, the impression that the British were far ahead of themselves in astronomy. [3]

## Virodha—Contradiction

Using a "divide and conquer" policy, Wilkinson et al attempted to show a contradiction (*virodha* in Sanskrit) between the Purāṇas and the more recent Siddhāntas or Jyoti Śāstras. The Gauḍīya Vaiṣṇavas in modern times headed by Śrīla Bhaktisiddhānta Sarasvatī Ṭhākura, accept both the Purāṇas and the Jyoti Śāstras.

> *A manvantara is the period controlled by one Manu. The reign of fourteen Manus equals the length of one day (twelve hours) in the life of Brahmā, and the night of Brahmā is of the same duration. These calculations are given in the authentic astronomy book known as the Sūrya-siddhānta. A Bengali translation of this book was compiled by the great professor of astronomy and mathematics Bimal Prasād Datta, later known as Bhaktisiddhānta Sarasvatī Gosvāmī, who was our merciful spiritual master. He was honored with the title Siddhānta Sarasvatī for translating the Sūrya-siddhānta, and the title Gosvāmī Mahārāja was added when he accepted sannyāsa, the renounced order of life.* (Caitanya-caritāmṛta: Ādi 3.9 Purport)

The Gauḍīya Vaiṣṇava school accept all Vedic wisdom yet categorize it within two educational systems:

*Śrīla Bhaktisiddhānta Sarasvatī Ṭhākura gives the following commentary on the words bhāgavata vicāra. As confirmed in the Muṇḍaka Upaniṣad (1.1.4–5), there are two kinds of educational systems:*
*dve vidye veditavya iti, ha sma yad brahma-vido vadanti—parā caivāparā ca. tatrāparā ṛg-vedo yajur-vedaḥ sāma-vedo 'tharva-vedaḥ śikṣā kalpo vyākaraṇaṁ niruktaṁ chando jyotiṣam iti. atha parā yayā tad akṣaram adhigamyate.*

*"There are two kinds of educational systems. One deals with transcendental knowledge [parā vidyā] and the other with material knowledge [aparā vidyā]. All the Vedas—the Ṛg Veda, Yajur Veda, Sāma Veda and Atharva Veda, along with their corollaries, known as śikṣā, kalpa, vyākaraṇa, nirukta, chanda and jyotiṣa—belong to the inferior system of material knowledge [aparā vidyā]. By parā vidyā one can understand the akṣara—Brahman or the Absolute Truth." As far as the Vedic literature is concerned, the Vedānta-sūtra is accepted as the parā vidyā. Śrīmad-Bhāgavatam is an explanation of that parā vidyā.* (Caitanya-caritāmṛta; Madhya 19.17 Purport)

The Jyotisa śāstras fall into the category of *aparā vidyā* or material knowledge whereas the Bhāgavata Purāṇa explains *parā vidyā*, or transcendental knowledge. The two systems are generally compatible, however if there should appear to be a *virodha* or contradiction between them, then the Śrīmad Bhāgavatam must be considered the most authoritative judgement. This is confirmed by Śrīla Jiva Goswāmī in his Sandarbhas as well as by numerous other Vaiṣṇava ācāryas.

धर्म" प्रोज्झितकैतवोऽत्र परमो निर्मत्सराणां सतां वेद्यं वास्तवमत्र वस्तु शिवदं
तापत्रयोन्मूलनम्
श्रीमद्भागवते महामुनिकृते किं वा परैरीश्वर" सद्यो हृद्यवरुध्यतेऽत्र कृतिभि"
शुश्रूषुभिस्तत्क्षणात्

*dharmaḥ projjhita-kaitavo 'tra paramo nirmatsarāṇāṁ satāṁ*
*vedyaṁ vāstavam atra vastu śivadaṁ tāpa-trayonmūlanam*
*śrīmad-bhāgavate mahā-muni-kṛte kiṁ vā parair īśvaraḥ*
*sadyo hṛdy avarudhyate 'tra kṛtibhiḥ śuśrūṣubhis tat-kṣaṇāt*

*Completely rejecting all religious activities which are materially motivated, this Bhāgavata Purāṇa propounds the highest truth, which is understandable by those devotees who are fully pure in heart. The highest*

> *truth is reality distinguished from illusion for the welfare of all. Such truth uproots the threefold miseries. This beautiful Bhāgavatam, compiled by the great sage Vyāsadeva [in his maturity], is sufficient in itself for God realization. What is the need of any other scripture? As soon as one attentively and submissively hears the message of Bhāgavatam, by this culture of knowledge the Supreme Lord is established within his heart.* (Śrīmad Bhāgavatam 1.1.2)

Vaiṣṇavas do not see any actual contradiction between the Jyotisa śāstras and the Bhāgavata Purāṇa since the Jyotisa śāstras are subordinate by nature. Yet the sabha of Wilkinson and Sastri put forward the view that Jyotisa śāstras outweigh the Bhāgavata Purāṇa. Moreover, scholars of this type tended to place high valuation upon reasoning from direct observation, especially as aided by technologically advanced instruments, and the corresponding devaluation, within the sphere of what is observable, of scriptural Vedic authority. Due to putting implicit faith in scientific discovery based on telescopes and his own interpretations, Wilkinson severely critiqued the Jyotisa śāstras as well as the Purāṇas.

It is understood from the statement of Mahābhārata that there are many *munis*, or speculators:

*tarko 'pratiṣṭhaḥ śrutayo vibhinnā*
*nāsāv ṛṣir yasya mataṁ na bhinnam*

> *All speculators must disagree with other speculators; otherwise, why should there be so many opposing parties concerned with ascertaining the supreme cause?* (Śrīmad Bhāgavatam 6.4.31 Purport)

On the other hand, pure devotees in the line of Śrī Caitanya Mahāprabhu, place full unreserved faith in the Bhāgavata Purāṇa knowing that any apparent discrepancies are there resolved.

> *Śrīmad-Bhāgavatam is Lord Śrī Kṛṣṇa Himself in the form of recorded knowledge. Therefore, it is the cream of all the Vedas, and it contains all historical facts of all times in relation with Śrī Kṛṣṇa. It is factually the essence of all histories.* (Śrīmad Bhāgavatam 1.3.42 Purport)

It is important to note that Sawai Jaisingh, the celebrated astronomer-king, in the 1720's commissioned Kevalarama, the Jyotiṣaraya in the court of

Amber, to write an independent work on the subject of the noncontradiction between the Purāṇic and Siddhāntic (Jyotisa) models. This work known as the *Bhāgavatajyotiṣayor Bhūgolavirodhaparihāra* was expanded by Nandarāma Miṇra in the 1780's in Kṛmakavana in southern Rājputana.[4] Some modern interpreters characterize Jaisingh as supporting a planisphere theory of the Bhāgavatam wherein Jambudvīpa occupies the northern hemisphere of this earth globe and the other six dvīpas as mentioned in the Śrīmad Bhāgavatam occupy the southern hemisphere, yet if that were factual how could he oppose the so-called theory of contradiction between Purāṇic and Siddhāntic (Jyotisa) models. The proposition of contradiction is difficult to accept by Vaiṣṇavas due to its non-literal interpretation of the Bhāgavatam texts.

In 1836-37 the learned Yajñeśvara wrote *Avirodhaprakāśa* wherein he showed that there is no contradiction between the Siddhāntic and Purāṇic models. He explained that the two models have different purposes and scopes of application. For instance, the small earth of 5000 yojanas in circumference assumed by the Siddhāntakāras is explained as being only a subsection of the Purāṇic bhūmaṇḍala; in fact it is a subportion of Bhāratakhaṇḍa which measures about 2000 yojanas on its longest side and 5000 yojanas in circumference. For Wilkinson, Sastri and Subbājī (another of Wilkinson's protoges) talk of *virodha* and *avirodha* activated doubts of the Brahminical cosmology, which they hoped to fragment and reassemble.

Regarding the supposed inconsistency between Purāṇic and Siddhāntic cosmologies and astronomical models, Yajñeśvara proposes that what appears to be the moon at the time of a solar eclipse is in fact the head of Rāhu, which has the same size as the moon. Rāhu is suspended from the wheel of the *nakṣatras* in such a way that his head hangs down to the level of the sun and blocks the sun's light from time to time, hence, the Siddhāntas predictions of eclipses can be used without rejecting the Purāṇic order of the planets.[5]

## Wilkinson and Sastri

Wilkinson's chief protege, Bāpu Deva Śāstrī, went on to teach both Indian and European astronomy at the Benares Sanskrit College beginning in 1841. Bāpu Deva published voluminously in Sanskrit and English, promoted a modernization of Indian astronomy and the knowledge of European astronomy in India. It was his pupils and intellectual descendants, including

Sudhākara Dvivedi, who dominated the intellectual scene in Benares for at least the rest of the century.[6]

> *"As the point where the equator cuts the horizon is the east, the sun therefore rises due east at, time of the equinoxes but on this ground, we cannot determine the direction at Meru (the north pole) because there the equator coincides with the horizon und consequently the sun moves at Meru under the horizon the whole day of the equinox."* (—Bāpu Deva Śāstrī commenting on Siddhānta-śiromaṇi chapter 3.36)

Throughout their translation of the Sanskrit text Siddhānta-śiromaṇi, Wilkinson and Sastri offer opinions such as the one above. This interpretation depicts Mount Meru to be in actuality the North Pole on this earth planet although no such mention is described in the Śrīmad Bhāgavatam and other Vedic literatures. Such an interpretation brings up questions since Mount Meru (also known as Sumeru) is (according to Śrīmad Bhāgavatam) situated in the center Ilāvṛta varsa which lies in the center of Jambūdvīpa comprising a lateral distance from the Bhārata-varṣa and its Himalaya Mountain range of approximately 31,000 yojanas (248,000 miles).

Commenting on Siddhānta-śiromaṇi verses 3. 21-43: Lancelot Wilkinson writes:

> *Bhaskaracarya has exercised his ingenuity in giving a locality on the earth to the poetical imaginations of Vyasa, at the same time that he has preserved his own principles in regard to the form and dimensions of the Earth. But he himself attached no credit to what he has discribed in these verses for he concludes his recital in his commentary with the words,*
>
> यदिद मुक्तं तत्सव पुराणाश्रितम्
>
> *yadida muktaṁ tat sava purāṇāśritam*
>
> *"What is stated here rests all on the authority of the Puranas."*
>
> *As much as to say "credat Judaeus."*

The Sanskrit verse quoted here is the only one from the entire book Siddhānta-śiromaṇi to be quoted in Devanāgarī or English diacritics.

*The Latin comment Wilkinson makes "credat Judaeus,"* is a phrase coming from a work by Horace which was ***Credat Iudaeus Apella, non ego*** ("Iudaeus" is sometimes spelled "Judaeus"); literally, "Let the Jew Apella believe it; I will not." Or according to Roget's Thesaurus entry #485: the phrase suggests "Let those believe who may." In other words, Mr. Wilkinson does not believe that the author of Siddhānta-śiromaṇi is consistent with the Purāṇas even though Bhaskaracarya testifies to his fidelity to the Purāṇas.

Throughout their commentary, Wilkinson and Sastri take the liberty to discredit all Vaiṣṇava parties related to the text namely the Siddhānta-śiromaṇi's author Bhaskaracarya; Siddhānta-śiromaṇi's speaker and Śrīla Vyāsadeva, the author of the Purāṇas and in particular the Bhāgavata Purāṇa.

## Burgess

In chapter 12 of this book we present an English translation for Sūrya-siddhānta done in 1860 A.D. by Reverend Ebeneezer Burgess, an American Christian missionary to India and published in the Journal of the American Oriental Society. Ebenezer Burgess was born at Grafton, Vermont, U.S.A., on June 25, in the year 1805. He graduated from Amherst College in 1831 and was a Tutor in the same College from 1833 to 1835. He then entered Andover Theological Seminary from which he graduated in 1837. After another year spent in advanced study at Andover, and after teaching Hebrew and Greek for sometime at Union Theological Seminary in New York City, he was ordained to the ministry.

In the year 1839 Burgess came to India as a Missionary to the Marathas. He lived in Bombay Presidency for fifteen years; first at Ahmednagar until 1851, and then at Satara. He returned to the United States in 1854. From 1857 to 1859 he acted as Pastor at Centreville, Mass., from 1861 to 1863 at Lanesville and from 1864 to 1867 at South Franklin. He was engaged in lecturing and literary work until his death. He died in Newton Centre, Mass., on January 1, 1870. He had written an elaborate essay on the history of astronomy among the Hindus and had nearly completed a treatise on the antiquity of man, but it remained unpublished.

Rev. Burgess was disturbed that a previous Western translator Mr. Bailly, had given too much value to the literal statements of Sūrya-siddhānta. He expresses his impetus for translating Sūrya-siddhānta in his Introductory Note:

*Soon after my entrance upon the Missionary field, in the Maratha country of western India, in the year 1839, my attention was directed to the preparation, in the Marathi language, of an astronomical text-book for schools. I was thus led to a study of the Hindu science of astronomy, as exhibited in the native text-books, and to an examination of what had been written respecting it by European scholars.*

*The Astronomie Indienne of Bailly, the first extended work upon its subject, had long been acknowledged to be founded upon insufficient data, to contain a greatly exaggerated estimate of the antiquity and value of the Hindu astronomy, and to have been written for the purpose of supporting an untenable theory.*

In his work of translation of Sūrya-siddhānta, no Devanāgarī is given nor are there any Sanskrit verses transcribed into English using diacritics, although key Sanskrit words are often supplied. Rev. Burgess held views adversarial toward Vedic texts as expressed in his Introductory Note:

*"In short, there was nothing in existence which showed the world how much and how little the Hindus know of astronomy [as Sūrya-siddhānta], as also their mode of presenting the subject in its totality, the intermixture in their science of old ideas with new, of astronomy with astrology, of observation and mathematical deduction with arbitrary theory, mythology, cosmogony, and pure imagination. It seemed to me that nothing would so well supply the deficiency as the translation and detailed explication of a complete treatise of Hindu astronomy : and this work I accordingly undertook to execute."*

The translator's strong bias against accepting the text he was translating at face value is nicely expressed in the following:

*"According to this [the actual text], the Sūrya-Siddhānta was revealed more than 2, 164,960 years ago, that amount of time having elapsed, according to Hindu reckoning, since the end of the Golden Age; see below, under verse 48, for the computation of the period. As regards the actual date of the treatise, it is, like all dates in Hindu history and history of Hindu literature, exceedingly difficult to ascertain. Bentley has endeavoured to show by internal evidence that the Sūrya-Siddhānta belongs to the end of the eleventh century; see below, under verse 29-34,*

*where his method and results are explained, and their value estimated." (Sūrya-siddhānta: 1.2-3 Comment by Rev. Burgess)*

In other words Reverend Burgess rejected the date given in the text and preferred the date assigned to the text given by a fellow scholar, Mr. Bentley at 1,000 A.D. The primary conjecture of the Western scholars was and is that Jyoti Śāstras such as Sūrya-siddhānta, Siddhānta-śiromaṇi, etc. were derived from Greek astronomy. Rev. Burgess for example thought the city Romaka as used in Jyoti śāstras must have been referring to the city of Rome.

> *"Go therefore to Romaka-city, thine own residence; there, undergoing incarnation as a barbarian, owning to a curse of Brahma, I will impart to thee this science."* (Sūrya-siddhānta: Chapter 1: verse deleted)
>
> *If this verse really formed a part of the text, it would be as clear an acknowledgement as the author could well convey indirectly, that the science displayed in his treatise was derived from the Greeks. Romaka-city is Rome, the great metropolis of the West.*

He then took the conjecture of Mr. Weber who thought that the Asura Maya as presented in the Sūrya-siddhānta as the student of Sūrya the sungod must be a Greek. Contrary to Mr. Weber's postulation, we find in the Mahābhārata (below) Asura Maya, also known as Maya Danava is described as a resident of Mainaka Mountain one of the prominent mountains in Bhārata-varṣa. Mainaka Mountain is not in Greece.

> *Maya explained that he had formerly been engaged by Vrishaparva, king of the Danavas, to construct sacrificial altars for the Asuras. He had gathered all kinds of celestial materials which he had stored at Vrishaparva's house high up on the Mainaka mountain. There was also a great club with which Vrishaparva had once withstood the gods in battle. Maya would bring that club, equal to one hundred thousand ordinary clubs, and give it to Bhima. He would also fetch from the depths of a lake on Mainaka the large celestial conch shell known as Devadatta for Arjuna. If Arjuna blew that conch on the battlefield, it would shatter his opponents' hearts.*
>
> *Having gained Yudhisthira's permission, the Asura left quickly for the north. He found all his wealth guarded by Yakshas and Rakshasas, and with their assistance he brought it back to Indraprastha. After presenting*

> *the club to Bhima and the conch shell to Arjuna, he commenced work.* (MAHĀBHĀRATA: At Mainaka Mountain P. 152)

Weber's method of speculation is not without bias since the commentators clearly wanted to diminish the value of the Indian culture and their astronomical treatises.

> *Is this verse, then, a fragment of a different, and perhaps more ancient, account of the orgin of the treatise, for which, as conveying too ingenuous a confession of the source of the Hindu astronomy, another has been substituted later? Such a supposition certainly does not lack plausibility. There is something which looks the same way in the selection of a demon, an Asura, to be the medium of the sun's revelation; as if, while the essential truth and value of the system was acknowledged, it were sought to affix a stigma to the source whence the Hindus derived it. Weber (Ind. Stud. ii. 242; Ind. Lit., p. 225), noticing that the name of the Egyptian sovereign Ptolemaios occurs in Indian inscriptions in the form Turamaya, conjectures that Asura Maya is an alteration of that name, and that the demon Maya accordingly represents the author of the Almagest himself; and the conjecture if; powerfully supported by the fact that al-Biruni (see Reinaud, as above) ascribes the Pāuliśa-Siddhānta, which the later Hindus attribute to a Pulica, to Paulus al-Yunani, Paulus the Greek.* (Comment by Reverend Burgess)

Rev. Burgess reveals his approach to translating and commenting upon theSūrya-siddhānta. Hespeculates that Romaka refers to the modern Italian city of Rome although according to the text the conversation was spoken millions of years before modern Rome appreared. Further he removes the text from the book because he felt it out of place. In the following statement Burgess expresses his frustration with "Hindu" writings.

> *[The former is] curiously illustrative of a fundamental trait of Hindu character: a fantastic imaginativeness, which delights itself with arbitrary theorizings, and is unrestrained by, and careless of, actual realities. Thus, having no instruments by which they could measure even seconds with any tolerable precision, they vied with one another in dividing the second down to the farthest conceivable limit of minuteness; thus, seeking infinity in the other direction also, while they were almost destitute of a chronology or a history, and could hardly fix with accuracy the date of any event beyond the memory of the living generation, they devised, and put forth*

> *as actual, a frame-work of chronology reaching for millions of millions of years back into the past and forward into the future.* (Comment by Rev. Burgess on Ch. 1.11)

Burgess finds it difficult to accept the concept of celestial days as described in the Vedas. Furthermore he conceives that the descriptions of Mount Meru as given in the Śrīmad Bhāgavatam must be referring to the north and south pole of this earth planet.

> *In the astronomical reconstruction of the Puranic system, however, a physical meaning has been given to this day of the gods: the gods are made to reside at the north pole, and the demons at the south; and then, of course, during the half year when the sun is north of the equator, it is day to the gods and night to the demons; and during the other half-year, the contrary. The subject is dwelt upon at some length in the twelfth chapter (xii. 45, etc.). To make such a division accurate, the year ought to be the tropical, and not the sidereal; but the author of the Sūrya-Siddhānta has not yet begun to take into account the precession. See what is said upon this subject in the third chapter (vv. 9-10).*(Sūrya-siddhānta Ch. 1.14 Comment by Burgess)

In fact Burgess, like the other Western commentators, has presented his own belief that Mount Meru could only mean the north pole.

> *The" seat of the gods" is Mount Meru, situated at the north pole.* (1.62 Comment by Burgess)

Burgess attributes Bhaskaracarya, author of Siddhānta-śiromaṇi, with inventing a theory which contradicts the Śrīmad Bhāgavatam.

> *The author of the Siddhānta-Śiromaṇi, more submissive than the writer of our chapter to the authority of tradition, accepts (Goladhy., chap. ii) the series of concentric continents and oceans, but gives them all a place in the unknown southern hemisphere, while he regards Jambudvipa as occupying the whole of the northern.* (Comment by Burgess on Sūrya-siddhānta 12.32-44)

> *As out of the Puranic Meru the new astronomical geography makes the axis and poles of the earth, so out of these mountains [Lokāloka] it makes the visible horizon.* (Comment by Burgess on Sūrya-siddhānta 13.15-16)

## Clark

Walter Eugene Clark, Wales Professor of Sanskrit at Harvard and Master of Kirkland House has translated and commented upon the Aryabhatiya of Aryabhatta in 1930. This book is very similar to Sūrya-siddhānta and Siddhānta-śiromaṇi—the commonly accepted date of this treatise is 499 A.D.

> *Stanza IV, 9, in spite of Parameavara [a previous commentator], must be interpreted as maintaining that the asterisms are stationary and that the Earth revolves. And yet the very next stanza (IV, 10) seems to describe a stationary Earth around which the asterisms revolve. Quotations by Bhaṭṭotpala, the Vāsanāvārttika, and the Madīci indicate that this stanza was known in its present form from the eleventh century on. Is it capable of some different interpretation? Is it intended merely as a statement of the popular view? Has its wording been changed as has been done with I, 4? I see at present no satisfactory solution of the problem.* (Walter Eugene Clark, Professor of Sanskrit in Harvard University; The University of Chicago Press; Chicago, Illinois Preface P. xiv)

From this we learn that tremendous confusion exists among scholars as to what the authors of Jyoti śāstras meant. For example Professor Clark cannot be sure whether Aryabhata subscribed to the idea that the earth revolves or is or is stationary nor whether the earth revolves around a stationary set of asterisms or visa versa. Thus one cannot gain any solid Vedic cosmological evidence from these books as translated and commented upon by secular scholars. Only original Sanskrit texts of the Jyoti Śāstras properly translated and explained by qualified Vaiṣṇavas, as was done by Śrīla Bhaktisiddhānta Sarasvatı Goswamı, should be accepted as bona fide.

Today's so-called Vedic scientists utilize the translations and commentaries of Jyoti śāstras given by secular scholars to substantiate their own imaginative cosmological theories. In my humble opinion, works by secular scholars or scientists should not be trusted.

*avaiṣṇava-mukhodgīrṇaṁ*
*pūtaṁ hari-kathāmṛtam*
*śravaṇaṁ naiva kartavyaṁ*
*sarpocchiṣṭaṁ yathā payaḥ*

> *"One should not hear anything about Kṛṣṇa from a non-Vaiṣṇava. Milk touched by the lips of a serpent has poisonous effects; similarly, talks about Kṛṣṇa given by a non-Vaiṣṇava are also poisonous."* (Hari-bhakti-vilāsa: Śrīla Sanātana Gosvāmī)

I have included the available English translations of Jyoti Śāstras by secular scholars in the later chapters of this book, therefore, merely to show how scholars and scientists, both past and present, enjoy tinkering and tampering with the Vedic literature.

It is the nature of secular scholars and scientists to ever postulate new theorems supposedly surpassing those of former reformers. Remember according to academic rules, if an academic puts complete faith in anything other than his own intellect, he at once ceases to be an academic. Thus a "true" academic cannot be a real Vaiṣṇava.

> *If the readings of our text are correct it is difficult t to see how the two stanzas can be brought into agreement. The ninth stanza states unequivocally that the asterisms are stationary and implies the rotation of the Earth. The tenth stanza seems to state that the asterisms, together with the planets, are driven by the provector wind. This would imply the ordinary point of view of most Indian astronomers that the Earth was stationary. But since several other stanzas (I, 1; I, 4; III, 5; IV, 48) and the testimony of later writers who quote Aryabhata prove that Aryabhata believed in the rotation of the Earth, it is impossible to follow Paramesvara. We might understand in stanza 10 the phrase "they seem to move" as stating a purvapaksa (the erroneous view), but in the absence of any word to suggest this interpretation it is a doubtful expedient. Stanza 10 cannot be regarded as an interpolation (unless one stanza has been dropped out in order to make room for it) because the last three sections of Aryabhata's work were known to Brahmagupta [an earlier jyoti sastri] as "The Hundred and Eight Stanzas" (and our text contains 108 stanzas). (Professor Young commenting on Aryabhatiya Chapter 4. 10)*

## Sengupta, Gangooly

Prabodh Chandra Sengupta (Sengupta) wrote a forty-two page foreword for Burgess's Sūrya-siddhānta.

*Some again of the Hindu astronomical works are alleged as revelations, which means that their authors have hid their names and their times with the definite motive of making their astronomical systems and calculations acceptable to the people of Hindu India, by representing them as direct transmission from their gods. To this class belong[eighteen] Siddhāntas. Their name is eighteen to match the Puranas of which also the name is eighteen; so revelation is eighteen ways stated; The versifier might have easily pushed up the number to twenty which is the number of the authors of the Dharma Sastras. The Sūrya Siddhānta is at the top of this class of revelations. It was revealed to Maya an Asura, in all probability an Assyrian or rather a Babylonian. The date is stated to be 2163102 B.C. to which no historical value may be attached.*

Like his colleagues, Sengupta cynically denounces Sūrya-siddhānta's textual date of transmission of two million years B.C. as absurd and assigns the asura Maya to be an Assyrian or Babylonian. And like others, he supposes that Hindu astronomy came from Babylonia and Greece.

*A peculiar feature of the knowledge of astronomy of this period was the idea that the moon was more distant than the sun, which is also noticeable in the Puranas. Thus from 100 to 400 A.D. we have a great gap of three hundred years in which astronomical knowledge from Babylonia and Greece came to India. (Sengupta: Introduction to Sūrya-siddhānta; xxxv January* 16,1935.*)*

Within the appendix at the end book we find editor, Phanindralal Gangooly adding his own views on Vedic astronomy and cosmography:

*It is now well known that Hindu culture cannot pretend to a remoter origin than 2000 B.C., and that, though marked by striking and eminent traits of intellect and character, the Hindus have ever been weak in positive science; metaphysics and grammar—with, perhaps, algebra and arithmetic, to them the mechanical part of mathematical science— being the only branches of knowledge in which they have independently won honorable distinction. We regard the Hindu science as an offshoot from the Greek, planted not far from the commencement of the Christian era, and attaining its fully developed form in the course of the fifth and sixth centuries.*

*The whole [astronomical and cosmological]system, in short, may be divided into two portions, whereof the one contains truth so successfully deduced that only the Greeks, among all other ancient nations, can show anything worthy to be compared with it; the other, the framework in which that truth is set, composed of arbitrary assumptions and absurd imaginings, which betray a close connection with the fictitious cosmogonies and geographies of the philosophical and Puranic literature of India. (Appendix to Sūrya Siddhānta p. 382-3)*

## Aftermath

With British (Wilkinson), American (Burgess, Clark), German (Weber) and Indian (Sastri, Sengupta, Gangooly) scholars each interpreting the Jyoti śāstras according to their overt Western bias, it is no wonder that today misconceptions remain. Individually and as a loose confederacy, the Western scholars have employed and perpetuated the method of accepting or rejecting Vedic evidence based on empirical so-called scientific paradigms. Consequently, they have employed non-literal interpretations contrary to the directions of the very śāstras they wish to explain.

In the course of their writings, the above-mentioned scholars have introduced four speculative interpretations in relationship to the Jyoti śāstras.

**Planisphere Interpretation.** Broadly speaking, in the planisphere interpretation, Jambūdvīpa as described in the Śrīmad Bhāgavatam represents the lands of the northern hemisphere such as North America, Europe and Asia whereas Africa, South America, Australia and Antarctica are supposed to be the other six dvīpas described in the Bhāgavatam. This interpretation claims Bhū-maṇḍala to be actually the earth globe mapped onto a plane by stereographic projection. We may note above that Wilkinson and Sastri in their commentary on Sūrya-siddhānta, treat the earth as a globe with Jambūdvīpa as the northern hemisphere and Mount Meru as the north pole.

Such an idea wreaks havoc on the Śrīmad Bhāgavatam's version of the universe by misconstruing six immense circular islands Plakṣa, Śālmali, Kuśa, Krauñca, Śāka and Puṣkara as existing on this globe. The six circular islands, said to cover millions of yojanas as well as the six oceans of sugarcane juice, liquor, clarified butter, milk, emulsified yogurt, and sweet drinking water are entirely disposed of as antiquated folklore. Rev. Burgess, for example, also interprets this passage from Sūrya-siddhānta Chapter 7.36 to support the planisphere interpretation: "Surrounding it [Meru] on every side is fixed next this great ocean, like a girdle about the earth, dividing the two hemispheres of the gods and the demons."

Indeed hemispheres exist in the Bhāgavata cosmology however they divide the entire universe not only this one planet. Rev. Burgess, of course considers many of the descriptions found in the Sūrya-siddhānta and the Purāṇas as mere mythology. (see Sūrya-siddhānta p. 286-288)

In order to try to substantiate the planesphere interpretation, the first step is to neglect certain features of the Bhāgavatam, in this case, the large dimensions assigned to the dvīpas of Bhū-maṇḍala in the Bhāgavatam. These begin with 100,000 yojanas for both the diameter of Jambūdvīpa and the width of the surrounding, annular Salt Ocean.
Yet Śrīla Prabhupāda says exactly the opposite:

> *The measurements given herein, such as 10,000 yojanas or 100,000 yojanas, should be considered correct because they have been given by Śukadeva Gosvāmī. (Śrīmad Bhāgavatam 5.16.10 Purport)*

One cannot remain neutral in such discussions. Either one sides with the so-called scientific secular scholars or with Śrīla Prabhupāda and the Gauḍīya Vaiṣṇava disciplic succession.

**Local Map of India Interpretation.** Due to wholesale rejection of the Purāṇas, the above-mentioned scholars, cannot imagine how the Bhū-maṇḍala could apply accurately to the vast universe using what is according to their view, "a primitive astronomical system." Thus they speculate that Jambudvīpa is a local map of India, the Himalayan region, and nearby territories in the south-central Asia. Burgess thought that the city Romaka described in the Sūrya-siddhānta was referring to the modern city of Rome and Mr. Weber speculated that the Maya Danava was a Greek. None of the scholars accept the dates given in the Vedas as realistic. If such secular theories are correct then the statements of Śrīla Śukadeva Goswāmī are fictitious and absurd. Both cannot stand. In order to make a valid case for the Local Map of India Interpretation, one would need to explain why Śrīla Śukadeva Goswāmī spoke so much nonsense to Mahārāja Parīkṣit. Then one would need to explain why Śrīla Vyāsadeva chose to include such erroneous accounts in the Śrīmad Bhāgavatam. At last one would then have to demonstrate why all of our ācāryas, most notably Śrīla Prabhupāda opposed such non-literal interpretations.

Śrīla Prabhupāda repeatedly states that formerly Bhārata-varṣa referred to the entire planet not just modern-day India. If at the time of Śrīmad Bhāgavatam's writing, Bhārata-varṣa referred to the entire earth globe we inhabit, which makes up less than an approximated seven per cent of the area of Jambūdvīpa, how can we suppose that Jambūdvīpa in actuality, refers to a mere part of modern day India.

**Celestial Map Interpretation.** The question may arise as to whether the Bhāgavatam is to be taken literally since Lord Brahmā is said to reside on Mount Meru yet elsewhere he is described as living on the highest planetary system Satyaloka. The answer is revealed in a discussion between Śrīla Prabhupāda and his disciples:

*Prabhupāda: Śivaloka? Śivaloka?*
*Prema-bhakti Swami: Yes, it is not actually Śivaloka, but residence of Lord Śiva.*
*Tamāla Kṛṣṇa: One of his... He described it that like during the summertime you go to a hill station. Each demigod has their place where they also go.*
*Prema-bhakti Swami: Śivaloka is different, but Lord Śiva is...*
*Prabhupāda: Summer residence.*
*Tamāla Kṛṣṇa: Yes.*
*Prabhupāda: Why not? If a man can have summer residence, Lord Śiva...*
*Prema-bhakti Swami: Actual residence is in Satyaloka. Brahmā is there in Satyaloka. That means 2,200,000,000 miles away from sun planet.*
*Prabhupāda: That... Then it is universe.*
*Prema-bhakti Swami: Yes, in the universe. It is very high.*
*Prabhupāda: And where is that cakra?*
*Prema-bhakti Swami: Cakra?*
*Prabhupāda: That polestar, center?*
*Tamāla Kṛṣṇa: That will be in another...*
*Prema-bhakti Swami: We will give another picture.*
*Prabhupāda: Oh.*
*Tamāla Kṛṣṇa: This picture only shows mainly part of Jambūdvīpa and especially Ilāvṛta-varṣa, the middle of...*
*Prabhupāda: So Brahmaloka is in Jambūdvīpa?*
*Prema-bhakti Swami: No, no, no. Brahmaloka is in Satyaloka.*

*Prabhupāda: Oh, then what is this?*
*Prema-bhakti Swami: This is Mount Meru.*
*Prabhupāda: But you said Brahmaloka.*
*Prema-bhakti Swami: Brahmapurī.*
*Prabhupāda: Brahmapurī, that.*
*Prema-bhakti Swami: Summer residence.*
*Tamāla Kṛṣṇa: The middle part.*
*Prema-bhakti Swami: And then there are eight different residences of different loka-pālas, Indra, Agni, Varuṇa...*
*Prabhupāda: Summer residence.*
*Tamāla Kṛṣṇa: Yes, on top of Meru, like a hill station. (laughter)* (Showing of Planetary Sketches — June 28, 1977, Vṛndāvana)

So again the Bhāgavatam's version of Bhū-maṇḍala is to be taken literally. Brahmā does in fact, reside atop Mount Meru but that residence is not Satyaloka but Brahmapurī, his "summer resort," if you will.

**Solar System Interpretation.** There is still another interpretation which attempts to take the distances of Bhū-maṇḍala as described in the Śrīmad Bhāgavatam and compare them to modern astronomy's estimates of the solar system. The logic here is that the orbits of the planets in our solar system are so big that they fill out the universe of four billion miles, therefore how can there be anything beyond them such as the galaxies and quasars to which we are so attached. Thus in order to hold on to the dear galaxy theories of modern science, we shall turn the Bhāgavatam's version of Bhū-maṇḍala into our local solar system. However let us not forget that the fantastic so-called accurate distances of galaxies of modern science are solely based on Big Bang Theory, Red Shift and Doppler Theories which have all yet to be proven true.

The Vaiṣṇava community faces a challenge. It is being asked to accept speculative interpretations the Śrīmad Bhāgavatam's fifth canto in a way which has never been attempted before by Gauḍīya Vaiṣṇava ācāryas. The unifying principle of all the interpretations mentioned above, is to unilaterally dismiss the distances supplied in the Śrīmad Bhāgavatam's cosmology. Intertwining Vedic statements with speculation has rendered a complex topic even more so thus confusing many a reader. Such misleading interpretation come from the Westernized commentators of the Jyoti śāstras, past and present who attempt to overrule Vedic wisdom with theories linked to modern science. These speculative scientists are tricky folks—they juggle yojanas as adeptly as Māyāvādīs wrangle scriptural suffixes and prefixes.

The Bhū-maṇḍala must not be taken as a "literary device" to be interpreted as one likes. It is real and realizable to those who are faithful devotees of Krishna. Having unceremoniously expelled the large distances for Bhū-maṇḍala's planetary systems, modern speculative scientists, now using Hubble telescopes and some probability math theory, assume that Bhū-maṇḍala doesn't really exist as described in the Śrīmad Bhāgavatam, but actually refers to the geocentric orbits of our nearest planets.

**FOOTNOTES:**

[1] (Richard F. Young "Political" Science: Astronomy in India, 1800-1850)
[2] Christopher Z. Minkowski; The Pandit as Public Intellectual: The Controversy over virodha or Inconsistency in the Astronomical Sciences; Cornell University
[3] (Richard F. Young "Political" Science: Astronomy in India, 1800-1850)
[4] Christopher Z. Minkowski; The Pandit as Public Intellectual: The Controversy over virodha or Inconsistency in the Astronomical Sciences; Cornell University
[5] Ibid.
[6] Ibid.

# 9

# Vedic Planetarium

One can argue that Śrīla Prabhupāda's insistence that the monumental Māyāpur temple contain a Vedic model of the universe or planetarium is meant to use astronomy to establish faith in the authority of the Vedas. Śrīla Prabhupāda said in a letter to Svarupa Damodara (now Goswāmī), dated April, 27,1976, "So now you all Ph.D.'s must carefully study the details of the 5th Canto and make a working model of the universe. If we can explain the passing seasons, eclipses, phases of the moon, passing of day and night, etc. then it will be very powerful propaganda."

The primary principle to note in this chapter is how His Divine Grace Śrīla Prabhupāda expected the planetarium to be designed. To the careful observant it is clear that his intent was for the planetarium to follow the literal descriptions of cosmology given in the Śrīmad Bhāgavatam. When his disciples prepared a working diagram it was on this principle and although not perfect, Śrīla Prabhupāda accepted the effort since it showed complete allegiance to the Bhāgavatam's statements. We pray that our Vedic Planetarium in Māyāpur will follow the path of the mahājana, Śrīla Prabhupāda in this regard.

## Śrīla Prabhupāda on the Vedic Planetarium

Prabhupāda: Planetarium. According to planetarium.
Dr. Patel: They have got a big planetarium in Calcutta.
Prabhupāda: Yes, they have got, but we have also got. So it is little difference. Just like we are calculating that the sun is the beginning of the planets, sun,

then moon. Therefore we say "Sunday, Monday." This is very quite reasonable. Tuesday.
Dr. Patel: And Tuesday is next word because Maṅgala.
Prabhupāda: Yes. And planet is 1,006,000 miles above, above. And Saturday, Saturn is the last. So now sun is calculated to be 93,000,000's miles from earth, and if the moon is still further 1,600,000, then it becomes about fifteen hundred thousand miles. No. Fifteen million miles. So how they are going, in four days, fifteen million?
Bhāgavata: …So we should display the real planetary system...
Prabhupāda: Yes.
Bhāgavata: As it really exists.
Prabhupāda: Yes.
Bhāgavata: And then we can put underneath explanations in Bengali and English.
Prabhupāda: Yes. Yes.
Dr. Patel: In Sanskrit and English because Sanskrit contains Bengali and Gujarati and all.
Prabhupāda: Yes. That's a good idea, good idea.
Prabhupāda:… The Vaikuṇṭha planet, where is the Goloka planet, where is this material world—in this way.
Prabhupāda:… So I think we shall have, according to our original plan, the temple, and in that temple we shall display all these dolls: "Here is Vaikuṇṭhaloka, here is this loka, this loka, this loka." (Morning Walk — November 7, 1975, Bombay)

Prabhupāda: ...from the Bhāgavatam. So this should be mentioned. It is not only a temple, but a planetarium according to Bhāgavatam, where which planet is situated, where is Vaikuṇṭhaloka, where is Goloka Vṛndāvana, where is Mahar...
Devotees: Wow!
Tamāla Kṛṣṇa: You want a real planetarium just like in the West.
Jayapatākā: That's what I told them, Prabhupāda...
Prabhupāda: Yes.
Jayapatākā: ...that this is not a temple; this is like a big cultural exposition, museum, planetarium.
Prabhupāda: Yes.
Tamāla Kṛṣṇa: Boy, Prabhupāda, the people will pay... Everyone will pay a rupee to go in and see that. One rupee for that, one rupee to ride the escalator.
Prabhupāda: And there will be escalator to take them to different planetary system. Mention there.

Bhavānanda: My father... In those planetariums, they use a machine in the center that shoots out light. My father helped to invent that, so we could probably...
Prabhupāda: So bring your father. Father and son, both together. He is... Where he is?
Bhavānanda: In America.
Prabhupāda: So call him. He'll not come?
Bhavānanda: I don't think so, Śrīla Prabhupāda. (Morning Walk — February 9, 1976, Māyāpura)

Prabhupāda: Yes, planetarium...
Devotee (4): Temple and planetarium.
Prabhupāda: Yes.
Devotee (4): "World's largest planetarium and Temple of Understanding."
Prabhupāda: No "Understanding" simply Vedic "Temple of Vedic Planetarium," That's all. We shall show the Vedic conception of planetary system within this material world and above the material world. [break] We are going to exhibit the Vedic culture throughout the whole world, and they'll come here.
Jayapatākā: The whole world will be coming here to...
Prabhupāda: Yes.
Jayapatākā: ...visit.
Prabhupāda: Just like they come to see the Taj Mahal...
Hṛdayānanda: Yes.
Prabhupāda: ...the architectural culture, they'll come to see the civilization culture, the philosophical culture, the religious culture by practical demonstration with dolls and other things. (Morning Walk — February 27, 1976, Māyāpura)

Jayapatākā: Farce of modern science, and every philosophy and science group, where they are just cheating the public. Then exhibit how Kṛṣṇa consciousness is showing the real way.
Prabhupāda: Yes. So do that in this planetarium.
Pañca-draviḍa: That will be very nice. (Morning Walk — March 18, 1976, Māyāpura)

Prabhupāda: Yes, available. And we have asked government to give us 350 acres of land in Māyāpura to construct Vedic planetarium, 350 feet high. It will require eight crores of rupees minimum. I shall exhibit there all the planetary system, bhurloka, Goloka...

goloka-nāmni nija-dhāmni tale ca tasya
devī-maheśa-hari-dhāmasu teṣu teṣu
te te prabhāva-nicayā vihitāś ca yena
govindam ādi-puruṣaṁ tam ahaṁ bhajāmi
Dr. Patel: Now the Russians also have come to this conclusion, the Russian scientists and astro-physicists, that there are some planets here in the cosmos which actually have people, I mean, has life like human beings. The other day I read a big article on that. Now they agree.
Prabhupāda: They will agree so many things. (Morning Walk — April 12, 1976, Bombay)

Tamāla Kṛṣṇa: The scientists are getting smashed to bits by your statements, Śrīla Prabhupāda. This destroys their whole theory. Orbs, round spheres. I think that this Māyāpura building, we must build a big planetarium in it.
Prabhupāda: Yes. That, that I am going to do, Vedic planetarium.
Tamāla Kṛṣṇa: Oh, boy. You're going to bring a lot of.... A lot of scientists will come here just to dispute this.
Prabhupāda: Yes.
Puṣṭa Kṛṣṇa: Wonderful attraction.
Prabhupāda: World people will come to see the way the planetary systems...
Tamāla Kṛṣṇa: We should advertise it very widely that this is the actual, factual explanation of the universe.
Prabhupāda: This will be automatically advertised. As soon as the temple is finished, people will come like anything. (Morning Walk March 18, 1976, Māyāpura)

Puṣṭa Kṛṣṇa: He wants to know something about the Vedic city which is being constructed in Māyāpura.
Prabhupāda: Yes, not Vedic city. Vedic planetarium. How you got this information? (Conversation with News Reporters — March 25, 1976, Delhi)

Now our PhD's must collaborate and study the 5th Canto to make a model for building the Vedic Planetarium. My final decision is that the universe is just like a tree, with root upwards. Just as a tree has branches and leaves so the universe is also composed of planets which are fixed up in the tree like the leaves, flowers, fruits, etc. of the tree. The pivot is the pole star, and the whole tree is rotating on this pivot. Mount Sumeru is the center, trunk, and is like a steep hill, like the alps mountains which also have very high peaks. I have seen in Switzerland one mountain peak which was so high that is penetrated through the clouds. The tree is turning and therefore, all the branches and leaves turn with the tree. The planets have their fixed orbits, but still they are turning with the turning of the great tree. There are pathways leading from one planet to another made of gold, copper, etc., and these are like the branches. Distances are also described in the 5th Canto just how far one planet is from another.

We can see that at night, how the whole planetary system is turning around, the pole star being the pivot. Each planet has its orbit fixed but the sun is moving up and down, north and south. It is not that we shall accept the theory that the sun is fixed up and the others are all going around the sun. That is not correct. Even the 7th star is rotating once around the pole star in each 24 hours. The whole universe is just like a big tree, that is a fact. I do not think that the modern astronomers have any such idea that the whole universe is like a big tree. The planets which are full of living entities are one after another, one above the other. The relative positions of the planets is fixed up but the whole thing is turning. The sun is going north and south, it has its own orbit below the moon. So now you all Ph.D.'s must carefully study the details of the 5th Canto and make a working model of the universe. If we can explain the passing seasons, eclipses, phases of the moon, passing of day and night, etc. then it will be very powerful propaganda.

I am sending this letter to you, and you can make photocopies of it and send to our other Ph.D.'s and begin serious research into the matter in detail. But one thing, I am convinced that the universe is just like a great tree as described therein. (Letter to: Svarupa Damodara — Auckland 27 April, 1976)

In the meantime, I have received one letter from Jayapataka Mahārāja in which it informs that Madhava Mahārāja and also possibly Damodara Mahārāja are making propaganda against our attempt to construct a township in Mayapur, with the centre of attraction being the "Vedic Planetarium." In this connection we have applied to the government for 350 acres of land and the matter is in the process.... As a matter of fact, I am trying to develop a township in Mayapur spending crores of rupees to give protection against the occasional inundation (flood) and construct a tall planetarium estimated to be 300 feet high. So why they are obstructing this program? (Letter to: B.R. Sridhara Mahārāja — Los Angeles 6 June, 1976)

Prabhupāda: Now you all together make this Vedic planetarium very nice, so that people will come and see. From the description of the Śrīmad-

Bhāgavatam, you prepare this Vedic planetarium. How do you like this idea, Vedic planetarium?
Ambarīṣa: It seems like a very nice idea.
Prabhupāda: You also like? So finance this project. (laughter) Vedic planetarium.
Ambarīṣa: Where will this be?
Prabhupāda: Māyāpura. My idea is to attract people of the whole world to Māyāpura. So we are just trying to acquire three hundred fifty acres of land from the government. It is going on under consideration. Some men are against and some are in favor. But those who are against, they have counted ninety?
Hari-śauri: Ninety against, two thousand for.
Prabhupāda: Yes.
Puṣṭa Kṛṣṇa: In Māyāpura itself.
Hari-śauri: The local farmers.
Prabhupāda: So all of you now make a complete idea how to make Vedic planetary, planetarium. (Room Conversation — June 15, 1976, Detroit)

To see the planetarium and how things are universally situated has nothing to do with sectarian ideas. It is a scientific presentation of spiritual life. (Letter to: Jayapataka — New Vrindaban 26 June, 1976)

Prabhupāda: ...and it is a fact, he did not fix it. I wanted both of you to take various detailed photographs of that Capitol.
Yadubara: The Capitol Building. For what purpose, Śrīla Prabhupāda?
Prabhupāda: We shall have picture, planetarium in Māyāpura. (aside:) That's better. [break] ...spiritual world, material world, and so on, so on. Planetary..., succession of the planetary systems, everything. A building like that.
Yadubara: That would be a separate building from the temple?
Prabhupāda: Yes. We are acquiring 350 acres of land for life for constructing a small township...
Yadubara: I think we...
Prabhupāda: ...to attract people from all the world to see the planetarium.
Svarūpa Dāmodara: Is this near the temple?
Prabhupāda: Yes. Planetarium name, actually it will be temple. But all round, things will be... Anyway.
Yadubara: I know before the idea was to have it inside the main temple.
Prabhupāda: Hm?

Yadubara: As you walked up the outside of the, or the inside of the main temple, inside that dome, they would have it on the walls. But that would... That original plan was to have it inside the main temple.
Prabhupāda: Yes. You take all details, inside, outside. That will be nice.
Svarūpa Dāmodara: Can you take inside? Is it allowed?
Yadubara: Yes, I think so.
Prabhupāda: Hare Kṛṣṇa, Hare Kṛṣṇa... (japa) And you prove that sun planet is first. It is stated in the Bible. (Room Conversation- July 6, 1976, Washington, D.C.)

Svarūpa Dāmodara: The other morning Śrīla Prabhupāda was mentioning about the Institute, the Bhaktivedanta Institute. So is there anything else that your Divine Grace wants to?
Prabhupāda: Institute, you have got sufficient subject matter as I was describing, this original source of life and the planetary system, as you are going to make planetarium. Who can say about so many planets in the sky? Who has got sufficient knowledge? They cannot even give... They think that moon is the nearest planet, but we do not think like that. But still they are unable to give sufficient knowledge about the moon. It is not vacant, it cannot be vacant. We do not find any part of the world vacant. There is living entities. This earth planet is part of the universe, and the moon is also part of the universe. If it is not vacant, how that can be vacant? You have got dust there, we have got dust here; you have got rocks there, here we have got rocks. And why it is vacant? We find in the dust also there is life. When we walk on the beach, it is simply sand, there are so many crabs.
(Room Conversation — July 7, 1976, Baltimore)

Tamāla Kṛṣṇa: There's a big planetarium here also at this museum.
Rāmeśvara: Biggest in the world, I think.
Tamāla Kṛṣṇa: Yes, this planetarium is the most famous one, Heyden Planetarium.
Prabhupāda: So see how the planetarium is done.
Tamāla Kṛṣṇa: Yes, we should go, Rāmeśvara. Have you ever gone?
Rāmeśvara: Many years ago.
Tamāla Kṛṣṇa: Should we go to see it?
Prabhupāda: Hmm?
Puṣṭa Kṛṣṇa: They use all kinds of lighting systems.
Prabhupāda: So you take the idea. We shall have to do that.
Tamāla Kṛṣṇa: Wow. All right. They have a special projector.
Prabhupāda: Whatever it may be, we may take the idea and utilize.

Puṣṭa Kṛṣṇa: In our planetarium, Śrīla Prabhupāda, we want to actually show the different qualities of life on the different planetary systems also?
Prabhupāda: Yes, as far as possible. There is Siddhaloka, they can go from one planet to another without any machine, without any vehicle. Siddhaloka means, just like the yogis, they can go.
Rāmeśvara: Their body have..., they have wings?
Prabhupāda: No wings. We think only our ideas, that without wings nobody can fly.
Tamāla Kṛṣṇa: That picture wasn't accurate.
Rāmeśvara: That was Gandharvaloka.
Tamāla Kṛṣṇa: Prabhupāda said they don't have wings either. (Morning Walk — July 13, 1976, New York)

Prabhupāda: Give him, give him, give him. No, that is the different. Oh, all right. We are just attempting a big planetarium in Māyāpur. We have asked government to acquire land, 350 acres. That is negotiation going on. We shall give a Vedic planetarium.
George Harrison: Is that the one you were talking about? With all the...
Prabhupāda: In the Fifth Canto.
Gurudāsa: The planetarium will be 350 feet high and show the cosmology of the spiritual world.
Prabhupāda: The construction will be like your Washington capital, like that.
George Harrison: A big dome.
Prabhupāda: Yes. Estimated eight crores of rupees. (Conversation with George Harrison — July 26, 1976, London)

Prabhupāda: Yes, actually this is fact. It is similar planet like the sun, but it is surrounded by cold atmosphere, therefore it is so pleasing. And because it is far away from the sun, the distance between moon and earth is more than the distance between the sun and the earth. Therefore sun looks bigger and it looks smaller. We are contemplating having a planetarium with electric arrangement. You'll have to work. The whole planetary system is moving from east to west, and the sun, moon and... They are up and down. (Room Conversation — August 8, 1976, Tehran)

Prabhupāda: This is the important point that my books are sold daily sixty thousand dollars all over the world. I have made the Trust so all the collections should be divided fifty percent for constructing temple and fifty percent for reprinting books. So we don't take a paisa profit. So far as an

author, they are selling sixty thousand dollars. And even if I would have taken ten percent royalty then it would have been six thousand dollars. Six thousand dollars means...
Gopāla Kṛṣṇa: Hundred and eighty thousand dollars a month.
Prabhupāda: No no no. Six thousand dollars means, say ten rupees.
Krishna Modi: Sixty thousand per day.
Prabhupāda: Sixty thousand per day. Actually. But we are all foregoing it for pushing on this movement, and if we bring that money and construct big, big temple or planetarium, what is the harm to India? (Meeting With Member of Parliament, Mr. Krishna Modi — August 31, 1976, Delhi)

Prabhupāda: We are trying for another big scheme in Bengal. We have applied to the government to acquire land, 350 acres, a big planetarium. Planetarium. We have described the planetarium in our Fifth Canto of Śrīmad-Bhāgavatam. In that planetarium it is said that the moon is above the sun planet. By one million six hundred thousand miles.
Indian man (3): What does astronomy say? The modern astronomy.
Prabhupāda: They say the moon planet is nearer to earth planet and they have gone there. (Evening Darśana — September 1, 1976, Delhi)

It will also claim the world's largest planetarium with 410 foot high Temple of Understanding..."
Dr. Kneupper: It sounds like a beautiful project. That is near Bombay, now?
Prabhupāda: No, that is near Calcutta.
Dr. Kneupper: Calcutta.
Prabhupāda: About sixty miles.
Dr. Kneupper: Have they progressed much now?
Prabhupāda: Yes, the enquiry is finished. Now government is considering to give that land. (indistinct) that land. There were many others, land offers in Hyderabad, but I wanted to start this institution in the, on the birth site of Lord Caitanya. Otherwise we have got better land. We have asked government for 350 acres but we have already bought in Hyderabad, 600 acres. Here also we can get immediately 500, 600 acres, like that. We have already got another plot of land, 100 acres in Ahmedabad. So there are so many lands available but I wanted to start this project in Māyāpura, the birthsite of Lord Caitanya. Therefore we are (indistinct). If they say no, then we shall attempt it somewhere else. (Room Conversation with Dr. Theodore Kneupper — November 6, 1976, Vṛndāvana)

Prabhupāda: One very big paṇḍita is coming. He'll help us about the Sanskrit language, how to form this...
Tamāla Kṛṣṇa: Planetarium.
Prabhupāda: ...planetarium.
Tamāla Kṛṣṇa: One big paṇḍita has been contacted, and he's very surrendered to Prabhupāda. He wants to help Prabhupāda and our movement to understand the meaning of the śāstras in regard to the layout of the universe, so that the planetary systems can be done in our planetarium.
Prabhupāda: This planetary system is rotating from east to west, and it is hanging like the chandelier, taking shelter of the polestar. That we can see every night. Now where is the situation, which planet, where is sun, where is moon—so that he has to assert.
Svarūpa Dāmodara: When he's coming, Śrīla Prabhupāda?
Prabhupāda: I have asked him to come in this ceremony. (Room Conversation with GBC members March 2-3, 1977, Māyāpura)

Now we are going to show this planetarium. These rascal scientists: "All desert. All rocks and desert." Simply this planet, for his father's property. This is now happening. "The moon planet is a desert." And from the desert such brilliant light is coming that is illuminating at night the whole universe. And we have to believe it because they are spoken by scientists. You see? All rascals, fools, rogues, thieves, they are leading. And our determination is to stop these rascals. That is our... It is not that "Let the rascals go on with their... Let us make our salvation." Prahlāda Mahārāja said, "No, no, no, I don't want. I don't want. If there is salvation, I must take them also." This is Vaiṣṇava. (Room Conversation — April 13, 1977, Bombay)

Prabhupāda: So we are presenting this planetarium...From Bhāgavatam. In the Fifth Canto there is description of the planetary system. So we want a diagram. So kindly help us. Hare Kṛṣṇa. Now find out that Fifth Canto. So this planetary system is hanging. Ūrdhva-mūlam adhah-śākham. Same thing is explained in the Śrīmad-Bhāgavatam. Now, how it is hanging and where are different situation of the planets and plateaus and hills and..., so as far as the book is concerned, kindly make a diagram so that we shall execute.
Indian Astronomer: Yes, yes. Ūrdhva-mūlam adhah-śākham aśvatthaṁ prāhur avyayam. That means, this, about the life.
Prabhupāda: Not life. This planetary system is like that.
Indian Astronomer: Oh.

Prabhupāda: Ūrdhva-mūlam adhah-śākham. That is... That is explained in the Śrīmad-Bhāgavatam. Which chapter it is? There is Jambudvīpa and... Pañcama-skandha, yes.
Indian Astronomer: There we find there about the Priyavrata. Priyavrata's sons, they divided the world...
Prabhupāda: No, persons we are not very much concerned. We are... Immediately we want the diagram how to fix it up so that people can see, "This is the situation." So you make this diagram. We have got very good scheme so that people from the world will come to see the Vedic idea of planetary system. This is the ambition. So you kindly help us. So you sit down and take paper and make it and..., so that he can understand. You have already studied that. So that you can do immediately. You do. Can you make a rough sketch immediately?
Indian Astronomer: First (indistinct). Because after he came there I told him that we must prepare a diagram which is acceptable to all, acceptable to all.
Prabhupāda: No, acceptable, the, I mean to say, Western astronomers, they...
Indian Astronomer: No, we... If you prefer Bhāgavatam and if you give only Bhāgavatam...
Prabhupāda: Yes.
Indian Astronomer: ...whether Western accepts or not, that is not worthwhile.
Prabhupāda: Eh?
Indian Astronomer: Whether Westerners or modern scientists accept or not, that is not worthwhile.
Prabhupāda: No, we want do it according to Bhāgavatam. Yes. So when you'll do it? No, my order... It is for this purpose I sent him, so you can begin immediately or...
Tamāla Kṛṣṇa: Surabhī has an office with all things ready, drawing board...
Prabhupāda: You can go to his office, can give him instruction how the planetary system is hanging. The polestar, dhruva-tārā, is the center, and it is moving. That at night you can see. It is moving. No, what is the first, beginning? You just give him. Read it. The Sanskrit reading. Sapta-sindhu... You have to make diagram where the sapta-sindhu are. No, all...So you can... Why not begin it in my presence? I have tried to translate it as far as possible, but I am not satisfied. What is your idea of the seven planets, sun...?
Indian Astronomer: You see, planets, in our system planets comes in the history very late.
Prabhupāda: Very?
Indian Astronomer: Late, very. Because you won't find any reference to grahas in the Vedas. As a student of Vedas, I can guarantee that there is no

reference to the grahas in the Vedas. Because nava-grahas are only important mantras, not they are (indistinct). So when because these things comes later on, Kalpa-sūtra, for those adopted the grahas and prepared the mantras from the Vedas for the grahas. So historically we must have come down. Now we'll find out in what date they are presented. Till now I was told all these things after only... I am also preaching Bhāgavata.
Prabhupāda: Now, where is the situation of the grahas in this planetary system? Whether sun is first or moon is first?
Indian Astronomer: Sun is first. Definitely sun is first because the moon is the reflection of sunshine. (quotes Sanskrit verse) Vedas say (Sanskrit).
Prabhupāda: Candrama has... Candrama has got his own light. It is not reflection.
Indian Astronomer: Not reflection. It has its own importance. But about the lights...
Prabhupāda: As sūrya is blazing, similarly, candrama is also blazing, but it is surrounded by cool cloud. Therefore it is so soothing. But it is above the sun planet, so far it is stated there.
Indian Astronomer: Natural... Naturally sūrya, our astronomy says, sūryaḥ agni (Sanskrit). Vedas say, (Sanskrit). Because the nakṣatras only.
Prabhupāda: So when you begin this diagram? I wanted to make it before me. What is the difficulty? The śāstra is there. You have to make drawing according to śāstra. That's all. (Conversation with Vedic Astronomer April 30, 1977, Bombay)

Prabhupāda: If you have got the name, so why not reply to each one, the letter...
Tamāla Kṛṣṇa: Oh, I see.
Prabhupāda: "We are intending to make a huge planetarium. If you can help us?"
Tamāla Kṛṣṇa: Mentioning the Fifth Canto, Bhāgavatam?
Prabhupāda: Exactly to the description of Fifth Canto, we want.
Tamāla Kṛṣṇa: How will we get their addresses?
Prabhupāda: Hm? Or I can write to Mr. Chandan(?). (Room Conversation — May 8, 1977, Hrishikesh)

After working on the planetarium sketches, I told Śrīla Prabhupāda that this planetarium would spell the downfall of Western civilization. Śrīla Prabhupāda very much appreciated this statement and added, "I want to expose that they are cheating. All these so-called scientists. I have said they have only stool in the head. Their only interest is to make money, and for this they cheat. If you can make this planetarium, then our movement is grand success, triumphant." (TKG's Diary; June 15, 1977)

"When we make our planetarium, they will see how frog-like they are. How they can measure such a huge distance as universe? They are simply speculating." (TKG's Diary; June 28, 1977)

Prabhupāda: Vedic planetarium.
Bhakti Prem: Next step will be planetary system.
*Prabhupāda: All right.*
Bhakti Prem:The third planetary system, above Mount Meru.
Prabhupāda: Jaya. You have taken lunch?
Upendra: Yes, Prabhupāda. [break]
Prabhupāda: ...there is sun, and above that, there is moon. And they are going to moon. They are going nowhere, simply taking laboratory photo, studio photo, and cheating. Why this cheating can go on? You do not know. That's all. Who is insisting that "We must know"? (end) (Conversation June 30, 1977, Vrindaban)

Tamāla Kṛṣṇa: So I think that Svarūpa Dāmodara will be helped by these drawings when the men come. 'Cause he said that even though they are scientists, they could not understand this volume. It's been a mystery practically. These drawings, one by one, should be able to help in the creation of that planetarium.
Prabhupāda: Thank you very much. Hare Kṛṣṇa. (July 2, 1977, Vṛndāvana)

Tamāla Kṛṣṇa: As we are conditioned, as everyone is conditioned, our planetarium will have to show the actual facts.
Prabhupāda: Yes. That fact we have learned from Bhāgavatam.
Tamāla Kṛṣṇa: So, so far we have not drawn this fact correctly.
Prabhupāda: That is your inability. That is another thing.
Tamāla Kṛṣṇa: Yeah, well that's why we're... That's the question that we've raised. This question that we've raised is due to that.
Prabhupāda: That is you are unable to, but the fact is that you are conditioned. You cannot go beyond that conditioned
Tamāla Kṛṣṇa: That's accepted.
Prabhupāda: So we are also conditioned. But as far as possible we take description from Bhāgavata, try to. That is our... Suppose here is India, here is Los Angeles. You start from India, Los Angeles..., or India, you'll come to

Los Angeles. And again return to India. Similarly you start from this again going.
Tamāla Kṛṣṇa: Yes. That's possible if you can fly this way, underneath.
Prabhupāda: But where is the underneath?
Tamāla Kṛṣṇa: See Prabhupāda, we weren't asking that issue. That's not the question we were asking.
Prabhupāda: What is that issue?
Tamāla Kṛṣṇa: Well, you've explained, if it's a lotus petal, then you can fly around it. That's all right. Then the answer is there. But if there's no lotus petal and it's simply flat, then it's a problem. That a problem.
Prabhupāda: No, we don't say flat.
Tamāla Kṛṣṇa: We've pictured it like that. There's some defect in our picture. If you recall, there's a picture we drew...
Prabhupāda: So rectify it.
Tamāla Kṛṣṇa: Yeah, well how?
Prabhupāda: That is the explanation. Petal.
Tamāla Kṛṣṇa: See it doesn't say petal anywhere in the Bhāgavatam. It doesn't say Bhārata-varṣa is a lotus petal.
Upendra: Neither does it say it's flat, though.
Tamāla Kṛṣṇa: Well I can get the Bhāgavatam right now if you allow me to. It doesn't say... I believe it says Jambūdvīpa is the second part of the lotus, and there aren't petals in the center.
Upendra: It's a transcendental lotus flower.
Prabhupāda: You are prejudiced. That's all. If I can understand, why don't you understand? You are prejudiced.
Tamāla Kṛṣṇa: Well I can understand if I accept it as... I'm only trying to think as the persons who are going to make this planetarium.
Prabhupāda: You are Western. You are prejudiced. That is reason. If I can understand why you...
(July 5, 1977, Vṛndāvana)

Prabhupāda: Oh, yes. And you can charge some fees, entrance fee. People will gladly pay. All around, this doll exhibition, and in the medium, a planetarium, small.
Tamāla Kṛṣṇa: Small?
Prabhupāda: Yes. If they're going to have it in Māyāpura, they can have a little... Not very small. Proportionate. (Conversation—Recent Mail — July 14, 1977, Vṛndāvana)

Tamāla Kṛṣṇa: One of our devotees, Ambarīṣa dāsa, his name is. He's very wealthy person.
Prabhupāda: He is grandson of Ford, great-grandson.
Dr. Kapoor: Yes, I was told. He came here once, I'm told.
Tamāla Kṛṣṇa: So he offered everything he has to Prabhupāda. So Śrīla Prabhupāda has requested him to build a large building and put in a Vedic planetarium in Washington, D.C., capital of the United States, right near the White House, with the Vedic planetary system according to the Śrīmad-Bhāgavatam, Fifth Canto. So he has agreed to do this, and now they are choosing the property where they will develop this program right in the heart of Washington, D.C. So it will be very good. And also the dolls will be... Doll exhibit.
Prabhupāda: How they are making dolls, show him. How nice dolls. (Room Conversation Gaurasundara, Dr. Kapoor July 26, 1977, Vṛndāvana)

## GBC on Vedic Planetarium

D. Motions : Re: Research institute

*1. That a research institute styled as "BHAKTIVEDANTA UNIVERSAL ORDER RESEARCH INSTITUTE" ( or "BHAKTIVEDANTA ASTRONOMICAL RESEARCH IMSTITUTE"),( or B.I.) shall be established and research on the vedic understanding of the Universal order shall be undertaken on a full time basis for the purpose of erecting an exhibit at Sri Mayapur, and there after in other places, for giving public and scientific circles of the real picture of the Universe.*

*2. That the above mentioned R.I. shall be financed under the auspices of the Śrī Mayapur Developement Project Committee's Fund*

*3. That the above mentioned Research Institute shall have the following devotees in the first committee and that other persons may be co-opted as per their decision.*

1) Swarup Damodar Das Brahmachari
2) Sadaputa Das Adhikari
3) Gurukripa Swami
4) Harikesh Swami
5) Yasodanandana Swami
6) Bhakti Prema Swami

7) Jagadish Das Adhikari
8) Pradyumna Das Adhikari
9) Hridayananda Goswami

RESOLUTIONS OF MEETING #6

*1. Jayapataka Swami's "Motions in regard to Mayapur Project" are accepted as amended above.*

*2. That above mentioned Research Institute shall be under the Bhaktivedanta Institute under the review of the SMPDC*

## Māyāpura Project on Vedic Planetarium

### The Temple of the Vedic Planetarium

As the name of the temple suggests, the planetarium is of central importance to this project. A major feature of the planetarium exhibition will be the representation of the planets rotating above the heads of the Deities. Śrīla Prabhupāda often requested that the seven great planets — the Sun, Moon, Mars, Mercury, Jupiter, Venus, and Saturn (which also correspond to the days of the week) — should rotate in orbit around the top of the Deity altars. Since Krishna is the Lord and source of the planets, *sarva loka mahesvaram*, this presentation will visibly impress upon the visitors that the Lord is in control of His creation, of which the entire universe is only a small part.

The planetarium will be housed in one of the main four domes surrounding the temple, just behind the Deities on the temple's western side. This will be a completely equipped, modern planetarium with full projection capacity, and it will seat at least 200 people at a time. A variety of shows on the history of the universe and other specific phenomena described in Vedic literature will be presented periodically throughout the day. High school and college students, both individually and with their classes, will be able to examine the scientific basis of Krishna consciousness and judge for themselves the merit of the Vedic version.

Above the planetarium will be exhibits. Dioramas will depict the elements of the universal creation, such as the different planetary systems and how they relate to each other, and a complete model of the Universe will be exhibited. Other exhibits will be decidedly more "hi-tech." Computerized

3D graphics on large high-definition screens will offer "tours" throughout the universe. Visitors will appreciate the immense variety of creatures, societies, and opulence of the different planetary systems — the upper heavenly planets, the earthly plane, and the lower planets. Demonstrations of special pastimes mentioned in the *Veda* will describe the relative differences of time in the universe and the cyclical nature of time through the manifestation of creation, maintenance, and annihilation again and again throughout the day of Lord Brahmā. This exhibition promises to be an eye opener — one of the most exciting presentations of the universe in the world today.

The special themes to be presented in the exhibitions include:

- The accurate astronomy of the Fifth Canto of the *Śrīmad-Bhāgavatam*
- Insights from the *Jyotisa Śāstra* (primarily the *Sūrya-siddhānta*)
- Stars and cosmic distances
- Higher dimensions and the middle planetary systems
- Cross-cultural references to Fifth Canto cosmology
- Living beings in the higher-dimensional universe
- Higher powers in the world of Maya

# 10

# Bhu-Mandala

## Showing of Planetary Sketches of Bhū-maṇḍala

Bhakti-prema: You like that.
Prabhupāda: Yes.
Bhakti-prema: Yes, it is nicer.
Prabhupāda: That I have seen. I told him in Switzerland. Yes. From Switzerland they'll...
Bhakti-prema: So big mountain.
Prabhupāda: Yes.
Bhakti-prema: And he took (indistinct) in other planets?
Prabhupāda: Yes.
Bhakti-prema: The same with this Meru Mountain, it spreads...
Prabhupāda: They cannot explain what is that. But I have seen it. Where it has gone, nobody knows. Just see. Down, hundred miles, and up... You cannot see more than hundred... But it appears like the instruction.
Tamāla Kṛṣṇa: This is Jambūdvīpa, part of Jambūdvīpa. That is Ilāvṛta-varṣa in the square. We have... These mountains are each two thousand yojanas high.
Bhakti-prema: After Mandara Meru, Mandara, that Mandara Mountains... And here he began to meditate. He meditated for six thousand years on this Mandara Mountain here. Then Pārvatī came, and she... Then next... These two mountains are like (describes sketch of planetary system, very difficult to hear) And here is the beautiful Indu stream(?). And it's heart is... Eight hundred miles.
Tamāla Kṛṣṇa: Where is it?

Bhakti-prema: On the Mandara Mountain. There are four different streams. This is Ilāvṛta-varṣa.
Yaśodānandana: This area?
Bhakti-prema: Yes.
Prabhupāda: Śivaloka? Śivaloka?
Bhakti-prema: Yes, it is not actually Śivaloka, but residence of Lord Śiva.
Tamāla Kṛṣṇa: One of his... He described it that like during the summertime you go to a hill station. Each demigod has their place where they also go.
Bhakti-prema: Śivaloka is different, but Lord Śiva is...
Prabhupāda: Summer residence.
Tamāla Kṛṣṇa: Yes.
Prabhupāda: Why not? If a man can have summer residence, Lord Śiva...
Bhakti-prema: Actual residence is in Satyaloka. Brahmā is there in Satyaloka. That means 2,200,000,000 miles away from sun planet.
Prabhupāda: That... Then it is universe.
Bhakti-prema: Yes, in the universe. It is very high.
Prabhupāda: And where is that cakra?
Bhakti-prema: Cakra?
Prabhupāda: That polestar, center?
Tamāla Kṛṣṇa: That will be in another...
Bhakti-prema: We will give another picture.
Prabhupāda: Oh.
Tamāla Kṛṣṇa: This picture only shows mainly part of Jambūdvīpa and especially Ilāvṛta-varṣa, the middle of...
Prabhupāda: So Brahmaloka is in Jambūdvīpa?
Bhakti-prema: No, no, no. Brahmaloka is in Satyaloka.
Prabhupāda: Oh, then what is this?
Bhakti-prema: This is Mount Meru.
Prabhupāda: But you said Brahmaloka.
Bhakti-prema: Brahmapurī.
Prabhupāda: Brahmapurī, that.
Bhakti-prema: Summer residence.
Tamāla Kṛṣṇa: The middle part.
Bhakti-prema: And then there are eight different residences of different lokapālas, Indra, Agni, Varuṇa...
Prabhupāda: Summer residence.
Tamāla Kṛṣṇa: Yes, on top of Meru, like a hill station. (laughter)
Bhakti-prema: You won't find many even like a (indistinct).
Prabhupāda: And they simply find rocks and sand. How nonsense...
Tamāla Kṛṣṇa: Because that's all there is in Arizona.

Prabhupāda: (laughs) That is... Do it nicely, carefully.
Bhakti-prema: So scientists cannot try to realize the height of this mountain...
Prabhupāda: Not advised.
Bhakti-prema: They come just here.
Prabhupāda: I have seen. They can go little higher than that earth. That's all. They cannot go even that peak. As soon as they go, finished.
Bhakti-prema: This is more than six lakhs' miles from the earth, the height of Mount Meru.
Tamāla Kṛṣṇa: The airplane goes about eight miles. Five thousand feet equals one mile. And they go forty thousand. So how much is forty thousand? Eight miles.
Devotee (1): Forty thousand feet divided by five means approximately eight miles.
Tamāla Kṛṣṇa: Airplane goes eight miles high, and this Meru Mandapa(?) is 800,000 total, a hundred thousand yojanas, one lakh of yojanas. So this is very high, nearly 600,000 miles high. An airplane only goes five miles.
Devotee (1): Eight miles.
Tamāla Kṛṣṇa: They have no idea about distances.
Bhakti-prema: If we rise above, man, six thousand miles, then we will arrive at sun planet.
Prabhupāda: Yes.
Bhakti-prema: And then more, eight lakhs miles, that means moon planet.
Prabhupāda: Then miles? They could not explain why Sunday first.
Tamāla Kṛṣṇa: Right. That question stumps.
Prabhupāda: Common sense. Sunday, Monday. Sunday. Sun must be first. Then... This is my commonsense interpretation.
Yaśodānandana: You have written in Bhāgavatam like that.
Prabhupāda: Yes. And these rascals say moon first.
Devotee (2): Śrīla Prabhupāda, previously we painted in the art department... Just like Varāha lifted the earth, and the earth was a globe, and we showed also a globe of the earth. How does that relate to this? Previously, when we painted, we showed the earth a ball. So now the artists will be very confused. How it fell in the Garbha Ocean as a ball?
Yaśodānandana: It depends on what we mean by earth. The Western conception of earth is just five continents and a few oceans, but according to Bhāgavatam, earth means Jambūdvīpa, because earth is connected with Jambūdvīpa.
Devotee (2): So whole Jambūdvīpa fell.
Bhakti-prema: Bhāgavata describes the height of Himalayas, eighty thousand miles.

Prabhupāda: No, about this earth globe.
Bhakti-prema: About the earth it describes four billion miles.
Tamāla Kṛṣṇa: That means Jambūdvīpa.
Bhakti-prema: No, that means complete earth, four billion miles. That is eight lakhs miles, Jambūdvīpa.
Tamāla Kṛṣṇa: Four billion is the universe.
Bhakti-prema: Bhū-maṇḍala.
Tamāla Kṛṣṇa: Bhū-maṇḍala. Oh, the Bhāgavatam describes Bhū-maṇḍala as earth.
Prabhupāda: Oh.
Yaśodānandana: Bhūmi.
Bhakti-prema: ...has come right from the ocean, from the bottom of the ocean.
Prabhupāda: So explained there. Then this question is solved.
Tamāla Kṛṣṇa: Yes. The whole Bhū-maṇḍala fell.
Bhakti-prema: So the universe is round, and from bottom of the Garbhodaka ocean to top of Satyaloka there is four billion miles. Then the again, from east to west again is another four billion miles. Because Bhū-maṇḍala is situated in between the earth.
Prabhupāda: Explain. He'll explain.
Yaśodānandana: (background talking)
Bhakti-prema: (background talking)
Tamāla Kṛṣṇa: And that is salt water ocean, behind it.
Yaśodānandana: Salt water ocean goes all around.
Bhakti-prema: All around.
Tamāla Kṛṣṇa: It's actually a circle, but we have shown one area, so it become flat.
Bhakti-prema: This is a circle. And this is (indistinct). And this is Brahma-varṣa(?). And this is (indistinct). This is now eighty thousand miles high and sixty thousand miles wide, these mountains. And these...
Prabhupāda: All marble. Mountain.
Tamāla Kṛṣṇa: What are they made of?
Yaśodānandana: Each mountain is made of a different kind of stone. Some are made of gold; some are made of coral; some of them, stones, are lapis lazuli, that blue stone that Kṛṣṇa wears. Each mountain has different variety.
Prabhupāda: They will be puzzled. These material scientists will be puzzled. (laughter)
Bhakti-prema: But according to Śrīmad-Bhāgavatam, in the beginning of creation we have relation with all this, in India. Even five thousand years ago Parīkṣit Mahārāja went and he conquered this area. I have translated this. This Tattvata-varṣa was conquered by... And this Ramya-varṣa was also con-

quered. And Vardhanya also... And then this is Bhārata-varṣa, this whole world.
Tamāla Kṛṣṇa: Where?
Bhakti-prema: This.
Tamāla Kṛṣṇa: That is Bhārata-varṣa.
Bhakti-prema: Yeah, and he was presiding here. He was living here. And then he crossed this mountain.
Prabhupāda: Crossed?
Bhakti-prema: Yes. And he conquered it. Not only he saw, he conquered it. And it is surrounded by five other oceans. And again this is surrounded by (indistinct). And is surrounded by (indistinct). Again there are ten mountains.
Tamāla Kṛṣṇa: That is what we're painting next.
Bhakti-prema: Next time we'll show you.
Tamāla Kṛṣṇa: Bhū-maṇḍala, the whole Bhū-maṇḍala, these different oceans and dvīpas. That will be in color.
Prabhupāda: Very nice.
Tamāla Kṛṣṇa: This Puṣkara, he did the...
Puṣkara: I made the mess. (laughs)
Prabhupāda: He'll do something wonderful in arts. Kṛṣṇa will be... No complaint. All right. Jaya.
Devotees: (obeisances)
Prabhupāda: Vedic planetarium.
Bhakti-prema: Next step will be planetary system.
Prabhupāda: All right.
Bhakti-prema: The third planetary system, above Mount Meru.
Prabhupāda: Jaya. You have taken lunch?
Upendra: Yes, Prabhupāda. [break]
Prabhupāda: ...there is sun, and above that, there is moon. And they are going to moon. They are going nowhere, simply taking laboratory photo, studio photo, and cheating. Why this cheating can go on? You do not know. That's all. Who is insisting that "We must know"? *(end)* (June 28, 1977, Vṛndāvana)

## Bhū-maṇḍala Diagram Discussion

Tamāla Kṛṣṇa: They have to stand up on the chairs because it's so big. Big project. This is only one drawing. Imagine the planetarium. [break]
Yaśodā-nandana: The directions are north, east and south and west. In the middle here, right in the middle, there is Mount Meru, which is very, very small. The scale, it comes to be one centimeter... What is this measurement

estimate?
Tamāla Kṛṣṇa: One twentieth of a centimeter.
Yaśodā-nandana: One twentieth of a centimeter is calculated 100,000 yojanas. One lakh of yojanas is one twentieth of a centimeter.
Tamāla Kṛṣṇa: Śrīla Prabhupāda, that one twentieth of a centimeter... One centimeter is about just the tip of the finger, so one twentieth of this is 100,000 yojanas. Actually you cannot even see Mount Meru on this picture. It is so small that we could not even draw it. It's just a pinpoint.
Yaśodā-nandana: Mount Meru is in the middle, and then, surrounding Mount Meru, is the whole Jambūdvīpa. Jambūdvīpa is 100,000 yojanas or (sic) 800 miles in length and width. These are the maps we have shown you already previously. And surrounding Jambūdvīpa in the salt ocean, this very little circle... The south ocean is the same width as Jambūdvīpa, or 100,000 yojanas. That is 800,000 miles. And it is all around Jambūdvīpa. Here it is. Maybe you can see. Then there is the south part of the ocean. You can see here?
Bhakti-prema: No, Plakṣadvīpa is the orange dot.
Yaśodā-nandana: Then surrounding Jambūdvīpa then there is Plakṣadvīpa, the next dvīpa, which is... Around the salt ocean there is Plakṣadvīpa. That is the planet beside(?) of the river we call ocean. That is 200,000 yojanas, or 1,600,000 miles. That is right in the middle.
Tamāla Kṛṣṇa: Actually you can hardly see it there. It's very small.
Yaśodā-nandana: Then, surrounding Plakṣadvīpa is another ocean, the sugarcane ocean. That sugarcane ocean is the same length as Plakṣadvīpa, or 200,000 yojanas, or 1,600,000 miles. And one each one of these dvīpas...
Prabhupāda: So in each ocean there are islands?
Bhakti-prema: No. These are all ocean.
Prabhupāda: Simply ocean.
Yaśodā-nandana: Now, each one of these dvīpas are managed by different sons of Mahārāja Priyavrata. Mahārāja Priyavrata had so many sons, and each one of their sons...
Prabhupāda: No, no, but the son is the king, but he is ruling over the ocean?
Tamāla Kṛṣṇa: No, he is ruling over the dvīpa.
Bhakti-prema: Yes, he rotated around the mountain area. Seven times he went. And then to (indistinct).
Prabhupāda: Now, oceans, but there is no inhabitant.
Bhakti-prema: No.
Prabhupāda: Simply ocean.
Bhakti-prema: There are no dvīpas, no islands. Islands are surrounded by oceans rather than(?) oceans are surrounded by islands.
Prabhupāda: So what is his kingdom?

Bhakti-prema: An island... Islands are so big, very big.
Prabhupāda: Where is the island?
Tamāla Kṛṣṇa: There.
Yaśodā-nandana: The islands are surrounding the oceans. It goes in a circle. There's a whole ocean surrounding one island. And each island is divided by seven mountains and seven rivers, and the main sons of Mahārāja Priyavrata, they also have sons, and each one of those seven divisions are being ruled by the seven grandsons of Mahārāja Priyavrata too.
Tamāla Kṛṣṇa: In other words, Śrīla Prabhupāda, Priyavrata had seven children on seven islands, and each one of them had seven children. So each son of Priyavrata divided his island in seven parts, and it was given to one of the grandsons of Priyavrata. We could not show that because it's so small. This is drawn exactly to scale. In order to include all of the Bhū-maṇḍala we had to make everything up very small because, as you'll see, most of Bhū-maṇḍala is made of Loka-varṣa and Lokāloka mountain. Everything else is very small compared to those two.
Yaśodā-nandana: Surrounding the sugarcane ocean, there is Śālmalidvīpa, where again there are seven divisions of land, seven rivers...
Prabhupāda: So island and surrounded by ocean, like that.
Bhakti-prema: Yes.
Yaśodā-nandana: Śālmalidvīpa is this purple here. Still, it is very difficult to see. And surrounding Śālmalidvīpa is an ocean of liquor, Surā-sāgara. There's an ocean of liquor. There is Kuśadvīpa.
Tamāla Kṛṣṇa: Which one is that?
Yaśodā-nandana: The little red circle is the Surā-sāgara. Then the little green here is the Kuśadvīpa. One particular feature of this land is that there's very..., kuśa grass, which is very, very cooling. It is all over the island. And then there is the ghee ocean which is surrounding Kuśadvīpa. That is the dark green.
Tamāla Kṛṣṇa: Dark green.
Yaśodā-nandana: No, the ghee ocean is the yellow color. The ghee ocean is surrounding the Kuśadvīpa. In fact, again surrounding the ghee ocean is Krauñcadvīpa, but Krauñcadvīpa is 1,600,000 yojanas in width. That means 12,800,000 miles. It is getting bigger, twice as big, as we go further. And after Krauñcadvīpa, surrounding Krauñcadvīpa is the milk ocean, and this milk ocean is surrounding the whole Krauñcadvīpa. And then again there is mention that the Śvetadvīpa Ocean, the Śvetadvīpa where Lord Viṣṇu... Lord Viṣṇu resides there in white island. This Your Divine Grace has described in the Caitanya-caritāmṛta.
Bhakti-prema: Kṣīrodakaśāyī Viṣṇu resides there.

Yaśodā-nandana: Yes, kṣīra. Kṣīra means milk. Śākadvīpa is surrounding the milk ocean.
Tamāla Kṛṣṇa: What color is it?
Yaśodā-nandana: Śākadvīpa, it is green, this green here. It is pale green. And Śākadvīpa is surrounded by yogurt ocean, dahi, dahi ocean. This is the white here. This yogurt ocean is 3,200,000 yojanas, or 25,600,000 miles of width. Then, surrounding this yogurt ocean is Puṣkaradvīpa. This is the reddish brown here. Puṣkaradvīpa is 6,400,000 yojanas in width, or 51,200,000 miles. And here, in the middle of Puṣkaradvīpa, all around is the Mānasottara range of mountains, which is situated in the middle and which has a width of 10,000 yojanas. Now, from the middle of Mount Meru until Puṣkaradvīpa it is calculated to be 15,750,000 yojanas.
Tamāla Kṛṣṇa: We calculated all these...
Yaśodā-nandana: This was all calculated according to the Bhāgavatam. And then, on the four corners, on top of this Mānasottara range of mountains, the loka-pālas, Maheśvara, Brahmā, Kuvera, they are residing in these four corners, guarding the directions of the universe. And...
Tamāla Kṛṣṇa: Now wait a moment. What goes there? The most important thing is that on top of that Mānasottara range...
Yaśodā-nandana: On top of the Mānasottara range... If this map could be placed on the ground and Mount Meru would be like this and all the oceans go around, the sun, which is again 16,000 yojanas above Mount Meru, goes all around Mount Mānasottara. It comes in this way every day.
Tamāla Kṛṣṇa: So where does it go now? Where does the sun go?
Yaśodā-nandana: The sun would go like this, all around. It would be above. It's constantly moving round the top, circumambulating the Mount Meru and going just above the Mānasottara range of mountains. This little brown line, gray line, within the circle is the Mānasottara range of mountains. And surrounding this Puṣkaradvīpa there is sweet water ocean throughout the whole... There's sweetwater ocean. It's 6,400,000 yojanas, or 51,200,000 miles. And past the sweet water ocean there is Kāñcana-bhūmi, or a golden land, where everything... This is a land that's described to be just like mirror, where no living being can be because as soon as something is dropped there, it disappears. That is a very peculiar feature of this land. This land, this golden land, is 15,750,000 yojanas. That means that this land here... The width of this land is the same as between Mānasottara Mountain and the middle of Mount Meru. It is very scientific like this. And past this golden land...
Tamāla Kṛṣṇa: We used a calculator to calculate all this.

Yaśodā-nandana: Past this golden land, until here, is the Lokāloka Mountain, which is the border between this Aloka-varṣa, or the uninhabited land, and the inhabited place. This Aloka-varṣa is constituted by a protective mountain that stops the rays of the sun to go beyond this portion. And it goes very, very high, it is described. It goes higher than Dhruvaloka. So the whole planetary system of Bhū-maṇḍala is like a big lotus flower with very high, high petals. It is very wonderful.
Tamāla Kṛṣṇa: Which is that Lokāloka?
Yaśodā-nandana: This is the Lokāloka mountain.
Tamāla Kṛṣṇa: Just see how much it is, Śrīla Prabhupāda. It's like the whole Bhū-maṇḍala is protected by a huge mountain.
Prabhupāda: Karach.(?)
Tamāla Kṛṣṇa: Yeah. And only in the very center are the inhabited lands of Bhū-maṇḍala.
Prabhupāda: The sun is above them?
Tamāla Kṛṣṇa: Not above Lokāloka. Inside, within.
Yaśodā-nandana: If this would be on the floor, that means Mount Meru would be 84,000 yojanas. Then, above Mount Meru, 16,000 yojanas above, is the sun. But then, so that the rays of the would not penetrate in that land all around, this great mountain, Lokāloka, extends all the way up to Svarga. So it is like a big, big cup, in which the middle of the cup, or the bottom, there is all of these planets, all of this Bhū-maṇḍala. And past this Lokāloka range of mountains is this Aloka-varṣa, which is described that there is no living being which can go to there. The only occasion where anyone went through there is when Kṛṣṇa and Arjuna went to see the Mahā-Viṣṇu in the spiritual sky. So Kṛṣṇa with His cakra penetrated through all of these regions, and then He went through all of the coverings of the universe who were there. That was the only occasion where anyone went to this land. So this is a general picture of Bhū-maṇḍala.
Tamāla Kṛṣṇa: To scale.
Prabhupāda: Then the sun... Above the sun there is moon.
Yaśodā-nandana: Yes, and above the sun there is the moon.
Tamāla Kṛṣṇa: That we have to...
Yaśodā-nandana: That is our next project.
Bhakti-prema: (indistinct)
Tamāla Kṛṣṇa: So far, we have stayed in the Bhū-maṇḍala. We've now just picked the cover of Bhū-maṇḍala. We were thinking to do one more. We can do it later on, one more drawing to give it more detail, because the centers Meru and Jambūdvīpa and salt ocean are so small on this map you cannot even see it. So we want to...

Prabhupāda: Hm, in detail.
Tamāla Kṛṣṇa: We want to blow up one area like they do on some maps. They show one city bigger than other places 'cause it's important. And after that, our next business is to now go upwards. Now we've gone outward this way. Now we have to show the sun, the moon, the sapta-ṛṣi, everything like that. That'll be very nice doing, how the sun is...
Yaśodā-nandana: Another unique feature of this map is that... Bhakti-prema Mahārāja found about... There is one space which is mentioned by Viśvanātha Cakravartī Ṭhākura in his commentary. The complete diameter from west to east and north to south, it is calculated to be 500,000,000 yojanas. In order to be able to adjust that, we needed 250,000,000 on each side.
Tamāla Kṛṣṇa: That means 4,000,000,000 miles.
Prabhupāda: That I have given there.
Tamāla Kṛṣṇa: Yes.
Yaśodā-nandana: Yes, that was mentioned. His Divine Grace has mentioned it. So the total distance of all the seas was coming to about 496. So we understood from the commentary of Viśvanātha Cakravartī that there is a little space, vacant space. Between this vacant land there is a little guarded space, between this Aloka-varṣa and the covering of the universe, which is 1,700,000 yojanas on each side of the universe. And Viśvanātha Cakravartī explains that the reason for this is that under all the planetary systems there is the eight elements, which are sustaining all the Bhū-maṇḍalas and all the lower planetary systems. And if this is touching the side of the universe, then these elements would have nothing to do. There would be no function for them. Therefore, that is one reason... And another reason he was giving is that when Varāha-avatāra lifted the earth, if the Bhū-maṇḍala is touching the side of the universe, then there's no place to lift it. Therefore, by his calculation... You have given then a very long purport with devotional senses(?). According to Jyotir Veda, like this, it was mentioned that there is little space between the outward part of the universe and Aloka-varṣa. That is the space, so that Varāha could lift up the earth from the Garbhodaka Ocean. (Prabhupāda chuckles)
Bhakti-prema: (indistinct)
Tamāla Kṛṣṇa: I think, Śrīla Prabhupāda, that you gave that right in the purport. You called them (indistinct). See, you gave this huge... It says here, "However , the technical terms used in the astronomical calculations given by the Jyotir Veda are difficult to translate into English. Therefore, to satisfy the reader, we may include the exact Sanskrit statement given by Śrīla Viśvanātha Cakravartī Ṭhākura, which records exact calculations regarding universal affairs." And then you give this huge Sanskrit quotation. And from

reading this, Bhakti-prema Mahārāja found out that there is space. Due to this purport we got that information. So it was perfectly put in here. It's very nice.
Prabhupāda: Yes.
Tamāla Kṛṣṇa: So I think that Svarūpa Dāmodara will be helped by these drawings when the men come. 'Cause he said that even though they are scientists, they could not understand this volume. It's been a mystery practically. These drawings, one by one, should be able to help in the creation of that planetarium.
Prabhupāda: Thank you very much. Hare Kṛṣṇa.
Tamāla Kṛṣṇa: Now we have to figure out how to preserve these, Śrīla Prabhupāda. Puṣkara Prabhu was concerned how to preserve these pictures, and we were thinking that maybe they should be mounted on canvas and stretched.
Prabhupāda: Yes.
Tamāla Kṛṣṇa: And that way, they can be preserved.
Prabhupāda: Like map.
Tamāla Kṛṣṇa: Oh, oh, like a map. Once they are put on canvas, can they be rolled, Puṣkara?
Puṣkara: Yeah. The best thing is to put them on a board and just permanently put these onto the best possible...
Tamāla Kṛṣṇa: He says the best thing is to mount them on something and then display them somewhere, keep them permanently that way, stretched out.
Puṣkara: Every time you roll it, it gets ruined more.
Tamāla Kṛṣṇa: Every time you have to roll it...
Prabhupāda: Take one room in the gurukula building.
Tamāla Kṛṣṇa: It will be very interesting. They can use this for teaching.
Prabhupāda: Yes. Explain there in bold and this picture.
Tamāla Kṛṣṇa: Yeah, explanation can be... This book is the explanation.
Prabhupāda: Hm.
Tamāla Kṛṣṇa: It's amazing how Kṛṣṇa has made each universe. It's just like all the fruits. When you get a fruit, it's protected by a skin. This universe is very much protected, first by uninhabited land, then by huge mountain, then by golden land. And then only in the center of these dvīpas there are all the inhabitants and all the activities are going on, very much protected.
Prabhupāda: And they are going to the moon. (devotees laugh) I... So far I remember, the moon is also like the sun, that it is fire blazing, but it is surrounded by a cool atmosphere. Therefore it is soothing. I think there is such description.

Bhakti-prema: Śrīmad-Bhāgavatam says the moon is also (indistinct).
Tamāla Kṛṣṇa: How far do they, the scientists, say the moon is from the earth? How long? I have a book which says it, and I'm bringing this book. It's very... You'll see it here.
Bhakti-prema: Twenty-four lakhs miles.
Tamāla Kṛṣṇa: Twenty-four lakhs miles, the scientists say?
Bhakti-prema: No.
Tamāla Kṛṣṇa: What do the scientists say? How...?
Prabhupāda: Two lakhs.
Tamāla Kṛṣṇa: Two lakhs miles, 250,000 miles. So that means about, in yojanas, very little, about 25-, 30,000 yojanas.
Yaśodā-nandana: They say the sun is 93,000,000 away.
Bhakti-prema: I think the difference of the (indistinct).
Tamāla Kṛṣṇa: They have froglike brains.
Prabhupāda: That, the microscope... What is called? Telescope.
Tamāla Kṛṣṇa: This is the book of the rascal scientists. They describe the solar system according to their nonsense. The solar system... Gives all the calculations. They calculated how much it weighs on each planet. (laughs) They haven't even been there. They say that each planet has moons. Says here... This is how scientific they are. "Pluto was discovered only in 1930, and as yet, little is known about this remote planet. Pluto is much smaller than Neptune and has a diameter probably about..."
Prabhupāda: "Probably."
Tamāla Kṛṣṇa: "...half the size..."
Prabhupāda: "Probably."
Tamāla Kṛṣṇa: "The orbit of Pluto is extremely elliptical, and the day is some 6,109 hours long. There is probably no atmosphere, and there are no known humans."
Prabhupāda: "Probably." Their science is "probably." Probably it is science; otherwise it is nonsense. (laughter)
Tamāla Kṛṣṇa: It says here, "The moon is 238,860 miles from the earth. It has no atmosphere, no weather and no wind."
Prabhupāda: "Probably." Everything "probably."
Tamāla Kṛṣṇa: It says here, "There is thick dust covering and no evidence to suggest that the moon has ever supported life." In that newspaper article the man who is exposing them said—because they say it is covered by dust—"How is it that no dust is shown on the astronauts' suits when they walked around?" He says, "If there's such a thick dust, then, when the rocket landed, it would have made a pocket within that dust." He says, "But there's no crater around the rocket. Then how it is possible that these things are like

that?" 'Cause actually they forgot. When they were making the stage setting in Arizona, they forgot these things.
Yaśodā-nandana: One argument Your Divine Grace gave in 1971 was that if they went to the moon and they found it was rock, how do they explain the moon is so shiny and gives such a cooling effect? They cannot explain that.
Tamāla Kṛṣṇa: Look at the earth. Now, this is a real question that we still have to answer. They picture the earth round, and we say, no. Bhū-maṇḍala is like a lotus, like this, and the earth is only one part of one island in Bhū-maṇḍala, and it's not, you know, it's not round(?). It doesn't look like that. And all the pictures they take of the earth when they go up in their satellites show round. And we're going to tell them that it's not. This is a very tricky question. In other words, if this is the picture of the world, like this, and we say that... If we take an airplane from here, from Los Angeles. Now, supposing we go to India, which is here. So there's two ways to go. One way, you can go like this, and the other way, you can go like that. But if the earth is not a round globe, then how is it sometimes people go from Los Angeles via Hawaii to Japan and then India? So we can't figure this out. We have experience, those of us who have flown, that actually the plane went from Los Angeles to Hawaii to Tokyo to Hong Kong and then to India. So it doesn't work out in our maps so far, right? We can't figure it out. This thing has to be very complete in its answers. Otherwise everyone will laugh at us. We can't leave any loopholes.
Prabhupāda: So are you thinking on this?
Bhakti-prema: In the Śrīmad-Bhāgavatam... According to Śrīmad-Bhāgavatam, it is (indistinct).
Prabhupāda: Find out from our side, according to Bhāgavatam.
Bhakti-prema: Scientists are lacking in the main points.
Prabhupāda: Yes. They "probably," so many theories.
Bhakti-prema: There is... Śrımad-Bhagavatam gives that about Himalaya. Himalaya is 80,000 miles high, 16,000 miles wide. That means covering more than earth, more area than (indistinct).
Tamāla Kṛṣṇa: Then it's not so much. In other words, the Himalaya Mountains are here, according to our, this yellow here is the Himalayas. This is a map, showing all the mountains. So according to Śrīmad-Bhāgavatam, how long do they stretch?
Bhakti-prema: Sixteen thousand miles wide.
Tamāla Kṛṣṇa: So the Himalayas are 16,000 miles wide. Sixteen thousand miles is a huge area, beyond this whole area. So according to the Bhāgavatam, this should be all Himalayas.
Bhakti-prema: Yes, then it is coming this side, up to Canada, all Himalayas.

Tamāla Kṛṣṇa: So what is the explanation?
Bhakti-prema: And bring this from there and there, there are nine islands. From each, divided one between (indistinct). They say that it's 8,000 miles. (indistinct)
Tamāla Kṛṣṇa: And how high is the Himalayan Mountains?
Bhakti-prema: Eighty thousand miles.
Tamāla Kṛṣṇa: Ten thousand yojanas, 80,000 miles. Here it says that "The highest mountain is Mount Everest, 29,000 feet." Not very high. That's about six miles. And we are saying 80,000 miles. So we want to know where is that. How high is Govardhana supposed to be?
Bhakti-prema: Govardhana (indistinct). Govardhana mountain is sinking.
Prabhupāda: Sinking.
Bhakti-prema: Crushed by (indistinct). [break]
Tamāla Kṛṣṇa: So the Rand McNally's Illustrated World Atlas, its special feature is that it gives maps describing all different subjects like language. What language is spoken throughout different places is the world is shown by map and the national areas, according to size, population, and cities, major populations, densities of population. Like China is supposed to be... China. This shows proportionately in population according to this and India, it says that it is... Everything else is very small compared to these two, India and China. Agriculture, what kind of agriculture, natural vegetation, climate... [break]
Bhakti-prema: But there was no Atlantic Ocean, Indian Oceans, no. But after that, there were sixteen thousand sons of Sagara Mahārāja. Then their sacrificial hearth was stolen by Indra. So it was put somewhere in the earth folds. They began to dig the earth to find out that hearth. So they dug other oceans, (indistinct) Kapila Muni (indistinct), and it explains the curse(?) that he is the chief. And then there's...
Tamāla Kṛṣṇa: It says here that "Only our knowledge of the crust of the earth is based on direct observation, but studies of paths of earthquake waves..." Then it goes... The only way they can understand is by direct observation. And that's very limited. Says, "All the planets were probably formed at much the same time." It doesn't sound like they have very much knowledge, Śrīla Prabhupāda. Everything is "probably," "maybe." Says that "Probably all the planets were formed at much the same time from the same great dust cloud." After you create this planetarium, Śrīla Prabhupāda, they'll have to rewrite all of these books. These businesses...
Bhakti-prema: Another book has to be written. Its name should be Easy-to-Read Geography or Advanced Geography. And also about history we have

to write. Your Divine Grace will write Advanced History, and there the complete lifetime of Manus and Indras should be given.
Tamāla Kṛṣṇa: According to the modern thinkers, any further back than about three or four thousand years ago, everyone was living in the caves. So they think that all of our books are mythology, some dreamt-up stories by some people...
Prabhupāda: So how they are writing of millions of years ago?
Tamāla Kṛṣṇa: This is all according to their mythology.
Prabhupāda: No, they are suggesting.
Tamāla Kṛṣṇa: And, of course, they say that there were no humans around, just dust and water and earth. There were no brains at that time.
Prabhupāda: Only brains are developed now.
Tamāla Kṛṣṇa: Yeah, especially now, this century. Before this, everybody was unintelligent, and now man's brain is developing to a higher and higher degree, and he can finally understand what is what. I don't think that... Your descriptions, especially this planetarium, will at first meet with a lot of heavy reaction. It is not going to be embraced immediately very favorably. It means that everyone who calls himself a Ph.D. is a fool, that students will laugh at their teachers, if what we say is correct. There will be chaos in educational circles. (Prabhupāda chuckles)
Prabhupāda: All right. )
Devotees: Jaya Śrīla Prabhupāda. *(July 2, 1977, Vṛndāvana)*

## Discussion about Bhū-maṇḍala

Prabhupāda: Lotus is also round. Lotus, that petal is also round. Lotus petal is round. So within one lotus petal this round or that round, there is.
Śatadhanya: And within their own limited sphere they can go this way or that way. But they don't know the whole shape.
Tamāla Kṛṣṇa: We have to draw the shape probably, because according to what we've drawn so far...
Prabhupāda: That you can do, but real thing is we can remain within the limit.
Tamāla Kṛṣṇa: That we accept.
Prabhupāda: So within the limit...Suppose the lotus petal this way, that way, or this way...
Tamāla Kṛṣṇa: You're asking us to draw the details and make a planetarium very exact.
Prabhupāda: Yes. Lotus petal, it is round. So in one lotus petal you are conditioned. You cannot go.

Tamāla Kṛṣṇa: So far that hasn't been said, but that's... You can say that Bhārata-varṣa is a petal of the lotus. But I think if you look at the Bhāgavatam, it may... I'd have to see it, what it says. There's a statement that it may be the inner portion of the lotus. I don't know...[break] And what we do, it has to agree with the Bhāgavatam.
Prabhupāda: Lotus petal... There are so many petals. You are conditioned with one petal.
Tamāla Kṛṣṇa: That's... If that's the explanation, then it can be somehow adjusted.
Prabhupāda: You do not know what is going on outside.
Devotee: There's so many round parts of a lotus petal.
Tamāla Kṛṣṇa: Yes, that part...That could be adjusted. But there's a place within it, you know, the center part, there's no petals. In the middle of the lotus there's like a... I don't know what you call it, but a... It's a flattish area. They show Kṛṣṇa standing sometimes when they draw a picture of Kṛṣṇa on a lotus.
Prabhupāda: That's all right. But your place is in.
Tamāla Kṛṣṇa: That we accept.
Prabhupāda: That we have to hear from authority.
Tamāla Kṛṣṇa: We accept that. I'm just thinking...
Prabhupāda: Unless you are obstinate, you have to accept if you are convinced.
Tamāla Kṛṣṇa: We accept.
Prabhupāda: There are so many millions of stars and moons that we cannot go.
Tamāla Kṛṣṇa: As we are conditioned, as everyone is conditioned, our planetarium will have to show the actual facts.
Prabhupāda: Yes. That fact we have learned from Bhāgavatam.
Tamāla Kṛṣṇa: So, so far we have not drawn this fact correctly.
Prabhupāda: That is your inability. That is another thing.
Tamāla Kṛṣṇa: Yeah, well that's why we're... That's the question that we've raised. This question that we've raised is due to that.
Prabhupāda: That is you are unable to, but the fact is that you are conditioned. You cannot go beyond that conditioned
Tamāla Kṛṣṇa: That's accepted.
Prabhupāda: So we are also conditioned. But as far as possible we take description from Bhāgavata, try to. That is our... Suppose here is India, here is Los Angeles. You start from India, Los Angeles..., or India, you'll come to Los Angeles. And again return to India. Similarly you start from this again going.

Tamāla Kṛṣṇa: Yes. That's possible if you can fly this way, underneath.
Prabhupāda: But where is the underneath?
Tamāla Kṛṣṇa: See Prabhupāda, we weren't asking that issue. That's not the question we were asking.
Prabhupāda: What is that issue?
Tamāla Kṛṣṇa: Well, you've explained, if it's a lotus petal, then you can fly around it. That's all right. Then the answer is there. But if there's no lotus petal and it's simply flat, then it's a problem. That a problem.
Prabhupāda: No, we don't say flat.
Tamāla Kṛṣṇa: We've pictured it like that. There's some defect in our picture. If you recall, there's a picture we drew...
Prabhupāda: So rectify it.
Tamāla Kṛṣṇa: Yeah, well how?
Prabhupāda: That is the explanation. Petal.
Tamāla Kṛṣṇa: See it doesn't say petal anywhere in the Bhāgavatam. It doesn't say Bhārata-varṣa is a lotus petal.
Upendra: Neither does it say it's flat, though.
Tamāla Kṛṣṇa: Well I can get the Bhāgavatam right now if you allow me to. It doesn't say... I believe it says Jambūdvīpa is the second part of the lotus, and there aren't petals in the center.
Upendra: It's a transcendental lotus flower.
Prabhupāda: You are prejudiced. That's all. If I can understand, why don't you understand? You are prejudiced.
Tamāla Kṛṣṇa: Well I can understand if I accept it as... I'm only trying to think as the persons who are going to make this planetarium.
Prabhupāda: You are Western. You are prejudiced. That is reason. If I can understand why you...
Tamāla Kṛṣṇa: I can understand blindly, but I don't want to do that.
Prabhupāda: I do not understand blindly.
Tamāla Kṛṣṇa: No.
Prabhupāda: Then.
Tamāla Kṛṣṇa: But I don't want to...
Prabhupāda: Still we differ. That means one of us is prejudiced.
Tamāla Kṛṣṇa: Well it's not Your Divine Grace. (laughter) I mean I'm sorry I have to take this thankless task to ask all these questions.
Prabhupāda: You are also prejudiced. It is written in the śāstra. In that sense we are also prejudiced.
Tamāla Kṛṣṇa: Yeah, but we're correctly prejudiced.
Prabhupāda: Just like I take Kṛṣṇa's word. Bas, fact. You can say that you are prejudiced, you see. This is the book.

Tamāla Kṛṣṇa: That is a nice explanation, the petals.
Prabhupāda: Everyone is prejudiced. But who is rightly prejudiced who is wrongly prejudiced. That is everywhere. Just like materialistic person will think, "Brainwash. These rascals, they have given all up material enjoyment, and after some phantasmagoria they sacrifice everything. Brainwash." And we are thinking, "Oh, these rascals, got this human form of body, he did not understand what is spiritual life." Both of them—he is rascal and he is rascal. This is going on. Yā niśā sarva-bhūtānāṁ tasyāṁ jāgarti saṁyamī. Is it not? They say, "Brainwash. Unnecessarily they've sacrificed everything." And we say that he got the human form and unnecessarily he is working like cats and...
Tamāla Kṛṣṇa: Both are saying the same thing. Obvious.
Prabhupāda: Yes. Now who is correct? Who will decide? I accuse you, you accuse me. But who is correct? Who will decide? That is śāstra. There is no question of argument. That is called pratiṣṭhā. You will never come to any conclusion by arguing. I think you are wrong, you think I am wrong. Somebody must decide. Judge. And that is śāstra.
Tamāla Kṛṣṇa: Supposing they say...
Prabhupāda: They say, we are saying so many things. In the court there are two parties. I say something, another party says. The judge is there. He has now decision.
Tamāla Kṛṣṇa: Then they'll ask for the verse to prove it, and we must quote the correct śloka to substantiate our claim.
Prabhupāda: Yes. That the judge gives. Judges according to lawbook. Tad-vijñānārthaṁ sa gurum evābhigacchet. So we have to approach the right person who can give the judgment. What is their answer about this, that by chemical combination they cannot produce life? Still why they are insisting?
Tamāla Kṛṣṇa: They are rascals.
Prabhupāda: Then what is the value of rascal's statement?
Tamāla Kṛṣṇa: No value. Lunatic asylum.
Prabhupāda: Why if one is settled up that he's a rascal, we should not hear anything. And why not? He has not gone to the moon. They are insisting, "Yes, we have gone," by false propaganda. Why the one who makes false propaganda and one who says nonsense, we have to believe? Immediately, whatever he says, reject. There's no argument. You have proved yourself rascal so we don't accept any statement. What do you think?
Tamāla Kṛṣṇa: Well I'm just trying to think of the description of Jambūdvīpa.
Prabhupāda: Whatever you say. But be aware you are rascal.
Tamāla Kṛṣṇa: But we have the practical task of...
Prabhupāda: No test.

Tamāla Kṛṣṇa: No—task.
Prabhupāda: No test. Mother says here is your father. That's all. Finish. You cannot test. Then you are rascal.
Tamāla Kṛṣṇa: Not test. I said...
Prabhupāda: No, these things cannot be tested.
Tamāla Kṛṣṇa: No, I'm not saying...
Prabhupāda: Mother says, "Here is your father." That's all. Finish. No test. No question. That is foolishness.
Tamāla Kṛṣṇa: I was using the word "task." That to... I'll have to bring in Bhakti-prema Swami. Maybe he can speak better. I'm not able to explain. But according to the description in the Bhāgavatam, everything that we've read, it is not petals. So let's... I don't know. On one hand you are asking us to follow the...
Prabhupāda: Bhāgavatam, one portion may not be exactly like... Just like if you are in the midst of the petal, you cannot understand it is petal. You see it is flat. You have shown in the map, there is a point so many miles. So that petal, middle portion, you can take it flat. It is not flat. It is curved. Whatever conception you make, that is defective. Rascals. First of all you have to understand this. So don't make it like this, like this, like this, like... Whatever it is said, that is all right. Therefore Kṛṣṇa said aśraddadhānāḥ. No faith. Faithless. We have challenged, by chemical combination make a small egg, sparrow's egg, and produce.
Tamāla Kṛṣṇa: They can't do anything.
Prabhupāda: Then what is the value of your experiment? And still they are saying. This one point will kill them. Why do they not take this challenge? Huh?
Tamāla Kṛṣṇa: They're rascals. There's no other answer. Only a rascal will go on maintaining that we can create life and they never have.
Prabhupada: And they are getting Nobel Prize. Just see. Such rascals. And we say on the authority of Bhagavad-gītā, apareyam. This is inferior and that is superior. So how you can make superior with inferior ingredients?
Tamāla Kṛṣṇa: Impossible. Without the superior touch the inferior cannot have any life.
Prabhupāda: That's a fact. They're seeing every moment.
Tamāla Kṛṣṇa: Everything has movement... If it's matter it only moves because of some superior force, living force, pushing it or entering it or something.
Prabhupāda: The superior energy is utilizing inferior energy for His purpose. You are utilizing this inferior, the lump of matter, for serving my purpose. That is superior energy. It cannot use me for its purpose.

Tamāla Kṛṣṇa: It has no purpose.
Prabhupāda: No. I can use it for my purpose. Therefore I am superior. (aside:) Come on.
Tamāla Kṛṣṇa: Śrīla Prabhupāda, it's about three o'clock now.
Prabhupāda: Shall we go?
Tamāla Kṛṣṇa: I may have to go for that... I told Bhakti-prema Mahārāja I would meet him at three.
Prabhupāda: Go on. [break] So what is the difficulty?
Bhakti-prema: First we are interested to know...
Prabhupāda: (aside:) Take from the middle. [break] It is something new?
Tamāla Kṛṣṇa: No. This is the first drawing we did. This is Jambūdvīpa. This is Bhārata....
Bhakti-prema: This is Bhārata-varṣa, Bhūrloka. Now, this is Himalayan mountain going from east to west. This is India.
Prabhupāda: First thing is, these people cannot cross the Himalayan mountain. (aside:) You can turn this light in this way, down. From my practical experience, I have seen Switzerland. It is so high, that so many accidents have taken place. Little inattention. They have experience. The plane goes thirty-two thousand feet high. Is it not?
Tamāla Kṛṣṇa: Yes. Thirty to forty thousand. Say forty thousand maximum.
Prabhupāda: Still they met with accident. But they are saying that Himalaya is twenty-eight thousand feet high.
Tamāla Kṛṣṇa: Maximum.
Prabhupāda: What is this? First of all answer this. Everything proposition is wrong. So how we can trust? What is the answer?
Tamāla Kṛṣṇa: We cannot trust them.
Prabhupāda: Then? But if you cannot trust them, they're rejected immediately. This is one argument. Another argument, (Bengali). The frog in the well, he's estimating, "This is the limit of water." What he'll understand about Pacific Ocean, Atlantic Ocean? There is so many things. On the whole we are imperfect, and although imperfect, they want to be perfect. That is cheating. Now for argument's sake, if they put some argument, we have to reply. We have to prepare for them. Otherwise they're useless. We know they're useless, but sometimes they put some argument. We have to give answer. This is the position. And our duty is to present picture of Śrīmad-Bhāgavatam correct. So what do you have to say? At least some answer you should give.
Yaśodā-nandana: The biggest problem we are discussing... We were thinking that the first question they will ask, this is their conception of the world.
Prabhupāda: We reject them.

Yaśodā-nandana: Yes.
Prabhupāda: "You are imperfect. So whatever you have written, that is nonsense. And everywhere problem. Why should we waste our time?" Now just try to settle from point of. He's showing this book, I am showing the Bhāgavatam. So you are imperfect, that's a fact. What is the value of your book? If you are basically a rascal, then what is the value of your book? Why shall I waste my time? First of all I take it and I prove it that you are a rascal. You show your book, I show my book. But you have been proved that you are a rascal. Because practically you cannot cross Himalaya. You have no idea what is the Himalaya. You are giving some conflict.
Bhakti-prema: This..., we have to reply to this question. They say if we go, we start from Los Angeles and arrive Japan, according to Śrīmad-Bhāgavatam...
Prabhupāda: Japan and Los Angeles and India, that is not the whole thing.
Bhakti-prema: Yeah, that is not the whole thing, but it is basic point.
Prabhupāda: Huh... Insignificant.
Bhakti-prema: According to Śrīmad-Bhāgavatam, if we start from London...
Prabhupāda: That argument also I have refuted. Just like animal, he is bound up. He is rotating around the log, and within that there may be Japan, there may be Calcutta, there may be Los Angeles. You can think this is there. But that is not all. Within that rounding circle, whatever is there, you may think this is all. But that is not all. He's limited condition. So within his limitation (Bengali). Within that limitation he's speaking. But Himalaya and other things, far beyond their limitation. That I have already explained. He's speaking within his limitation. Our position should be, correctly represent what is described in the Śrīmad-Bhāgavatam. But if there is some question, we should ask. Just like I am answering to the reasonable point. That if you are conditioned, within your condition you can see, you can experience, but beyond that you have no right to see. What is Los Angeles, Calcutta, Japan, this is very insignificant space. And they're talking of that. We are talking that Himalaya mountain, we have crossed over Himalaya, we conquered the outer space. How they can think of it?
Bhakti-prema: How to take them to (indistinct).
Prabhupāda: There is no question of take them. If you take, he's all right. If you don't take, we cannot change.
Bhakti-prema: As long as we cannot show them.
Prabhupāda: First of all you have to prove that you are imperfect.
Bhakti-prema: That has already been proved.
Prabhupāda: Yes. You're simply insisting on things by "probably." "Probably beyond the Himalaya there is something." That is one thing. We say "Definitely, here is. Beyond the Himalayas like this, that." That is the

difference. You say "probably," I say "definitely." Father, mother said, "Here is your father." You can say "probably," and mother is saying "definitely." She knows perfectly. You may say probably he may be your father, but mother knows that definitely. Therefore we take Vedas as mother, Purāṇas as sister. Śukadeva Gosvāmī is explaining it. Iti śuśruma. But śuśruma, why he should waste his valuable time? He knows it is definite. So unless you come to this standpoint that whatever is spoken in the Vedic literature, that is definite, you cannot be convinced by argument. Tarko 'pratiṣṭhaḥ śrutayo vibhinnā nāsāv ṛṣir yasya mataṁ na bhinnam, mahājano yena gataḥ sa panthāḥ. Śukadeva Gosvāmī said—that's all. Mahājano yena. Vaiyāsakiḥ.

svayambhūr nāradaḥ śambhuḥ
kumāraḥ kapilo manuḥ
prahlādo janako bhīṣmo
balir vaiyāsakir vayam

These mahājanas. This is our argument. And for common-sense argument, the Himalaya is very, very high. Very, very broad. You have never crossed and you have met with so many accidents. They avoid that portion, flying plane. And I have seen how high has it gone, then it will be in the clouds. Still they say twenty-eight thousand. Huh? Twenty-eight thousand?
Tamāla Kṛṣṇa: Twenty-nine thousand.
Prabhupāda: So I conclude like that. When you show this book, accept this authority.
Tamāla Kṛṣṇa: When we show this book it means we're accepting the authority?
Yaśodā-nandana: Not accepting the authority. The problem is the Pacific Ocean, according to Jambūdvīpa, for us it is the salt ocean. So the first question that Mahārāja was raising, that they will ask, how do we go between the west coast of America, which is very tiny, and Japan, as this according to their calculation we go west and we arrive around. And you keep still going further and you arrive back in America. The point about the Himalayas, that we did not raise, because that we could well understood. The Himalayas is much broader and much bigger than they think. We can fully understand it. But that point, from common-sense point of view we couldn't exactly understand. We thought of it all afternoon, and we came up with a few ideas, but we wanted to hear what Your Divine Grace...
Tamāla Kṛṣṇa: Our question was mostly coming out of how to draw what the Śrīmad-Bhāgavatam is saying. We're not supporting any kind of mundane

argument, nor do we have any doubt in Bhāgavatam. We're simply trying to understand the Bhāgavatam.
Prabhupāda: And that is your credit.
Tamāla Kṛṣṇa: That's why we were meeting.
Prabhupāda: Where is the difficulty?
Tamāla Kṛṣṇa: Well, I think that one of the difficulties arose from my seeing this map originally, because in this map it shows the continents that we today speak of as various continents. So when we all looked at this map, our immediate question was, as I said the other day, how do we go from one place to the next? It's not such an unreasonable question. So we're just trying to answer it from the Bhāgavatam—not to give our own speculations, because they're imperfect.
Prabhupāda: That is already answered. You are limited within certain area. So you can conclude in so many ways. Sara bhuri kara, kara bhuri sara.(?) There is a vegetable prepared, the following vegetables are taken: sara, bhuri, and kara. And again you say kara bhuri sara. This way or that way. Sara bhuri kara, kara bhuri sara. That three things, either you take this way consecutively or that way consecutively. But you have got only that kara bhuri sara. You have no experience. There is potato, there is (indistinct), there is (indistinct—Sanskrit). You know these three things. Sara bhuri kara. Kara bhuri sara. Just like they are making arrangement in the laboratory. Companies are there. They present in such a way, that he was flying in the sky, and big sputnik and very small town. Do you know that man came?
Tamāla Kṛṣṇa: Man came and said he could do anything. Make any trick or illusion. (indistinct)
Prabhupāda: And they have done so in the matter of moon planet. They've never gone there by laboratory arrangement. Arizona?
Tamāla Kṛṣṇa: Arizona.
Prabhupada: That's all. That is their business. They might have presented such devices in a book, a different color. Just like medical science. The preparation is nothing, but they'll employ big, big medical students to write in such a way that it will be presented as very important. They are doing this. We have done in Dr. Bose's laboratory. They employ. (speaks garbled words to sound like big words) Go on speaking. (laughter) It sounds very nice. What is the meaning of? (more garbled phrases) So I have got my own common sense. I understand. (indistinct) in the Vedic literatures. (indistinct comments by devotees)
Prabhupāda: That I already told. Prejudiced.
Yaśodā-nandana: I'm not prejudiced. I'm just trying to understand. I'm trying to understand what is the facts according to Bhāgavatam.

Prabhupāda: How you can understand which is beyond your understanding?
Tamāla Kṛṣṇa: He is saying he wants to understand according to the Bhāgavatam.
Yaśodā-nandaṇa: Not we're challenging.
Prabhupāda: That you can describe.
Tamāla Kṛṣṇa: That is what we're having trouble with. We're not trying to do our own thing. We're trying to understand Bhāgavatam. That we're a little stuck on some point.
Prabhupāda: Bhāgavata is there. You try to understand.
Tamāla Kṛṣṇa: We're not siding with the preconceptions that we had before. We'll throw them away. We're trying to accept the Bhāgavatam. Everything has to be according to the Bhāgavatam.
Prabhupāda: Make it nice. We are going to spend so much money, people may not reject it. [break]
Tamāla Kṛṣṇa: Just like as we're talking, we're a little... Sometimes it's difficult to understand the Bhāgavatam.
Prabhupāda: But I take it simply—that there is ocean, and it was churned. So there is no difficulty.
Tamāla Kṛṣṇa: No. Mostly we have understood. Only in one place we are a little...
Prabhupāda: But nobody has seen that ocean. And nobody can believe that ocean can be churned. Would you believe it? Because it is. And the Vāsuki was taken as rope.
Bhakti-prema: And this Mandara mountain, fifty thousand miles high, was taken there, carried by.
Prabhupāda: And it was born by tortoise incarnation.
Bhakti-prema: This is combined with description.
Prabhupāda: Acintyāḥ khalu ye bhāvā na tāṁs tarkeṇa yojayet. What is beyond your conception, don't try to. So that is Vedic civilization. They were satisfied with information received from the Vedas.
Bhakti-prema: The first and last thing we have to prove logically that this is...
Prabhupāda: No, no. Logically you cannot. Acintya. Logic comes when it is conceivable; but it is inconceivable. Where is your logic?
Bhakti-prema: But first to prove them right understanding we have to bring them.
Prabhupāda: It is very difficult. Acintya.
Yaśodā-nandana: If such a thing as going from one place to the other on earth, this is cintya, this is conceivable.

Prabhupāda: Argument is there when it is conceivable. It is inconceivable. Kṛṣṇa lifted Govardhana. Seven years old, six years old. How it is conceivable? But devotees, they are crying, "Oh, Giri-vara-dhārī." And the Māyāvādīs they say kalpanāyā. The Akhandananda was saying. And Śukadeva Gosvāmī is narrating something kalpanāyā and wasting his time?
Bhakti-prema: (indistinct)
Prabhupāda: This is acintya for the Māyāvādīs. They say kalpanāyā. These Akhandananda and other Māyāvādīs, they explain Bhāgavata-kalpanāyā. They are making some imagination that "I am God," but they are alleging us, that "You are in illusion." God, as soon as there is some toothache, he goes to the doctor. He's such a rascal hypocrite. There was some heart attack going on, so Akhandananda, immediately he called one of his chief disciples, that Mishra, Jasri, and he was taken to Bombay hospital. And he's God.
Bhakti-prema: He's expired?
Prabhupāda: Expired? No, no. He's living. But talking all nonsense, reading Bhāgavatam, and hundreds of people go to the...
Bhakti-prema: Today we were discussing...
Prabhupāda: That difference of opinion will continue. You cannot stop.
Bhakti-prema: This Pacific Ocean is saltwater...
Prabhupāda: Why you are trying to adjust Pacific Ocean? Pacific Ocean, any ocean, it is just like kūpa-maṇḍūka. It is very big for you, but you are a very teeny identity. Take the universe. What is the Pacific? Is it not a drop. There are so many Pacific Oceans floating in the sky. Everything is acintya. Yasyaika-niśvasita-kālam athāvalambya jīvanti loma-vilajā jagad-aṇḍa-nāthāḥ. This is acintya. So acintyāḥ khalu ye bhāvā na tāṁs tarkeṇa yojayet. You cannot make an experiment or see it. Take some information from the authority and be satisfied. With your limited knowledge, if you want to bring it to experiment, that is not possible. That is not possible.
Bhakti-prema: If this is the background I can give a lot of material. But they want logic.
Prabhupāda: Where is the logic?
Yaśodā-nandana: I don't want any logic. The scientists will come to the planetarium and ask...
Prabhupāda: How they can? Scientist is rascal. That is proved. They are insisting that chemical can produce life. He's a rascal. They have gone to moon planet. That's a rascal. So what is the value of so-called scientist. Why should we give any importance? I'm not giving any importance. If you become scientist, that so much ghee and so much āṭā makes puri, and we can eat very nicely, all right, you are a scientist. But so much chemicals, make it life—prove that. The confectioner is also scientist. He knows very well how

to do his business. A carpenter is also scientist. Here is some work nice done by the carpenter. I cannot do it. You may be a great scientist, but me? It is not possible for me to do a carpentry work. In this way it is going on. (Bengali) You have learned something, you can do it very nicely. But I cannot do it. For me it will be beating by the rod if I am given this work. I can translate, my work. So everyone is scientist, his own field of activities, to some extent. You cannot make everyone agree. That is not... Vox populi you cannot. That is not possible. What is that vox populi?
Tamāla Kṛṣṇa: General population.
Prabhupāda: Mohitaṁ nābhijānāti mām ebhyaḥ param avyayam. So what is the value of vox populi? I concluded. Munayāḥ sādhu te 'ham. Huh? What is that? Munayāḥ sādhu pṛṣṭo 'ham. Loka-maṅgalam. Yenātmā suprasīdati. So Kṛṣṇa's childhood pastime, enjoyed by mother Yaśodā, others will think, "What is this nonsense? Kṛṣṇa is breaking the butterpot and it is taken as sublime?" They'll think like that.
Bhakti-prema: Kṛṣṇa opened His mouth, His mother Yaśodā saw all this universe.
Prabhupāda: And first of all she became surprised. And next moment, "Whatever it is, my dear Kṛṣṇa, You come on my lap." So it is not for all.
Bhakti-prema: Therefore we should name it Esoteric Geography.
Prabhupāda: Journal?
Bhakti-prema: Esoteric.
Prabhupāda: Yes, esoteric I know. Journal?
Bhakti-prema: Geography.
Prabhupāda: Yes. Esoteric and exoteric. But one thing is that it is folly to be wise where ignorance is bliss. If you think that ignorance is bliss, then why should we waste our time and money?
Yaśodā-nandana: Prabhupāda, what is the shape of this tiny portion of earth or whatever place we are on? What is the shape of this, whatever you call...
Prabhupāda: Ask them. Why don't you ask them? Sometimes they say flat, sometimes they say it is round. Why don't you ask them, the scientists?
Yaśodā-nandana: We don't accept what they say.
Prabhupāda: No, no. Formerly they were under the impression the world is flat. And now they are saying round. So what they'll say after few years?
Yaśodā-nandana: They are not consistent. That's a fact. They're very inconsistent in their theories.
Prabhupāda: Ask them which is correct. "Probably" this is correct.
Bhakti-prema: When someone asks this question, first one would reply.
Prabhupāda: I answered it. You people say like that, so which is correct? Flat or round?

Tamāla Kṛṣṇa: They will say "What does the Bhāgavatam say?"
Prabhupāda: Huh?
Tamāla Kṛṣṇa: They may reply to us, "All right, we are rascals. So please tell us what is the fact."
Prabhupāda: That is, we are...
Tamāla Kṛṣṇa: That's what he's asking. What should be shown? Actually we're a little stumped by... I mean...
Prabhupāda: Yes, you must have proper answer as far as possible.
Tamāla Kṛṣṇa: We just read... We got a version from South India, and we've even found that there are different conceptions of what the Bhāgavata is saying. But the Purāṇas, they give some Puranic references.
Bhakti-prema: It is written the world... The earth is round and flat.
Prabhupāda: Hm?
Bhakti-prema: Earth is round and flat both, together.
Prabhupāda: Yes.
Bhakti-prema: First we should reply it is acintya. This should be the reply. "Inconceivable."
Yaśodā-nandana: If it is inconceivable, then they will say how we can conceive it?
Prabhupāda: Take the version of Bhāgavatam.
Bhakti-prema: Everything we conceived, that is wrong?
Prabhupāda: Everything you conceived, that is wrong. Yes. Therefore inconceivable.
Bhakti-prema: The Lord is inconceivable always and any (indistinct), it is inconceivable.
Prabhupāda: But we have to accept śāstra.
Bhakti-prema: This outer structure of the Lord is one with the Lord. It is inconceivable; it is not conceivable.
Prabhupada: Acintya-guṇa-svarūpam. Acintya-guṇa-svarūpam. Yaṁ śyāmasundaram acintya-guṇa. Acintya. Govindam ādi-puruṣaṁ tam ahaṁ bhajāmi. Acintya.
Bhakti-prema: (indistinct)
Prabhupāda: Give me that pineapple juice. [break]
Bhakti-prema: If it is inconceivable, then don't try to...
Prabhupāda: We are not lying to you.
Bhakti-prema: ...draw.
Tamāla Kṛṣṇa: We're drawing according to the Bhāgavatam.
Prabhupāda: We're not conceiving it. It is already there.
Bhakti-prema: Then it is conceived, he says.
Prabhupāda: Conceived not by me.

Tamāla Kṛṣṇa: But not by us. By the Unlimited.
Prabhupāda: I am imperfect. That is the difference.
Bhakti-prema: Yes. It is conceivable by you.
Tamāla Kṛṣṇa: By Kṛṣṇa.
Prabhupāda: By Śukadeva... Not even Śukadeva Gosvāmī. He says śuśruma.
Tamāla Kṛṣṇa: "I heard."
Prabhupāda: Janmādy asya yataḥ anvayād itarataś cārtheṣu abhijñaḥ svarāṭ tene brahma hṛdā. Brahma hṛdā. This is Brahmā. Ādi-kavaye. The Brahmā is very significant. Ādi-kavaye. So it is coming that way.
Bhakti-prema: (indistinct)
Prabhupāda: Yes.
Bhakti-prema: Scholars
Prabhupāda: This is their... Apauruṣeyam. Actually India's culture is going on that way. Mass of people, they are going to Prayāga for taking bath. What do they know? They have received it from authorities that if you take bath in such and such place... Ah, lakhs of people will go. That is India's culture. Without any advertisement, without any means, walking hundreds of miles they are coming, yes, that is their culture. And the government is perturbed that people are so prejudiced. So how to make them forget? This is going on. But they don't listen. They just, "If I take bath I'll..." That is the difference between Western and Eastern. And as soon as there is interpretation, it is Māyāvāda. And Caitanya Mahāprabhu has rejected—māyāvādi-bhāṣya śunile haya sarva-nāśa. You see in the Kumbhamela how peacefully they are sitting. They are accepting Vedic culture. So nice atmosphere. Simply by going there you'll be satisfied. That is the difference between East and West.

manuṣyāṇāṁ sahasreṣu
kaścid yatati siddhaye
yatatām api siddhānāṁ
kaścin māṁ vetti...

Therefore tad viddhi praṇipātena paripraśnena sevayā. There is no other. Submissive. [break] Guru. Tasyaitc kathitā hy arthāḥ prakāśante. The author is revealed to him. Yasya deve parā bhaktir yathā deve. Otherwise not. So do it as far as possible to your capacity. But things are inconceivable. You cannot adjust within the limitation of your understanding. That is not...
Tamāla Kṛṣṇa: That's... We're trying to impose some preconceived idea onto this. As soon as that happens, this knowledge is blocked. Because the

whole attitude shouldn't be like that. One should come out of service and devotion, not with some mental, materialistic speculations.
Prabhupāda: Did you know that story, the Nārada was going to Vaikuṇṭha? Nārada came back and replied to a cobbler... Cobbler asked him what Nārāyaṇa is doing. "He has taken one elephant and He's drawing through the hole of a needle like this and again taking." The learned brāhmaṇa, he began to laugh. "These are all stories." And the cobbler began to cry, "Oh, Nārāyaṇa, Kṛṣṇa, can..." Nārada inquired, "How do you believe that elephant is being drawn through the hole of needle?" "No, why not? I'm daily seeing by sitting under this banyan tree, and within a fruit there are thousands of seeds. And each seed contains the big tree." Can the scientists make such small seed contain a big banyan tree? So it is acintya. That's a fact. [break] ...thing is inconceivable. And these rascals want to bring them as conceivable. He's conditioned, and he's trying to bring inconceivable thing to his conception. Useless, futile attempt. How the scientist will answer? We take a fruit. There are hundreds of seeds, and each seed contains a big tree. How you can explain? Is it not inconceivable?
Tamāla Kṛṣṇa: Yes.
Prabhupāda: So what is the use of arguing?
Tamāla Kṛṣṇa: It's better to take the fruit and offer it to Kṛṣṇa.
Prabhupāda: That's right. We take it as accepted, mahā-muni kṛte. Dharmaḥ projjhita-kaitavo 'tra paramo nirmatsarāṇāṁ satām.
Bhakti-prema: Vedyaṁ vāstavam atra vastu śivadaṁ tāpa-trayonmūlanam, śrīmad-bhāgavate mahā-muni-kṛte kiṁ vā parair īśvaraḥ sadyo hṛdy avarudhyate 'tra kṛtibhiḥ śuśrūṣubhis tat-kṣaṇāt.
Prabhupāda: Śuśrūṣubhis tat-kṣaṇāt. There is no other way.
Tamāla Kṛṣṇa: This is a very good lesson, Śrīla Prabhupāda.
Prabhupāda: You do not require to be puffed-up with your so-called education. It has no value. *(July 5, 1977, Vṛndāvana)*

# 11

# Fifth Canto Cosmology

## *(Śrīmad Bhāgavatam )*

### Part I: Bhū-Maṇḍala Created

5.1.30: yāvad avabhāsayati sura-girim anuparikrāman bhagavān ādityo vasudhā-talam ardhenaiva pratapaty ardhenāvacchādayati tadā hi bhagavad-upāsanopacitāti-puruṣa-prabhāvas tad anabhinandan samajavena rathena jyotirmayena rajanīm api dinaṁ kariṣyāmīti sapta-kṛt vastaraṇim anuparyakrāmad dvitīya iva pataṅgaḥ.

SYNONYMS: yāvat—so long; avabhāsayati—illuminates; sura-girim—the Sumeru Hill; anuparikrāman—by circumambulating; bhagavān—the most powerful; ādityaḥ—sun-god; vasudhā-talam—the lower planetary system; ardhena—by half; eva—certainly; pratapati—makes dazzling; ardhena—by half; avacchādayati—covers with darkness; tadā—at that time; hi—certainly; bhagavat-upāsanā—by worshiping the Supreme Personality of Godhead; upacita—by satisfying Him perfectly; ati-puruṣa—superhuman; prabhāvaḥ—influence; tat—that; anabhinandan—without appreciating; samajavena—by equally powerful; rathena—on a chariot; jyotiḥ-mayena—dazzling; rajanīm—night; api—also; dinam—day; kariṣyāmi—I shall make it; iti—thus; sapta-kṛt—seven times; vastaraṇim—exactly following the orbit of the sun; anuparyakrāmat—circumambulated; dvitīyaḥ—second; iva—like; pataṅgaḥ—sun.

TRANSLATION: While so excellently ruling the universe, King Priyavrata once became dissatisfied with the circumambulation of the most powerful sun-god. Encircling Sumeru Hill on his chariot, the sun-god illuminates all the surrounding planetary systems. However, when the sun is on the northern side of the hill, the south receives less light, and when the sun is in the south, the north receives less.

King Priyavrata disliked this situation and therefore decided to make daylight in the part of the universe where there was night. He followed the orbit of the sun-god on a brilliant chariot and thus fulfilled his desire. He could perform such wonderful activities because of the power he had achieved by worshiping the Supreme Personality of Godhead.

PURPORT: From the description in this verse, it appears that the sun moves. According to modern astronomers, the sun is fixed in one place, surrounded by the solar system, but here we find that the sun is not stationary: it is rotating in a prescribed orbit. This fact is corroborated by Brahma-saṁhitā (5.52). Yasyājñayā bhramati saṁbhṛta-kāla-cakraḥ: the sun is rotating in its fixed orbit in accordance with the order of the Supreme Personality of Godhead. According to Jyotir Veda, the science of astronomy in the Vedic literature, the sun moves for six months on the northern side of the Sumeru Hill and for six months on the southern side. We have practical experience on this planet that when there is summer in the north there is winter in the south and vice versa. Modern materialistic scientists sometimes present themselves as knowing all the ingredients of the sun, yet they are unable to offer a second sun like Mahārāja Priyavrata's.

Śrīla Viśvanātha Cakravartī Ṭhākura remarks that in the months of April and May the rays of Mahārāja Priyavrata's brilliant sun were as pleasing as the rays of the moon, and in October and November, both morning and evening, that sun provided more warmth than the sunshine. In short, Mahārāja Priyavrata was extremely powerful, and his actions extended his power in all directions.

5.1.31: ye vā u ha tad-ratha-caraṇa-nemi-kṛta-parikhātās te sapta sindhava āsan
yata eva kṛtāḥ sapta bhuvo dvīpāḥ.

SYNONYMS: ye—that; vā u ha—certainly; tat-ratha—of his chariot; caraṇa—of the wheels; nemi—by the rims; kṛta—made; parikhātāḥ—trenches; te—those; sapta—seven; sindhavaḥ—oceans; āsan—became; yataḥ—because of which; eva—certainly; kṛtāḥ—were made; sapta—seven; bhuvaḥ—of the Bhū-maṇḍala; dvīpāḥ—islands.

TRANSLATION: When Priyavrata drove his chariot behind the sun, the rims of his chariot wheels created impressions that later became seven oceans, dividing the planetary system known as Bhū-maṇḍala into seven islands.

PURPORT: Sometimes the planets in outer space are called islands. We have experience of various types of islands in the ocean, and similarly the various planets, divided into fourteen lokas, are islands in the ocean of space. As Priyavrata drove his chariot behind the sun, he created seven different types of oceans and planetary systems, which altogether are known as Bhū-maṇḍala, or Bhūloka. In the Gāyatrī mantra, we chant, oṁ bhūr bhuvaḥ svaḥ tat savitur vareṇyam. Above the Bhūloka planetary system is Bhuvarloka, and above that is Svargaloka, the heavenly planetary system. All these planetary systems are controlled by Savitā, the sun-god. By chanting the Gāyatrī mantra just after rising early in the morning, one worships the sun-god.

5.1.32: jambū-plakṣa-śālmali-kuśa-krauñca-śāka-puṣkara-saṁjñās teṣāṁ parimāṇaṁ pūrvasmāt pūrvasmād uttara uttaro yathā-saṅkhyaṁ dvi-guṇa-mānena bahiḥ samantata upakḷptāḥ.

SYNONYMS: jambū—Jambū; plakṣa—Plakṣa; śālmali—Śālmali; kuśa—Kuśa; krauñca—Krauñca; śāka—Śāka; puṣkara—Puṣkara; saṁjñāḥ—known as; teṣām—of them; parimāṇam—measurement; pūrvasmāt pūrvasmāt—from the former; uttaraḥ uttaraḥ—the following; yathā—according to; saṅkhyam—number; dvi-guṇa—twice as much; mānena—with a measure; bahiḥ—outside; samantataḥ—all around; upakḷptāḥ—produced.

TRANSLATION: The names of the islands are Jambū, Plakṣa, Śālmali, Kuśa, Krauñca, Śāka and Puṣkara. Each island is twice as large as the one preceding it, and each is surrounded by a liquid substance, beyond which is the next island.

PURPORT: The ocean in each planetary system has a different type of liquid. How they are situated is explained in the next verse.

5.1.33: kṣārodekṣu-rasoda-suroda-ghṛtoda-kṣīroda-dadhi-maṇḍoda-śuddhodāḥ sapta jaladhayaḥ sapta dvīpa-parikhā ivābhyantara-dvīpa-samānā ekaikaśyena yathānupūrvaṁ saptasv api bahir dvīpeṣu pṛthak parita upakalpitās teṣu jambv-ādiṣu barhiṣmatī-patir anuvratānātmajān āgnīdhredhmajihva-yajñabāhu-hiraṇyareto-ghṛtapṛṣṭha-medhātithi-vītihotra-saṁjñān yathā-saṅkhyenaikaikasminn ekam evādhi-patiṁ vidadhe.

SYNONYMS: kṣāra—salt; uda—water; ikṣu-rasa—the liquid extract from sugarcane; uda—water; surā—liquor; uda—water; ghṛta—clarified butter; uda—water; kṣīra—milk; uda—water; dadhi-maṇḍa—emulsified yogurt; uda—water; śuddha-udāḥ—and drinking water; sapta—seven; jala-dhayaḥ—oceans; sapta—seven; dvīpa—islands; parikhāḥ—trenches; iva—like; abhyantara—internal; dvīpa—islands; samānāḥ—equal to; eka-ekaśyena—one after another; yathā-anupūrvam—in chronological order; saptasu—seven; api—although; bahiḥ—outside; dvīpeṣu—in islands; pṛthak—separate; paritaḥ—all around; upakalpitāḥ—situated; teṣu—within them; jambū-ādiṣu—beginning with Jambū; barhiṣmatī—of Barhiṣmatī; patiḥ—the husband; anuvratān—who were actually followers of the father's principles; ātma-jān—sons; āgnīdhra-idhmajihva-yajñabāhu-hiraṇyaretaḥ-ghṛtapṛṣṭha-medhātithi-vītihotra-saṁjñān—named Āgnīdhra, Idhmajihva, Yajñabāhu, Hiraṇyaretā, Ghṛtapṛṣṭha, Medhātithi and Vītihotra;

yathā-saṅkhyena—by the same number; eka-ekasmin—in each island; ekam—one; eva—certainly; adhi-patim—king; vidadhe—he made.

TRANSLATION: The seven oceans respectively contain salt water, sugarcane juice, liquor, clarified butter, milk, emulsified yogurt, and sweet drinking water. All the islands are completely surrounded by these oceans, and each ocean is equal in breadth to the island it surrounds. Mahārāja Priyavrata, the husband of Queen Barhiṣmatī, gave sovereignty over these islands to his respective sons, namely Āgnīdhra, Idhmajihva, Yajñabāhu, Hiraṇyaretā, Ghṛtapṛṣṭha, Medhātithi and Vītihotra. Thus they all became kings by the order of their father.

PURPORT: It is to be understood that all the dvīpas, or islands, are surrounded by different types of oceans, and it is said herein that the breadth of each ocean is the same as that of the island it surrounds. The length of the oceans, however, cannot equal the length of the islands. According to Vīrarāghava Ācārya, the breadth of the first island is 100,000 yojanas. One yojana equals eight miles, and therefore the breadth of the first island is calculated to be 800,000 miles. The water surrounding it must have the same breadth, but its length must be different.

5.1.39: tasya ha vā ete ślokāḥ——/priyavrata-kṛtaṁ karma/ko nu kuryād vineśvaram/yo nemi-nimnair akaroc/chāyāṁ ghnan sapta vāridhīn

SYNONYMS: tasya—his; ha vā—certainly; ete—all these; ślokāḥ—verses; priyavrata—by King Priyavrata; kṛtam—done; karma—activities; kaḥ—who; nu—then; kuryāt—can execute; vinā—without; īśvaram—the Supreme Personality of Godhead; yaḥ—one who; nemi—of the rim of the wheels of his chariot; nimnaiḥ—by the depressions; akarot—made; chāyām—darkness; ghnan—dissipating; sapta—seven; vāridhīn—oceans.

TRANSLATION: There are many famous verses regarding Mahārāja Priyavrata's activities:
"No one but the Supreme Personality of Godhead could do what Mahārāja Priyavrata has done. Mahārāja Priyavrata dissipated the darkness of night, and with the rims of his great chariot, he excavated seven oceans."

5.1.40: bhū-saṁsthānaṁ kṛtaṁ yena/sarid-giri-vanādibhiḥ/sīmā ca bhūta-nirvṛtyai/dvīpe dvīpe vibhāgaśaḥ

SYNONYMS: bhū-saṁsthānam—the situation of the earth; kṛtam—done; yena—by whom; sarit—by rivers; giri—by hills and mountains; vana-ādibhiḥ—by forests and so on; sīmā—boundaries; ca—also; bhūta—of different nations; nirvṛtyai—to stop fighting; dvīpe dvīpe—on the various islands; vibhāgaśaḥ—separately.

TRANSLATION: "To stop the quarreling among different peoples, Mahārāja Priyavrata marked boundaries at rivers and at the edges of mountains and forests so that no one would trespass upon another's property."

PURPORT: The example set by Mahārāja Priyavrata in marking off different states is still followed. As indicated here, different classes of men are destined to live in different areas, and therefore the boundaries of various tracts of land, which are described here as islands, should be defined by different rivers, forests and hills...Mahārāja Priyavrata divided the surface of the globe into different islands so that each class of men would live peacefully and not clash with the others. The modern idea of nationhood has gradually developed from the divisions made by Mahārāja Priyavrata.

## Part II: Ruling Jambūdvīpa

5.2.1: śrī-śuka uvāca/ evaṁ pitari sampravṛtte tad-anuśāsane vartamāna āgnīdhro jambūdvīpaukasaḥ prajā aurasavad dharmāvekṣamāṇaḥ paryagopāyat.

SYNONYMS: śrī-śukaḥ—Śrī Śukadeva Gosvāmī; uvāca—said; evam—thus; pitari—when his father; sampravṛtte—took to the path of liberation; tat-anuśāsane—according to his order; vartamānaḥ—situated; āgnīdhraḥ—King Āgnīdhra; jambū-dvīpa-okasaḥ—the inhabitants of Jambūdvīpa; prajāḥ—citizens; aurasa-vat—as if they were his sons; dharma—religious principles; avekṣamāṇaḥ—strictly observing; paryagopāyat—completely protected.

TRANSLATION: Śrī Śukadeva Gosvāmī continued: After his father, Mahārāja Priyavrata, departed to follow the path of spiritual life by undergoing austerities, King Āgnīdhra completely obeyed his order. Strictly observing the principles of religion, he gave full protection to the inhabitants of Jambūdvīpa as if they were his own begotten sons.

5.2.19: tasyām u ha vā ātmajān sa rāja-vara āgnīdhro nābhi-kimpuruṣa-harivarṣelāvṛta-ramyaka-hiraṇmaya-kuru-bhadrāśva-ketumāla-saṁjñān nava putrān ajanayat.

SYNONYMS: tasyām—in her; u ha vā—certainly; ātma-jān—sons; saḥ—he; rāja-varaḥ—the best of kings; āgnīdhraḥ—Āgnīdhra; nābhi—Nābhi; kiṁpuruṣa—Kiṁpuruṣa; hari-varṣa—Harivarṣa; ilāvṛta—Ilāvṛta; ramyaka—Ramyaka; hiraṇmaya—Hiraṇmaya; kuru—Kuru; bhadrāśva—Bhadrāśva; ketu-māla—Ketumāla; saṁjñān—named; nava—nine; putrān—sons; ajanayat—begot.

TRANSLATION: In the womb of Pūrvacitti, Mahārāja Āgnīdhra, the best of kings, begot nine sons, named Nābhi, Kiṁpuruṣa, Harivarṣa, Ilāvṛta, Ramyaka, Hiraṇmaya, Kuru, Bhadrāśva and Ketumāla.

5.2.21: āgnīdhra-sutās te mātur anugrahād autpattikenaiva saṁhanana-balopetāḥ pitrā vibhaktā ātma-tulya-nāmāni yathā-bhāgaṁ jambūdvīpa-varṣāṇi bubhujuḥ.

SYNONYMS: āgnīdhra-sutāḥ—the sons of Mahārāja Āgnīdhra; te—they; mātuḥ—of the mother; anugrahāt—by the mercy or by drinking the breast milk; autpattikena—naturally; eva—certainly; saṁhanana—well-built body; bala—strength; upetāḥ—obtained; pitrā—by the father; vibhaktāḥ—divided; ātma-tulya—following their own; nāmāni—possessing names; yathā-bhāgam—divided properly; jambūdvīpa-varṣāṇi—different parts of Jambūdvīpa (probably Asia and Europe combined together); bubhujuḥ—ruled.

TRANSLATION: Because of drinking the breast milk of their mother, the nine sons of Āgnīdhra naturally had strong, well-built bodies. Their father gave them each a kingdom in a different part of Jambūdvīpa. The kingdoms were named according to the names of the sons. Thus the sons of Āgnīdhra ruled the kingdoms they received from their father.

5.4.5: viditānurāgam āpaura-prakṛti jana-pado rājā nābhir ātmajaṁ samaya-setu-rakṣāyām abhiṣicya brāhmaṇeṣūpanidhāya saha merudevyā viśālāyāṁ prasanna-nipuṇena tapasā samādhi-yogena nara-nārāyaṇākhyaṁ bhagavantaṁ vāsudevam upāsīnaḥ kālena tan-mahimānam avāpa.

SYNONYMS: vidita—known very well; anurāgam—popularity; āpaura-prakṛti—among all the citizens and government officers; jana-padaḥ—desiring to serve the people in general; rājā—the King; nābhiḥ—Nābhi; ātmajam—his son; samaya-setu-rakṣāyām—to protect the people strictly according to the Vedic principles of religious life; abhiṣicya—enthroning; brāhmaṇeṣu—to the learned brāhmaṇas; upanidhāya—entrusting; saha—with; merudevyā—his wife. Merudevī; viśālāyām—in Badarikāśrama; prasanna-nipuṇena—performed with great satisfaction and expertise; tapasā—by austerities and penances; samādhi-yogena—by full samādhi; nara-nārāyaṇa-ākhyam—named Nara-Nārāyaṇa; bhagavantam—the Supreme Personality of Godhead; vāsudevam—Kṛṣṇa; upāsīnaḥ—worshiping; kālena—in due course of time; tat-mahimānam—His glorious abode, the spiritual world, Vaikuṇṭha; avāpa—achieved.

TRANSLATION: King Nābhi understood that his son, Ṛṣabhadeva, was very popular among the citizens and among government officers and ministers. Understanding the popularity of his son, Mahārāja Nābhi enthroned Him as the emperor of the world to give protection to the general populace in terms of the Vedic religious system. To do this, he entrusted Him into the hands of learned brāhmaṇas, who would guide Him in administrating the government. Then Mahārāja Nābhi and his wife, Merudevī, went to Badarikāśrama in the Himalaya Mountains, where the King engaged Himself very expertly in austerities and penances with great jubilation. In full samādhi he worshiped the Supreme Personality of Godhead, Nara-Nārāyaṇa, who is Kṛṣṇa in His plenary expansion. By doing so, in course of time Mahārāja Nābhi was elevated to the spiritual world known as Vaikuṇṭha.

5.4.9: yeṣāṁ khalu mahā-yogī bharato jyeṣṭhaḥ śreṣṭha-guṇa āsīd yenedaṁ varṣaṁ bhāratam iti vyapadiśanti.

SYNONYMS: yeṣām—of whom; khalu—indeed; mahā-yogī—a very highly exalted devotee of the Lord; bharataḥ—Bharata; jyeṣṭhaḥ—the eldest; śreṣṭha-guṇaḥ—qualified with the best attributes; āsīt—was; yena—by whom; idam—this; varṣam—planet; bhāratam—Bhārata; iti—thus; vyapadiśanti—people call.

TRANSLATION: Of Ṛṣabhadeva's one hundred sons, the eldest, named Bharata, was a great, exalted devotee qualified with the best attributes. In his honor, this planet has become known as Bhārata-varṣa.

PURPORT: This planet known as Bhārata-varṣa is also called puṇya-bhūmi, the pious land. At the present moment Bhārata-bhūmi, or Bhārata-varṣa, is a small piece of land extending from the Himalaya Mountains to Cape Comorin. Sometimes this peninsula is called puṇya-bhūmi. Śrī Caitanya Mahāprabhu has given special importance to the people of this land. bhārata-bhūmite haila manuṣya-janma yāra janma sārthaka kari' kara para-upakāra "One who has taken his birth as a human being in the land of India (Bhārata-varṣa) should make his life successful and work for the benefit of all other people." (Cc. Ādi 9.41) The inhabitants of this piece of land are very fortunate. They can purify their existence by accepting this Kṛṣṇa consciousness movement and go outside Bhārata-bhūmi (India) and preach this cult to benefit the whole world.

5.6.13: tasyānuguṇān ślokān gāyanti——/aho bhuvaḥ sapta-samudravatyā dvīpeṣu varṣeṣv adhipuṇyam etat/gāyanti yatratya-janā murāreḥ karmāṇi bhadrāṇy avatāravanti

SYNONYMS: tasya—of Him (Lord Ṛṣabhadeva); anuguṇān—conforming to the instructions for liberation; ślokān—verses; gāyanti—chant; aho—oh; bhuvaḥ—of this earthly planet; sapta-samudra-vatyāḥ—possessing seven seas; dvīpeṣu—among the islands; varṣeṣu—among the lands; adhipuṇyam—more pious than any other island; etat—this (Bhārata-varṣa); gāyanti—sing about; yatratya-janāḥ—the people of this tract of land; murāreḥ—of Murāri, the Supreme Personality of

Godhead; karmāṇi—the activities; bhadrāṇi—all-auspicious; avatāravanti—in many incarnations such as Lord Ṛṣabhadeva.

TRANSLATION: Learned scholars chant about the transcendental qualities of Lord Ṛṣabhadeva in this way: "Oh, this earthly planet contains seven seas and many islands and lands, of which Bhārata-varṣa is considered the most pious. People of Bhārata-varṣa are accustomed to glorifying the activities of the Supreme Personality of Godhead in His incarnations as Lord Ṛṣabhadeva and others. All these activities are very auspicious for the welfare of humanity.

5.7.1: śrī-śuka uvāca/ bharatas tu mahā-bhāgavato yadā bhagavatāvani-tala-paripālanāya sañcintitas tad-anuśāsana-paraḥ pañcajanīṁ viśvarūpa-duhitaram upayeme.

SYNONYMS: śrī-śukaḥ uvāca—Śukadeva Gosvāmī said; bharataḥ—Mahārāja Bharata; tu—but; mahā-bhāgavataḥ—a mahā-bhāgavata, most exalted devotee of the Lord; yadā—when; bhagavatā—by the order of his father, Lord Ṛṣabhadeva; avani-tala—the surface of the globe; paripālanāya—for ruling over; sañcintitaḥ—made up his mind; tat-anuśāsana-paraḥ—engaged in governing the globe; pañcajanīm—Pañcajanī; viśvarūpa-duhitaram—the daughter of Viśvarūpa; upayeme—married.

TRANSLATION: Śukadeva Gosvāmī continued speaking to Mahārāja Parīkṣit: My dear King, Bharata Mahārāja was a topmost devotee. Following the orders of his father, who had already decided to install him on the throne, he began to rule the earth accordingly. When Bharata Mahārāja ruled the entire globe, he followed the orders of his father and married Pañcajanī, the daughter of Viśvarūpa.

5.7.3: ajanābhaṁ nāmaitad varṣaṁ bhāratam iti yata ārabhya vyapadiśanti.

SYNONYMS: ajanābham—Ajanābha; nāma—by the name; etat—this; varṣam—island; bhāratam—Bhārata; iti—thus; yataḥ—from whom; ārabhya—beginning; vyapadiśanti—they celebrate.

TRANSLATION: Formerly this planet was known as Ajanābha-varṣa, but since Mahārāja Bharata's reign, it has become known as Bhārata-varṣa.

PURPORT: This planet was formerly known as Ajanābha because of the reign of King Nābhi. After Bharata Mahārāja ruled the planet, it became celebrated as Bhārata-varṣa.

## Part III: Characteristics of Bhū-Maṇḍala

5.16.1: rājovāca/ uktas tvayā bhū-maṇḍalāyāma-viśeṣo yāvad ādityas tapati yatra cāsau jyotiṣāṁ gaṇaiś candramā vā saha dṛśyate.

SYNONYMS: rājā uvāca—Mahārāja Parīkṣit said; uktaḥ—already been said; tvayā—by you; bhū-maṇḍala—of the planetary system known as Bhū-maṇḍala; āyāma-viśeṣaḥ—the specific length of the radius; yāvat—as far as; ādityaḥ—the sun; tapati—heats; yatra—wherever; ca—also; asau—that; jyotiṣām—of the luminaries; gaṇaiḥ—with hordes; candramā—the moon; vā—either; saha—with; dṛśyate—is seen.

TRANSLATION: King Parīkṣit said to Śukadeva Gosvāmī: O brāhmaṇa, you have already informed me that the radius of Bhū-maṇḍala extends as far as the sun spreads its light and heat and as far as the moon and all the stars can be seen.

PURPORT: In this verse it is stated that the planetary system known as Bhū-maṇḍala extends to the limits of the sunshine. According to modern science, the sunshine reaches earth from a distance of 93,000,000 miles. If we calculate according to this modern information, 93,000,000 miles can be considered the radius of Bhū-maṇḍala. In the Gāyatrī mantra, we chant oṁ bhūr bhuvaḥ svaḥ. The word bhūr refers to Bhū-maṇḍala. Tat savitur vareṇyam: the sunshine spreads throughout Bhū-maṇḍala. Therefore the sun is worshipable. The stars, which are known as nakṣatra, are not different suns, as modern astronomers suppose. From Bhagavad-gītā (10.21) we understand that the stars are similar to the moon (nakṣatrāṇām ahaṁ śaśī). Like the moon, the stars reflect the sunshine. Apart from our modern distinguished estimations of where the planetary systems are located, we can understand that the sky and its various planets were studied long, long before Śrīmad-Bhāgavatam was compiled. Śukadeva Gosvāmī explained the location of the planets, and this indicates that the information was known long, long before Śukadeva Gosvāmī related it to Mahārāja Parīkṣit. The location of the various planetary systems was not unknown to the sages who flourished in the Vedic age.

5.16.2: tatrāpi priyavrata-ratha-caraṇa-parikhātaiḥ saptabhiḥ sapta sindhava upakḷptā yata etasyāḥ sapta-dvīpa-viśeṣa-vikalpas tvayā bhagavan khalu sūcita etad evākhilam ahaṁ mānato lakṣaṇataś ca sarvaṁ vi-jijñāsāmi.

SYNONYMS: tatra api—in that Bhū-maṇḍala; priyavrata-ratha-caraṇa-parikhātaiḥ—by the ditches made by the wheels of the chariot used by Priyavrata Mahārāja while circumambulating Sumeru behind the sun; saptabhiḥ—by the seven; sapta—seven; sindhavaḥ—oceans; upakḷptāḥ—created; yataḥ—because of which; etasyāḥ—of this Bhū-maṇḍala; sapta-dvīpa—of the seven islands; viśeṣa-vikalpaḥ—the mode of the construction; tvayā—by you; bhagavan—O great saint;

khalu—indeed; sūcitaḥ—described; etat—this; eva—certainly; akhilam—whole subject; aham—I; mānataḥ—from the point of view of measurement; lakṣaṇataḥ—and from symptoms; ca—also; sarvam—everything; vijijñāsāmi—wish to know.

TRANSLATION: My dear Lord, the rolling wheels of Mahārāja Priyavrata's chariot created seven ditches, in which the seven oceans came into existence. Because of these seven oceans, Bhū-maṇḍala is divided into seven islands. You have given a very general description of their measurement, names and characteristics. Now I wish to know of them in detail. Kindly fulfill my desire.

5.16.3: bhagavato guṇamaye sthūla-rūpa āveśitaṁ mano hy aguṇe 'pi sūkṣmatama ātma-jyotiṣi pare brahmaṇi bhagavati vāsudevākhye kṣamam āveśituṁ tad u haitad guro 'rhasy anuvarṇayitum iti.

SYNONYMS: bhagavataḥ—of the Supreme Personality of Godhead; guṇa-maye—into the external features, consisting of the three modes of material nature; sthūla-rūpe—the gross form; āveśitam—entered; manaḥ—the mind; hi—indeed; aguṇe—transcendental; api—although; sūkṣmatame—in His smaller form as Paramātmā within the heart; ātma-jyotiṣi—who is full of Brahman effulgence; pare—the supreme; brahmaṇi—spiritual entity; bhagavati—the Supreme Personality of Godhead; vāsudeva-ākhye—known as Bhagavān Vāsudeva; kṣamam—suitable; āveśitum—to absorb; tat—that; u ha—indeed; etat—this; guro—O my dear spiritual master; arhasi anuvarṇayitum—please describe factually; iti—thus.

TRANSLATION: When the mind is fixed upon the Supreme Personality of Godhead in His external feature made of the material modes of nature—the gross universal form—it is brought to the platform of pure goodness. In that transcendental position, one can understand the Supreme Personality of Godhead, Vāsudeva, who in His subtler form is self-effulgent and beyond the modes of nature. O my lord, please describe vividly how that form, which covers the entire universe, is perceived.

PURPORT: Mahārāja Parīkṣit had already been advised by his spiritual master, Śukadeva Gosvāmī, to think of the universal form of the Lord, and therefore, following the advice of his spiritual master, he continuously thought of that form. The universal form is certainly material, but because everything is an expansion of the energy of the Supreme Personality of Godhead, ultimately nothing is material. Therefore Parīkṣit Mahārāja's mind was saturated with spiritual consciousness...When Mahārāja Parīkṣit was thinking of the universal form of the Lord, his mind was certainly situated on the transcendental platform. Therefore although he might not have had any reason to be concerned with detailed information of the universe, he was thinking of it in relationship with the Supreme Lord, and therefore such geographical knowledge was not material but transcendental...Elsewhere in Śrīmad-Bhāgavatam (1.5.20) Nārada Muni has said, idaṁ hi viśvaṁ bhagavān ivetaraḥ: the entire universe is also the Supreme Personality of Godhead, although it appears different from Him. Therefore although Parīkṣit Mahārāja had no need for geographical knowledge of this universe, that knowledge was also spiritual and transcendental because he was thinking of the entire universe as an expansion of the energy of the Lord.

5.16.4: ṛṣir uvāca/ na vai mahārāja bhagavato māyā-guṇa-vibhūteḥ kāṣṭhāṁ manasā vacasā vādhigantum alaṁ vibudhāyuṣāpi puruṣas tasmāt prādhān-yenaiva bhū-golaka-viśeṣaṁ nāma-rūpa-māna-lakṣaṇato vyākhyāsyāmaḥ.

SYNONYMS: ṛṣiḥ uvāca—Śrī Śukadeva Gosvāmī continued to speak; na—not; vai—indeed; mahā-rāja—O great King; bhagavataḥ—of the Supreme Personality of Godhead; māyā-guṇa-vibhūteḥ—of the transformation of the qualities of the material energy; kāṣṭhām—the end; manasā—by the mind; vacasā—by words; vā—either; adhigantum—to understand fully; alam—capable; vibudha-āyuṣā—with a duration of life like that of Brahmā; api—even; puruṣaḥ—a person; tasmāt—therefore; prādhānyena—by a general description of the chief places; eva—certainly; bhū-golaka-viśeṣam—the particular description of Bhūloka; nāma-rūpa—names and forms; māna—measurements; lakṣaṇataḥ—according to symptoms; vyākhyāsyāmaḥ—I shall try to explain.

TRANSLATION: The great ṛṣi Śukadeva Gosvāmī said: My dear King, there is no limit to the expansion of the Supreme Personality of Godhead's material energy. This material world is a transformation of the material qualities [sattva-guṇa, rajo-guṇa and tamo-guṇa], yet no one could possibly explain it perfectly, even in a lifetime as long as that of Brahmā No one in the material world is perfect, and an imperfect person could not describe this material universe accurately, even after continued speculation. O King, I shall nevertheless try to explain to you the principal regions, such as Bhū-goloka [Bhūloka], with their names, forms, measurements and various symptoms.

PURPORT: The material world is only one fourth of the Supreme Personality of Godhead's creation, but it is unlimited and impossible for anyone to know or describe, even with the qualification of a life as long as that of Brahmā, who lives for millions and millions of years. Modern scientists and astronomers try to explain the cosmic situation and the vastness of space, and some of them believe that all the glittering stars are different suns. From Bhagavad-gītā, however, we understand that all these stars (nakṣatras) are like the moon, in that they reflect the sunshine. They are not independent luminaries. Bhūloka is explained to be that portion of outer space through which the heat and light of the sun extend. Therefore it is natural to conclude that this universe extends in space as far as we can see and encompasses the glittering stars. Śrīla Śukadeva Gosvāmī admitted that to give full details of this expansive material universe would be impossible, but nevertheless he wanted to give the King as much knowledge as he had received through the paramparā system. We should conclude that if one cannot comprehend the material expansions of the Supreme Personality of Godhead, one certainly cannot estimate the expansiveness of the spiritual world. The Brahma-saṁhitā (5.33) confirms this: advaitam acyutam anādim ananta-rūpam ādyaṁ purāṇa-puruṣaṁ nava-yauvanaṁ ca. The limits of the expansions of Govinda, the Supreme Personality of Godhead, cannot be estimated by anyone, even a person as perfect as Brahmā, not to speak of tiny scientists whose senses and instruments are all imperfect and who cannot give us information of even this one universe. We should therefore be satisfied with the information obtainable from Vedic sources as spoken by authorities like Śukadeva Gosvāmī.

5.16.5: yo vāyaṁ dvīpaḥ kuvalaya-kamala-kośābhyantara-kośo niyuta-yojana-viśālaḥ samavartulo yathā puṣkara-patram.

SYNONYMS: yaḥ—which; vā—either; ayam—this; dvīpaḥ—island; kuvalaya—the Bhūloka; kamala-kośa—of the whorl of a lotus flower; abhyantara—inner; kośaḥ—whorl; niyuta-yojana-viśālaḥ—one million yojanas (eight million miles) wide; samavartulaḥ—equally round, or having a length and breadth of the same measurement; yathā—like; puṣkara-patram—a lotus leaf.

TRANSLATION: The planetary system known as Bhū-maṇḍala resembles a lotus flower, and its seven islands resemble the whorl of that flower. The length and breadth of the island known as Jambūdvīpa, which is situated in the middle of the whorl, are one million yojanas [eight million miles]. Jambūdvīpa is round like the leaf of a lotus flower.

## Part IV: A Description of Jambūdvīpa

5.16.6: yasmin nava varṣāṇi nava-yojana-sahasrāyāmāny aṣṭabhir maryādā-giribhiḥ suvibhaktāni bhavanti.

SYNONYMS: yasmin—in that Jambūdvīpa; nava—nine; varṣāṇi—divisions of land; nava-yojana-sahasra—72,000 miles in length; āyāmāni—measuring; aṣṭabhiḥ—by eight; maryādā—indicating the boundaries; giribhiḥ—by mountains; suvibhaktāni—nicely divided from one another; bhavanti—are.

TRANSLATION: In Jambūdvīpa there are nine divisions of land, each with a length of 9,000 yojanas [72,000 miles]. There are eight mountains that mark the boundaries of these divisions and separate them nicely.

PURPORT: Śrīla Viśvanātha Cakravartī Ṭhākura gives the following quotation from the Vāyu Purāṇa, wherein the locations of the various mountains, beginning with the Himalayas, are described.
dhanurvat saṁsthite jñeye dve varṣe dakṣiṇottare; dīrghāṇi tatra catvāri caturasram ilāvṛtam iti dakṣiṇottare bhāratottara-kuru-varṣe catvāri kiṁpuruṣa-harivarṣa-ramyaka-hiraṇmayāni varṣāṇi nīla-niṣadhayos tiraścinībhūya samudra-praviṣṭayoḥ saṁlagnatvam aṅgīkṛtya bhadrāśva-ketumālayor api dhanur-ākṛtitvam; atas tayor dairghyata eva madhye saṅkucitatvena nava-sahasrāyāmatvam; ilāvṛtasya tu meroḥ sakāśāt catur-dikṣu nava-sahasrāyama-tvaṁ saṁbhavet vastutas tv ilāvṛta-bhadrāśva-ketumālānāṁ catus-triṁśat-sahasrāyāmatvaṁ jñeyam.

5.16.7: eṣāṁ madhye ilāvṛtaṁ nāmābhyantara-varṣaṁ yasya nābhyām avasthitaḥ sarvataḥ sauvarṇaḥ kula-giri-rājo merur dvīpāyāma-samunnāhaḥ karṇikā-bhūtaḥ kuvalaya-kamalasya mūrdhani dvā-triṁśat sahasra-yojana-vitato mūle ṣoḍaśa-sahasraṁ tāvat āntar-bhūmyāṁ praviṣṭaḥ.

SYNONYMS: eṣām—all these divisions of Jambūdvīpa; madhye—among; ilāvṛtam nāma—named Ilāvṛta-varṣa; abhyantara-varṣam—the inner division; yasya—of which; nābhyām—in the navel; avasthitaḥ—situated; sarvataḥ—entirely; sauvarṇaḥ—made of gold; kula-giri-rājaḥ—the most famous among famous mountains; meruḥ—Mount Meru; dvīpa-āyāma-samunnāhaḥ—whose height is the same measurement as the width of Jambūdvīpa; karṇikā-bhūtaḥ—existing as the pericarp; kuvalaya—of this planetary system; kamalasya—like a lotus flower; mūrdhani—on the top; dvā-triṁśat—thirty-two; sahasra—thousand; yojana—yojanas (eight miles each; vitataḥ—expanded; mūle—at the base; ṣoḍaśa-sahasram—sixteen thousand yojanas; tāvat—so much; āntaḥ-bhūmyām—within the earth; praviṣṭaḥ—entered.

TRANSLATION: Amidst these divisions, or varṣas, is the varṣa named Ilāvṛta, which is situated in the middle of the whorl of the lotus. Within Ilāvṛta-varṣa is Sumeru Mountain, which is made of gold. Sumeru Mountain is like the pericarp of the lotuslike Bhū-maṇḍala planetary system. The mountain's height is the same as the width of Jambūdvīpa—or, in other words, 100,000 yojanas [800,000 miles]. Of that, 16,000 yojanas [128,000 miles] are within the earth, and therefore the mountain's height above the earth is 84,000 yojanas [672,000 miles]. The mountain's width is 32,000 yojanas [256,000 miles] at its summit and 16,000 yojanas at its base.

5.16.8: uttarottareṇelāvṛtaṁ nīlaḥ śvetaḥ śṛṅgavān iti trayo ramyaka-hiraṇmaya-kurūṇāṁ varṣāṇāṁ maryādā-girayaḥ prāg-āyatā ubhayataḥ kṣārodāvadhayo dvi-sahasra-pṛthava ekaikaśaḥ pūrvasmāt pūrvasmād uttara uttaro daśāṁśādhikāṁśena dairghya eva hrasanti.

SYNONYMS: uttara-uttareṇa ilāvṛtam—further and further north of Ilāvṛta-varṣa; nīlaḥ—Nīla; śvetaḥ—Śveta; śṛṅgavān—Śṛṅgavān; iti—thus; trayaḥ—three mountains; ramyaka—Ramyaka; hiraṇmaya—Hiraṇmaya; kurūṇām—of the Kuru division; varṣāṇām—of the varṣas; maryādā-girayaḥ—the mountains marking the borders; prāk-āyatāḥ—extended on the eastern side; ubhayataḥ—to the east and the west; kṣāroda—the ocean of salt water; avadhayaḥ—extending to; dvi-sahasra-pṛthavaḥ—which are two thousand yojanas wide; eka-ekaśaḥ—one after another; pūrvasmāt—than the former; pūrvasmāt—than the former; uttaraḥ—further north; uttaraḥ—further north; daśa-aṁśa-adhika-aṁśena—by one tenth of the former; dairghyaḥ—in length; eva—indeed; hrasanti—become shorter.

TRANSLATION: Just north of Ilāvṛta-varṣa—and going further northward, one after another—are three mountains named Nīla, Śveta and Śṛṅgavān. These mark the borders of the three varṣas named Ramyaka, Hiraṇmaya and Kuru and separate them from one another. The width of these mountains is 2,000 yojanas [16,000 miles]. Lengthwise, they extend east and west to the beaches of the ocean of salt water. Going from south to north, the length of each mountain is one tenth that of the previous mountain, but the height of them all is the same.

PURPORT: In this regard, Madhvācārya quotes the following verses from the Brahmāṇḍa Purāṇa: yathā bhāgavate tūktaṁ/ bhauvanaṁ kośa-lakṣaṇam/ tasyāvirodhato yojyam/ anya-granthāntare sthitam/ maṇḍode puraṇaṁ caiva/ vyatyāsaṁ kṣīra-sāgare/ rāhu-soma-ravīṇāṁ ca/ maṇḍalād dvi-guṇoktitām/ vinaiva sarvam unneyaṁ/ yojanābhedato 'tra tu. It appears from these verses that aside from the sun and moon, there is an invisible planet called Rāhu. The movements of Rāhu cause both solar and lunar eclipses. We suggest that the modern expeditions attempting to reach the moon are mistakenly going to Rāhu.

5.16.9: evaṁ dakṣiṇenelāvṛtaṁ niṣadho hemakūṭo himālaya iti prāg-āyatā yathā nīlādayo 'yuta-yojanotsedhā hari-varṣa-kimpuruṣa-bhāratānāṁ yathā-saṅkhyam.

SYNONYMS: evam—thus; dakṣiṇena—by degrees to the southern side; ilāvṛtam—of Ilāvṛta-varṣa; niṣadhaḥ hema-kūṭaḥ himālayaḥ—three mountains named Niṣadha, Hemakūṭa and Himālaya; iti—thus; prāk-āyatāḥ—extended to the east; yathā—just as; nīla-ādayaḥ—the mountains headed by Nīla; ayuta-yojana-utsedhāḥ—ten thousand yojanas high; hari-varṣa—the division named Hari-varṣa; kimpuruṣa—the division named Kimpuruṣa; bhāratānām—the division named Bhārata-varṣa; yathā-saṅkhyam—according to number.

TRANSLATION: Similarly, south of Ilāvṛta-varṣa and extending from east to west are three great mountains named (from north to south) Niṣadha, Hemakūṭa and Himālaya. Each of them is 10,000 yojanas [80,000 miles] high. They mark the boundaries of the three varṣas named Hari-varṣa, Kimpuruṣa-varṣa and Bhārata-varṣa [India].

5.16.10: tathaivelāvṛtam apareṇa pūrveṇa ca mālyavad-gandhamādanāv ānīla-niṣadhāyatau dvi-sahasraṁ paprathatuḥ ketumāla-bhadrāśvayoḥ sīmānaṁ vidadhāte.

SYNONYMS: tathā eva—exactly like that; ilāvṛtam apareṇa—on the western side of Ilāvṛta-varṣa; pūrveṇa ca—and on the eastern side; mālyavad-gandha-mādanau—the demarcation mountains of Mālyavān on the west and Gandhamādana on the east; ā-nīla-niṣadha-āyatau—on the northern side up to the mountain known as Nīla and on the southern side up to the mountain known as Niṣadha; dvi-sahasram—two thousand yojanas; paprathatuḥ—they extend; ketumāla-bhadrāśvayoḥ—of the two varṣas named Ketumāla and Bhadrāśva; sīmānam—the border; vidadhāte—establish.

TRANSLATION: In the same way, west and east of Ilāvṛta-varṣa are two great mountains named Mālyavān and Gandhamādana respectively. These two mountains, which are 2,000 yojanas [16,000 miles] high, extend as far as Nīla Mountain in the north and Niṣadha in the south. They indicate the borders of Ilāvṛta-varṣa and also the varṣas known as Ketumāla and Bhadrāśva.

PURPORT: There are so many mountains, even on this planet earth. We do not think that the measurements of all of them have actually been calculated. While passing over the mountainous region from Mexico to Caracas, we actually saw so many mountains that we doubt whether their height, length and breadth have been properly measured. Therefore, as indicated in Śrīmad-Bhāgavatam by Śukadeva Gosvāmī, we should not try to comprehend the greater mountainous areas of the universe merely by our calculations. Śukadeva Gosvāmī has already stated that such calculations would be very difficult even if one had a duration of life like that of Brahmā. We should simply be satisfied with the statements of authorities like Śukadeva Gosvāmī and appreciate how the entire cosmic manifestation has been made possible by the external energy of the Supreme Personality of Godhead. The measurements given herein, such as 10,000 yojanas or 100,000 yojanas, should be considered correct because they have been given by Śukadeva Gosvāmī. Our experimental knowledge can neither verify nor disprove the statements of Śrīmad-Bhāgavatam. We should simply hear these statements from the authorities. If we can appreciate the extensive energy of the Supreme Personality of Godhead, that will benefit us.

5.16.11: mandaro merumandaraḥ supārśvaḥ kumuda ity ayuta-yojana-vistāronnāhā meroś catur-diśam avaṣṭambha-giraya upakḷptāḥ.

SYNONYMS: mandaraḥ—the mountain named Mandara; meru-mandaraḥ—the mountain named Merumandara; supārśvaḥ—the mountain named Supārśva; kumudaḥ—the mountain named Kumuda; iti—thus; ayuta-yojana-vistāra-unnāhāḥ—which measure ten thousand yojanas high and wide; meroḥ—of Sumeru; catuḥ-diśam—the four sides; avaṣṭambha-girayaḥ—mountains that are like the belts of Sumeru; upakḷptāḥ—situated.

TRANSLATION: On the four sides of the great mountain known as Sumeru are four mountains—Mandara, Merumandara, Supārśva and Kumuda—which are like its belts. The length and height of these mountains are calculated to be 10,000 yojanas [80,000 miles].

5.16.12: caturṣv eteṣu cūta-jambū-kadamba-nyagrodhāś catvāraḥ pādapa-pravarāḥ parvata-ketava ivādhi-sahasra-yojanonnāhās tāvad viṭapa-vitatayaḥ śata-yojana-pariṇāhāḥ.

SYNONYMS: caturṣu—on the four; eteṣu—on these mountains, beginning with Mandara; cūta-jambū-kadamba—of trees such as the mango, rose apple and kadamba; nyagrodhāḥ—and the banyan tree; catvāraḥ—four kinds; pādapa-pravarāḥ—the best of trees; parvata-ketavaḥ—the flagstaffs on the mountains; iva—like; adhi—over; sahasra-yojana-unnāhāḥ—one thousand yojanas high; tāvat—so much also; viṭapa-vitatayaḥ—the length of the branches; śata-yojana—one hundred yojanas; pariṇāhāḥ—wide.

TRANSLATION: Standing like flagstaffs on the summits of these four mountains are a mango tree, a rose apple tree, a kadamba tree and a banyan tree. Those trees are calculated to have a width of 100 yojanas [800 miles] and a height of 1,100 yojanas [8,800 miles]. Their branches also spread to a radius of 1,100 yojanas.

5.16.13-14: hradāś catvāraḥ payo-madhv-ikṣurasa-mṛṣṭa-jalā yad-upasparśina upadeva-gaṇā yogaiśvaryāṇi svābhāvikāni bharatarṣabha dhārayanti; devodyānāni ca bhavanti catvāri nandanaṁ caitrarathaṁ vaibhrājakaṁ sarvatobhadram iti.

SYNONYMS: hradāḥ—lakes; catvāraḥ—four; payaḥ—milk; madhu—honey; ikṣu-rasa—sugarcane juice; mṛṣṭa-jalāḥ—filled with pure water; yat—of which; upasparśinaḥ—those who use the liquids; upadeva-gaṇāḥ—the demigods; yoga-aiśvaryāṇi—all the perfections of mystic yoga; svābhāvikāni—without being tried for; bharata-ṛṣabha—O best of the Bharata dynasty; dhārayanti—possess; deva-udyānāni—celestial gardens; ca—also; bhavanti—there are; catvāri—four; nandanam—of the Nandana garden; caitra-ratham—Caitraratha garden; vaibhrājakam—Vaibhrājaka garden; sarvataḥ-bhadram—Sarvatobhadra garden; iti—thus.

TRANSLATION: O Mahārāja Parīkṣit, best of the Bharata dynasty, between these four mountains are four huge lakes. The water of the first tastes just like milk; the water of the second, like honey; and that of the third, like sugarcane juice. The fourth lake is filled with pure water. The celestial beings such as the Siddhas, Cāraṇas and Gandharvas, who are also known as demigods, enjoy the facilities of those four lakes. Consequently they have the natural perfections of mystic yoga, such as the power to become smaller than the smallest or greater than the greatest. There are also four celestial gardens named Nandana, Caitraratha, Vaibhrājaka and Sarvatobhadra.

5.16.15: yeṣv amara-parivṛḍhāḥ saha sura-lalanā-lalāma-yūtha-pataya upadeva-gaṇair upagīyamāna-mahimānaḥ kila viharanti.

SYNONYMS: yeṣu—in which; amara-parivṛḍhāḥ—the best of the demigods; saha—with; sura-lalanā—of the wives of all the demigods and semidemigods; lalāma—of those women who are like ornaments; yūtha-patayaḥ—the husbands; upadeva-gaṇaiḥ—by the semi-demigods (the Gandharvas); upagīyamāna—being chanted; mahimānaḥ—whose glories; kila—indeed; viharanti—they enjoy sports.

TRANSLATION: The best of the demigods, along with their wives, who are like ornaments of heavenly beauty, meet together and enjoy within those gardens, while their glories are sung by lesser demigods known as Gandharvas.

5.16.16: mandarotsaṅga ekādaśa-śata-yojanottuṅga-devacūta-śiraso giri-śikhara-sthūlāni phalāny amṛta-kalpāni patanti.

SYNONYMS: mandara-utsaṅge—on the lower slopes of Mandara Mountain; ekādaśa-śata-yojana-uttuṅga—1,100 yojanas high; devacūta-śirasaḥ—from the top of a mango tree named Devacūta; giri-śikhara-sthūlāni—which are as fat as mountain peaks; phalāni—fruit; amṛta-kalpāni—as sweet as nectar; patanti—fall down.

TRANSLATION: On the lower slopes of Mandara Mountain is a mango tree named Devacūta. It is 1,100 yojanas high. Mangoes as big as mountain peaks and as sweet as nectar fall from the top of this tree for the enjoyment of the denizens of heaven.

PURPORT: In the Vāyu Purāṇa there is also a reference to this tree by great learned sages:
aratnīnāṁ śatāny aṣṭāv/eka-ṣaṣṭy-adhikāni ca/phala-pramāṇam ākhyātam/ṛṣibhis tattva-darśibhiḥ

5.16.17: teṣāṁ viśīryamāṇānām ati-madhura-surabhi-sugandhi-bahulāruṇa-rasodenāruṇodā nāma nadī mandara-giri-śikharān nipatantī pūrveṇelāvṛtam upaplāvayati.

SYNONYMS: teṣām—of all the mangoes; viśīryamāṇānām—being broken because of falling from the top; ati madhura—very sweet; surabhi—fragrant; sugandhi—scented with other aromas; bahula—large quantities; aruṇa-rasa-udena—by reddish juice; aruṇodā—Aruṇodā; nāma—named; nadī—the river; mandara-giri-śikharāt—from the top of Mandara Mountain; nipatantī—falling down; pūrveṇa—on the eastern side; ilāvṛtam—through Ilāvṛta-varṣa; upaplāvayati—flows.

TRANSLATION: When all those solid fruits fall from such a height, they break, and the sweet, fragrant juice within them flows out and becomes increasingly more fragrant as it mixes with other scents. That juice cascades from the mountain in waterfalls and becomes a river called Aruṇodā, which flows pleasantly through the eastern side of Ilāvṛta.

5.16.18: yad-upajoṣaṇād bhavānyā anucarīṇāṁ puṇya-jana-vadhūnām avayava-sparśa-sugandha-vāto daśa-yojanaṁ samantād anuvāsayati.

SYNONYMS: yat—of which; upajoṣaṇāt—because of using the fragrant water; bhavānyāḥ—of Bhavānī, the wife of Lord Śiva; anucarīṇām—of attendant maidservants; puṇya-jana-vadhūnām—who are wives of the most pious Yakṣas; avayava—of the bodily limbs; sparśa—from contact; sugandha-vātaḥ—the wind, which becomes fragrant; daśa-yojanam—up to ten yojanas (about eighty miles); samantāt—all around; anuvāsayati—makes fragrant.

TRANSLATION: The pious wives of the Yakṣas act as personal maidservants to assist Bhavānī, the wife of Lord Śiva. Because they drink the water of the River Aruṇodā, their bodies become fragrant, and as the air carries away that fragrance, it perfumes the entire atmosphere for eighty miles around.

5.16.19: evaṁ jambū-phalānām atyucca-nipāta-viśīrṇānām anasthi-prāyāṇām ibha-kāya-nibhānāṁ rasena jambū nāma nadī meru-mandara-śikharād ayuta-yojanād avani-tale nipatantī dakṣiṇenātmānaṁ yāvad ilāvṛtam upasyandayati.

SYNONYMS: evam—similarly; jambū-phalānām—of the fruits called jambū (the rose apple); ati-ucca-nipāta—because of falling from a great height; viśīrṇānām—which are broken to pieces; anasthi-prāyāṇām—having very small seeds; ibha-kāya-nibhānām—and which are as large as the bodies of elephants; rasena—by the juice; jambū nāma nadī—a river named Jambū-nadī; meru-mandara-śikharāt—from the top of Merumandara Mountain; ayuta-yojanāt—ten thousand yojanas high; avani-tale—on the ground; nipatantī—falling; dakṣiṇena—on the southern side; ātmānam—itself; yāvat—the whole; ilāvṛtam—Ilāvṛta-varṣa; upasyandayati—flows through.

TRANSLATION: Similarly, the fruits of the jambū tree, which are full of pulp and have very small seeds, fall from a great height and break to pieces. Those fruits are the size of elephants, and the juice gliding from them becomes a river named Jambū-nadī. This river falls a distance of 10,000 yojanas, from the summit of Merumandara to the southern side of Ilāvṛta, and floods the entire land of Ilāvṛta with juice.

PURPORT: We can only imagine how much juice there might be in a fruit that is the size of an elephant but has a very tiny seed. Naturally the juice from the broken jambū fruits forms waterfalls and floods the entire land of Ilāvṛta. That juice produces an immense quantity of gold, as will be explained in the next verses.

5.16.20-21: tāvad ubhayor api rodhasor yā mṛttikā tad-rasenānuvidhyamānā vāyv-arka-saṁyoga-vipākena sadāmara-lokābharaṇaṁ jāmbū-nadaṁ nāma suvarṇaṁ bhavati; yad u ha vāva vibudhādayaḥ saha yuvatibhir mukuṭa-kaṭaka-kaṭi-sūtrādy-ābharaṇa-rūpeṇa khalu dhārayanti.

SYNONYMS: tāvat—entirely; ubhayoḥ api—of both; rodhasoḥ—of the banks; yā—which; mṛttikā—the mud; tat-rasena—with the juice of the jambū fruits that flows in the river; anuvidhyamānā—being saturated; vāyu-arka-saṁyoga-vipākena—because of a chemical reaction with the air and sunshine; sadā—always; amara-loka-ābharaṇam—which is used for the ornaments of the demigods, the denizens of the heavenly planets; jāmbū-nadam nāma—named Jāmbū-nada; suvarṇam—gold; bhavati—becomes; yat—which; u ha vāva—indeed; vibudha-ādayaḥ—the great demigods; saha—with; yuvatibhiḥ—their everlastingly youthful wives; mukuṭa—crowns; kaṭaka—bangles; kaṭi-sūtra—belts; ādi—and so on; ābharaṇa—of all kinds of ornaments; rūpeṇa—in the form; khalu—indeed; dhārayanti—they possess.

TRANSLATION: The mud on both banks of the River Jambū-nadī, being moistened by the flowing juice and then dried by the air and the sunshine, produces huge quantities of gold called Jāmbū-nada. The denizens of heaven use this gold for various kinds of ornaments. Therefore all the inhabitants of the heavenly planets and their youthful wives are fully decorated with golden helmets, bangles and belts, and thus they enjoy life.

PURPORT: By the arrangement of the Supreme Personality of Godhead, the rivers on some planets produce gold on their banks. The poor inhabitants of this earth, because of their incomplete knowledge, are captivated by a so-called bhagavān who can produce a small quantity of gold. However, it is understood that in a higher planetary system in this material world, the mud on the banks of the Jambū-nadī mixes with jambū juice, reacts with the sunshine in the air, and automatically produces huge quantities of gold. Thus the men and women are decorated there by various golden ornaments, and they look very nice.

5.16.22: yas tu mahā-kadambaḥ supārśva-nirūḍho yās tasya koṭarebhyo viniḥsṛtāḥ pañcāyāma-pariṇāhāḥ pañca madhu-dhārāḥ supārśva-śikharāt patantyo 'pareṇātmānam ilāvṛtam anumodayanti.

SYNONYMS: yaḥ—which; tu—but; mahā-kadambaḥ—the tree named Mahākadamba; supārśva-nirūḍhaḥ—which stands on the side of the mountain known as Supārśva; yāḥ—which; tasya—of that; koṭarebhyaḥ—from the hollows; viniḥsṛtāḥ—flowing; pañca—five; āyāma—vyāma, a unit of measurement of about eight feet; pariṇāhāḥ—whose measurement; pañca—five; madhu-dhārāḥ—flows of honey; supārśva-śikharāt—from the top of Supārśva Mountain; patantyaḥ—flowing down; apareṇa—on the western side of Sumeru Mountain; ātmānam—the whole of; ilāvṛtam—Ilāvṛta-varṣa; anumodayanti—make fragrant.

TRANSLATION: On the side of Supārśva Mountain stands a big tree called Mahākadamba, which is very celebrated. From the hollows of this tree flow five rivers of honey, each about five vyāmas wide. This flowing honey falls incessantly from the top of Supārśva Mountain and flows all around Ilāvṛta-varṣa, beginning from the western side. Thus the whole land is saturated with the pleasing fragrance.

PURPORT: The distance between one hand and another when one spreads both his arms is called a vyāma. This comes to about eight feet. Thus each of the rivers was about forty feet wide, making a total of about two hundred feet.

5.16.23: yā hy upayuñjānānāṁ mukha-nirvāsito vāyuḥ samantāc chata-yojanam anuvāsayati.

SYNONYMS: yāḥ—which (those flows of honey); hi—indeed; upayuñjānānām—of those who drink; mukha-nirvāsitaḥ vāyuḥ—the air emanating from the mouths; samantāt—all around; śata-yojanam—up to one hundred yojanas (eight hundred miles); anuvāsayati—makes sweetly flavored.
TRANSLATION: The air carrying the scent from the mouths of those who drink that honey perfumes the land for a hundred yojanas around.

5.16.24: evaṁ kumuda-nirūḍho yaḥ śatavalśo nāma vaṭas tasya skandhebhyo nīcīnāḥpayo-dadhi-madhu-ghṛta-guḍānnādy-ambara-śayyāsanābharaṇādayaḥ sarva eva kāma-dughā nadāḥ kumudāgrāt patantas tam uttareṇelāvṛtam upayojayanti.

SYNONYMS: evam—thus; kumuda-nirūḍhaḥ—having grown on Kumuda Mountain; yaḥ—that; śata-valśaḥ nāma—the tree named Śatavalśa (because of having hundreds of trunks); vaṭaḥ—a banyan tree; tasya—of it; skandhebhyaḥ—from the thick branches; nīcīnāḥ—flowing down; payaḥ—milk; dadhi—yogurt; madhu—honey; ghṛta—clarified butter; guḍa—molasses; anna—food grains; ādi—and so on; ambara—clothing; śayyā—bedding; āsana—sitting places; ābharaṇa-ādayaḥ—carrying ornaments and so on; sarve—everything; eva—certainly; kāma-dughāḥ—fulfilling all desires; nadāḥ—big rivers; kumuda-agrāt—from the top of Kumuda Mountain; patantaḥ—flowing; tam—to that; uttareṇa—on the northern side; ilāvṛtam—the land known as Ilāvṛta-varṣa; upayojayanti—give happiness.

TRANSLATION: Similarly, on Kumuda Mountain there is a great banyan tree, which is called Śatavalśa because it has a hundred main branches. From those branches come many roots, from which many rivers are flowing. These rivers flow down from the top of the mountain to the northern side of Ilāvṛta-varṣa for the benefit of those who live there. Because of these flowing rivers, all the people have ample supplies of milk, yogurt, honey, clarified butter [ghee], molasses, food grains, clothes, bedding, sitting places and ornaments. All the objects they desire are sufficiently supplied for their prosperity, and therefore they are very happy.

5.16.25: yān upajuṣāṇānāṁ na kadācid api prajānāṁ valī-palita-klama-sveda-daurgandhya-jarāmaya-mṛtyu-śītoṣṇa-vaivarṇyopasargādayas tāpa-viśeṣā bhavanti yāvaj jīvaṁ sukhaṁ niratiśayam eva.

SYNONYMS: yān—which (all the products produced because of the flowing rivers mentioned above); upajuṣāṇānām—of persons who are fully utilizing; na—not; kadācit—at any time; api—certainly; prajānām—of the citizens; valī—wrinkles; palita—grey hair; klama—fatigue; sveda—perspiration; daurgandhya—bad odors because of unclean perspiration; jarā—old age; āmaya—disease; mṛtyu—untimely death; śīta—severe cold; uṣṇa—scorching heat; vaivarṇya—fading of the luster of the body; upasarga—troubles; ādayaḥ—and so on; tāpa—of sufferings; viśeṣāḥ—varieties; bhavanti—are; yāvat—as long as; jīvam—life; sukham—happiness; niratiśayam—unlimited; eva—only.

TRANSLATION: The residents of the material world who enjoy the products of these flowing rivers have no wrinkles on their bodies and no grey hair. They never feel fatigue, and perspiration does not give their bodies a bad odor. They are not afflicted by old age, disease or untimely death, they do not suffer from chilly cold or scorching heat, nor do their bodies lose their luster. They all live very happily, without anxieties, until death.

5.16.26: kuraṅga-kurara-kusumbha-vaikaṅka-trikūṭa-śiśira-pataṅga-rucaka-niṣadha-śinīvāsa-kapila-śaṅkha-vaidūrya-jārudhi-haṁsa-ṛṣabha-nāga-kālañjara-nāradādayo viṁśati-girayo meroḥ karṇikāyā iva kesara-bhūtā mūla-deśe parita upakḷptāḥ.

SYNONYMS: kuraṅga—Kuraṅga; kurara—Kurara; kusumbha-vaikaṅka-trikūṭa-śiśira-pataṅga-rucaka-niṣadha-śinīvāsa-kapila-śaṅkha-vaidūrya-jārudhi-haṁsa-ṛṣabha-nāga-kālañjara-nārada—the names of mountains; ādayaḥ—and so on; viṁśati-girayaḥ—twenty mountains; meroḥ—of Sumeru Mountain; karṇikāyāḥ—of the whorl of the lotus; iva—like; kesara-bhūtāḥ—as filaments; mūla-deśe—at the base; paritaḥ—all around; upakḷptāḥ—arranged by the Supreme Personality of Godhead.

TRANSLATION: There are other mountains beautifully arranged around the foot of Mount Meru like the filaments around the whorl of a lotus flower. Their names are Kuraṅga, Kurara, Kusumbha, Vaikaṅka, Trikūṭa, Śiśira, Pataṅga, Rucaka, Niṣadha, Sinīvāsa, Kapila, Śaṅkha, Vaidūrya, Jārudhi, Haṁsa, Ṛṣabha, Nāga, Kālañjara and Nārada.

5.16.27: jaṭhara-devakūṭau meruṁ pūrveṇāṣṭādaśa-yojana-sahasram udagāyatau dvi-sahasraṁ pṛthu-tuṅgau bhavataḥ; evam apareṇa pavana-pāriyātrau dakṣiṇena kailāsa-karavīrau prāg-āyatāv evam uttaratas triśṛṅga-makarāv aṣṭabhir etaiḥ parisṛto 'gnir iva paritaś cakāsti kāñcana-giriḥ.

SYNONYMS: jaṭhara-devakūṭau—two mountains named Jaṭhara and Devakūṭa; merum—Sumeru Mountain; pūrveṇa—on the eastern side; aṣṭādaśa-yojana-sahasram—eighteen thousand yojanas; udagāyatau—stretching from north to south; dvi-sahasram—two thousand yojanas; pṛthu-tuṅgau—in width and height; bhavataḥ—there are; evam—similarly; apareṇa—on the western side; pavana-pāriyātrau—two mountains named Pavana and Pāriyātra; dakṣiṇena—on the southern side; kailāsa-karavīrau—two mountains named Kailāsa and Karavīra; prāk-āyatau—expanding east and west; evam—similarly; uttarataḥ—on the northern side; triśṛṅga-makarau—two mountains named Triśṛṅga and Makara; aṣṭabhiḥ etaiḥ—by these eight mountains; parisṛtaḥ—surrounded; agniḥ iva—like fire; paritaḥ—all over; cakāsti—brilliantly shines; kāñcana-giriḥ—the golden mountain named Sumeru, or Meru.

TRANSLATION: On the eastern side of Sumeru Mountain arc two mountains named Jaṭhara and Devakūṭa, which extend to the north and south for 18,000 yojanas [144,000 miles]. Similarly, on the western side of Sumeru are two mountains named Pavana and Pāriyātra, which also extend north and south for the same distance. On the southern side of Sumeru are two mountains named Kailāsa and Karavīra, which extend east and west for 18,000 yojanas, and on the northern side of Sumeru, extending for the same distance east and west, are two mountains named Triśṛṅga and Makara. The width and height of all these mountains is 2,000 yojanas [16,000 miles]. Sumeru, a mountain of solid gold shining as brilliantly as fire, is surrounded by these eight mountains.

5.16.28: meror mūrdhani bhagavata ātma-yoner madhyata upakḷptāṁ purīm ayuta-yojana-sāhasrīṁ sama-caturasrāṁ śātakaumbhīṁ vadanti.

SYNONYMS: meroḥ—of Sumeru Mountain; mūrdhani—on the head; bhagavataḥ—of the most powerful being; ātma-yoneḥ—of Lord Brahmā; madhyataḥ—in the middle; upakḷptām—situated; purīm—the great township; ayuta-yojana—ten thousand yojanas; sāhasrīm—one thousand; sama-caturasrām—of the same length on all sides; śāta-kaumbhīm—made entirely of gold; vadanti—the great learned sages say.

TRANSLATION: In the middle of the summit of Meru is the township of Lord Brahmā. Each of its four sides is calculated to extend for ten million yojanas [eighty million miles]. It is made entirely of gold, and therefore learned scholars and sages call it Śātakaumbhī.

5.16.29: tām anuparito loka-pālānām aṣṭānāṁ yathā-diśaṁ yathā-rūpaṁ turīya-mānena puro 'ṣṭāv upakḷptāḥ.

SYNONYMS: tām—that great township named Brahmapurī; anuparitaḥ—surrounding; loka-pālānām—of the governors of the planets; aṣṭānām—eight; yathā-diśam—according to the directions; yathā-rūpam—in exact conformity with the township of Brahmapurī; turīya-mānena—by measurement only one fourth; puraḥ—townships; aṣṭau—eight; upakḷptāḥ—situated.

TRANSLATION: Surrounding Brahmapurī in all directions are the residences of the eight principal governors of the planetary systems, beginning with King Indra. These abodes are similar to Brahmapurī but are one fourth the size.

PURPORT: Śrīla Viśvanātha Cakravartī Ṭhākura confirms that the townships of Lord Brahmā and the eight subordinate governors of the planetary systems, beginning with Indra, are mentioned in other Purāṇas.

merau nava-pūrāṇi syur/manovaty amarāvatī/tejovatī saṁyamanī/tathā kṛṣṇāṅganā parā/śraddhāvatī gandhavatī/tathā cānyā mahodayā/yaśovatī ca brahmendra/bahyādīnāṁ yathā-kramam//

Brahmā's township is known as Manovatī, and those of his assistants such as Indra and Agni are known as Amarāvatī, Tejovatī, Saṁyamanī, Kṛṣṇāṅganā, Śraddhāvatī, Gandhavatī, Mahodayā and Yaśovatī. Brahmapurī is situated in the middle, and the other eight purīs surround it in all directions.

## Part V: The Descent of the River Ganges

5.17.1: śrī-śuka uvāca/ tatra bhagavataḥ sākṣād yajña-liṅgasya viṣṇor vikramato vāma-pādāṅguṣṭha-nakha-nirbhinnordhvāṇḍa-kaṭāha-vivareṇāntaḥ-praviṣṭā yā bāhya-jala-dhārā tac-caraṇa-paṅkajāvanejanāruṇa-kiñjalkoparañjitākhila-jagad-agha-malāpahopasparśanāmalā sākṣād bhagavat-padīty anupalakṣita-vaco 'bhidhīyamānāti-mahatā kālena yuga-sahasropalakṣaṇena divo mūrdhany avatatāra yat tad viṣṇu-padam āhuḥ.

SYNONYMS: śrī-śukaḥ uvāca—Śrī Śukadeva Gosvāmī said; tatra—at that time; bhagavataḥ—of the incarnation of the Supreme Personality of Godhead; sākṣāt—directly; yajña-liṅgasya—the enjoyer of the results of all sacrifices; viṣṇoḥ—of Lord Viṣṇu; vikramataḥ—while taking His second step; vāma-pāda—of His left leg; aṅguṣṭha—of the big toe; nakha—by the nail; nirbhinna—pierced; ūrdhva—upper; aṇḍa-kaṭāha—the covering of the universe (consisting of seven layers—earth, water, fire, etc.); vivareṇa—through the bole; antaḥ-praviṣṭā—having entered the universe; yā—which; bāhya-jala-dhārā—the flow of water from the Causal Ocean outside the universe; tat—of Him; caraṇa-paṅkaja—of the lotus feet; avanejana—by the washing; aruṇa-kiñjalka—by reddish powder; uparañjitā—being colored; akhila-jagat—of the whole world; agha-mala—the sinful activities; apahā—destroys; upasparśana—the touching of which; amalā—completely pure; sākṣāt—directly; bhagavat-padī—emanating from the lotus feet of the Supreme Personality of Godhead; iti—thus; anupalakṣita—described; vacaḥ—by the name; abhidhīyamānā—being called; ati-mahatā kālena—after a long time; yuga-sahasra-upalakṣaṇena—consisting of one thousand millenniums; divaḥ—of the sky; mūrdhani—on the head (Dhruvaloka); avatatāra—descended; yat—which; tat—that; viṣṇu-padam—the lotus feet of Lord Viṣṇu; āhuḥ—they call.

TRANSLATION: Śukadeva Gosvāmī said: My dear King, Lord Viṣṇu, the enjoyer of all sacrifices, appeared as Vāmanadeva in the sacrificial arena of Bali Mahārāja. Then He extended His left foot to the end of the universe and pierced a hole in its covering with the nail of His big toe. Through the hole, the pure water of the Causal Ocean entered this universe as the Ganges River. Having washed the lotus feet of the Lord, which are covered with reddish powder, the water of the Ganges acquired a very beautiful pink color. Every living being can immediately purify his mind of material contamination by touching the transcendental water of the Ganges, yet its waters remain ever pure. Because the Ganges directly touches the lotus feet of the Lord before descending within this universe, she is known as Viṣṇupadī. Later she received other names like Jāhnavī and Bhāgīrathī. After one thousand millenniums, the water of the Ganges descended on Dhruvaloka, the topmost planet in this universe. Therefore all learned sages and scholars proclaim Dhruvaloka to be Viṣṇupada ["situated on Lord Viṣṇu's lotus feet"].

PURPORT: Many sages, including Śaṅkarācārya, have composed prayers in praise of the Ganges, and the land of India itself has become glorious because such rivers as the Ganges, Yamunā, Godāvarī, Kāverī, Kṛṣṇā and Narmadā flow there. Anyone living on the land adjacent to these rivers is naturally advanced in spiritual consciousness. Śrīla Madhvācārya says: vārāhe vāma-pādaṁ tu/tad-anyeṣu tu dakṣiṇam/pādaṁ kalpeṣu bhagavān/ujjahāra trivikramaḥ. Standing on His right foot and extending His left to the edge of the universe, Lord Vāmana became known as Trivikrama, the incarnation who performed three heroic deeds.

5.17.2: yatra ha vāva vīra-vrata auttānapādiḥ parama-bhāgavato 'smat-kula-devatā-caraṇāravindodakam iti yām anusavanam utkṛṣyamāṇa-bhagavad-bhakti-yogena dṛḍhaṁ klidyamānāntar-hṛdaya autkaṇṭhya-vivaśāmīlita-locana-yugala-kuḍmala-vigalitāmala-bāṣpa-kalayābhivyajyamāna-roma-pulaka-kulako 'dhunāpi paramādareṇa śirasā bibharti.

SYNONYMS: yatra ha vāva—in Dhruvaloka; vīra-vrataḥ—firmly determined; auttānapādiḥ—the famous son of Mahārāja Uttānapāda; parama-bhāgavataḥ—the most exalted devotee; asmat—our; kula-devatā—of the family Deity; caraṇa-aravinda—of the lotus feet; udakam—in the water; iti—thus; yām—which; anusavanam—constantly; utkṛṣyamāṇa—being increased; bhagavat-bhakti-yogena—by devotional service unto the Lord; dṛḍham—greatly; klidyamāna-antaḥ-hṛdayaḥ—being softened within the core of his heart; autkaṇṭhya—by great anxiety; vivaśa—spontaneously; amīlita—slightly open; locana—of eyes; yugala—pair; kuḍmala—from the flowerlike; vigalita—emanating; amala—uncontaminated; bāṣpa-kalayā—with tears; abhivyajyamāna—being manifested; roma-pulaka-kulakaḥ—whose symptoms of ecstasy on the body; adhunā api—even now; parama-ādareṇa—with great reverence; śirasā—by the head; bibharti—he bears.

TRANSLATION: Dhruva Mahārāja, the famous son of Mahārāja Uttānapāda, is known as the most exalted devotee of the Supreme Lord because of his firm determination in executing devotional service. Knowing that the sacred Ganges water washes the lotus feet of Lord Viṣṇu, Dhruva Mahārāja, situated on his own planet, to this very day accepts that water on his head with great devotion. Because he constantly thinks of Kṛṣṇa very devoutly within the core of his heart, he is overcome with ecstatic anxiety. Tears flow from his half-open eyes, and eruptions appear on his entire body.

5.17.3: tataḥ sapta ṛṣayas tat prabhāvābhijñā yāṁ nanu tapasa ātyantikī siddhir etāvatī bhagavati sarvātmani vāsudeve 'nuparata-bhakti-yoga-lābhenaivopekṣitānyārthātma-gatayo muktim ivāgatāṁ mumukṣava iva sabahu-mānam adyāpi jaṭā-jūṭair udvahanti.

SYNONYMS: tataḥ—thereafter; sapta ṛṣayaḥ—the seven great sages (beginning with Marīci); tat prabhāva-abhijñāḥ—who knew very well the influence of the Ganges River; yām—this Ganges water; nanu—indeed; tapasaḥ—of our austerities; ātyantikī—the ultimate; siddhiḥ—perfection; etāvatī—this much; bhagavati—the Supreme Personality of Godhead; sarva-ātmani—in the all-pervading; vāsudeve—Kṛṣṇa; anuparata—continuous; bhakti-yoga—of the mystic process of devotional service; lābhena—simply by achieving this platform; eva—certainly; upekṣita—neglected; anya—other; artha-ātma-gatayaḥ—all other means of perfection (namely religion, economic development, sense gratification and liberation); muktim—liberation from material bondage; iva—like; āgatām—obtained; mumukṣavaḥ—persons desiring liberation; iva—like; sa-bahu-mānam—with great honor; adya api—even now; jaṭā-jūṭaiḥ—with matted locks of hair; udvahanti—they carry.

TRANSLATION: The seven great sages [Marīci, Vasiṣṭha, Atri and so on] reside on planets beneath Dhruvaloka. Well aware of the influence of the water of the Ganges, to this day they keep Ganges water on the tufts of hair on their heads. They have concluded that this is the ultimate wealth, the perfection of all austerities, and the best means of prosecuting transcendental life. Having obtained uninterrupted devotional service to the Supreme Personality of Godhead, they neglect all other beneficial processes like religion, economic development, sense gratification and even merging into the Supreme. Just as jñānīs think that merging into the existence of the Lord is the highest truth, these seven exalted personalities accept devotional service as the perfection of life.

PURPORT: In the upper planetary system, the topmost planet is Dhruvaloka, and beneath Dhruvaloka are the seven planets occupied by the great sages, beginning with Marīci, Vasiṣṭha and Atri. All these sages regard devotional service as the highest perfection of life. Therefore they all carry the holy water of the Ganges on their heads.

5.17.4: tato 'neka-sahasra-koṭi-vimānānīka-saṅkula-deva-yānenāvatar-antīndu maṇḍalam āvārya brahma-sadane nipatati.

SYNONYMS: tataḥ—after purifying the seven planets of the seven great sages; aneka—many; sahasra—thousands; koṭi—of millions; vimāna-anīka—with contingents of airplanes; saṅkula—congested; deva-yānena—by the spaceways of the demigods; avatarantī—descending; indu-maṇḍalam—the moon planet; āvārya—inundated; brahma-sadane—to the abode of Lord Brahmā atop Sumeru-parvata; nipatati—falls down.

TRANSLATION: After purifying the seven planets near Dhruvaloka [the polestar], the Ganges water is carried through the spaceways of the demigods in billions of celestial airplanes. Then it inundates the moon [Candraloka] and finally reaches Lord Brahmā's abode atop Mount Meru.

PURPORT: We should always remember that the Ganges River comes from the Causal Ocean, beyond the covering of the universe. After the water of the Causal Ocean leaks through the hole created by Lord Vāmanadeva, it flows down to Dhruvaloka (the polestar) and then to the seven planets beneath Dhruvaloka. Then it is carried to the moon by innumerable celestial airplanes, and then it falls to the top of Mount Meru, which is known as Sumeru-parvata. In this way, the water of the Ganges finally reaches the lower planets and the peaks of the Himalayas, and from there it flows through Hardwar and throughout the plains of India, purifying the entire land. How the Ganges water reaches the various planets from the top of the universe is explained herein. Celestial airplanes carry the water from the planets of the sages to other planets. So-called advanced scientists of the modern age are trying to go to the higher planets, but at the same time they are experiencing a power shortage on earth. If they were actually capable scientists, they could personally go by airplane to other planets, but this they are unable to do. Having now given up their moon excursions, they are attempting to go to other planets, but without success.

5.17.5: tatra caturdhā bhidyamānā caturbhir nāmabhiś catur-diśam abhispandantī nada-nadī-patim evābhiniviśati sītālakanandā cakṣur bhadreti.

SYNONYMS: tatra—there (on top of Mount Meru); caturdhā—into four branches; bhidyamānā—being divided; caturbhiḥ—with four; nāmabhiḥ—names; catuḥ-diśam—the four directions (east, west, north and south); abhispandantī—profusely flowing; nada-nadī-patim—the reservoir of all great rivers (the ocean); eva—certainly; abhiniviśati—enters; sītā-alakanandā—Sītā and Alakanandā; cakṣuḥ—Cakṣu; bhadrā—Bhadrā; iti—known by these names.

TRANSLATION: On top of Mount Meru, the Ganges divides into four branches, each of which gushes in a different direction [east, west, north and south]. These branches, known by the names Sītā, Alakanandā, Cakṣu and Bhadrā, flow down to the ocean.

5.17.6: sītā tu brahma-sadanāt kesarācalādi-giri-śikharebhyo 'dho 'dhaḥ prasravantī gandhamādana-mūrdhasu patitvāntareṇa bhadrāśva-varṣaṁ prācyāṁ diśi kṣāra-samudram abhipraviśati.

SYNONYMS: sītā—the branch known as Sītā; tu—certainly; brahma-sadanāt—from Brahmapurī; kesarācala-ādi—of Kesarācala and of other great mountains; giri—hills; śikharebhyaḥ—from the tops; adhaḥ adhaḥ—downward; prasravantī—flowing; gandhamādana—of Gandhamādana Mountain; mūrdhasu—on the top;

patitvā—falling down; antareṇa—within; bhadrāśva-varṣam—the province known as Bhadrāśva; prācyām—in the western; diśi—direction; kṣāra-samudram—the ocean of salt water; abhipraviśati—enters.

TRANSLATION: The branch of the Ganges known as the Sītā flows through Brahmapurī atop Mount Meru, and from there it runs down to the nearby peaks of the Kesarācala Mountains, which stand almost as high as Mount Meru itself. These mountains are like a bunch of filaments around Mount Meru. From the Kesarācala Mountains, the Ganges falls to the peak of Gandhamādana Mountain and then flows into the land of Bhadrāśva-varṣa. Finally it reaches the ocean of salt water in the west.

5.17.7: evaṁ mālyavac-chikharān niṣpatantī tato 'nuparata-vegā ketumālam abhi cakṣuḥ pratīcyāṁ diśi sarit-patiṁ praviśati.

SYNONYMS: evam—in this way; mālyavat-śikharāt—from the top of Mālyavān Mountain; niṣpatantī—falling down; tataḥ—thereafter; anuparata-vegā—whose force is uninterrupted; ketumālam abhi—into the land known as Ketumāla-varṣa; cakṣuḥ—the branch known as Cakṣu; pratīcyām—in the West; diśi—direction; sarit-patim—the ocean; praviśati—enters into.

TRANSLATION: The branch of the Ganges known as Cakṣu falls onto the summit of Mālyavān Mountain and from there cascades onto the land of Ketumāla-varṣa. The Ganges flows incessantly through Ketumāla-varṣa and in this way also reaches the ocean of salt water in the West.

5.17.8: bhadrā cottarato meru-śiraso nipatitā giri-śikharād giri-śikharam atihāya śṛṅgavataḥ śṛṅgād avasyandamānā uttarāṁs tu kurūn abhita udīcyāṁ diśi jaladhim abhipraviśati.

SYNONYMS: bhadrā—the branch known as Bhadrā; ca—also; uttarataḥ—to the northern side; meru-śirasaḥ—from the top of Mount Meru; nipatitā—having fallen; giri-śikharāt—from the peak of Kumuda Mountain; giri-śikharam—to the peak of Nīla Mountain; atihāya—passing over as if not touching; śṛṅgavataḥ—of the mountain known as Śṛṅgavān; śṛṅgāt—from the peak; avasyandamānā—flowing; uttarān—the northern; tu—but; kurūn—the land known as Kuru; abhitaḥ—on all sides; udīcyām—in the northern; diśi—direction; jaladhim—the ocean of salt water; abhipraviśati—enters into.

TRANSLATION: The branch of the Ganges known as Bhadrā flows from the northern side of Mount Meru. Its waters fall onto the peaks of Kumuda Mountain, Mount Nīla, Śveta Mountain and Śṛṅgavān Mountain in succession. Then it runs down into the province of Kuru and, after crossing through that land, flows into the saltwater ocean in the north.

5.17.9: tathaivālakanandā dakṣiṇena brahma-sadanād bahūni giri-kūṭāny atikramya hemakūṭād dhaimakūṭāny ati-rabhasatara-raṁhasā luṭhayantī bhāratam abhivarṣaṁ dakṣiṇasyāṁ diśi jaladhim abhipraviśati yasyāṁ snānārthaṁ cāgacchataḥ puṁsaḥ pade pade 'śvamedha-rājasūyādīnāṁ phalaṁ na durlabham iti.

SYNONYMS: tathā eva—similarly; alakanandā—the branch known as Alakanandā; dakṣiṇena—by the southern side; brahma-sadanāt—from the city known as Brahmapurī; bahūni—many; giri-kūṭāni—the tops of mountains; atikramya—crossing over; hemakūṭāt—from Hemakūṭa Mountain; haimakūṭāni—and Himakūṭa; ati-rabhasatara—more fiercely; raṁhasā—with great force; luṭhayantī—plundering; bhāratam abhivarṣam—on all sides of Bhārata-varṣa; dakṣiṇasyām—in the southern; diśi—direction; jaladhim—the ocean of salt water; abhipraviśati—enters into; yasyām—in which; snāna-artham—for bathing; ca—and; āgacchataḥ—of one who is coming; puṁsaḥ—a person; pade pade—at every step; aśvamedha-rājasūya-ādīnām—of great sacrifices like the Aśvamedha yajña and Rājasūya yajña; phalam—the result; na—not; durlabham—very difficult to obtain; iti—thus.

TRANSLATION Similarly, the branch of the Ganges known as Alakanandā flows from the southern side of Brahmapurī [Brahma-sadana]. Passing over the tops of mountains in various lands, it falls down with fierce force upon the peaks of the mountains Hemakūṭa and Himakūṭa. After inundating the tops of those mountains, the Ganges falls down onto the tract of land known as Bhārata-varṣa, which she also inundates. Then the Ganges flows into the ocean of salt water in the south. Persons who come to bathe in this river are fortunate. It is not very difficult for them to achieve with every step the results of performing great sacrifices like the Rājasūya and Aśvamedha yajñas.

PURPORT: The place where the Ganges flows into the salt water of the Bay of Bengal is still known as Gaṅgā-sāgara, or the meeting place of the Ganges and the Bay of Bengal. On Makara-saṅkrānti, in the month of January–February, thousands of people still go there to bathe, hoping to be liberated. That they can actually be liberated in this way is confirmed herein.

5.17.10: anye ca nadā nadyaś ca varṣe varṣe santi bahuśo merv-ādi-giri-duhitaraḥ śataśaḥ.

SYNONYMS: anye—many others; ca—also; nadāḥ—rivers; nadyaḥ—small rivers; ca—and; varṣe varṣe—in each tract of land; santi—are; bahuśaḥ—of many varieties; meru-ādi-giri-duhitaraḥ—daughters of the mountains beginning with Meru; śataśaḥ—in the hundreds.

TRANSLATION: Many other rivers, both big and small, flow from the top of Mount Meru. These rivers are like daughters of the mountain, and they flow to the various tracts of land in hundreds of branches.

## Part VI: Nine Varṣas of Jambūdvīpa

5.17.11: tatrāpi bhāratam eva varṣaṁ karma-kṣetram anyāny aṣṭa varṣāṇi svargiṇāṁ puṇya-śeṣopabhoga-sthānāni bhaumāni svarga-padāni vyapadiśanti.

SYNONYMS: tatra api—out of all of them; bhāratam—known as Bhārata-varṣa; eva—certainly; varṣam—the tract of land; karma-kṣetram—the field of activities; anyāni—the others; aṣṭa varṣāṇi—eight tracts of land; svargiṇām—of the living entities elevated to the heavenly planets by extraordinary pious activities; puṇya—of the results of pious activities; śeṣa—of the remainder; upabhoga-sthānāni—the places for material enjoyment; bhaumāni svarga-padāni—as the heavenly places on earth; vyapadiśanti—they designate.

TRANSLATION: Among the nine varṣas, the tract of land known as Bhārata-varṣa is understood to be the field of fruitive activities. Learned scholars and saintly persons declare the other eight varṣas to be meant for very highly elevated pious persons. After returning from the heavenly planets, they enjoy the remaining results of their pious activities in these eight earthly varṣas.

PURPORT: The heavenly places of enjoyment are divided into three groups: the celestial heavenly planets, the heavenly places on earth, and the bila heavenly places, which are found in the lower regions. Among these three classes of heavenly places (bhauma-svarga-pada-ni), the heavenly places on earth are the eight varṣas other than Bhārata-varṣa. In Bhagavad-gītā (9.21) Kṛṣṇa says, kṣīṇe puṇye martya-lokaṁ viśanti: when the persons living in the heavenly planets exhaust the results of their pious activities, they return to this earth. In this way, they are elevated to the heavenly planets, and then they again fall to the earthly planets. This process is known as brahmāṇḍa bhramaṇa, wandering up and down throughout the universes.

5.17.12: eṣu puruṣāṇām ayuta-puruṣāyur-varṣāṇāṁ deva-kalpānāṁ nāgāyuta-prāṇānāṁ vajra-saṁhanana-bala-vayo-moda-pramudita-mahā-saurata-mithuna-vyavāyāpavarga-varṣa-dhṛtaika-garbha-kalatrāṇāṁ tatra tu tretā-yuga-samaḥ kālo vartate.

SYNONYMS: eṣu—in these (eight) varṣas, or tracts of land; puruṣāṇām—of all the men; ayuta—ten thousand; puruṣa—by the measure of men; āyuḥ-varṣāṇām—of those whose years of life; deva-kalpānām—who are like the demigods; nāga-ayuta-prāṇānām—having the strength of ten thousand elephants; vajra-saṁhanana—by bodies as solid as thunderbolts; bala—by bodily strength; vayaḥ—by youth; moda—by abundant sense enjoyment; pramudita—being excited; mahā-saurata—a great deal of sexual; mithuna—combinations of man and woman; vyavāya-apavarga—at the end of their period of sexual enjoyment; varṣa—in the last year; dhṛta-eka-garbha—who conceive one child; kalatrāṇām—of those who have wives; tatra—there; tu—but; tretā-yuga-samaḥ—exactly like the Tretā-yuga (when there is no tribulation); kālaḥ—time; vartate—exists.

TRANSLATION: In these eight varṣas, or tracts of land, human beings live ten thousand years according to earthly calculations. All the inhabitants are almost like demigods. They have the bodily strength of ten thousand elephants. Indeed, their bodies are as sturdy as thunderbolts. The youthful duration of their lives is very pleasing, and both men and women enjoy sexual union with great pleasure for a long time. After years of sensual pleasure—when a balance of one year of life remains—the wife conceives a child. Thus the standard of pleasure for the residents of these heavenly regions is exactly like that of the human beings who lived during Tretā-yuga.

5.17.13: yatra ha deva-patayaḥ svaiḥ svair gaṇa-nāyakair vihita-mahārhaṇāḥ sarvartu-kusuma-stabaka-phala-kisalaya-śriyānamyamāna-viṭapa-latā-viṭapibhir upaśumbhamāna-rucira-kānanāśramāyatana-varṣa-giri-droṇīṣu tathā cāmala-jalāśayeṣu vikaca-vividha-nava-vanaruhāmoda-mudita-rāja-haṁsa-jala-kukkuṭa-kāraṇḍava-sārasa-cakravākādibhir madhukara-nikarākṛtibhir upakūjiteṣu jala-krīḍādibhir vicitra-vinodaiḥ sulalita-sura-sundarīṇāṁ kāma-kalila-vilāsa-hāsa-līlāvalokākṛṣṭa-mano-dṛṣṭayaḥ svairaṁ viharanti.

SYNONYMS: yatra ha—in those eight tracts of land; deva-patayaḥ—the lords of the demigods, such as Lord Indra; svaiḥ svaiḥ—by their own respective; gaṇa-nāyakaiḥ—leaders of the servants; vihita—furnished with; mahā-arhaṇāḥ—valuable gifts, such as sandalwood pulp and garlands; sarva-ṛtu—in all seasons; kusuma stabaka of bunches of flowers; phala—of fruits; kisalaya-śriyā—by the opulences of shoots; ānamyamāna—being bent down; viṭapa—whose branches; latā—and creepers; viṭapibhiḥ—by many trees; upaśumbhamāna—being fully decorated; rucira—beautiful; kānana—gardens; āśrama-āyatana—and many hermitages; varṣa-giri-droṇīṣu—the valleys between the mountains designating the borders of the tracts of land; tathā—as well as; ca—also; amala-jala-āśayeṣu—in lakes with clear water; vikaca—just fructified; vividha—varieties; nava-vanaruha-āmoda—by the fragrance of lotus flowers; mudita—enthused; rāja-haṁsa—great swans; jala-kukkuṭa—water fowl; kāraṇḍava—aquatic birds called kāraṇḍavas; sārasa—cranes; cakravāka-ādibhiḥ—by birds known as cakravākas and so on; madhukara-nikara-ākṛtibhiḥ—by the bumblebees; upakūjiteṣu—which were made to resound; jala-krīḍā-ādibhiḥ—such as water sports; vicitra—various; vinodaiḥ—by pastimes; su-lalita—attractive; sura-sundarīṇām—of the women of the demigods; kāma—from lust; kalila—born; vilāsa—pastimes; hāsa—smiling; līlā-avaloka—by playful glances; ākṛṣṭa-manaḥ—whose minds are attracted; dṛṣṭayaḥ—and whose vision is attracted; svairam—very freely; viharanti—engage in sportive enjoyment.

TRANSLATION: In each of those tracts of land, there are many gardens filled with flowers and fruits according to the season, and there are beautifully decorated hermitages as well. Between the great mountains demarcating the borders of those lands lie enormous lakes of clear water filled with newly grown lotus flowers. Aquatic birds such as swans, ducks, water chickens, and cranes become greatly excited by the fragrance of lotus flowers, and the charming sound of bumblebees fills the air. The inhabitants of those lands are important leaders among the demigods. Always attended by their respective servants, they enjoy life in gardens alongside the lakes. In this pleasing situation, the wives of the demigods smile playfully at their husbands and look upon them with lusty desires. All the demigods and their wives are constantly supplied with sandalwood pulp and flower garlands by their servants. In this way, all the residents of the eight heavenly varṣas enjoy, attracted by the activities of the opposite sex.

PURPORT: Here is a description of the lower heavenly planets. The inhabitants of those planets enjoy life in a pleasing atmosphere of clear lakes filled with newly grown lotus flowers and gardens filled with fruits, flowers, various kinds of birds and humming bees. In that atmosphere they enjoy life with their very beautiful wives, who are always sexually stimulated.

5.17.14: navasv api varṣeṣu bhagavān nārāyaṇo mahā-puruṣaḥ puruṣāṇāṁ tad-anugrahāyātma-tattva-vyūhenātmanādyāpi sannidhīyate.

SYNONYMS: navasu—in the nine; api—certainly; varṣeṣu—tracts of land known as varṣas; bhagavān—the Supreme Personality of Godhead; nārāyaṇaḥ—Lord Viṣṇu; mahā-puruṣaḥ—the Supreme Person; puruṣāṇām—unto His various devotees; tat-anugrahāya—to show His mercy; ātma-tattva-vyūhena—by expansions of Himself in the quadruple forms Vāsudeva, Saṅkarṣaṇa, Pradyumna and Aniruddha; ātmanā—personally; adya api—until now; sannidhīyate—is near the devotees for accepting their services.

TRANSLATION: To show mercy to His devotees in each of these nine tracts of land, the Supreme Personality of Godhead known as Nārāyaṇa expands Himself in His quadruple principles of Vāsudeva, Saṅkarṣaṇa, Pradyumna and Aniruddha. In this way He remains near His devotees to accept their service.

PURPORT: In this connection, Viśvanātha Cakravartī Ṭhākura informs us that the demigods worship the Supreme Lord in His various Deity forms (arcā-vigraha) because except in the spiritual world, the Supreme Personality of Godhead cannot be directly worshiped in person. In the material world, the Lord is always worshiped as the arcā-vigraha, or Deity in the temple. There is no difference between the arcā-vigraha and the original person, and therefore those who are engaged in worshiping the Deity in the temple in full opulence, even on this planet, should be understood to be directly in touch with the Supreme Personality of Godhead without a doubt... In this regard, the following quotation appears in the Laghu-bhāgavatāmṛta:

pādme tu parama-vyomnaḥ/pūrvādye dik-catuṣṭaye/vāsudevādayo vyūhaś/ catvāraḥ kathitāḥ kramāt/tathā pāda-vibhūtau ca nivasanti kramādi me/jalāvṛti-stha-vaikuṇṭha-/sthita vedavatī-pure/satyordhve vaiṣṇave loke/nityākhye dvārakā-pure śuddhodād uttare śveta-/dvīpe cairāvatī-pure/kṣīrāmbudhi-sthitānte/kroḍa-paryaṅka-dhāmani/sātvatīye kvacit tantre nava vyūhāḥ prakīrtitāḥ/catvāro vāsudevādyā/nārāyaṇa-nṛsiṁhakau/hayagrīvo mahā-kroḍo/brahmā ceti navoditāḥ tatra brahmā tu vijñeyaḥ/pūrvokta-vidhayā hariḥ//

"In the Padma Purāṇa it is said that in the spiritual world the Lord personally expands in all directions and is worshiped as Vāsudeva, Saṅkarṣaṇa, Pradyumna and Aniruddha. The same God is represented by the Deity in this material world, which is only one quarter of His creation. Vāsudeva, Saṅkarṣaṇa, Pradyumna and Aniruddha are also present in the four directions of this material world. There is a Vaikuṇṭhaloka covered with water in this material world, and on that planet is a place called Vedavatī, where Vāsudeva is located. Another planet known as Viṣṇuloka is situated above Satyaloka, and there Saṅkarṣaṇa is present. Similarly, in Dvārakā-purī, Pradyumna is the predominator. On the island known as Śvetadvīpa, there is an ocean of milk, and in the midst of that ocean is a place called Airāvatī-pura, where Aniruddha lies on Ananta. In some of the sātvata-tantras, there is a description of the nine varṣas and the predominating Deity worshiped in each: (1) Vāsudeva, (2) Saṅkarṣaṇa, (3) Pradyumna, (4) Aniruddha, (5) Nārāyaṇa, (6) Nṛsiṁha, (7) Hayagrīva, (8) Mahāvarāha, and (9) Brahmā. "The Lord Brahmā mentioned in this connection is the Supreme Personality of Godhead. When there is no fit human being to empower as Lord Brahma, the Lord Himself takes the position of Lord Brahmā. Tatra brahmā tu vijñeyaḥ pūrvokta-vidhayā hariḥ. That Brahmā mentioned here is Hari Himself.

5.17.15: ilāvṛte tu bhagavān bhava eka eva pumān na hy anyas tatrāparo nirviśati bhavānyāḥ śāpa-nimitta-jño yat-pravekṣyataḥ strī-bhāvas tat paścād vakṣyāmi.

SYNONYMS: ilāvṛte—in the tract of land known as Ilāvṛta-varṣa; tu—but; bhagavān—the most powerful; bhavaḥ—Lord Śiva; eka—only; eva—certainly; pumān—male person; na—not; hi—certainly; anyaḥ—any other; tatra—there; aparaḥ—besides; nirviśati—enters; bhavānyāḥ śāpa-nimitta-jñaḥ—who knows the cause of the curse by Bhavānī, the wife of Lord Śiva; yat-pravekṣyataḥ—of one who forcibly enters that tract of land; strī-bhāvaḥ—transformation into a female; tat—that; paścāt—later; vakṣyāmi—I shall explain.

TRANSLATION: Śukadeva Gosvāmī said: In the tract of land known as Ilāvṛta-varṣa, the only male person is Lord Śiva, the most powerful demigod. Goddess Durgā, the wife of Lord Śiva, does not like any man to enter that land. If any foolish man dares to do so, she immediately turns him into a woman. I shall explain this later [in the Ninth Canto of Śrīmad-Bhāgavatam].

5.17.16: bhavānī-nāthaiḥ strī-gaṇārbuda-sahasrair avarudhyamāno bhagavataś catur-mūrter mahā-puruṣasya turīyāṁ tāmasīṁ mūrtiṁ prakṛtim ātmanaḥ saṅkarṣaṇa-saṁjñām ātma-samādhi-rūpeṇa sannidhāpyaitad abhigṛṇan bhava upadhāvati.

SYNONYMS: bhavānī-nāthaiḥ—by the company of Bhavānī; strī-gaṇa—of females; arbuda-sahasraiḥ—by ten billion; avarudhyamānaḥ—always being served; bhagavataḥ catuḥ-mūrteḥ—the Supreme Personality of Godhead, who is expanded in four; mahā-puruṣasya—of the Supreme Person; turīyām—the fourth expansion; tāmasīm—related to the mode of ignorance; mūrtim—the form; prakṛtim—as the source; ātmanaḥ—of himself (Lord Śiva); saṅkarṣaṇa-saṁjñām—known as Saṅkarṣaṇa; ātma-samādhi-rūpeṇa—by meditating upon Him in trance; sannidhāpya—bringing Him near; etat—this; abhigṛṇan—clearly chanting; bhavaḥ—Lord Śiva; upadhāvati—worships.

TRANSLATION: In Ilāvṛta-varṣa, Lord Śiva is always encircled by ten billion maidservants of goddess Durgā, who minister to him. The quadruple expansion of the Supreme Lord is composed of Vāsudeva, Pradyumna, Aniruddha and Saṅkarṣaṇa. Saṅkarṣaṇa, the fourth expansion, is certainly transcendental, but because his activities of destruction in the material world are in the mode of ignorance, He is known as tāmasī, the Lord's form in the mode of ignorance. Lord Śiva knows that Saṅkarṣaṇa is the original cause of his own existence, and thus he always meditates upon Him in trance by chanting the following mantra.

5.17.21: yam āhur asya sthiti-janma-saṁyamaṁ/tribhir vihīnaṁ yam anantam ṛṣayaḥna veda siddhārtham iva kvacit sthitaṁ/bhū-maṇḍalaṁ mūrdha-sahasra-dhāmasu

SYNONYMS: yam—whom; āhuḥ—they said; asya—of the material world; sthiti—the maintenance; janma—creation; saṁyamam—annihilation; tribhiḥ—these three; vihīnam—without; yam—which; anantam—unlimited; ṛṣayaḥ—all the great sages; na—not; veda—feels; siddha-artham—a mustard seed; iva—like; kvacit—where; sthitam—situated; bhū-maṇḍalam—the universe; mūrdha-sahasra-dhāmasu—on the hundreds and thousands of hoods of the Lord.

TRANSLATION: Lord Śiva continued: All the great sages accept the Lord as the source of creation, maintenance and destruction, although He actually has nothing to do with these activities. Therefore the Lord is called unlimited. Although the Lord in His incarnation as Śeṣa holds all the universes on His hoods, each universe feels no heavier than a mustard seed to Him. Therefore, what person desiring perfection will not worship the Lord?

PURPORT: The incarnation of the Supreme Personality of Godhead known as Śeṣa or Ananta has unlimited strength, fame, wealth, knowledge, beauty and renunciation. As described in this verse, Ananta's strength is so great that the innumerable universes rest on His hoods. He has the bodily features of a snake with thousands of hoods, and since His strength is unlimited, all the universes resting on His hoods feel no heavier than mustard seeds. We can just imagine how insignificant a mustard seed is on the hood of a serpent. In this connection, the reader is referred to Śrī Caitanya-caritāmṛta, Ādi-līlā, Chapter Five, verses 117–125. There it is stated that Lord Viṣṇu's incarnation as the serpentine Ananta Śeṣa Nāga supports all the universes on His hoods. By our calculation, a universe may be very, very heavy, but because the Lord is ananta (unlimited), He feels the weight to be no heavier than a mustard seed.

5.18.1: śrī-śuka uvāca/ tathā ca bhadraśravā nāma dharma-sutas tat-kula-patayaḥ puruṣā bhadrāśva-varṣe sākṣād bhagavato vāsudevasya priyāṁ tanuṁ dharmamayīṁ hayaśīrṣābhidhānāṁ paramena samādhinā sannidhāpyedam abhigṛṇanta upadhāvanti.

SYNONYMS: śrī-śukaḥ uvāca—Śukadeva Gosvāmī said; tathā ca—similarly (just as Lord Śiva worships Saṅkarṣaṇa in Ilāvṛta-varṣa); bhadra-śravā—Bhadraśravā; nāma—known as; dharma-sutaḥ—the son of Dharmarāja; tat—of him; kula-patayaḥ—the chiefs of the dynasty; puruṣāḥ—all the residents; bhadrāśva-varṣe—in the land known as Bhadrāśva-varṣa; sākṣāt—directly; bhagavataḥ—of the Supreme Personality of Godhead; vāsudevasya—of Lord Vāsudeva; priyām tanum—very dear form; dharma-mayīm—the director of all religious principles; hayaśīrṣa-abhidhānām—the Lord's incarnation named Hayaśīrṣa (also called Hayagrīva); paramena samādhinā—with the highest form of trance; sannidhāpya—coming near; idam—this; abhigṛṇantaḥ—chanting; upadhāvanti—they worship.

TRANSLATION: Śrī Śukadeva Gosvāmī said: Bhadraśravā, the son of Dharmarāja, rules the tract of land known as Bhadrāśva-varṣa. Just as Lord Śiva worships Saṅkarṣaṇa in Ilāvṛta-varṣa, Bhadraśravā, accompanied by his intimate servants and all the residents of the land, worships the plenary expansion of Vāsudeva known as Hayaśīrṣa. Lord Hayaśīrṣa is very dear to the devotees, and He is the director of all religious principles. Fixed in the topmost trance,

Bhadraśravā and his associates offer their respectful obeisances to the Lord and chant the following prayers with careful pronunciation.

5.18.6: vedān yugānte tamasā tiraskṛtān/rasātalād yo nṛ-turaṅga-vigrahaḥ
pratyādade vai kavaye 'bhiyācate/tasmai namas te 'vitathehitāya iti

SYNONYMS: vedān—the four Vedas; yuga-ante—at the end of the millennium; tamasā—by the demon of ignorance personified; tiraskṛtān—stolen away; rasātalāt—from the lowest planetary system (Rasātala); yaḥ—who (the Supreme Personality of Godhead); nṛ-turaṅga-vigrahaḥ—assuming the form of half-horse, half-man; pratyādade—returned; vai—indeed; kavaye—to the supreme poet (Lord Brahmā); abhiyā-cate—when he asked for them; tasmai—unto Him (the form of Hayagrīva); namaḥ—my respectful obeisances; te—to You; avitatha-īhitāya—whose resolution never fails; iti—thus.

TRANSLATION: At the end of the millennium, ignorance personified assumed the form of a demon, stole all the Vedas and took them down to the planet of Rasātala. The Supreme Lord, however, in His form of Hayagrīva retrieved the Vedas and returned them to Lord Brahmā when he begged for them. I offer my respectful obeisances unto the Supreme Lord, whose determination never fails.

5.18.7: hari-varṣe cāpi bhagavān nara-hari-rūpeṇāste; tad-rūpa-grahaṇa-nimittam uttaratrābhidhāsye; tad dayitaṁ rūpaṁ mahā-puruṣa-guṇa-bhājano mahā-bhāgavato daitya-dānava-kula-tīrthīkaraṇa-śīlā-caritaḥ prahlādo 'vyavadhānānanya-bhakti-yogena saha tad-varṣa-puruṣair upāste idaṁ codāharati.

SYNONYMS: hari-varṣe—in the tract of land known as Harivarṣa; ca—also; api—indeed; bhagavān—the Supreme Personality of Godhead; nara-hari-rūpeṇa—His form of Nṛsiṁhadeva; āste—is situated; tat-rūpa-grahaṇa-nimittam—the reason why Lord Kṛṣṇa (Keśava) assumed the form of Nṛsiṁha; uttaratra—in later chapters; abhidhāsye—I shall describe; tat—that; dayitam—most pleasing; rūpam—form of the Lord; mahā-puruṣa-guṇa-bhājanaḥ—Prahlāda Mahārāja, who is the abode of all the good qualities of great personalities; mahā-bhāgavataḥ—the topmost devotee; daitya-dānava-kula-tīrthī-karaṇa-śīlā-caritaḥ—whose activities and character were so exalted that he delivered all the daityas (demons) born in his family; prahlādaḥ—Mahārāja Prahlāda; avyavadhāna-ananya-bhakti-yogena—by uninterrupted and unflinching devotional service; saha—with; tat-varṣa-puruṣaiḥ—the inhabitants of Hari-varṣa; upāste—offers obeisances to and worships; idam—this; ca—and; udāharati—chants.

TRANSLATION: Śukadeva Gosvāmī continued: My dear King, Lord Nṛsiṁhadeva resides in the tract of land known as Hari-varṣa. In the Seventh Canto of Śrīmad-Bhāgavatam, I shall describe to you how Prahlāda Mahārāja caused the Lord to assume the form of Nṛsiṁhadeva. Prahlāda Mahārāja, the topmost devotee of the Lord, is a reservoir of all the good qualities of great personalities. His character and activities have delivered all the fallen members of his demoniac family. Lord Nṛsiṁhadeva is very dear to this exalted personality. Thus Prahlāda Mahārāja, along with his servants and all the denizens of Harivarṣa, worships Lord Nṛsiṁhadeva by chanting the following mantra.

5.18.15: ketumāle 'pi bhagavān kāmadeva-svarūpeṇa lakṣmyāḥ priya-cikīrṣayā prajāpater duhitṝṇāṁ putrāṇāṁ tad-varṣa-patīnāṁ puruṣāyuṣāho-rātra-parisaṅkhyānānāṁ yāsāṁ garbhā mahā-puruṣa-mahāstra-tejasodvejita-manasāṁ vidhvastā vyasavaḥ saṁvatsarānte vinipatanti.

SYNONYMS: ketumāle—in the tract of land known as Ketumāla-varṣa; api—also; bhagavān—the Supreme Personality of Godhead, Lord Viṣṇu; kāmadeva-svarūpeṇa—in the form of Kāmadeva (Cupid or Pradyumna); lakṣmyāḥ—of the goddess of fortune; priya-cikīrṣayā—with a desire to bring about the satisfaction; prajāpateḥ—of Prajāpati; duhitṝṇām—of the daughters; putrāṇām—of the sons; tat-varṣa-patīnām—the ruler of that land; puruṣa-āyuṣā—in a human lifetime (about one hundred years); ahaḥ-rātra—the days and nights; parisaṅkhyānānām—which equal in number; yāsām—of whom (the daughters); garbhāḥ—fetuses; mahā-puruṣa—of the Supreme Personality of Godhead; mahā-astra—of the great weapon (the disc); tejasā—by the effulgence; udvejita-manasām—whose minds are agitated; vidhvastāḥ—ruined; vyasavaḥ—dead; saṁvatsara-ante—at the end of the year; vinipatanti—fall down.

TRANSLATION: Śukadeva Gosvāmī continued: In the tract of land called Ketumāla-varṣa, Lord Viṣṇu lives in the form of Kāmadeva, only for the satisfaction of His devotees. These include Lakṣmījī [the goddess of fortune], the Prajāpati Saṁvatsara and all of Saṁvatsara's sons and daughters. The daughters of Prajāpati are considered the controlling deities of the nights, and his sons are considered the controllers of the days. The Prajāpati's offspring number 36,000, one for each day and each night in the lifetime of a human being. At the end of each year, the Prajāpati's daughters become very agitated upon seeing the extremely effulgent disc of the Supreme Personality of Godhead, and thus they all suffer miscarriages.

5.18.17: tad bhagavato māyāmayaṁ rūpaṁ parama-samādhi-yogena ramā devī saṁvatsarasya rātriṣu prajāpater duhitṛbhir upetāhaḥsu ca tad-bhartṛbhir upāste idaṁ codāharati.

SYNONYMS: tat—that; bhagavataḥ—of the Supreme Personality of Godhead; māyā-mayam—full of affection for the devotees; rūpam—form; parama—highest; samādhi-yogena—by absorption of the mind in the service of the Lord; ramā—the goddess of fortune; devī—divine woman; saṁvatsarasya—known as Saṁvatsara; rātriṣu—during the nights; prajāpateḥ—of Prajāpati; duhitṛbhiḥ—

with the daughters; upeta—combined; ahaḥsu—during the days; ca—also; tat-bhartṛbhiḥ—with the husbands; upāste—worships; idam—this; ca—also; udāharati—chants.

TRANSLATION: Accompanied during the daytime by the sons of the Prajāpati [the predominating deities of the days] and accompanied at night by his daughters [the deities of the nights], Lakṣmīdevī worships the Lord during the period known as the Saṁvatsara in His most merciful form as Kāmadeva. Fully absorbed in devotional service, she chants the following mantras.

5.18.24: ramyake ca bhagavataḥ priyatamaṁ mātsyam avatāra-rūpaṁ tad-varṣa-puruṣasya manoḥ prāk-pradarśitaṁ sa idānīm api mahatā bhakti-yogenārādhayatīdaṁ codāharati.

SYNONYMS: ramyake ca—also in Ramyaka-varṣa; bhagavataḥ—of the Supreme Personality of Godhead; priya-tamam—the foremost; mātsyam—fish; avatāra-rūpam—the form of the incarnation; tat-varṣa-puruṣasya—of the ruler of that land; manoḥ—Manu; prāk—previously (at the end of the Cākṣuṣa-manvantara); pradarśitam—exhibited; saḥ—that Manu; idānīm api—even until now; mahatā bhakti-yogena—by dint of advanced devotional service; ārādhayati—worships the Supreme Personality of Godhead; idam—this; ca—and; udāharati—chants.

TRANSLATION: Śukadeva Gosvāmī continued: In Ramyaka-varṣa, where Vaivasvata Manu rules, the Supreme Personality of Godhead appeared as Lord Matsya at the end of the last era [the Cākṣuṣa-manvantara]. Vaivasvata Manu now worships Lord Matsya in pure devotional service and chants the following mantra.

5.18.28: bhavān yugāntārṇava ūrmi-mālini/kṣoṇīm imām oṣadhi-vīrudhāṁ nidhim
mayā sahoru kramate 'ja ojasā/tasmai jagat-prāṇa-gaṇātmane nama iti

SYNONYMS: bhavān—Your Lordship; yuga-anta-arṇave—in the water of devastation at the end of the millennium; ūrmi-mālini—possessing rows of big waves; kṣoṇīm—the planet earth; imām—this; oṣadhi-vīrudhām—of all kinds of herbs and drugs; nidhim—the storehouse; mayā—me; saha—with; uru—great; kramate—You roamed; aja—O unborn one; ojasā—with speed; tasmai—unto Him; jagat—of the entire universe; prāṇa-gaṇa-ātmane—the ultimate source of life; namaḥ—my respectful obeisances; iti—thus.

TRANSLATION: O almighty Lord, at the end of the millennium this planet earth, which is the source of all kinds of herbs, drugs and trees, was inundated by water and drowned beneath the devastating waves. At that time, You protected me along with the earth and roamed the sea with great speed. O unborn one, You are the actual maintainer of the entire universal creation, and therefore You are the cause of all living entities. I offer my respectful obeisances unto You.

5.18.29: hiraṇmaye 'pi bhagavān nivasati kūrma-tanuṁ bibhrāṇas tasya tat priyatamāṁ tanum aryamā saha varṣa-puruṣaiḥ pitṛ-gaṇādhipatir upadhāvati mantram imaṁ cānujapati.

SYNONYMS: hiraṇmaye—in Hiraṇmaya-varṣa; api—indeed; bhagavān—the Supreme Personality of Godhead; nivasati—resides; kūrma-tanum—the body of a tortoise; bibhrāṇaḥ—manifesting; tasya—of the Supreme Personality of Godhead; tat—that; priya-tamām—dearmost; tanum—body; aryamā—Aryamā, the chief resident of Hiraṇmaya-varṣa; saha—with; varṣa-puruṣaiḥ—the people of that tract of land; pitṛ-gaṇa-adhipatiḥ—who is the chief of the pitās; upadhāvati—worships in devotional service; mantram—hymn; imam—this; ca—also; anujapati—chants.

TRANSLATION: Śukadeva Gosvāmī continued: In Hiraṇmaya-varṣa, the Supreme Lord, Viṣṇu, lives in the form of a tortoise [kūrma-śarīra]. This most dear and beautiful form is always worshiped there in devotional service by Aryamā, the chief resident of Hiraṇmaya-varṣa, along with the other inhabitants of that land. They chant the following hymns.

5.18.34: uttareṣu ca kuruṣu bhagavān yajña-puruṣaḥ kṛta-varāha-rūpa āste taṁ tu devī haiṣā bhūḥ saha kurubhir askhalita-bhakti-yogenopadhāvati imāṁ ca paramām upaniṣadam āvartayati.

SYNONYMS: uttareṣu—on the northern side; ca—also; kuruṣu—in the tract of land known as Kuru; bhagavān—the Supreme Personality of Godhead; yajña-puruṣaḥ—who accepts all results of sacrifices; kṛta-varāha-rūpaḥ—having accepted the form of a boar; āste—exists eternally; tam—Him; tu—certainly; devī—the goddess; ha—certainly; eṣā—this; bhūḥ—planet earth; saha—along with; kurubhiḥ—the inhabitants of the land known as Kuru; askhalita—unfailing; bhakti-yogena—by devotional service; upadhāvati—worship; imām—this; ca—also; paramām upaniṣadam—the supreme Upaniṣad (the process by which one can approach the Lord); āvartayati—chants again and again for the purpose of practice.

TRANSLATION: Śukadeva Gosvāmī said: Dear King, the Supreme Lord in His boar incarnation, who accepts all sacrificial offerings, lives in the northern part of Jambūdvīpa. There, in the tract of land known as Uttarakuru-varṣa, mother earth and all the other inhabitants worship Him with unfailing devotional service by repeatedly chanting the following Upaniṣad mantra.

5.18.39: pramathya daityaṁ prativāraṇaṁ mṛdhe/yo māṁ rasāyā jagad-ādi-sūkaraḥ kṛtvāgra-daṁṣṭre niragād udanvataḥ/krīḍann ivebhaḥ praṇatāsmi taṁ vibhum iti

SYNONYMS: pramathya—after killing; daityam—the demon; prativāraṇam—most formidable opponent; mṛdhe—in the fight; yaḥ—He who; mām—me (the earth); rasāyāḥ—fallen to the bottom of the universe; jagat—in this material world; ādi-sūkaraḥ—the original form of a boar; kṛtvā—keeping it; agra-daṁṣṭre—on the end of the tusk; niragāt—came out of the water; udanvataḥ—from the Garbhodaka Ocean; krīḍan—playing; iva—like; ibhaḥ—elephant; praṇatā asmi—I bow down; tam—to Him; vibhum—the Supreme Lord; iti—thus.

TRANSLATION: My Lord, as the original boar within this universe, You fought and killed the great demon Hiraṇyakṣa. Then You lifted me [the earth] from the Garbhodaka Ocean on the end of Your tusk, exactly as a sporting elephant plucks a lotus flower from the water. I bow down before You.

5.19.1: śrī-śuka uvāca/ kimpuruṣe varṣe bhagavantam ādi-puruṣaṁ lakṣmaṇāgrajaṁ sītābhirāmaṁ rāmaṁ tac-caraṇa-sannikarṣābhirataḥ parama-bhāgavato hanumān saha kimpuruṣair avirata-bhaktir upāste.

SYNONYMS: śrī-śukaḥ uvāca—Śrī Śukadeva Gosvāmī continued to speak; kimpuruṣe varṣe—the tract of land known as Kimpuruṣa; bhagavantam—the Supreme Personality of Godhead; ādi-puruṣam—the original cause of all causes; lakṣmaṇa-agra-jam—the elder brother of Lakṣmaṇa; sītā-abhirāmam—who is very pleasing to mother Sītā, or who is the husband of Sītādevī; rāmam—Lord Rāmacandra; tat-caraṇa-sannikarṣa-abhirataḥ—one always engaged in service at the lotus feet of Lord Rāmacandra; parama-bhāgavataḥ—the great devotee celebrated throughout the universe; hanumān—His Grace Hanumānjī; saha—with; kimpuruṣaiḥ—the inhabitants of the tract of land known as Kimpuruṣa; avirata—continuous; bhaktiḥ—who possesses devotional service; upāste—worships.

TRANSLATION: Śrīla Śukadeva Gosvāmī said: My dear King, in Kimpuruṣa-varṣa the great devotee Hanumān is always engaged with the inhabitants of that land in devotional service to Lord Rāmacandra, the elder brother of Lakṣmaṇa and dear husband of Sītādevī.

5.19.10: taṁ bhagavān nārado varṇāśramavatībhir bhāratībhiḥ prajābhir bhagavat-proktābhyāṁ sāṅkhya-yogābhyāṁ bhagavad-anubhāvopavarṇanaṁ sāvarṇer upadekṣyamāṇaḥ parama-bhakti-bhāvenopasarati idaṁ cābhigṛṇāti.

SYNONYMS: tam—Him (Nara-Nārāyaṇa); bhagavān—the most powerful saintly person; nāradaḥ—the great sage Nārada; varṇa-āśrama-vatībhiḥ—by followers of the institution of the four varṇas and four āśramas; bhāratībhiḥ—of the land known as Bhārata-varṣa (India); prajābhiḥ—who are the inhabitants; bhagavat-proktābhyām—which was stated by the Supreme Personality of Godhead; sāṅkhya—by the sāṅkhya-yoga system (the analytical study of material conditions); yogābhyām—by practice of the yoga system; bhagavat-anubhāva-upavarṇanam—which describes the process of God realization; sāvarṇeḥ—unto Sāvarṇi Manu; upadekṣyamāṇaḥ—instructing; parama-bhakti-bhāvena—in greatly ecstatic loving service to the Lord; upasarati—serves the Lord; idam—this; ca—and; abhigṛṇāti—chants.

TRANSLATION: In his own book, known as Nārada Pañcarātra, Bhagavān Nārada has very vividly described how to work to achieve the ultimate goal of life—devotion—through knowledge and through execution of the mystic yoga system. He has also described the glories of the Lord, the Supreme Personality of Godhead. The great sage Nārada instructed the tenets of this transcendental literature to Sāvarṇi Manu in order to teach those inhabitants of Bhārata-varṣa who strictly follow the principles of varṇāśrama-dharma how to achieve the devotional service of the Lord. Thus Nārada Muni, along with the other inhabitants of Bhārata-varṣa, always engages in the service of Nara-Nārāyaṇa, and he chants as follows.

## Part VII: Bhārata-Varṣa

5.19.16: bhārate 'py asmin varṣe saric-chailāḥ santi bahavo malayo maṅgala-prastho maināka s trikūṭa ṛṣabhaḥ kūṭakaḥ kollakaḥ sahyo devagirir ṛṣyamūkaḥ śrī-śailo veṅkaṭo mahendro vāridhāro vindhyaḥ śuktimān ṛkṣagiriḥ pāriyātro droṇaś citrakūṭo govardhano raivatakaḥ kakubho nīlo gokāmukha indrakīlaḥ kāmagirir iti cānye ca śata-sahasraśaḥ śailās teṣāṁ nitamba-prabhavā nadā nadyaś ca santy asaṅkhyātāḥ.

SYNONYMS:bhārate—in the land of Bhārata-varṣa; api—also; asmin—in this; varṣe—tract of land; sarit—rivers; śailāḥ—mountains; santi—there are; bahavaḥ—many; malayaḥ—Malaya; maṅgala-prasthaḥ—Maṅgala-prastha; mainākaḥ—Maināka; tri-kūṭaḥ—Trikūṭa; ṛṣabhaḥ—Ṛṣabha; kūṭakaḥ—Kūṭaka; kollakaḥ—Kollaka; sahyaḥ—Sahya; devagiriḥ—Devagiri; ṛṣya-mūkaḥ—Ṛṣyamūka; śrī-śailaḥ—Śrī-śaila; veṅkaṭaḥ—Veṅkaṭa; mahendraḥ—Mahendra; vāri-dhāraḥ—Vāridhāra; vindhyaḥ—Vindhya; śuktimān—Śuktimān; ṛkṣa-giriḥ—Ṛkṣagiri; pāriyātraḥ—Pāriyātra; droṇaḥ—Droṇa; citra-kūṭaḥ—Citrakūṭa; govardhanaḥ—Govardhana; raivatakaḥ—Raivataka; kakubhaḥ—Kakubha; nīlaḥ—Nīla; gokāmukhaḥ—Gokāmukha; indrakīlaḥ—Indrakīla; kāma-giriḥ—Kāmagiri; iti—thus; ca—and; anye—others; ca—also; śata-sahasraśaḥ—many hundreds and thousands; śailāḥ—mountains; teṣām—of them; nitamba-prabhavāḥ—born of the slopes; nadāḥ—big rivers; nadyaḥ—small rivers; ca—and; santi—there are; asaṅkhyātāḥ—innumerable.

TRANSLATION: In the tract of land known as Bhārata-varṣa, as in Ilāvṛta-varṣa, there are many mountains and rivers. Some of the mountains are known as Malaya, Maṅgala-prastha, Maināka, Trikūṭa, Ṛṣabha, Kūṭaka, Kollaka, Sahya, Devagiri, Ṛṣyamūka, Śrī-śaila, Veṅkaṭa, Mahendra, Vāridhāra, Vindhya, Śuktimān, Ṛkṣagiri, Pāriyātra, Droṇa, Citrakūṭa, Govardhana, Raivataka, Kakubha, Nīla, Gokāmukha, Indrakīla and Kāmagiri. Besides these, there are many other hills, with many large and small rivers flowing from their slopes.

5.19.17-18: etāsām apo bhāratyaḥ prajā nāmabhir eva punantīnām ātmanā copaspṛśanti; candravasā tāmraparṇī avaṭodā kṛtamālā vaihāyasī kāverī veṇī payasvinī śarkarāvartā tuṅgabhadrā kṛṣṇāveṇyā bhīmarathī godāvarī nirvindhyā payoṣṇī tāpī revā surasā narmadā carmaṇvatī sindhur andhaḥ śoṇaś ca nadau mahānadī vedasmṛtir ṛṣikulyā trisāmā kauśikī mandākinī yamunā sarasvatī dṛṣadvatī gomatī sarayū rodhasvatī saptavatī suṣomā śatadrūś candrabhāgā marudvṛdhā vitastā asiknī viśveti mahā-nadyaḥ.

SYNONYMS: etāsām—of all these; apaḥ—water; bhāratyaḥ—of Bhārata-varṣa (India); prajāḥ—the residents; nāmabhiḥ—by the names; eva—only; punantīnām—are purifying; ātmanā—by the mind; ca—also; upaspṛśanti—touch; candra-vasā—Candravasā; tāmra-parṇī—Tāmraparṇī; avaṭodā—Avaṭodā; kṛta-mālā—Kṛtamālā; vaihāyasī—Vaihāyasī; kāverī—Kāverī; veṇī—Veṇī; payasvinī—Payasvinī; śarkarāvartā—Śarkarāvartā; tuṅga-bhadrā—Tuṅgabhadrā; kṛṣṇā-veṇyā—Kṛṣṇāveṇyā; bhīma-rathī—Bhīmarathī; godāvarī—Godāvarī; nirvindhyā—Nirvindhyā; payoṣṇī—Payoṣṇī; tāpī—Tāpī; revā—Revā; surasā—Surasā; narmadā—Narmadā; carmaṇvatī—Carmaṇvatī; sindhuḥ—Sindhu; andhaḥ—Andha; śoṇaḥ—Śoṇa; ca—and; nadau—two rivers; mahā-nadī—Mahānadī; veda-smṛtiḥ—Vedasmṛti; ṛṣi-kulyā—Ṛṣikulyā; tri-sāmā—Trisāmā; kauśikī—Kauśikī; mandākinī—Mandākinī; yamunā—Yamunā; sarasvatī—Sarasvatī; dṛṣadvatī—Dṛṣadvatī; gomatī—Gomatī; sarayū—Sarayū; rodhasvatī—Rodhasvatī; saptavatī—Saptavatī; suṣomā—Suṣomā; śata-drūḥ—Śatadrū; candrabhāgā—Candrabhāgā; marudvṛdhā—Marudvṛdhā; vitastā—Vitastā; asiknī—Asiknī; viśvā—Viśvā; iti—thus; mahā-nadyaḥ—big rivers.

TRANSLATION: Two of the rivers—the Brahmaputra and the Śoṇa—are called nadas, or main rivers. These are other great rivers that are very prominent: Candravasā, Tāmraparṇī, Avaṭodā, Kṛtamālā, Vaihāyasī, Kāverī, Veṇī, Payasvinī, Śarkarāvartā, Tuṅgabhadrā, Kṛṣṇāveṇyā, Bhīmarathī, Godāvarī, Nirvindhyā, Payoṣṇī, Tāpī, Revā, Surasā, Narmadā, Carmaṇvatī, Mahānadī, Vedasmṛti, Ṛṣikulyā, Trisāmā, Kauśikī, Mandākinī, Yamunā, Sarasvatī, Dṛṣadvatī, Gomatī, Sarayū, Rodhasvatī, Saptavatī, Suṣomā, Śatadrū, Candrabhāgā, Marudvṛdhā,

Vitastā, Asiknī and Viśvā. The inhabitants of Bhārata-varṣa are purified because they always remember these rivers. Sometimes they chant the names of these rivers as mantras, and sometimes they go directly to the rivers to touch them and bathe in them. Thus the inhabitants of Bhārata-varṣa become purified.

PURPORT: All these rivers are transcendental. Therefore one can be purified by remembering them, touching them or bathing in them. This practice is still going on.

5.19.21: etad eva hi devā gāyanti——/aho amīṣāṁ kim akāri śobhanaṁ/prasanna eṣāṁ svid uta svayaṁ hariḥ yair janma labdhaṁ nṛṣu bhāratājire/ mukunda-sevaupayikaṁ spṛhā hi naḥ

SYNONYMS: etat—this; eva—indeed; hi—certainly; devāḥ—all the demigods; gāyanti—chant; aho—oh; amīṣām—of these inhabitants of Bhāratavarṣa; kim—what; akāri—was done; śobhanam—pious, beautiful activities; prasannaḥ—pleased; eṣām—upon them; svit—or; uta—it is said; svayam—personally; hariḥ—the Supreme Personality of Godhead; yaiḥ—by whom; janma—birth; labdham—obtained; nṛṣu—in human society; bhārata-ajire—in the courtyard of Bhārata-varṣa; mukunda—the Supreme Personality of Godhead, who can offer liberation; sevā-aupayikam—which is the means of serving; spṛhā—desire; hi—indeed; naḥ—our.

TRANSLATION: Since the human form of life is the sublime position for spiritual realization, all the demigods in heaven speak in this way: How wonderful it is for these human beings to have been born in the land of Bhārata-varṣa. They must have executed pious acts of austerity in the past, or the Supreme Personality of Godhead Himself must have been pleased with them. Otherwise, how could they engage in devotional service in so many ways? We demigods can only aspire to achieve human births in Bhārata-varṣa to execute devotional service, but these human beings are already engaged there.

PURPORT:These facts are further explained in Caitanya-caritāmṛta (Ādi 9.41): bhārata-bhūmite haila manuṣya-janma yāra janma sārthaka kari' kara para-upakāra. "One who has taken his birth as a human being in the land of India [Bhārata-varṣa] should make his life successful and work for the benefit of all other people." There are many facilities in India, Bhārata-varṣa, for executing devotional service. In Bhārata-varṣa, all the ācāryas contributed their experience, and Śrī Caitanya Mahāprabhu personally appeared to teach the people of Bhārata-varṣa how to progress in spiritual life and be fixed in devotional service to the Lord. From all points of view, Bhārata-varṣa is the special land where one can very easily understand the process of devotional service and adopt it to make his life successful. If one makes his life successful in devotional service and then preaches devotional service in other parts of the world, people throughout the world will actually benefit.

5.19.22: kiṁ duṣkarair naḥ kratubhis tapo-vratair/dānādibhir vā dyujayena
phalgunā na yatra nārāyaṇa-pāda-paṅkaja-/smṛtiḥ
pramuṣṭātiśayendriyotsavāt

SYNONYMS: kim—what is the value; duṣkaraiḥ—very difficult to perform; naḥ—our; kratubhiḥ—with performances of sacrifice; tapaḥ—with austerities; vrataiḥ—vows; dāna-ādibhiḥ—with executing charitable activities and so on; vā—or; dyujayena—with achieving the heavenly kingdom; phalgunā—which is insignificant; na—no; yatra—where; nārāyaṇa-pāda-paṅkaja—of the lotus feet of Lord Nārāyaṇa; smṛtiḥ—the remembrance; pramuṣṭa—lost; atiśaya—excessive; indriya-utsavāt—due to material sense gratification.

TRANSLATION: The demigods continue: After performing the very difficult tasks of executing Vedic ritualistic sacrifices, undergoing austerities, observing vows and giving charity, we have achieved this position as inhabitants of the heavenly planets. But what is the value of this achievement? Here we are certainly very engaged in material sense gratification, and therefore we can hardly remember the lotus feet of Lord Nārāyaṇa. Indeed, because of our excessive sense gratification, we have almost forgotten His lotus feet.

PURPORT: The land of Bhārata-varṣa is so exalted that by taking birth there one can not only attain the heavenly planets but also go directly back home, back to Godhead...Therefore the demigods even regret having been elevated to the higher planetary systems. The denizens of the heavenly planets regret that they could not take full advantage of being born in the land of Bhārata-varṣa. Instead, they became captivated by a higher standard of sense gratification, and therefore they forgot the lotus feet of Lord Nārāyaṇa at the time of death.

5.19.23: kalpāyuṣāṁ sthānajayāt punar-bhavāt/kṣaṇāyuṣāṁ bhārata-bhūjayo
varam kṣaṇena martyena kṛtaṁ manasvinaḥ/sannyasya saṁyānty
abhayaṁ padaṁ hareḥ

SYNONYMS: kalpa-āyuṣām—of those who have a life duration of many millions of years, like Lord Brahmā; sthāna-jayāt—than achievement of the place or planetary systems; punaḥ-bhavāt—which is liable to birth, death and old age; kṣaṇa-āyuṣām—of persons who have only one hundred years of life; bhārata-bhū-jayaḥ—a birth in the land of Bhārata-varṣa; varam—more valuable; kṣaṇena—for such a short life; martyena—by the body; kṛtam—the work executed; manasvinaḥ—those actually understanding the value of life; sannyasya—surrendering unto the lotus feet of Kṛṣṇa; saṁyānti—they achieve; abhayam—

where there is no anxiety; padam—the abode; hareḥ—of the Supreme Personality of Godhead.

TRANSLATION: A short life in the land of Bharata-varṣa is preferable to a life achieved in Brahmaloka for millions and billions of years because even if one is elevated to Brahmaloka, he must return to repeated birth and death. Although life in Bhārata-varṣa, in a lower planetary system, is very short, one who lives there can elevate himself to full Kṛṣṇa consciousness and achieve the highest perfection, even in this short life, by fully surrendering unto the lotus feet of the Lord. Thus one attains Vaikuṇṭhaloka, where there is neither anxiety nor repeated birth in a material body.

5.19.25: prāptā nṛ-jātiṁ tv iha ye ca jantavo/jñāna-kriyā-dravya-kalāpa-sambhṛtām na vai yaterann apunar-bhavāya te/bhūyo vanaukā iva yānti bandhanam

SYNONYMS: prāptāḥ—who have obtained; nṛ-jātim—a birth in human society; tu—certainly; iha—in this land of Bhārata-varṣa; ye—those who; ca—also; jantavaḥ—the living beings; jñāna—with knowledge; kriyā—with activities; dravya—of ingredients; kalāpa—with a collection; sambhṛtām—full; na—not; vai—certainly; yateran—endeavor; apunaḥ-bhavāya—for the position of immortality; te—such persons; bhūyaḥ—again; vanaukāḥ—birds; iva—like; yānti—go; bandhanam—to bondage.

TRANSLATION: Bhārata-varṣa offers the proper land and circumstances in which to execute devotional service, which can free one from the results of jñāna and karma. If one obtains a human body in the land of Bhārata-varṣa, with clear sensory organs with which to execute the saṅkīrtana-yajña, but in spite of this opportunity he does not take to devotional service, he is certainly like liberated forest animals and birds that are careless and are therefore again bound by a hunter.

PURPORT: In the land of Bhārata-varṣa one can very easily perform the saṅkīrtana-yajña, which consists of śravaṇaṁ kīrtanaṁ viṣṇoḥ, or one can perform other methods of devotional service, such as smaraṇaṁ vandanam arcanaṁ dāsyaṁ sakhyam and ātma-nivedanam. In Bhārata-varṣa one has the opportunity to visit many holy places, especially Lord Caitanya's birthsite and Lord Kṛṣṇa's birthsite—Navadvīpa and Vṛndāvana—where there are many pure devotees who have no desire other than to execute devotional service (anyābhilāṣitā-śūnyaṁ jñāna-karmādy-anāvṛtam), and one may thus become free from the bondage of material conditions...The inhabitants of places other than Bhārata-varṣa have facilities for material enjoyment, but they do not have the same facility to take to Kṛṣṇa consciousness. Therefore Śrī Caitanya Mahāprabhu has advised that one who has taken birth as a human being in Bhārata-varṣa must first realize himself as a part and parcel of Kṛṣṇa, and after taking to Kṛṣṇa consciousness, he must distribute this knowledge all over the world.

5.19.26: yaiḥ śraddhayā barhiṣi bhāgaśo havir/niruptam iṣṭaṁ vidhi-mantra-vastutaḥ ekaḥ pṛthaṅ-nāmabhir āhuto mudā/gṛhṇāti pūrṇaḥ svayam āśiṣāṁ prabhuḥ

SYNONYMS: yaiḥ—by whom (the inhabitants of Bhārata-varṣa); śraddhayā—faith and confidence; barhiṣi—in the performance of Vedic ritualistic sacrifices; bhāgaśaḥ—by division; haviḥ—oblations; niruptam—offered; iṣṭam—to the desired deity; vidhi—by the proper method; mantra—by reciting mantras; vastutaḥ—with the proper ingredients; ekaḥ—that one Supreme Personality of Godhead; pṛthak—separate; nāmabhiḥ—by names; āhutaḥ—called; mudā—with great happiness; gṛhṇāti—He accepts; pūrṇaḥ—the Supreme Lord. who is full in Himself; svayam—personally; āśiṣām—of all benedictions; prabhuḥ—the bestower.

TRANSLATION: In India [Bhārata-varṣa], there are many worshipers of the demigods, the various officials appointed by the Supreme Lord, such as Indra, Candra and Sūrya, all of whom are worshiped differently. The worshipers offer the demigods their oblations, considering the demigods part and parcel of the whole, the Supreme Lord. Therefore the Supreme Personality of Godhead accepts these offerings and gradually raises the worshipers to the real standard of devotional service by fulfilling their desires and aspirations. Because the Lord is complete, He offers the worshipers the benedictions they desire even if they worship only part of His transcendental body.

5.19.28: yady atra naḥ svarga-sukhāvaśeṣitaṁ/sviṣṭasya sūktasya kṛtasya śobhanam tenājanābhe smṛtimaj janma naḥ syād/varṣe harir yad bhajatāṁ śaṁ tanoti

SYNONYMS: yadi—if; atra—in this heavenly planet; naḥ—of us; svarga-sukha-avaśeṣitam—whatever remains after the enjoyment of heavenly happiness; su-iṣṭasya—of a perfect sacrifice; su-uktasya—of diligently studying the Vedic literature; kṛtasya—of having performed a kind act; śobhanam—the resultant actions; tena—by such a resultant action; ajanābhe—in the land of Bhārata-varṣa; smṛti-mat janma—a birth enabling one to remember the lotus feet of the Lord; naḥ—of us; syāt—let there be; varṣe—in the land; hariḥ—the Supreme Personality of Godhead; yat—wherein; bhajatām—of the devotees; śam tanoti—expands the auspiciousness.

TRANSLATION: We are now living in the heavenly planets, undoubtedly as a result of our having performed ritualistic ceremonies, pious activities and yajñas and having studied the Vedas. However, our lives here will one day be finished. We pray that at that time, if any merit remains from our pious activities, we may again take birth in Bhārata-varṣa as human beings able to remember the lotus feet of the Lord. The Lord is so kind that He personally comes to the land of Bhārata-varṣa and expands the good fortune of its people.

PURPORT: It is certainly as a result of pious activities that one takes birth in the heavenly planets, but from those planets one must nevertheless come down again to earth, as stated in Bhagavad-gītā (kṣīṇe puṇye martya-lokaṁ viśanti). Even the demigods must return to earth to work like ordinary men when the results of their pious activities expire. Nevertheless, the demigods desire to come to the land of Bhārata-varṣa if even a small portion of the merits of their pious activities remains. In other words, to take birth in Bhārata-varṣa, one must perform more pious activities than the demigods. In Bhārata-varṣa one is naturally Kṛṣṇa conscious, and if one further cultivates his Kṛṣṇa consciousness, by the grace of Kṛṣṇa he certainly expands his good fortune by becoming perfect in Kṛṣṇa consciousness and very easily going back home, back to Godhead. In many other places in the Vedic literature it is found that even the demigods want to come to this land of Bhārata-varṣa. A foolish person may desire to be promoted to the heavenly planets as a result of his pious activities, but even the demigods from the heavenly planets want to come to Bhārata-varṣa and achieve bodies that may be very easily used to cultivate Kṛṣṇa consciousness...Therefore those already born in Bhārata-varṣa should take lessons from the śāstras and guru and should fully take advantage of the mercy of Śrī Caitanya Mahāprabhu in order to be fully equipped with Kṛṣṇa consciousness.

5.19.29-30: śrī-śuka uvāca/ jambūdvīpasya ca rājann upadvīpān aṣṭau haika upadiśanti sagarātmajair aśvānveṣaṇa imāṁ mahīṁ parito nikhanadbhir upakalpitān; tad yathā svarṇaprasthaś candraśukla āvartano ramaṇako mandarahariṇaḥ pāñcajanyaḥ siṁhalo laṅketi.

SYNONYMS: śrī-śukaḥ uvāca—Śrī Śukadeva Gosvāmī continued to speak; jambūdvīpasya—of the island known as Jambūdvīpa; ca—also; rājan—O King; upadvīpān aṣṭau—eight subordinate islands; ha—certainly; eke—some; upadiśanti—learned scholars describe; sagara-ātma-jaiḥ—by the sons of Mahārāja Sagara; aśva-anveṣaṇe—while trying to find their lost horse; imām—this; mahīm—tract of land; paritaḥ—all around; nikhanadbhiḥ—digging; upakalpitān—created; tat—that; yathā—as follows; svarṇa-prasthaḥ—Svarṇaprastha; candra-śuklaḥ—Candraśukla; āvartanaḥ—Āvartana; ramaṇakaḥ—Ramaṇaka; mandara-hariṇaḥ—Mandara-hariṇa; pāñcajanyaḥ—Pāñcajanya; siṁhalaḥ—Siṁhala; laṅkā—Laṅkā; iti—thus.

TRANSLATION: Śrī Śukadeva Gosvāmī said: My dear King, in the opinion of some learned scholars, eight smaller islands surround Jambūdvīpa. When the sons of Mahārāja Sagara were searching all over the world for their lost horse, they dug up the earth, and in this way eight adjoining islands came into existence. The names of these islands are Svarṇaprastha, Candraśukla, Āvartana, Ramaṇaka, Mandara-hariṇa, Pāñcajanya, Siṁhala and Laṅkā.

PURPORT: In the Kūrma Purāṇa there is this statement about the desires of the demigods: anadhikāriṇo devāḥ/svarga-sthā bhāratodbhavam/vāñchanty ātma-vimokṣārtha-/mudrekārthe 'dhikāriṇaḥ// Although the demigods are situated in exalted positions in the heavenly planets, they nevertheless desire to descend to the land of Bhārata-varṣa on the planet earth. This indicates that even the demigods are unfit to reside in Bhārata-varṣa. Therefore if persons already born in Bhāratavarṣa live like cats and dogs, not taking full advantage of their birth in this land, they are certainly unfortunate.

5.19.31: evaṁ tava bhāratottama jambūdvīpa-varṣa-vibhāgo yathopadeśam upavarṇita iti.

SYNONYMS: evam—thus; tava—unto you; bhārata-uttama—O best of the descendents of Bharata; jambūdvīpa-varṣa-vibhāgaḥ—the divisions of the island of Jambūdvīpa; yathā-upadeśam—as much as I am instructed by the authorities; upavarṇitaḥ—explained; iti—thus.

TRANSLATION: My dear King Parīkṣit, O best of the descendants of Bharata Mahārāja, I have thus described to you, as I myself have been instructed, the island of Bhārata-varṣa and its adjoining islands. These are the islands that constitute Jambūdvīpa.

## Part VIII: Plakṣadvīpa

5.20.1: śrī-śuka uvāca/ ataḥ paraṁ plakṣādīnāṁ pramāṇa-lakṣaṇa-saṁsthānato varṣa-vibhāga upavarṇyate.

SYNONYMS: śrī-śukaḥ uvāca—Śukadeva Gosvāmī said; ataḥ param—after this; plakṣa-ādīnām—of the island named Plakṣa and others; pramāṇa-lakṣaṇa-saṁsthānataḥ—from the angle of dimensions, particular characteristics and form; varṣa-vibhāgaḥ—the division of the island; upavarṇyate—is described.

TRANSLATION: The great sage Śukadeva Gosvāmī said: Hereafter I shall describe the dimensions, characteristics and forms of the six islands beginning with the island of Plakṣa.

5.20.2: jambūdvīpo 'yaṁ yāvat-pramāṇa-vistāras tāvatā kṣārodadhinā pariveṣṭito yathā merur jambv-ākhyena lavaṇodadhir api tato dvi-guṇa-viśālena plakṣākhyena parikṣipto yathā parikhā bāhyopavanena; plakṣo jambū-pramāṇo dvīpākhyākaro hiraṇmaya utthito yatrāgnir upāste sapta-jihvas tasyādhipatiḥ priyavratātmaja idhmajihvaḥ svaṁ dvīpaṁ sapta-varṣāṇi vibhajya sapta-varṣa-nāmabhya ātmajebhya ākalayya svayam ātma-yogenopararāma.

SYNONYMS: jambū-dvīpaḥ—Jambūdvīpa, the island named Jambū; ayam—this; yāvat-pramāṇa-vistāraḥ—as much as the measure of its width, namely 100,000 yojanas (one yojana equals eight miles); tāvatā—so much; kṣāra-udadhinā—by the ocean of salt water; pariveṣṭitaḥ—surrounded; yathā—just as; meruḥ—Sumeru Mountain; jambū-ākhyena—by the island named Jambū; lavaṇa-udadhiḥ—the ocean of salt water; api—certainly; tataḥ—thereafter; dvi-guṇa-viśālena—which is twice as wide; plakṣa-ākhyena—by the island named Plakṣa; parikṣiptaḥ—surrounded; yathā—like; parikhā—a moat; bāhya—external; upavanena—by a gardenlike forest; plakṣaḥ—a plakṣa tree; jambū-pramāṇaḥ—having the height of the jambū tree; dvīpa-ākhyā-karaḥ—causing the name of the island; hiraṇmayaḥ—magnificently splendorous; utthitaḥ—rising; yatra—where; agniḥ—a fire; upāste—is situated; sapta-jihvaḥ—having seven flames; tasya—of that island; adhipatiḥ—the king or master; priyavrata-ātmajaḥ—the son of King Priyavrata; idhma-jihvaḥ—named Idhmajihva; svam—own; dvīpam—island; sapta—seven; varṣāṇi—tracts of land; vibhajya—dividing into; sapta-varṣa-nāmabhyaḥ—for whom the seven tracts of land were named; ātmajebhyaḥ—to his own sons; ākalayya—offering; svayam—personally; ātma-yogena—by the devotional service of the Lord; upararāma—he retired from all material activities.

TRANSLATION: As Sumeru Mountain is surrounded by Jambūdvīpa, Jambūdvīpa is also surrounded by an ocean of salt water. The breadth of Jambūdvīpa is 100,000 yojanas [800,000 miles], and the breadth of the saltwater ocean is the same. As a moat around a fort is sometimes surrounded by gardenlike forest, the saltwater ocean surrounding Jambūdvīpa is itself surrounded by Plakṣadvīpa. The breadth of Plakṣadvīpa is twice that of the saltwater ocean—in other words 260,000 yojanas [1,600,000 miles]. On Plakṣadvīpa there is a tree shining like gold and as tall as the jambū tree on Jambūdvīpa. At its root is a fire with seven flames. It is because this tree is a plakṣa tree that the island is called Plakṣadvīpa. Plakṣadvīpa was governed by Idhmajihva, one of the sons of Mahārāja Priyavrata. He endowed the seven islands with the names of his seven sons, divided the islands among the sons, and then retired from active life to engage in the devotional service of the Lord.

5.20.3-4: śivaṁ yavasaṁ subhadraṁ śāntaṁ kṣemam amṛtam abhayam iti varṣāṇi teṣu girayo nadyaś ca saptaivābhijñātāḥ; maṇikūṭo vajrakūṭa indraseno jyotiṣmān suparṇo hiraṇyaṣṭhīvo meghamāla iti setu-śailāḥ aruṇā nṛmṇāṅgirasī sāvitrī suptabhātā ṛtambharā satyambharā iti mahā-nadyaḥ; yāsāṁ jalopasparśana-vidhūta-rajas-tamaso haṁsa-pataṅgordhvāyana-satyāṅga-saṁjñāś catvāro varṇāḥ sahasrāyuṣo vibudhopamasandarśana-prajananāḥ svarga-dvāraṁ trayyā vidyayā bhagavantaṁ trayīmayaṁ sūryam ātmānaṁ yajante.

SYNONYMS: śivam—Śiva; yavasam—Yavasa; subhadram—Subhadra; śāntam—Śānta; kṣemam—Kṣema; amṛtam—Amṛta; abhayam—Abhaya; iti—thus; varṣāṇi—the tracts of land according to the names of the seven sons; teṣu—in them; girayaḥ—mountains; nadyaḥ ca—and rivers; sapta—seven; eva—indeed; abhijñātāḥ—are known; maṇi-kūṭaḥ—Maṇikūṭa; vajra-kūṭaḥ—Vajrakūṭa; indra-senaḥ—Indrasena; jyotiṣmān—Jyotiṣmān; suparṇaḥ—Suparṇa; hiraṇya-ṣṭhīvaḥ—Hiraṇyaṣṭhīva; megha-mālaḥ—Meghamāla; iti—thus; setu-śailāḥ—the ranges of mountains marking the borders of the varṣas; aruṇā—Aruṇā; nṛmṇā—Nṛmṇā; āṅgirasī—Āṅgirasī; sāvitrī—Sāvitrī; supta-bhātā—Suptabhātā; ṛtambharā—Ṛtambharā; satyambharā—Satyambharā; iti—thus; mahā-nadyaḥ—very large rivers; yāsām—of which; jala-upasparśana—simply by touching the water; vidhūta—washed off; rajaḥ-tamasaḥ—whose modes of passion and ignorance; haṁsa—Haṁsa; pataṅga—Pataṅga; ūrdhvāyana—Ūrdhvāyana; satyāṅga—Satyāṅga; saṁjñāḥ—named; catvāraḥ—four; varṇāḥ—castes or divisions of men; sahasra-āyuṣaḥ—living for one thousand years; vibudha-upama—similar to the demigods; sandarśana—in having very beautiful forms; prajananāḥ—and in producing children; svarga-dvāram—the gateway to the heavenly planets; trayyā vidyayā—by performing ritualistic ceremonies according to Vedic principles; bhagavantam—the Supreme Personality of Godhead; trayī-mayam—established in the Vedas; sūryam ātmānam—the Supersoul, represented by the sun-god; yajante—they worship.

TRANSLATION: The seven islands [varṣas] are named according to the names of those seven sons—Śiva, Yavasa, Subhadra, Śānta, Kṣema, Amṛta and Abhaya. In those seven tracts of land, there are seven mountains and seven rivers. The mountains are named Maṇikūṭa, Vajrakūṭa, Indrasena, Jyotiṣmān, Suparṇa, Hiraṇyaṣṭhīva and Meghamāla, and the rivers are named Aruṇā, Nṛmṇā, Āṅgirasī, Sāvitrī, Suptabhātā, Ṛtambharā and Satyambharā. One can immediately be free from material contamination by touching or bathing in those rivers, and the four castes of people who live in Plakṣadvīpa—the Haṁsas, Pataṅgas, Ūrdhvāyanas and Satyāṅgas—purify themselves in that way. The inhabitants of Plakṣadvīpa live for one thousand years. They are beautiful like the demigods, and they also beget children like the demigods. By completely performing the ritualistic ceremonies mentioned in the Vedas and by worshiping the Supreme Personality of Godhead as represented by the sun-god, they attain the sun, which is a heavenly planet.

5.20.5: pratnasya viṣṇo rūpaṁ yat/satyasyartasya brahmaṇaḥ/amṛtasya ca mṛtyoś ca/sūryam ātmānam īmahīti

SYNONYMS: pratnasya—of the oldest person; viṣṇoḥ—Lord Viṣṇu; rūpam—the form; yat—which; satyasya—of the Absolute Truth; ṛtasya—of dharma; brahmaṇaḥ—of the Supreme Brahman; amṛtasya—of the auspicious result; ca—and; mṛtyoḥ—of death (the inauspicious result); ca—and; sūryam—the demigod Sūrya; ātmānam—the Supersoul or origin of all souls; īmahi—we approach for shelter; iti—thus.

TRANSLATION: [This is the mantra by which the inhabitants of Plakṣadvīpa worship the Supreme Lord.] Let us take shelter of the sun-god, who is a reflection of Lord Viṣṇu, the all-expanding Supreme Personality of Godhead, the oldest of all persons. Viṣṇu is the only worshipable Lord. He is the Vedas, He is religion, and He is the origin of all auspicious and inauspicious results.

PURPORT: As described in this chapter, the inhabitants of the five islands beginning with Plakṣadvīpa worship the sun-god, the moon-god, the fire-god, the air-god and Lord Brahmā respectively. Although they engage in the worship of these five demigods, however, they actually worship Lord Viṣṇu, the Supersoul of all living entities, as indicated in this verse by the words pratnasya viṣṇo rūpam. Viṣṇu is brahma, amṛta, mṛtyu—the Supreme Brahman and the origin of everything, auspicious and inauspicious.

5.20.6: plakṣādiṣu pañcasu puruṣāṇām āyur indriyam ojaḥ saho balaṁ buddhir vikrama iti ca sarveṣām autpattikī siddhir aviśeṣeṇa vartate.

SYNONYMS: plakṣa-ādiṣu—in the islands headed by Plakṣa; pañcasu—five; puruṣāṇām—of the inhabitants; āyuḥ—long duration of life; indriyam—soundness of the senses; ojaḥ—bodily strength; sahaḥ—mental strength; balam—physical strength; buddhiḥ—intelligence; vikramaḥ—bravery; iti—thus; ca—also; sarveṣām—of all of them; autpattikī—inborn; siddhiḥ—perfection; aviśeṣeṇa—without distinction; vartate—exists.

TRANSLATION: O King, longevity, sensory prowess, physical and mental strength, intelligence and bravery are naturally and equally manifested in all the inhabitants of the five islands headed by Plakṣadvīpa.

## Part IX: Śālmalīdvīpa

5.20.7: plakṣaḥ sva-samānenekṣu-rasodenāvṛto yathā tathā dvīpo 'pi śālmalo dvi-guṇa-viśālaḥ samānena surodenāvṛtaḥ parivṛṅkte.

SYNONYMS: plakṣaḥ—the land known as Plakṣadvīpa; sva-samānena—equal in width; ikṣu-rasa—of sugarcane juice; udena—by an ocean; āvṛtaḥ—surrounded; yathā—just as; tathā—similarly; dvīpaḥ—another island; api—also; śālmalaḥ—

known as Śālmala; dvi-guṇa-viśālaḥ—twice as big; samānena—equal in width; surā-udena—by an ocean of liquor; āvṛtaḥ—surrounded; parivṛṅkte—exists.

TRANSLATION: Plakṣadvīpa is surrounded by an ocean of sugarcane juice, equal in breadth to the island itself. Similarly, there is then another island—Sālmalīdvīpa—twice as broad as Plakṣadvīpa [400,000 yojanas, or 3,200,000 miles] and surrounded by an equally broad body of water called Surāsāgara, the ocean that tastes like liquor.

5.20.8: yatra ha vai śālmalī plakṣāyāmā yasyāṁ vāva kila nilayam āhur bhagavataś chandaḥ-stutaḥ patattri-rājasya sā dvīpa-hūtaye upalakṣyate.

SYNONYMS: yatra—where; ha vai—certainly; śālmalī—a śālmalī tree; plakṣa-āyāmā—as big as the plakṣa tree (one hundred yojanas broad and eleven hundred yojanas high); yasyām—in which; vāva kila—indeed; nilayam—rest or living place; āhuḥ—they say; bhagavataḥ—of the most powerful; chandaḥ-stutaḥ—who worships the Lord by Vedic prayers; patattri-rājasya—of Garuḍa, the carrier of Lord Viṣṇu; sā—that tree; dvīpa-hūtaye—for the name of the island; upalakṣyate—is distinguished.

TRANSLATION: On Sālmalīdvīpa there is a śālmalī tree, from which the island takes its name. That tree is as broad and tall as the plakṣa tree—in other words 100 yojanas [800 miles] broad and 1,100 yojanas [8,800 miles] tall. Learned scholars say that this gigantic tree is the residence of Garuḍa, the king of all birds and carrier of Lord Viṣṇu. In that tree, Garuḍa offers Lord Viṣṇu his Vedic prayers.

5.20.9: tad-dvīpādhipatiḥ priyavratātmajo yajñabāhuḥ sva-sutebhyaḥ saptabhyas tan-nāmāni sapta-varṣāṇi vyabhajat surocanaṁ saumanasyaṁ ramaṇakaṁ deva-varṣaṁ pāribhadram āpyāyanam avijñātam iti.

SYNONYMS: tat-dvīpa-adhipatiḥ—the master of that island; priyavrata-ātmajaḥ—the son of Mahārāja Priyavrata; yajña-bāhuḥ—named Yajñabāhu; sva-sutebhyaḥ—unto his sons; saptabhyaḥ—seven in number; tat-nāmāni—having names according to their names; sapta-varṣāṇi—seven tracts of land; vyabhajat—divided; surocanam—Surocana; saumanasyam—Saumanasya; ramaṇakam—Ramaṇaka; deva-varṣam—Deva-varṣa; pāribhadram—Pāribhadra; āpyāyanam—Āpyāyana; avijñātam—Avijñāta; iti—thus.

TRANSLATION: The son of Mahārāja Priyavrata named Yajñabāhu, the master of Sālmalīdvīpa, divided the island into seven tracts of land, which he gave to his seven sons. The names of those divisions, which correspond to the names of the sons, are Surocana, Saumanasya, Ramaṇaka, Deva-varṣa, Pāribhadra, Āpyāyana and Avijñāta.

5.20.10: teṣu varṣādrayo nadyaś ca saptaivābhijñātāḥ svarasaḥ śataśṛṅgo vāmadevaḥ kundo mukundaḥ puṣpa-varṣaḥ sahasra-śrutir iti; anumatiḥ sinīvālī sarasvatī kuhū rajanī nandā rāketi.

SYNONYMS: teṣu—in those tracts of land; varṣa-adrayaḥ—mountains; nadyaḥ ca—as well as rivers; sapta eva—seven in number; abhijñātāḥ—understood; svarasaḥ—Svarasa; śata-śṛṅgaḥ—Śataśṛṅga; vāma-devaḥ—Vāmadeva; kundaḥ—Kunda; mukundaḥ—Mukunda; puṣpa-varṣaḥ—Puṣpa-varṣa; sahasra-śrutiḥ—Sahasra-śruti; iti—thus; anumatiḥ—Anumati; sinīvālī—Sinīvālī; sarasvatī—Sarasvatī; kuhū—Kuhū; rajanī—Rajanī; nandā—Nandā; rākā—Rākā; iti—thus.

TRANSLATION: In those tracts of land there are seven mountains—Svarasa, Śataśṛṅga, Vāmadeva, Kunda, Mukunda, Puṣpa-varṣa and Sahasra-śruti. There are also seven rivers—Anumati, Sinīvālī, Sarasvatī, Kuhū, Rajanī, Nandā and Rākā. They are still existing.

5.20.11: tad-varṣa-puruṣāḥ śrutadhara-vīryadhara-vasundhareṣandhara-saṁjñā bhagavantaṁ vedamayaṁ somam ātmānaṁ vedena yajante.

SYNONYMS: tat-varṣa-puruṣāḥ—the residents of those tracts of land; śrutadhara—Śrutadhara; vīryadhara—Vīryadhara; vasundhara—Vasundhara; iṣandhara—Iṣandhara; saṁjñāḥ—known as; bhagavantam—the Supreme Personality of Godhead; veda-mayam—fully conversant with the Vedic knowledge; somam ātmānam—represented by the living entity known as Soma; vedena—by following the Vedic rules and regulations; yajante—they worship.

TRANSLATION: Strictly following the cult of varṇāśrama-dharma, the inhabitants of those islands, who are known as Śrutidharas, Vīryadharas, Vasundharas and Iṣandharas, all worship the expansion of the Supreme Personality of Godhead named Soma, the moon-god.

5.20.12: sva-gobhiḥ pitṛ-devebhyo/vibhajan kṛṣṇa-śuklayoḥ/prajānāṁ sarvāsāṁ rājā-/ndhaḥ somo na āstv iti

SYNONYMS: sva-gobhiḥ—by expansion of his own illuminating rays; pitṛ-devebhyaḥ—unto the pitās and the demigods; vibhajan—dividing; kṛṣṇa-śuklayoḥ—into the two fortnights, dark and light; prajānām—of the citizens; sarvāsām—of all; rājā—the king; andhaḥ—food grains; somaḥ—the moon-god; naḥ—toward us; āstu—let him remain favorable; iti—thus.

TRANSLATION: [The inhabitants of Śālmalīdvīpa worship the demigod of the moon in the following words.] By his own rays, the moon-god has divided the month into two fortnights, known as śukla and kṛṣṇa, for the distribution of food grains to the pitās and the demigods. The demigod of the moon is he who divides time, and he is the king of all the residents of the universe. We therefore pray that he may remain our king and guide, and we offer him our respectful obeisances.

## Part X: Kuśadvīpa

5.20.13: evaṁ surodād bahis tad-dvi-guṇaḥ samānenāvṛto ghṛtodena yathā-pūrvaḥ kuśa-dvīpo yasmin kuśa-stambo deva-kṛtas tad-dvīpākhyākaro jvalana ivāparaḥ sva-śaṣpa-rociṣā diśo virājayati.

SYNONYMS: evam—thus; surodāt—from the ocean of liquor; bahiḥ—outside; tat-dvi-guṇaḥ—twice that; samānena—equal in width; āvṛtaḥ—surrounded; ghṛta-udena—an ocean of clarified butter; yathā-pūrvaḥ—as previously with Sālmalīdvīpa; kuśa-dvīpa—the island called Kuśadvīpa; yasmin—in which; kuśa-stambaḥ—kuśa grass; deva-kṛtaḥ—created by the supreme will of the Supreme Personality of Godhead; tat-dvīpa-ākhyā-karaḥ—giving the island its name; jvalanaḥ—fire; iva—like; aparaḥ—another; sva-śaṣpa-rociṣā—by the effulgence of the young sprouting grass; diśaḥ—all directions; virājayati—illuminates.

TRANSLATION: Outside the ocean of liquor is another island, known as Kuśadvīpa, which is 800,000 yojanas [6,400,000 miles] wide, twice as wide as the ocean of liquor. As Śālmalīdvīpa is surrounded by a liquor ocean, Kuśadvīpa is surrounded by an ocean of liquid ghee as broad as the island itself. On Kuśadvīpa there are clumps of kuśa grass, from which the island takes its name. This kuśa grass, which was created by the demigods by the will of the Supreme Lord, appears like a second form of fire, but with very mild and pleasing flames. Its young shoots illuminate all directions.

PURPORT: From the descriptions in this verse, we can make an educated guess about the nature of the flames on the moon. Like the sun, the moon must also be full of flames because without flames there cannot be illumination. The flames on the moon, however, unlike those on the sun, must be mild and pleasing. This is our conviction. The modern theory that the moon is full of dust is not accepted in the verses of Śrīmad-Bhāgavatam. In regard to this verse, Śrīla Viśvanātha Cakravartī Ṭhākura says, suśaṣpāṇi sukomala-śikhās teṣāṁ rociṣā: the kuśa grass illuminates all directions, but its flames are very mild and pleasing. This gives some idea of the flames existing on the moon.

5.20.14: tad-dvīpa-patiḥ praiyavrato rājan hiraṇyaretā nāma svaṁ dvīpaṁ saptabhyaḥ sva-putrebhyo yathā-bhāgaṁ vibhajya svayaṁ tapa ātiṣṭhata vasu-vasudāna-dṛḍharuci-nābhigupta-stutyavrata-vivikta-vāmadeva-nāmabhyaḥ.

SYNONYMS: tat-dvīpa-patiḥ—the master of that island; praiyavrataḥ—the son of Mahārāja Priyavrata; rājan—O King; hiraṇyaretā—Hiraṇyaretā; nāma—named;

svam—his own; dvīpam—island; saptabhyaḥ—unto seven; sva-putrebhyaḥ—his own sons; yathā-bhāgam—according to division; vibhajya—dividing; svayam—himself; tapaḥ ātiṣṭhata—engaged in austerities; vasu—unto Vasu; vasudāna—Vasudāna; dṛḍharuci—Dṛḍharuci; nābhi-gupta—Nābhigupta; stutya-vrata—Stutyavrata; vivikta—Vivikta; vāma-deva—Vāmadeva; nāmabhyaḥ—named.

TRANSLATION: O King, another son of Mahārāja Priyavrata, Hiraṇyaretā, was the king of this island. He divided it into seven parts, which he delivered to his seven sons according to the rights of inheritance. The King then retired from family life to engage in austerities. The names of those sons are Vasu, Vasudāna, Dṛḍharuci, Stutyavrata, Nābhigupta, Vivikta and Vāmadeva.

5.20.15: teṣāṁ varṣeṣu sīmā-girayo nadyaś cābhijñātāḥ sapta saptaiva cakraś
catuḥśṛṅgaḥ kapilaś citrakūṭo devānīka ūrdhvaromā draviṇa iti rasakulyā
madhukulyā mitravindā śrutavindā devagarbhā ghṛtacyutā mantramāleti.

SYNONYMS: teṣām—all those sons; varṣeṣu—in the tracts of land; sīmā-girayaḥ—border mountains; nadyaḥ ca—as well as rivers; abhijñātāḥ—known; sapta—seven; sapta—seven; eva—certainly; cakraḥ—Cakra; catuḥ-śṛṅgaḥ—Catuḥ-śṛṅga; kapilaḥ—Kapila; citra-kūṭaḥ—Citrakūṭa; devānīkaḥ—Devānīka; ūrdhva-romā—Ūrdhvaromā; draviṇaḥ—Draviṇa; iti—thus; rasa-kulyā—Ramakulyā; madhu-kulyā—Madhukulyā; mitra-vindā—Mitravindā; śruta-vindā—Śrutavindā; deva-garbhā—Devagarbhā; ghṛta-cyutā—Ghṛtacyutā; mantra-mālā—Mantramālā; iti—thus.

TRANSLATION: In those seven islands there are seven boundary mountains, known as Cakra, Catuḥśṛṅga, Kapila, Citrakūṭa, Devānīka, Ūrdhvaromā and Draviṇa. There are also seven rivers, known as Ramakulyā, Madhukulyā, Mitravindā, Śrutavindā, Devagarbhā, Ghṛtacyutā and Mantramālā.

5.20.16: yāsāṁ payobhiḥ kuśadvīpaukasaḥ kuśala-kovidābhiyukta-kulaka-
saṁjñā bhagavantaṁ jātaveda-sarūpiṇaṁ karma-kauśalena yajante.

SYNONYMS: yāsām—of which; payobhiḥ—by the water; kuśa-dvīpa-okasaḥ—the inhabitants of the island known as Kuśadvīpa; kuśala—Kuśala; kovida—Kovida; abhiyukta—Abhiyukta; kulaka—Kulaka; saṁjñāḥ—named; bhagavantam—unto the Supreme Personality of Godhead; jātaveda—the demigod of fire; sa-rūpiṇam—manifesting the form; karma-kauśalena—by expertise in ritualistic ceremonies; yajante—they worship.

TRANSLATION: The inhabitants of the island of Kuśadvīpa are celebrated as the Kuśalas, Kovidas, Abhiyuktas and Kulakas. They are like the brāhmaṇas, kṣatriyas, vaiśyas and śūdras respectively. By bathing in the waters of those rivers, they all become purified. They are expert in performing ritualistic ceremonies according to the orders of the Vedic scriptures. Thus they worship the Lord in His aspect as the demigod of fire.

5.20.17: parasya brahmaṇaḥ sākṣāj/jāta-vedo 'si havyavāṭ/devānāṁ puruṣāṅgānāṁ/yajñena puruṣaṁ yajeti

SYNONYMS: parasya—of the Supreme; brahmaṇaḥ—Brahman; sākṣāt—directly; jāta-vedaḥ—O fire-god; asi—you are; havyavāṭ—the carrier of Vedic offerings of grains and ghee; devānām—of all the demigods; puruṣa-aṅgānām—who are limbs of the Supreme Person; yajñena—by performing the ritualistic sacrifices; puruṣam—to the Supreme person; yaja—please carry oblations; iti—thus.

TRANSLATION: [This is the mantra by which the inhabitants of Kuśadvīpa worship the fire-god.] O fire-god, you are a part of the Supreme Personality of Godhead, Hari, and you carry to Him all the offerings of sacrifices. Therefore we request you to offer to the Supreme Personality of Godhead the yajñic ingredients we are offering the demigods, for the Lord is the real enjoyer.

## Part XI: Krauñcadvīpa

5.20.18: tathā ghṛtodād bahiḥ krauñcadvīpo dvi-guṇaḥ sva-mānena kṣīrodena parita upakḷpto vṛto yathā kuśadvīpo ghṛtodena yasmin krauñco nāma parvata-rājo dvīpa-nāma-nirvartaka āste.

SYNONYMS: tathā—so also; ghṛta-udāt—from the ocean of clarified butter; bahiḥ—outside; krauñca-dvīpaḥ—another island, known as Krauñcadvīpa; dvi-guṇaḥ—twice as big; sva-mānena—as the same measurement; kṣīra-udena—by an ocean of milk; paritaḥ—all around; upakḷptaḥ—surrounded; vṛtaḥ—surrounded; yathā—like; kuśa-dvīpaḥ—the island known as Kuśadvīpa; ghṛta-udena—by an ocean of clarified butter; yasmin—in which; krauñcaḥ nāma—named Krauñca; parvata-rājaḥ—a king of mountains; dvīpa-nāma—the name of the island; nirvartakaḥ—bringing about; āste—exists.

TRANSLATION: Outside the ocean of clarified butter is another island, known as Krauñcadvīpa, which has a width of 1,600,000 yojanas [12,800,000 miles], twice the width of the ocean of clarified butter. As Kuśadvīpa is surrounded by an ocean of clarified butter, Krauñcadvīpa is surrounded by an ocean of milk as broad as the island itself. On Krauñcadvīpa there is a great mountain known as Krauñca, from which the island takes its name.

5.20.19: yo 'sau guha-praharaṇonmathita-nitamba-kuñjo 'pi kṣīrodenā-sicyamāno bhagavatā varuṇenābhigupto vibhayo babhūva.

SYNONYMS: yaḥ—which; asau—that (mountain); guha-praharaṇa—by the weapons of Kārttikeya, the son of Lord Śiva; unmathita—shaken; nitamba-kuñjaḥ—whose trees and vegetables along the slopes; api—although; kṣīra-udena—by the ocean of milk; āsicyamānaḥ—being always bathed; bhagavatā—by the greatly powerful; varuṇena—the demigod known as Varuṇa; abhiguptaḥ—protected; vibhayaḥ babhūva—has become fearless.

TRANSLATION: Although the vegetables living on the slopes of Mount Krauñca were attacked and devastated by the weapons of Kārttikeya, the mountain has become fearless because it is always bathed on all sides by the ocean of milk and protected by Varuṇadeva.

5.20.20: tasminn api praiyavrato ghṛtapṛṣṭho nāmādhipatiḥ sve dvīpe varṣāṇi sapta vibhajya teṣu putra-nāmasu sapta rikthādān varṣapān niveśya svayaṁ bhagavān bhagavataḥ parama-kalyāṇa-yaśasa ātma-bhūtasya hareś caraṇāravindam upajagāma.

SYNONYMS: tasmin—in that island; api—also; praiyavrataḥ—the son of Mahārāja Priyavrata; ghṛta-pṛṣṭhaḥ—Ghṛtapṛṣṭha; nāma—named; adhipatiḥ—the king of that island; sve—his own; dvīpe—in the island; varṣāṇi—tracts of land; sapta—seven; vibhajya—dividing; teṣu—in each of them; putra-nāmasu—possessing the names of his sons; sapta—seven; rikthā-dān—sons; varṣa-pān—the masters of the varṣas; niveśya—appointing; svayam—himself; bhagavān—very powerful; bhagavataḥ—of the Supreme Personality of Godhead; parama-kalyāṇa-yaśasaḥ—whose glories are so auspicious; ātma-bhūtasya—the soul of all souls; hareḥ caraṇa-aravindam—the lotus feet of the Lord; upajagāma—took shelter at.

TRANSLATION: The ruler of this island was another son of Mahārāja Priyavrata. His name was Ghṛtapṛṣṭha, and he was a very learned scholar. He also divided his own island among his seven sons. After dividing the island into seven parts, named according to the names of his sons, Ghṛtapṛṣṭha Maharaja completely retired from family life and took shelter at the lotus feet of the Lord, the soul of all souls, who has all auspicious qualities. Thus he attained perfection.

5.20.21: āmo madhuruho meghapṛṣṭhaḥ sudhāmā bhrājiṣṭho lohitārṇo vanaspatir iti ghṛtapṛṣṭha-sutās teṣāṁ varṣa-girayaḥ sapta saptaiva nadyaś cābhikhyātāḥ śuklo vardhamāno bhojana upabarhiṇo nando nandanaḥ sarvatobhadra iti abhayā amṛtaughā āryakā tīrthavatī rūpavatī pavitravatī śukleti.

SYNONYMS: āmaḥ—Āma; madhu-ruhaḥ—Madhuruha; megha-pṛṣṭhaḥ—Meghapṛṣṭha; sudhāmā—Sudhāmā; bhrājiṣṭhaḥ—Bhrājiṣṭha; lohitārṇaḥ—Lohitārṇa; vanaspatiḥ—Vanaspati; iti—thus; ghṛtapṛṣṭha-sutāḥ—the sons of

Ghṛtapṛṣṭha; teṣām—of those sons; varṣa-girayaḥ—boundary hills of the tracts of land; sapta—seven; sapta—seven; eva—also; nadyaḥ—rivers; ca—and; abhikhyātāḥ—celebrated; śuklaḥ vardhamānaḥ—Śukla and Vardhamāna; bhojanaḥ—Bhojana; upabarhiṇaḥ—Upabarhiṇa; nandaḥ—Nanda; nandanaḥ—Nandana; sarvataḥ-bhadraḥ—Sarvatobhadra; iti—thus; abhayā—Abhayā; amṛtaughā—Amṛtaughā; āryakā—Āryakā; tīrthavatī—Tīrthavatī; rūpavatī—Rūpavatī; pavitravatī—Pavitravatī; śuklā—Śuklā; iti—thus.

TRANSLATION: The sons of Mahārāja Ghṛtapṛṣṭha were named Āma, Madhuruha, Meghapṛṣṭha, Sudhāmā, Bhrājiṣṭha, Lohitārṇa and Vanaspati. In their island there are seven mountains, which indicate the boundaries of the seven tracts of land, and there are also seven rivers. The mountains are named Śukla, Vardhamāna, Bhojana, Upabarhiṇa, Nanda, Nandana and Sarvatobhadra. The rivers are named Abhayā, Amṛtaughā, Āryakā, Tīrthavatī, Rūpavatī, Pavitravatī and Śuklā.

5.20.22: yāsām ambhaḥ pavitram amalam upayuñjānāḥ puruṣa-ṛṣabha-draviṇa-devaka-saṁjñā varṣa-puruṣā āpomayaṁ devam apāṁ pūrṇenāñjalinā yajante.

SYNONYMS: yāsām—of all the rivers; ambhaḥ—the water; pavitram—very sanctified; amalam—very clean; upayuñjānāḥ—using; puruṣa—Puruṣa; ṛṣabha—Ṛṣabha; draviṇa—Draviṇa; devaka—Devaka; saṁjñāḥ—endowed with the names; varṣa-puruṣāḥ—the inhabitants of those varṣas; āpaḥ-mayam—Varuṇa, the lord of water; devam—as the worshipable deity; apām—of water; pūrṇena—with full; añjalinā—folded palms; yajante—do worship.

TRANSLATION: The inhabitants of Krauñcadvīpa are divided into four castes, called the Puruṣas, Ṛṣabhas, Draviṇas and Devakas. Using the waters of those sanctified rivers, they worship the Supreme Personality of Godhead by offering a palmful of water at the lotus feet of Varuṇa, the demigod who has a form of water.

PURPORT: Viśvanātha Cakravartī Ṭhākura says, āpomayaḥ asmayam: with joined palms the inhabitants of the various sections of Krauñcadvīpa offer the sanctified waters of the rivers to a deity made of stone or iron.

5.20.23: āpaḥ puruṣa-vīryāḥ stha/punantīr bhūr-bhuvaḥ-suvaḥ/tā naḥ punītāmīva-ghnīḥ/spṛśatām ātmanā bhuva iti

SYNONYMS: āpaḥ—O water; puruṣa-vīryāḥ—endowed with the energy of the Supreme Personality of Godhead; stha—you are; punantīḥ—sanctifying; bhūḥ—of the planetary system known as Bhūḥ; bhuvaḥ—of the Bhuvaḥ planetary system; suvaḥ—of the Svaḥ planetary system; tāḥ—that water; naḥ—of us;

punīta—purify; amīva-ghnīḥ—who destroys sin; spṛśatām—of those touching; ātmanā—by your constitutional position; bhuvaḥ—the bodies; iti—thus.

TRANSLATION: [The inhabitants of Krauñcadvīpa worship with this mantra.] O water of the rivers, you have obtained energy from the Supreme Personality of Godhead. Therefore you purify the three planetary systems, known as Bhūloka, Bhuvarloka and Svarloka. By your constitutional nature, you take away sins, and that is why we are touching you. Kindly continue to purify us.

PURPORT: The specific rivers mentioned in the śāstras are also energies of the Supreme Personality of Godhead, and people who regularly bathe in them are purified. It can actually be seen that many people are cured of diseases simply by bathing in the Ganges. Similarly, the inhabitants of Krauñcadvīpa purify themselves by bathing in the rivers there.

## Part XII: Śākadvīpa

5.20.24: evaṁ purastāt kṣīrodāt parita upaveśitaḥ śākadvīpo dvātriṁśal-lakṣa-yojanāyāmaḥ samānena ca dadhi-maṇḍodena parīto yasmin śāko nāma mahīruhaḥ sva-kṣetra-vyapadeśako yasya ha mahā-surabhi-gandhas taṁ dvīpam anuvāsayati.

SYNONYMS: evam—thus; purastāt—beyond; kṣīra-udāt—from the ocean of milk; paritaḥ—all around; upaveśitaḥ—situated; śāka-dvīpaḥ—another island, known as Śākadvīpa; dvā-triṁśat—thirty-two; lakṣa—100,000; yojana—yojanas; āyāmaḥ—whose measure; samānena—of equal length; ca—and; dadhi-maṇḍa-udena—by an ocean containing water resembling churned yogurt; parītaḥ—surrounded; yasmin—in which land; śākaḥ—śāka; nāma—named; mahīruhaḥ—a fig tree; sva-kṣetra-vyapadeśakaḥ—giving the island its name; yasya—of which; ha—indeed; mahā-surabhi—a greatly fragrant; gandhaḥ—aroma; tam dvīpam—that island; anuvāsayati—makes fragrant.

TRANSLATION: Outside the ocean of milk is another island, Śākadvīpa, which has a width of 3,200,000 yojanas [25,600,000 miles]. As Krauñcadvīpa is surrounded by its own ocean of milk, Śākadvīpa is surrounded by an ocean of churned yogurt as broad as the island itself. In Śākadvīpa there is a big śāka tree, from which the island takes its name. This tree is very fragrant. Indeed, it lends its scent to the entire island.

5.20.25: tasyāpi praiyavrata evādhipatir nāmnā medhātithiḥ so 'pi vibhajya sapta varṣāṇi putra-nāmāni teṣu svātmajān purojava-manojava-pavamāna-dhūmrānīka-citrarepha-bahurūpa-viśvadhāra-saṁjñān nidhāpyādhipatīn svayaṁ bhagavaty ananta ā-veśita-matis tapovanaṁ praviveśa.

SYNONYMS: tasya api—of that island also; praiyavrataḥ—a son of Mahārāja Priyavrata; eva—certainly; adhipatiḥ—the ruler; nāmnā—by the name; medhātithiḥ—Medhātithi; saḥ api—he also; vibhajya—dividing; sapta varṣāṇi—seven divisions of the island; putra-nāmāni—possessing the names of his sons; teṣu—in them; sva-ātmajān—his own sons; purojava—Purojava; manojava—Manojava; pavamāna—Pavamāna; dhūmrānīka—Dhūmrānīka; citra-repha—Citrarepha; bahu-rūpa—Bahurūpa; viśvadhāra—Viśvadhāra; saṁjñān—having as names; nidhāpya—establishing as; adhipatīn—the rulers; svayam—himself; bhagavati—in the Supreme Personality of Godhead; anante—in the unlimited; āveśita-matiḥ—whose mind was fully absorbed; tapaḥ-vanam—in the forest where meditation is performed; praviveśa—he entered.

TRANSLATION: The master of this island, also one of the sons of Priyavrata, was known as Medhātithi. He also divided his island into seven sections, named according to the names of his own sons, whom he made the kings of that island. The names of those sons are Purojava, Manojava, Pavamāna, Dhūmrānīka, Citrarepha, Bahurūpa and Viśvadhāra. After dividing the island and situating his sons as its rulers, Medhātithi personally retired, and to fix his mind completely upon the lotus feet of the Supreme Personality of Godhead, he entered a forest suitable for meditation.

5.20.26: teṣāṁ varṣa-maryādā-girayo nadyaś ca sapta saptaiva īśāna uruśṛṅgo balabhadraḥ śatakesaraḥ sahasrasroto devapālo mahānasa iti anaghāyurdā ubhayaspṛṣṭir aparājitā pañcapadī sahasrasrutir nijadhṛtir iti.

SYNONYMS: eteṣām—of all these divisions; varṣa-maryādā—acting as the boundary limits; girayaḥ—the big hills; nadyaḥ ca—and the rivers also; sapta—seven; sapta—seven; eva—indeed; īśānaḥ—Īśāna; uruśṛṅgaḥ—Uruśṛṅga; balabhadraḥ—Balabhadra; śata-kesaraḥ—Śatakesara; sahasra-srotaḥ—Sahasrasrota; deva-pālaḥ—Devapāla; mahānasaḥ—Mahānasa; iti—thus; anaghā—Anaghā; āyurdā—Āyurdā; ubhayaspṛṣṭiḥ—Ubhayaspṛṣṭi; aparājitā—Aparājitā; pañcapadī—Pañcapadī; sahasra-srutiḥ—Sahasra-śruti; nija-dhṛtiḥ—Nijadhṛti; iti—thus.

TRANSLATION: For these lands also, there are seven boundary mountains and seven rivers. The mountains are Īśāna, Uruśṛṅga, Balabhadra, Śatakesara, Sahasrasrota, Devapāla and Mahānasa. The rivers are Anaghā, Āyurdā, Ubhayaspṛṣṭi, Aparājitā, Pañcapadī, Sahasra-śruti and Nijadhṛti.

5.20.27: tad-varṣa-puruṣā ṛtavrata-satyavrata-dānavratānuvrata-nāmāno bhagavantaṁ vāyv-ātmakaṁ prāṇāyāma-vidhūta-rajas-tamasaḥ parama-samādhinā yajante.

SYNONYMS: tat-varṣa-puruṣāḥ—the inhabitants of those tracts of land; ṛta-vrata—Ṛtavrata; satya-vrata—Satyavrata; dāna-vrata—Dānavrata; anuvrata—Anuvrata; nāmānaḥ—having the four names; bhagavantam—the Supreme Personality of Godhead; vāyu-ātmakam—represented by the demigod Vāyu; prāṇāyāma—by the practice of regulating the airs within the body; vidhūta—cleansed away; rajaḥ-tamasaḥ—whose passion and ignorance; parama—sublime; samādhinā—by trance; yajante—they worship.

TRANSLATION: The inhabitants of those islands are also divided into four castes—Ṛtavrata, Satyavrata, Dānavrata and Anuvrata—which exactly resemble brāhmaṇa, kṣatriya, vaiśya and śūdra. They practice prāṇāyāma and mystic yoga, and in trance they worship the Supreme Lord in the form of Vāyu.

5.20.28: antaḥ-praviśya bhūtāni/yo bibharty ātma-ketubhiḥ/antaryāmīśvaraḥ sākṣāt/pātu no yad-vaśe sphuṭam

SYNONYMS: antaḥ-praviśya—entering within; bhūtāni—all living entities; yaḥ—who; bibharti—maintains; ātma-ketubhiḥ—by the functions of the inner airs (prāṇa, apāna, etc.); antaryāmī—the Supersoul within; īśvaraḥ—the Supreme person; sākṣāt—directly; pātu—please maintain; naḥ—us; yat-vaśe—under whose control; sphuṭam—the cosmic manifestation.

TRANSLATION: [The inhabitants of Śākadvīpa worship the Supreme Personality of Godhead in the form of Vāyu in the following words.] O Supreme Person, situated as the Supersoul within the body, You direct the various actions of the different airs, such as prāṇa, and thus You maintain all living entities. O Lord, O Supersoul of everyone, O controller of the cosmic manifestation under whom everything exists, may You protect us from all dangers.

## Part XIII: Puṣkaradvīpa

5.20.29 evam eva dadhi-maṇḍodāt parataḥ puṣkaradvīpas tato dvi-guṇāyāmaḥ samantata upakalpitaḥ samānena svādūdakena samudreṇa bahir āvṛto yasmin bṛhat-puṣkaraṁ jvalana-śikhāmala-kanaka-patrāyutāyutaṁ bhagavataḥ kamalāsanasyādhyāsanaṁ parikalpitam.

SYNONYMS: evam eva—thus; dadhi-maṇḍa-udāt—the ocean of yogurt; parataḥ—beyond; puṣkara-dvīpaḥ—another island, named Puṣkaradvīpa; tataḥ—than that (Śākadvīpa); dvi-guṇa-āyāmaḥ—whose measurement is twice as great; samantataḥ—on all sides; upakalpitaḥ—surrounded; samānena—equal in width; svādu-udakena—possessing sweet water; samudreṇa—by an ocean; bahiḥ—outside; āvṛtaḥ—surrounded; yasmin—in which; bṛhat—very big; puṣkaram—lotus flower; jvalana-śikhā—like the flames of a blazing fire; amala—pure;

kanaka—gold; patra—leaves; ayuta-ayutam—possessing 100,000,000; bhagavataḥ—greatly powerful; kamala āsanasya—of Lord Brahmā, whose sitting place is on the lotus flower; adhyāsanam—sitting place; parikalpitam—considered.

TRANSLATION: Outside the ocean of yogurt is another island, known as Puṣkaradvīpa, which is 6,400,000 yojanas [51,200,000 miles] wide, twice as wide as the ocean of yogurt. It is surrounded by an ocean of very tasteful water as broad as the island itself. On Puṣkaradvīpa there is a great lotus flower with 100,000,000 pure golden petals, as effulgent as the flames of fire. That lotus flower is considered the sitting place of Lord Brahmā, who is the most powerful living being and who is therefore sometimes called bhagavān.

5.20.30: tad-dvīpa-madhye mānasottara-nāmaika evārvācīna-parācīna-varṣayor maryādācalo 'yuta-yojanocchrāyāyāmo yatra tu catasṛṣu dikṣu catvāri purāṇi loka-pālānām indrādīnāṁ yad-upariṣṭāt sūrya-rathasya meruṁ paribhramataḥ saṁvatsarātmakaṁ cakraṁ devānām aho-rātrābhyāṁ paribhramati.

SYNONYMS: tat-dvīpa-madhye—within that island; mānasottara—Mānasottara; nāma—named; ekaḥ—one; eva—indeed; arvācīna—on this side; parācīna—and beyond, or outside; varṣayoḥ—of tracts of land; maryādā—indicating the boundary; acalaḥ—a great mountain; ayuta—ten thousand; yojana—eight miles; ucchrāya-āyāmaḥ—whose height and width; yatra—where; tu—but; catasṛṣu—in the four; dikṣu—directions; catvāri—four; purāṇi—cities; loka-pālānām—of the directors of planetary systems; indra-ādīnām—headed by Indra; yat—of which; upariṣṭāt—on the top; sūrya-rathasya—of the chariot of the sun-god; merum—Meru Mountain; paribhramataḥ—while circumambulating; saṁvatsara-ātmakam—consisting of one saṁvatsara; cakram—wheel or orbit; devānām—of the demigods; ahaḥ-rātrābhyām—by the day and night; paribhramati—moves around.

TRANSLATION: In the middle of that island is a great mountain named Mānasottara, which forms the boundary between the inner side and the outer side of the island. Its breadth and height are 10,000 yojanas [80,000 miles]. On that mountain, in the four directions, are the residential quarters of demigods such as Indra. In the chariot of the sun-god, the sun travels on the top of the mountain in an orbit called the Saṁvatsara, encircling Mount Meru. The sun's path on the northern side is called Uttarāyaṇa, and its path on the southern side is called Dakṣiṇāyana. One side represents a day for the demigods, and the other represents their night.

PURPORT: The movement of the sun is confirmed in the Brahma-saṁhitā (5.52): yasyājñāya bhramati saṁbhṛta-kāla-cakraḥ. The sun orbits around Mount Sumeru, for six months on the northern side and for six months on the southern.

This adds up to the duration of a day and night of the demigods in the upper planetary systems.

5.20.31: tad-dvīpasyāpy adhipatiḥ praiyavrato vītihotro nāmaitasyātmajau ramaṇaka-dhātaki-nāmānau varṣa-patī niyujya sa svayaṁ pūrvajavad-bhagavat-karma-śīla evāste.

SYNONYMS: tat-dvīpasya—of that island; api—also; adhipatiḥ—the ruler; praiyavrataḥ—a son of Mahārāja Priyavrata; vītihotraḥ nāma—named Vītihotra; etasya—of him; ātma-jau—unto the two sons; ramaṇaka—Ramaṇaka; dhātaki—and Dhātaki; nāmānau—having the names; varṣa-patī—the rulers of the two tracts of land; niyujya—appointing; saḥ svayam—himself; pūrvaja-vat—like his other brothers; bhagavat-karma-śīlaḥ—being absorbed in activities to satisfy the Supreme Personality of Godhead; eva—indeed; āste—remains.

TRANSLATION: The ruler of this island, the son of Mahārāja Priyavrata named Vītihotra, had two sons named Ramaṇaka and Dhātaki. He granted the two sides of the island to these two sons and then personally engaged himself in activities for the sake of the Supreme Personality of Godhead like his elder brother Medhātithi.

5.20.32: tad-varṣa-puruṣā bhagavantaṁ brahma-rūpiṇaṁ sakarmakeṇa karmaṇārādhayantīdaṁ codāharanti.

SYNONYMS: tat-varṣa-puruṣāḥ—the inhabitants of that island; bhagavantam—the Supreme Personality of Godhead; brahma-rūpiṇam—exhibited as Lord Brahmā being seated on the lotus; sa-karmakeṇa—for fulfillment of material desires; karmaṇā—by performing ritualistic activities according to the Vedas; ārādhayanti—worship; idam—this; ca—and; udāharanti—they chant.

TRANSLATION: For the fulfillment of material desires, the inhabitants of this tract of land worship the Supreme Personality of Godhead as represented by Lord Brahmā. They offer prayers to the Lord as follows.

5.20.33: yat tat karmamayaṁ liṅgaṁ/brahma-liṅgaṁ jano 'rcayet/ekāntam advayaṁ śāntaṁ/tasmai bhagavate nama iti

SYNONYMS: yat—which; tat—that; karma-mayam—obtainable by the Vedic ritualistic system; liṅgam—the form; brahma-liṅgam—which makes known the Supreme Brahman; janaḥ—a person; arcayet—must worship; ekāntam—who has full faith in the one Supreme; advayam—nondifferent; śāntam—peaceful; tasmai—unto him; bhagavate—the most powerful; namaḥ—our respects; iti—thus.

TRANSLATION: Lord Brahmā is known as karma-maya, the form of ritualistic ceremonies, because by performing ritualistic ceremonies one may attain his position and because the Vedic ritualistic hymns become manifest from him. He is devoted to the Supreme Personality of Godhead without deviation, and therefore in one sense he is not different from the Lord. Nevertheless, he should be worshiped not as the monists worship him, but in duality. One should always remain a servitor of the Supreme Lord, the supreme worshipable Deity. We therefore offer our respectful obeisances unto Lord Brahmā, the form of manifest Vedic knowledge.

## Part XIV: Lokāloka Mountain

5.20.34: tataḥ parastāl lokāloka-nāmācalo lokālokayor antarāle parita upakṣiptaḥ.

SYNONYMS: tataḥ—from that ocean of sweet drinking water; parastāt—beyond; lokāloka-nāma—named Lokāloka; acalaḥ—a mountain; loka-alokayoḥ antarāle—between the countries full of sunlight and those without sunlight; paritaḥ—all around; upakṣiptaḥ—exists.

TRANSLATION: Thereafter, beyond the ocean of sweet water and fully surrounding it, is a mountain named Lokāloka, which divides the countries that are full of sunlight from those not lit by the sun.

5.20.35: yāvan mānasottara-mervor antaraṁ tāvatī bhūmiḥ kāñcany anyādarśa-talopamā yasyāṁ prahitaḥ padārtho na kathañcit punaḥ pratyupalabhyate tasmāt sarva-sattva-parihṛtāsīt.

SYNONYMS: yāvat—as much as; mānasottara-mervoḥ antaram—the land between Mānasottara and Meru (beginning from the middle of Mount Sumeru); tāvatī—that much; bhūmiḥ—land; kāñcanī—made of gold; anyā—another; ādarśa-tala-upamā—whose surface is just like the surface of a mirror; yasyām—on which; prahitaḥ—dropped; padārthaḥ—a thing; na—not; kathañcit—in any way; punaḥ—again; pratyupalabhyate—is found; tasmāt—therefore; sarva-sattva—by all living entities; parihṛtā—abandoned; āsīt—was.

TRANSLATION: Beyond the ocean of sweet water is a tract of land as broad as the area between the middle of Mount Sumeru and the boundary of Mānasottara Mountain. In that tract of land there are many living beings. Beyond it, extending to Lokāloka Mountain, is another land, which is made of gold. Because of its golden surface, it reflects light like the surface of a mirror, and any physical article that falls on that land can never be perceived again. All living entities, therefore, have abandoned that golden land.

5.20.36: lokāloka iti samākhyā yad anenācalena lokālokasyāntarvartināvasthāpyate.

SYNONYMS: loka—with light (or with inhabitants); alokaḥ—without light (or without inhabitants); iti—in this way; samākhyā—designation; yat—which; anena—by this; acalena—mountain; loka—of the land inhabited by living entities; alokasya—and of the land not inhabited by living entities; antarvartinā—which is in the middle; avasthāpyate—is established.

TRANSLATION: Between the lands inhabited by living entities and those that are uninhabited stands the great mountain which separates the two and which is therefore celebrated as Lokāloka.

5.20.37: sa loka-trayānte parita īśvareṇa vihito yasmāt sūryādīnāṁ dhruvāpavargāṇāṁ jyotir-gaṇānāṁ gabhastayo 'rvācīnāṁs trīû lokān āvitanvānā na kadācit parācīnā bhavitum utsahante tāvad unnahanāyāmaḥ.

SYNONYMS: saḥ—that mountain; loka-traya-ante—at the end of the three lokas (Bhūrloka, Bhuvarloka and Svarloka); paritaḥ—all around; īśvareṇa—by the Supreme Personality of Godhead, Kṛṣṇa; vihitaḥ—created; yasmāt—from which; sūrya-ādīnām—of the sun planet; dhruva-apavargāṇām—up to Dhruvaloka and other, inferior luminaries; jyotiḥ-gaṇānām—of all the luminaries; gabhastayaḥ—the rays; arvācīnān—on this side; trīn—the three; lokān—planetary systems; āvitanvānāḥ—spreading throughout; na—not; kadācit—at any time; parācīnāḥ—beyond the jurisdiction of that mountain; bhavitum—to be; utsahante—are able; tāvat—that much; unnahana-āyāmaḥ—the measure of the height of the mountain.

TRANSLATION: By the supreme will of Kṛṣṇa, the mountain known as Lokāloka has been installed as the outer border of the three worlds—Bhūrloka, Bhuvarloka and Svarloka—to control the rays of the sun throughout the universe. All the luminaries, from the sun up to Dhruvaloka, distribute their rays throughout the three worlds, but only within the boundary formed by this mountain. Because it is extremely high, extending even higher than Dhruvaloka, it blocks the rays of the luminaries, which therefore can never extend beyond it.

PURPORT: When we speak of loka-traya, we refer to the three primary planetary systems—Bhūḥ, Bhuvaḥ and Svaḥ—into which the universe is divided. Surrounding these planetary systems are the eight directions, namely east, west, north, south, northeast, southeast, northwest and southwest. Lokāloka Mountain has been established as the outer boundary of all the lokas to distribute the rays of the sun and other luminaries equally throughout the universe.

This vivid description of how the rays of the sun are distributed throughout the different planetary systems of the universe is very scientific. Śukadeva Gosvāmī described these universal affairs to Mahārāja Parīkṣit as he had heard about them from his predecessor. He explained these facts five thousand years ago, but the knowledge existed long, long before because Śukadeva Gosvāmī received it through disciplic succession. Because this knowledge is accepted through the disciplic succession, it is perfect. The history of modern scientific knowledge, on the contrary, does not go back more than a few hundred years. Therefore, even if modern scientists do not accept the other factual presentations of Śrīmad-Bhāgavatam, how can they deny the perfect astronomical calculations that existed long before they could imagine such things? There is so much information to gather from Śrīmad-Bhāgavatam. Modern scientists, however, have no information of other planetary systems and, indeed, are hardly conversant with the planet on which we are now living.

5.20.38: etāvāû loka-vinyāso māna-lakṣaṇa-saṁsthābhir vicintitaḥ kavibhiḥ sa tu pañcāśat-koṭi-gaṇitasya bhū-golasya turīya-bhāgo 'yaṁ lokālokācalaḥ.

SYNONYMS: etāvān—this much; loka-vinyāsaḥ—the placing of the different planets; māna—with measurements; lakṣaṇa—symptoms; saṁsthābhiḥ—as well as with their different situations; vicintitaḥ—established by scientific calculations; kavibhiḥ—by learned scholars; saḥ—that; tu—but; pañcāśat-koṭi—500,000,000 yojanas; gaṇitasya—which is measured at; bhū-golasya—of the planetary system known as Bhūgolaka; turīya-bhāgaḥ—one fourth; ayam—this; lokāloka-acalaḥ—the mountain known as Lokāloka.

TRANSLATION: Learned scholars who are free from mistakes, illusions and propensities to cheat have thus described the planetary systems and their particular symptoms, measurements and locations. With great deliberation, they have established the truth that the distance between Sumeru and the mountain known as Lokāloka is one fourth of the diameter of the universe—or, in other words, 125,000,000 yojanas [1 billion miles].

PURPORT: Śrīla Viśvanātha Cakravartī Ṭhākura has given accurate astronomical information about the location of Lokāloka Mountain, the movements of the sun globe and the distance between the sun and the circumference of the universe. However, the technical terms used in the astronomical calculations given by the Jyotir Veda are difficult to translate into English. Therefore to satisfy the reader, we may include the exact Sanskrit statement given by Śrīla Viśvanātha Cakravartī Ṭhākura, which records exact calculations regarding universal affairs.

sa tu lokālokas tu bhū-golakasya bhū-sambandhāṇḍa-golakasyety arthaḥ; sūryasy eva bhuvo 'py aṇḍa-golakayor madhya-vartitvāt kha-golam iva bhū-golam api pañcāśat-koṭi-yojana-pramāṇaṁ tasya turīya-bhāgaḥ sārdha-dvādaśa-koṭi-yojana-vistārocchrāya ity arthaḥ bhūs tu catus-triṁśal-lakṣonapañcāśat-koṭi-pramāṇā jñeyā; yathā meru-madhyān mānasottara-madhya-paryantaṁ sārdha-sapta-pañcāśal-lakṣottara-koṭi-yojana-pramāṇam; mānasottara-madhyāt svādūdaka-samudra-paryantaṁ ṣaṇ-ṇavati-lakṣa-yojana-pramāṇaṁ tataḥ kāñcanī-bhūmiḥ sārdha-sapta-pañcāśal-lakṣottara-koṭi-yojana-pramāṇā evam ekato meru-lokālokayor antarālam ekādaśa-śal-lakṣādhika-catuṣ-koṭi-parimitam anyato 'pi tathatyeto lokālokāl loka-paryantaṁ sthānaṁ dvāviṁśati-lakṣottarāṣṭa-koṭi-parimitaṁ lokālokād bahir apy ekataḥ etāvad eva anyato 'py etāvad eva yad vakṣyate, yo 'ntar-vistāra etena hy aloka-parimāṇaṁ ca vyākhyātaṁ yad-bahir lokālokācalād ity ekato lokālokaḥ sārdha-dvādaśa-koṭi-yojana-parimāṇaḥ anyato 'pi sa tathety evaṁ catus-triṁśal-lakṣonapañcāśat-koṭi-pramāṇā bhūḥ sābdhi-dvīpa-parvatā jñeyā; ata evāṇḍa-golakāt sarvato dikṣu sapta-daśa-lakṣa-yojanāvakāśe vartamāne sati pṛthivyāḥ śeṣa-nāgena dhāraṇaṁ dig-gajaiś ca niścalī-karaṇaṁ sārthakaṁ bhaved anyathā tu vyākhyāntare pañcāśat-koṭi-pramāṇatvād aṇḍa-golaka-lagnatve tat tat sarvam akiñcit-karaṁ syāt cākṣuṣe manvantare cākasmāt majjanaṁ śrī-varāha-devenotthāpanaṁ ca durghaṭaṁ syād ity adikaṁ vivecanīyam.

5.20.39: tad-upariṣṭāc catasṛṣv āśāsvātma-yoninākhila-jagad-guruṇādhiniveśitā ye dvirada-pataya ṛṣabhaḥ puṣkaracūḍo vāmano 'parājita iti sakala-loka-sthiti-hetavaḥ.

SYNONYMS: tat-upariṣṭāt—on the top of Lokāloka Mountain; catasṛṣu āśāsu—in the four directions; ātma-yoninā—by Lord Brahmā; akhila-jagat-guruṇā—the spiritual master of the whole universe; adhiniveśitāḥ—established; ye—all those; dvirada-patayaḥ—the best of elephants; ṛṣabhaḥ—Ṛṣabha; puṣkara-cūḍaḥ—Puṣkaracūḍa; vāmanaḥ—Vāmana; aparājitaḥ—Aparājita; iti—thus; sakala-loka-sthiti-hetavaḥ—the causes of the maintenance of the different planets within the universe.

TRANSLATION: On the top of Lokāloka Mountain are the four gaja-patis, the best of elephants, which were established in the four directions by Lord Brahmā, the supreme spiritual master of the entire universe. The names of those elephants are Ṛṣabha, Puṣkaracūḍa, Vāmana and Aparājita. They are responsible for maintaining the planetary systems of the universe.

5.20.40: teṣāṁ sva-vibhūtīnāṁ loka-pālānāṁ ca vividha-vīryopabṛṁhaṇāya bhagavān parama-mahā-puruṣo mahā-vibhūti-patir antaryāmy ātmano viśuddha-sattvaṁ dharma-jñāna-vairāgyaiśvaryādy-aṣṭa-mahā-siddhy-upalakṣaṇaṁ viṣvaksenādibhiḥ sva-pārṣada-pravaraiḥ parivārito nija-varāyudhopaśobhitair nija-bhuja-daṇḍaiḥ sandhārayamāṇas tasmin giri-vare samantāt sakala-loka-svastaya āste.

SYNONYMS: teṣām—of all of them; sva-vibhūtīnām—who are his personal expansions and assistants; loka-pālānām—who are entrusted with looking after the universal affairs; ca—and; vividha—varieties; vīrya-upabṛṁhaṇāya—for expanding the powers; bhagavān—the Supreme Personality of Godhead; parama-mahā-puruṣaḥ—the foremost master of all kinds of opulence, the Supreme Personality of Godhead; mahā-vibhūti-patiḥ—the master of all inconceivable potencies; antaryāmī—the Supersoul; ātmanaḥ—of Himself; viśuddha-sattvam—having an existence without contamination by the material modes of nature; dharma-jñāna-vairāgya—of religion, pure knowledge and renunciation; aiśvarya-ādi—of all kinds of opulence; aṣṭa—eight; mahā-siddhi—and of great mystic perfections; upalakṣaṇam—having the characteristics; viṣvaksena-ādibhiḥ—by His expansion known as Viṣvaksena and others; sva-pārṣada-pravaraiḥ—the best of personal assistants; parivāritaḥ—surrounded; nija—His own; vara-āyudha—by different types of weapons; upaśobhitaiḥ—being decorated; nija—own; bhuja-daṇḍaiḥ—with stout arms; sandhārayamāṇaḥ—manifesting this form; tasmin—on that; giri-vare—great mountain; samantāt—all around; sakala-loka-svastaye—for the benefit of all the planetary systems; āste—exists.

TRANSLATION: The Supreme Personality of Godhead is the master of all transcendental opulences and the master of the spiritual sky. He is the Supreme Person, Bhagavān, the Supersoul of everyone. The demigods, led by Indra, the King of heaven, are entrusted with seeing to the affairs of the material world. To benefit all living beings in all the varied planets and to increase the power of those elephants and of the demigods, the Lord manifests Himself on top of that mountain in a spiritual body, uncontaminated by the modes of material nature. Surrounded by His personal expansions and assistants like Viṣvaksena, He exhibits all His perfect opulences, such as religion and knowledge, and His mystic powers such as aṇimā, laghimā and mahimā. He is beautifully situated, and He is decorated by the different weapons in His four hands.

5.20.41: ākalpam evaṁ veṣaṁ gata eṣa bhagavān ātma-yogamāyayā viracita-vividha-loka-yātrā-gopīyāyety arthaḥ.

SYNONYMS: ā-kalpam—for the duration of the time of creation; evam—thus; veṣam—appearance; gataḥ—has accepted; eṣaḥ—this; bhagavān—the Supreme Personality of Godhead; ātma-yoga-māyayā—by His own spiritual potency; viracita—perfected; vividha-loka-yātrā—the livelihood of the various planetary systems; gopīyāya—just to maintain; iti—thus; arthaḥ—the purpose.

TRANSLATION: The various forms of the Supreme Personality of Godhead, such as Nārāyaṇa and Viṣṇu, are beautifully decorated with different weapons. The Lord exhibits those forms to maintain all the varied planets created by His personal potency, yogamāyā.

## Part XV: Aloka-Varṣa

5.20.42: yo 'ntar-vistāra etena hy aloka-parimāṇaṁ ca vyākhyātaṁ yad bahir lokālokācalāt; tataḥ parastād yogeśvara-gatiṁ viśuddhām udāharanti.

SYNONYMS: yaḥ—that which; antaḥ-vistāraḥ—the distance inside Lokāloka Mountain; etena—by this; hi—indeed; aloka-parimāṇam—the width of the tract of land known as Aloka-varṣa; ca—and; vyākhyātam—described; yat—which; bahiḥ—outside; lokāloka-acalāt—beyond Lokāloka Mountain; tataḥ—that; parastāt—beyond; yogeśvara-gatim—the path of Yogeśvara (Kṛṣṇa) in penetrating the coverings of the universe; viśuddhām—without material contamination; udāharanti—they say.

TRANSLATION: My dear King, outside Lokāloka Mountain is the tract of land known as Aloka-varṣa, which extends for the same breadth as the area within the mountain—in other words, 125,000,000 yojanas [one billion miles]. Beyond Aloka-varṣa is the destination of those who aspire for liberation from the material world. It is beyond the jurisdiction of the material modes of nature, and therefore it is completely pure. Lord Kṛṣṇa took Arjuna through this place to bring back the sons of the brāhmaṇa.

## Part XVI: The Sun

5.20.43: aṇḍa-madhya-gataḥ sūryo/dyāv-ābhūmyor yad antaram/sūryāṇḍa-golayor madhye/koṭyaḥ syuḥ pañca-viṁśatiḥ

SYNONYMS: aṇḍa-madhya-gataḥ—situated in the center of the universe; sūryaḥ—the sun globe; dyāv-ābhūmyoḥ—the two planetary systems Bhūrloka and Bhuvarloka; yat—which; antaram—in between; sūrya—of the sun; aṇḍa-golayoḥ—and the globe of the universe; madhye—in the middle; koṭyaḥ—groups of ten million; syuḥ—are; pañca-viṁśatiḥ—twenty-five.

TRANSLATION: The sun is situated [vertically] in the middle of the universe, in the area between Bhūrloka and Bhuvarloka, which is called antarikṣa, outer space. The distance between the sun and the circumference of the universe is twenty-five koṭi yojanas [two billion miles].

PURPORT: The word koṭi means ten million, and a yojana is eight miles. The diameter of the universe is fifty koṭi yojanas (four billion miles). Therefore, since the sun is in the middle of the universe, the distance between the sun and the edge of the universe is calculated to be twenty-five koṭi yojanas (two billion miles).

5.20.44: mṛte 'ṇḍa eṣa etasmin yad abhūt tato mārtaṇḍa iti vyapadeśaḥ; hiraṇyagarbha iti yad dhiraṇyāṇḍa-samudbhavaḥ.

SYNONYMS: mṛte—dead; aṇḍe—in the globe; eṣaḥ—this; etasmin—in this; yat—which; abhūt—entered personally at the time of creation; tataḥ—from that; mārtaṇḍa—Mārtaṇḍa; iti—thus; vyapadeśaḥ—the designation; hiraṇya-garbhaḥ—known as Hiraṇyagarbha; iti—thus; yat—because; hiraṇya-aṇḍa-samudbhavaḥ—his material body was created from Hiraṇyagarbha.

TRANSLATION: The sun-god is also known as Vairāja, the total material body for all living entities. Because he entered this dull egg of the universe at the time of creation, he is also called Mārtaṇḍa. He is also known as Hiraṇyagarbha because he received his material body from Hiraṇyagarbha [Lord Brahmā].

PURPORT: The post of Lord Brahmā is meant for very highly elevated, spiritually advanced living beings. When such living beings are unavailable, Lord Viṣṇu, the Supreme Personality of Godhead, expands Himself as Lord Brahmā. This takes place very rarely. Consequently there are two kinds of Brahmās. Sometimes Brahmā is an ordinary living entity, and at other times Brahmā is the Supreme Personality of Godhead. The Brahmā spoken of here is an ordinary living being. Whether Brahmā is the Supreme Personality of Godhead or an ordinary living being, he is known as Vairāja Brahmā and Hiraṇyagarbha Brahmā. Therefore the sun-god is also accepted as Vairāja Brahmā.

5.20.45: sūryeṇa hi vibhajyante/diśaḥ khaṁ dyaur mahī bhidā/svargāpavargau narakā/rasaukāṁsi ca sarvaśaḥ

SYNONYMS: sūryeṇa—by the sun-god within the sun planet; hi—indeed; vibhajyante—are divided; diśaḥ—the directions; kham—the sky; dyauḥ—the heavenly planets; mahī—the earthly planets; bhidā—other divisions; svarga—the heavenly planets; apavargau—and the places for liberation; narakāḥ—the hellish planets; rasaukāṁsi—such as Atala; ca—also; sarvaśaḥ—all.

TRANSLATION: O King, the sun-god and the sun planet divide all the directions of the universe. It is only because of the presence of the sun that we can understand what the sky, the higher planets, this world and the lower planets are. It is also only because of the sun that we can understand which places are for material enjoyment, which are for liberation, which are hellish and subterranean.

5.20.46: deva-tiryaṅ-manuṣyāṇāṁ/sarīsṛpa-savīrudhām/sarva-jīva-nikāyānāṁ/ sūrya ātmā dṛg-īśvaraḥ

SYNONYMS: deva—of the demigods; tiryak—the lower animals; manuṣyāṇām—and the human beings; sarīsṛpa—the insects and the serpents; sa-vīrudhām—and the plants and trees; sarva-jīva-nikāyānām—of all groups of living entities; sūryaḥ—the sun-god; ātmā—the life and soul; dṛk—of the eyes; īśvaraḥ—the Personality of Godhead.

TRANSLATION: All living entities, including demigods, human beings, animals, birds, insects, reptiles, creepers and trees, depend upon the heat and light given by the sun-god from the sun planet. Furthermore, it is because of the sun's presence that all living entities can see, and therefore he is called dṛg-īśvara, the Personality of Godhead presiding over sight.

PURPORT: In this regard, Śrīla Viśvanātha Cakravartī Ṭhākura says, sūrya ātmā ātmatvenopāsyaḥ. The actual life and soul of all living entities within this universe is the sun. He is therefore upāsya, worshipable. We worship the sun-god by chanting the Gāyatrī mantra (oṁ bhūr bhuvaḥ svaḥ tat savitur vareṇyaṁ bhargo devasya dhīmahi). Sūrya is the life and soul of this universe, and there are innumerable universes for which a sun-god is the life and soul, just as the Supreme Personality of Godhead is the life and soul of the entire creation. We have information that Vairāja, Hiraṇyagarbha, entered the great, dull, material globe called the sun. This indicates that the theory held by so-called scientists that no one lives there is wrong. Bhagavad-gītā also says that Kṛṣṇa first instructed Bhagavad-gītā to the sun-god (imaṁ vivasvate yogaṁ proktavān aham avyayam). Therefore the sun is not vacant. It is inhabited by living entities, and the predominating deity is Vairāja, or Vivasvān. The difference between the sun and earth is that the sun is a fiery planet, but everyone there has a suitable body and can live there without difficulty.

5.21 Summary: This chapter informs us of the movements of the sun. The sun is not stationary; it is also moving like the other planets. The sun's movements determine the duration of night and day. When the sun travels north of the equator, it moves slowly during the day and very quickly at night, thus increasing the duration of the daytime and decreasing the duration of night. Similarly, when the sun travels south of the equator, the exact opposite is true—the duration of the day decreases, and the duration of night increases. When the sun enters Karkaṭa-rāśi (Cancer) and then travels to Siṁha-rāśi (Leo) and so on through Dhanuḥ-rāśi (Sagittarius), its course is called Dakṣiṇāyana, the southern way, and when the sun enters Makara-rāśi (Capricorn) and thereafter travels through Kumbharāśi (Aquarius) and so on through Mithuna-rāśi (Gemini), its course is called Uttarāyaṇa, the northern way. When the sun is in Meṣa-rāśi (Aries) and Tulā-rāśi (Libra), the duration of day and night are equal.

On Mānasottara Mountain are the abodes of four demigods. East of Sumeru Mountain is Devadhānī, where King Indra lives, and south of Sumeru is Saṁyamanī, the abode of Yamarāja, the superintendent of death. Similarly, west of Sumeru is Nimlocanī, the abode of Varuṇa, the demigod who controls the water, and north of Sumeru is Vibhāvarī, where the demigod of the moon lives. Sunrise, noon, sunset and midnight occur in all these places because of the movements of the sun. Diametrically opposite the place where the sunrise takes places and the sun is seen by human eyes, the sun will be setting and passing away from human vision. Similarly, the people residing diametrically opposite the point where it is midday will be experiencing midnight. The sun rises and sets with all the other planets, headed by the moon and other luminaries.

The entire kāla-cakra, or wheel of time, is established on the wheel of the sun-god's chariot. This wheel is known as Saṁvatsara. The seven horses pulling the chariot of the sun are known as Gāyatrī, Bṛhatī, Uṣṇik, Jagatī, Triṣṭup, Anuṣṭup and Paṅkti. They are harnessed by a demigod known as Aruṇadeva to a yoke 900,000 yojanas wide. Thus the chariot carries Ādityadeva, the sun-god. Always staying in front of the sun-god and offering their prayers are sixty thousand sages known as Vālikhilyas. There are fourteen Gandharvas, Apsarās and other demigods, who are divided into seven parties and who perform ritualistic activities every month to worship the Supersoul through the sun-god according to different names. Thus the sun-god travels through the universe for a distance of 95,100,000 yojanas (760,800,000 miles) at a speed of 16,004 miles at every moment.

5.21.1: śrī-śuka uvāca/ etāvān eva bhū-valayasya sanniveśaḥ pramāṇa-lakṣaṇato vyākhyātaḥ.

SYNONYMS: śrī-śukaḥ uvāca—Śrī Śukadeva Gosvāmī said; etāvān—so much; eva—certainly; bhū-valayasya sanniveśaḥ—the arrangement of the whole universe; pramāṇa-lakṣaṇataḥ—according to measurement (fifty crores of yojanas, or four billion miles in width and length) and characteristics; vyākhyātaḥ—estimated.

TRANSLATION: Śukadeva Gosvāmī said: My dear King, I have thus far described the diameter of the universe [fifty crores of yojanas, or four billion miles] and its general characteristics, according to the estimations of learned scholars.

5.21.2: etena hi divo maṇḍala-mānaṁ tad-vida upadiśanti yathā dvi-dalayor niṣpāvādīnāṁ te antareṇāntarikṣaṁ tad-ubhaya-sandhitam.

SYNONYMS: etena—by this estimation; hi—indeed; divaḥ—of the upper planetary system; maṇḍala-mānam—the measurement of the globe; tat-vidaḥ—the experts who know about it; upadiśanti—instruct; yathā—just as; dvi-dalayoḥ—in the two halves; niṣpāva-ādīnām—of grain such as wheat; te—of the two divisions; antareṇa—in the intervening space; antarikṣam—the sky or outer space; tat—by the two; ubhaya—on both sides; sandhitam—where the two parts join.

TRANSLATION: As a grain of wheat is divided into two parts and one can estimate the size of the upper part by knowing that of the lower, so, expert geographers instruct, one can understand the measurements of the upper part of the universe by knowing those of the lower part. The sky between the earthly sphere and heavenly sphere is called antarikṣa, or outer space. It adjoins the top of the sphere of earth and the bottom of that of heaven.

5.21.3: yan-madhya-gato bhagavāṁs tapatāṁ patis tapana ātapena tri-lokīṁ pratapaty avabhāsayaty ātma-bhāsā sa eṣa udagayana-dakṣiṇāyana-vaiṣuvata-saṁjñābhir māndya-śaighrya-samānābhir gatibhir ārohaṇāvarohaṇa-samāna-sthāneṣu yathā-savanam abhipadyamāno makarādiṣu rāśiṣv aho-rātrāṇi dīrgha-hrasva-samānāni vidhatte.

SYNONYMS: yat—of which (the intermediate space); madhya-gataḥ—being situated in the middle; bhagavān—the most powerful; tapatām patiḥ—the master of those that heat the whole universe; tapanaḥ—the sun; ātapena—by heat; tri-lokīm—the three worlds; pratapati—heats; avabhāsayati—lights; ātma-bhāsā—by its own illuminating rays; saḥ—that; eṣaḥ—the sun globe; udagayana—of passing to the northern side of the equator; dakṣiṇa-ayana—of passing to the southern side of the equator; vaiṣuvata—or of passing through the equator; saṁjñābhiḥ—by different names; māndya—characterized by slowness; śaighrya—swiftness; samānābhiḥ—and by equality; gatibhiḥ—by movement; ārohaṇa—of rising; avarohaṇa—of going down; samāna—or of staying in the middle; sthāneṣu—in positions; yathā-savanam—according to the order of the Supreme Personality of Godhead; abhipadyamānaḥ—moving; makara-ādiṣu—headed by the sign Makara (Capricorn); rāśiṣu—in different signs; ahaḥ-rātrāṇi—the days and nights; dīrgha—long; hrasva—short; samānāni—equal; vidhatte—makes.

TRANSLATION: In the midst of that region of outer space [antarikṣa] is the most opulent sun, the king of all the planets that emanate heat, such as the moon. By the influence of its radiation, the sun heats the universe and maintains its proper order. It also gives light to help all living entities see. While passing toward the north, toward the south or through the equator, in accordance with the order of the Supreme Personality of Godhead, it is said to move slowly, swiftly or moderately. According to its movements in rising above, going beneath or passing through the equator—and correspondingly coming in touch with various signs of the zodiac, headed by Makara [Capricorn]—days and nights are short, long or equal to one another.

PURPORT: Lord Brahmā prays in his Brahma-saṁhitā (5.52): yac cakṣur eṣa savitā sakala-grahāṇāṁ/rājā samasta-sura-mūrtir aśeṣa-tejāḥ/yasyājñayā bhramati saṁbhṛta-kāla-cakro/govindam ādi-puruṣaṁ tam ahaṁ bhajāmi
"I worship Govinda, the primeval Lord, the Supreme Personality of Godhead under whose control even the sun, which is considered to be the eye of the Lord,

rotates within the fixed orbit of eternal time. The sun is the king of all planetary systems and has unlimited potency in heat and light." Although the sun is described as bhagavān, the most powerful, and although it is actually the most powerful planet within the universe, it nevertheless has to carry out the order of Govinda, Kṛṣṇa. The sun-god cannot deviate even an inch from the orbit designated to him. Therefore in every sphere of life, the supreme order of the Supreme Personality of Godhead is carried out. The entire material nature carries out His orders. However, we foolishly see the activities of material nature without understanding the supreme order and Supreme Person behind them. As confirmed in Bhagavad-gītā, mayādhyakṣeṇa prakṛtiḥ: material nature carries out the orders of the Lord, and thus everything is maintained in an orderly way.

5.21.4: yadā meṣa-tulayor vartate tadāho-rātrāṇi samānāni bhavanti yadā vṛṣabhādiṣu pañcasu ca rāśiṣu carati tadāhāny eva vardhante hrasati ca māsi māsy ekaikā ghaṭikā rātriṣu.

SYNONYMS: yadā—when; meṣa-tulayoḥ—in Meṣa (Aries) and Tulā (Libra); vartate—the sun exists; tadā—at that time; ahaḥ-rātrāṇi—the days and nights; samānāni—equal in duration; bhavanti—are; yadā—when; vṛṣabha-ādiṣu—headed by Vṛṣabha (Taurus) and Mithuna (Gemini); pañcasu—in the five; ca—also; rāśiṣu—signs; carati—moves; tadā—at that time; ahāni—the days; eva—certainly; vardhante—increase; hrasati—is diminished; ca—and; māsi māsi—in every month; eka-ekā—one; ghaṭikā—half hour; rātriṣu—in the nights.

TRANSLATION: When the sun passes through Meṣa [Aries] and Tulā [Libra], the durations of day and night are equal. When it passes through the five signs headed by Vṛṣabha [Taurus], the duration of the days increases [until Cancer], and then it gradually decreases by half an hour each month, until day and night again become equal [in Libra].

5.21.5: yadā vṛścikādiṣu pañcasu vartate tadāho-rātrāṇi viparyayāṇi bhavanti.

SYNONYMS: yadā—when; vṛścika-ādiṣu—headed by Vṛścika (Scorpio); pañcasu—five; vartate—remains; tadā—at that time; ahaḥ-rātrāṇi—the days and nights; viparyayāṇi—the opposite (the duration of the day decreases, and that of night increases); bhavanti—are.

TRANSLATION: When the sun passes through the five signs beginning with Vṛścika [Scorpio], the duration of the days decreases [until Capricorn], and then gradually it increases month after month, until day and night become equal [in Aries].

5.21.6: yāvad dakṣiṇāyanam ahāni vardhante yāvad udagayanaṁ rātrayaḥ.

SYNONYMS: yāvat—until; dakṣiṇa-ayanam—the sun passes to the southern side; ahāni—the days; vardhante—increase; yāvat—until; udagayanam—the sun passes to the northern side; rātrayaḥ—the nights.

TRANSLATION: Until the sun travels to the south the days grow longer, and until it travels to the north the nights grow longer.

5.21.7: evaṁ nava koṭaya eka-pañcāśal-lakṣāṇi yojanānāṁ mānasottara-giri-parivartanasyopadiśanti tasminn aindrīṁ purīṁ pūrvasmān meror devadhānīṁ nāma dakṣiṇato yāmyāṁ saṁyamanīṁ nāma paścād vāruṇīṁ nimlocanīṁ nāma uttarataḥ saumyāṁ vibhāvarīṁ nāma tāsūdaya-madhyāhnāstamaya-niśīthānīti bhūtānāṁ pravṛtti-nivṛtti-nimittāni samaya-viśeṣeṇa meroś catur-diśam.

SYNONYMS: evam—thus; nava—nine; koṭayaḥ—ten millions; eka-pañcāśat—fifty-one; lakṣāṇi—hundred thousands; yojanānām—of the yojanas; mānasottara-giri—of the mountain known as Mānasottara; parivartanasya—of the circumambulation; upadiśanti—they (learned scholars) teach; tasmin—on that (Mānasottara Mountain); aindrīm—of King Indra; purīm—the city; pūrvasmāt—on the eastern side; meroḥ—of Sumeru Mountain; devadhānīm—Devadhānī; nāma—of the name; dakṣiṇataḥ—on the southern side; yāmyām—of Yamarāja; saṁyamanīm—Saṁyamanī; nāma—named; paścāt—on the western side; vāruṇīm—of Varuṇa; nimlocanīm—Nimlocanī; nāma—named; uttarataḥ—on the northern side; saumyām—of the moon; vibhāvarīm—Vibhāvarī; nāma—named; tāsu—in all of them; udaya—rising; madhyāhna—midday; astamaya—sunset; niśīthāni—midnight; iti—thus; bhūtānām—of the living entities; pravṛtti—of activity; nivṛtti—and cessation of activity; nimittāni—the causes; samaya-viśeṣeṇa—by the particular times; meroḥ—of Sumeru Mountain; catuḥ-diśam—the four sides.

TRANSLATION: Śukadeva Gosvāmī continued; My dear King, as stated before, the learned say that the sun travels over all sides of Manasottara Mountain in a circle whose length is 95,100,000 yojanas [760,800,000 miles]. On Mānasottara Mountain, due east of Mount Sumeru, is a place known as Devadhānī, possessed by King Indra. Similarly, in the south is a place known as Saṁyamanī, possessed by Yamarāja, in the west is a place known as Nimlocanī, possessed by Varuṇa, and in the north is a place named Vibhāvarī, possessed by the moon-god. Sunrise, midday, sunset and midnight occur in all those places according to specific times, thus engaging all living entities in their various occupational duties and also making them cease such duties.

5.21.8-9: tatratyānāṁ divasa-madhyaṅgata eva sadādityas tapati savyenācalaṁ dakṣiṇena karoti; yatrodeti tasya ha samāna-sūtra-nipāte nimlocati yatra kvacana syandenābhitapati tasya haiṣa samāna-sūtra-nipāte prasvāpayati tatra gataṁ na paśyanti ye taṁ samanupaśyeran.

SYNONYMS: tatratyānām—for the living entities residing on Mount Meru; divasa-madhyaṅgataḥ—being positioned as at midday; eva—indeed; sadā—always; ādityaḥ—the sun; tapati—heats; savyena—to the left side; acalam—Sumeru Mountain; dakṣiṇena—to the right (being forced by wind blowing to the right, the sun moves to the right); karoti—moves; yatra—the point where; udeti—it rises; tasya—of that position; ha—certainly; samāna-sūtra-nipāte—at the diametrically opposite point; nimlocati—the sun sets; yatra—where; kvacana—somewhere; syandena—with perspiration; abhitapati—heats (at midday); tasya—of that; ha—certainly; eṣaḥ—this (the sun); samāna-sūtra-nipāte—at the diametrically opposite point; prasvāpayati—the sun causes to sleep (as at midnight); tatra—there; gatam—gone; na paśyanti—do not see; ye—who; tam—the sunset; samanupaśyeran—seeing.

TRANSLATION: The living entities residing on Sumeru Mountain are always very warm, as at midday, because for them the sun is always overhead. Although the sun moves counterclockwise, facing the constellations, with Sumeru Mountain on its left, it also moves clockwise and appears to have the mountain on its right because it is influenced by the dakṣiṇāvarta wind. People living in countries at points diametrically opposite to where the sun is first seen rising will see the sun setting, and if a straight line were drawn from a point where the sun is at midday, the people in countries at the opposite end of the line would be experiencing midnight. Similarly, if people residing where the sun is setting were to go to countries diametrically opposite, they would not see the sun in the same condition.

5.21.10: yadā caindryāḥ puryāḥ pracalate pañcadaśa-ghaṭikābhir yāmyāṁ sapāda-koṭi-dvayaṁ yojanānāṁ sārdha-dvādaśa-lakṣāṇi sādhikāni copayāti.

SYNONYMS: yadā—when; ca—and; aindryāḥ—of Indra; puryāḥ—from the residence; pracalate—it moves; pañcadaśa—by fifteen; ghaṭikābhiḥ—half hours (actually twenty-four minutes); yāmyām—to the residence of Yamarāja; sapāda-koṭi-dvayam—two and a quarter crores (22,500,000); yojanānām—of yojanas; sārdha—and one half; dvādaśa-lakṣāṇi—twelve hundred thousand; sādhikāni—twenty-five thousand more; ca—and; upayāti—he passes over.

TRANSLATION: When the sun travels from Devadhānī, the residence of Indra, to Saṁyamanī, the residence of Yamarāja, it travels 23,775,000 yojanas [190,200,000 miles] in fifteen ghaṭikās [six hours].

PURPORT: The distance indicated by the word sādhikāni is pañca-viṁśati-sahasrādhikāni, or 25,000 yojanas. That plus two and a half crores and twelve and a half lakṣa of yojanas is the distance the sun travels between each two cities. This comes to 23,775,000 yojanas, or 190,200,000 miles. The total orbit of the sun is four times that distance, or 95,100,000 yojanas (760,800,000 miles).

5.21.11: evaṁ tato vāruṇīṁ saumyām aindrīṁ ca punas tathānye ca grahāḥ somādayo nakṣatraiḥ saha jyotiś-cakre samabhyudyanti saha vā nimlocanti.

SYNONYMS: evam—in this way; tataḥ—from there; vāruṇīm—to the quarters where Varuṇa lives; saumyām—to the quarters where the moon lives; aindrīṁ ca—and to the quarters where Indra lives; punaḥ—again; tathā—so also; anye—the others; ca—also; grahāḥ—planets; soma-ādayaḥ—headed by the moon; nakṣatraiḥ—all the stars; saha—with; jyotiḥ-cakre—in the celestial sphere; samabhyudyanti—rise; saha—along with; vā—or; nimlocanti—set.

TRANSLATION: From the residence of Yamarāja the sun travels to Nimlocanī, the residence of Varuṇa, from there to Vibhāvarī, the residence of the moon-god, and from there again to the residence of Indra. In a similar way, the moon, along with the other stars and planets, becomes visible in the celestial sphere and then sets and again becomes invisible.

PURPORT: In Bhagavad-gītā (10.21) Kṛṣṇa says, nakṣatrāṇām ahaṁ śaśī: "Of stars I am the moon." This indicates that the moon is similar to the other stars. The Vedic literature informs us that within this universe there is one sun, which is moving. The Western theory that all the luminaries in the sky are different suns is not confirmed in the Vedic literature. Nor can we assume that these luminaries are the suns of other universes, for each universe is covered by various layers of material elements, and therefore although the universes are clustered together, we cannot see from one universe to another. In other words, whatever we see is within this one universe. In each universe there is one Lord Brahmā, and there are other demigods on other planets, but there is only one sun.

## Part XVII: The Chariot of The Sun

5.21.12: evaṁ muhūrtena catus-triṁśal-lakṣa-yojanāny aṣṭa-śatādhikāni sauro rathas trayīmayo 'sau catasṛṣu parivartate purīṣu.

SYNONYMS: evam—thus; muhūrtena—in a muhūrta (forty-eight minutes); catuḥ-triṁśat—thirty-four; lakṣa—hundred thousands; yojanāni—yojanas; aṣṭa-śatādhikāni—increased by eight hundred; sauraḥ rathaḥ—the chariot of the sun-god; trayī-mayaḥ—which is worshiped by the Gāyatrī mantra (oṁ bhūr bhuvaḥ svaḥ tat savitur, etc.); asau—that; catasṛṣu—to the four; parivartate—he moves; purīṣu—through different residential quarters.

TRANSLATION: Thus the chariot of the sun-god, which is trayīmaya, or worshiped by the words oṁ bhūr bhuvaḥ svaḥ, travels through the four residences mentioned above at a speed of 3,400,800 yojanas [27,206,400 miles] in a muhūrta.

5.21.13: yasyaikaṁ cakraṁ dvādaśāraṁ ṣaṇ-nemi tri-ṇābhi saṁvatsarātmakaṁ samāmananti tasyākṣo meror mūrdhani kṛto mānasottare kṛtetara-bhāgo yatra protaṁ ravi-ratha-cakraṁ taila-yantra-cakravad bhraman mānasottara-girau paribhramati.

SYNONYMS: yasya—of which; ekam—one; cakram—wheel; dvādaśa—twelve; aram—spokes; ṣaṭ—six; nemi—the segments of the rim; tri-ṇābhi—the three pieces of the hub; saṁvatsara-ātmakam—whose nature is a saṁvatsara; samāmananti—they fully describe; tasya—the chariot of the sun-god; akṣaḥ—the axle; meroḥ—of Sumeru Mountain; mūrdhani—on the top; kṛtaḥ—fixed; mānasottare—on the mountain known as Mānasottara; kṛta—fixed; itara-bhāgaḥ—the other end; yatra—where; protam—fixed on; ravi-ratha-cakram—the wheel of the chariot of the sun-god; taila-yantra-cakra-vat—like the wheel of an oil-pressing machine; bhramat—moving; mānasottara-girau—on Mānasottara Mountain; paribhramati—turns.

TRANSLATION: The chariot of the sun-god has only one wheel, which is known as Saṁvatsara. The twelve months are calculated to be its twelve spokes, the six seasons are the sections of its rim, and the three cātur-māsya periods are its three-sectioned hub. One side of the axle carrying the wheel rests upon the summit of Mount Sumeru, and the other rests upon Mānasottara Mountain. Affixed to the outer end of the axle, the wheel continuously rotates on Mānasottara Mountain like the wheel of an oil-pressing machine.

5.21.14: tasminn akṣe kṛtamūlo dvitīyo 'kṣas turyamānena sammitas taila-yantrākṣavad dhruve kṛtopari-bhāgaḥ.

SYNONYMS: tasmin akṣe—in that axle; kṛta-mūlaḥ—whose base is fixed; dvitīyaḥ—a second; akṣaḥ—axle; turyamānena—by one fourth; sammitaḥ—measured; taila-yantra-akṣa-vat—like the axle of an oil-pressing machine; dhruve—to Dhruvaloka; kṛta—fixed; upari-bhāgaḥ—upper portion.

TRANSLATION: As in an oil-pressing machine, this first axle is attached to a second axle, which is one-fourth as long [3,937,500 yojanas, or 31,500,000 miles]. The upper end of this second axle is attached to Dhruvaloka by a rope of wind.

5.21.15: ratha-nīḍas tu ṣaṭ-triṁśal-lakṣa-yojanāyatas tat-turīya-bhāga-viśālas tāvān ravi-ratha-yugo yatra hayāś chando-nāmānaḥ saptāruṇa-yojitā vahanti devam ādityam.

SYNONYMS: ratha-nīḍaḥ—the interior of the chariot; tu—but; ṣaṭ-triṁśat-lakṣa-yojana-āyataḥ—3,600,000 yojanas long; tat-turīya-bhāga—one quarter of that measure (900,000 yojanas); viśālaḥ—having a width; tāvān—so much also; ravi-ratha-yugaḥ—the yoke for the horses; yatra—where; hayāḥ—horses; chandaḥ-nāmānaḥ—having the different names of Vedic meters; sapta—seven; aruṇa-yojitāḥ—hooked up by Aruṇadeva; vahanti—carry; devam—the demigod; ādityam—the sun-god.

TRANSLATION: My dear King, the carriage of the sun-god's chariot is estimated to be 3,600,000 yojanas [28,800,000 miles] long and one-fourth as wide [900,000 yojanas, or 7,200,000 miles]. The chariot's horses, which are named after Gāyatrī and other Vedic meters, are harnessed by Aruṇadeva to a yoke that is also 900,000 yojanas wide. This chariot continuously carries the sun-god.

PURPORT: In the Viṣṇu Purāṇa it is stated: gāyatrī ca bṛhaty uṣṇig/jagatī triṣṭup eva ca/anuṣṭup paṅktir ity uktāś/chandāṁsi harayo raveḥ// The seven horses yoked to the sun-god's chariot are named Gāyatrī, Bṛhati, Uṣṇik, Jagatī, Triṣṭup, Anuṣṭup and Paṅkti. These names of various Vedic meters designate the seven horses that carry the sun-god's chariot.

5.21.16: purastāt savitur aruṇaḥ paścāc ca niyuktaḥ sautye karmaṇi kilāste.

SYNONYMS: purastāt—in front; savituḥ—of the sun-god; aruṇaḥ—the demigod named Aruṇa; paścāt—looking backward; ca—and; niyuktaḥ—engaged; sautye—of a charioteer; karmaṇi—in the work; kila—certainly; āste—remains.

TRANSLATION: Although Aruṇadeva sits in front of the sun-god and is engaged in driving the chariot and controlling the horses, he looks backward toward the sun-god.

PURPORT: In the Vāyu Purāṇa the position of the horses is described: saptāśva-rūpa-cchandāṁsī/vahante vāmato ravim/cakra-pakṣa-nibaddhani/cakre vakṣaḥ samāhitaḥ// Although Aruṇadeva is in the front seat, controlling the horses, he looks back toward the sun-god from his left side.

5.21.17: tathā vālikhilyā ṛṣayo 'ṅguṣṭha-parva-mātrāḥ ṣaṣṭi-sahasrāṇi purataḥ sūryaṁ sūkta-vākāya niyuktāḥ saṁstuvanti.

SYNONYMS: tathā—there; vālikhilyāḥ—Vālikhilyas; ṛṣayaḥ—great sages; aṅguṣṭha-parva-mātrāḥ—whose size is that of a thumb; ṣaṣṭi-sahasrāṇi—sixty thousand; purataḥ—in front; sūryam—the sun-god; su-ukta-vākāya—for speaking eloquently; niyuktāḥ—engaged; saṁstuvanti—offer prayers.

TRANSLATION: There are sixty thousand saintly persons named Vālikhilyas, each the size of a thumb, who are located in front of the sun-god and who offer him eloquent prayers of glorification.

5.21.18: tathānye ca ṛṣayo gandharvāpsaraso nāgā grāmaṇyo yātudhānā devā ity ekaikaśo gaṇāḥ sapta caturdaśa māsi māsi bhagavantaṁ sūryam ātmānaṁ nānā-nāmānaṁ pṛthaṅ-nānā-nāmānaḥ pṛthak-karmabhir dvandvaśa upāsate.

SYNONYMS: tathā—similarly; anye—others; ca—also; ṛṣayaḥ—saintly persons; gandharva-apsarasaḥ—Gandharvas and Apsarās; nāgāḥ—Nāga snakes; grāmaṇyaḥ—Yakṣas; yātudhānāḥ—Rākṣasas; devāḥ—demigods; iti—thus; eka-ekaśaḥ—one by one; gaṇāḥ—groups; sapta—seven; caturdaśa—fourteen in number; māsi māsi—in every month; bhagavantam—unto the most powerful demigod; sūryam—the sun-god; ātmānam—the life of the universe; nānā—various; nāmānam—who possesses names; pṛthak—separate; nānā-nāmānaḥ—having various names; pṛthak—separate; karmabhiḥ—by ritualistic ceremonies; dvandvaśaḥ—in groups of two; upāsate—worship.

TRANSLATION: Similarly, fourteen other saints, Gandharvas, Apsarās, Nāgas, Yakṣas, Rākṣasas and demigods, who are divided into groups of two, assume different names every month and continuously perform different ritualistic ceremonies to worship the Supreme Lord as the most powerful demigod Sūryadeva, who holds many names.

PURPORT: In the Viṣṇu Purāṇa it is said: stuvanti munayaḥ sūryaṁ/gandharvair gīyate puraḥ/nṛtyanto 'psaraso yānti/sūryasyānu niśācarāḥ/vahanti pannagā yakṣaiḥ/kriyate 'bhiṣusaṅgrahaḥ/vālikhilyās tathaivainaṁ/parivārya samāsate/so 'yaṁ sapta-gaṇaḥ sūrya-/maṇḍale muni-sattama/himoṣṇa vāri-vṛṣṭīṇāṁ/hetutve samayaṁ gataḥ// Worshiping the most powerful demigod Sūrya, the Gandharvas sing in front of him, the Apsarās dance before the chariot, the Niśācaras follow the chariot, the Pannagas decorate the chariot, the Yakṣas guard the chariot, and the saints called the Vālikhilyas surround the sun-god and offer prayers. The seven groups of fourteen associates arrange the proper times for regular snow, heat and rain throughout the universe.

5.21.19: lakṣottaraṁ sārdha-nava-koṭi-yojana-parimaṇḍalaṁ bhū-valayasya kṣaṇena sagavyuty-uttaraṁ dvi-sahasra-yojanāni sa bhuṅkte.

SYNONYMS: lakṣa-uttaram—increased by 100,000; sārdha—with 5,000,000; nava-koṭi-yojana—of 90,000,000 yojanas; parimaṇḍalam—circumference; bhū-valayasya—of the earthly sphere; kṣaṇena—in one moment; sagavyūti-uttaram—augmented by two krośas (four miles); dvi-sahasra-yojanāni—2,000 yojanas; saḥ—the sun-god; bhuṅkte—traverses.

TRANSLATION: My dear King, in his orbit through Bhū-maṇḍala, the sun-god traverses a distance of 95,100,000 yojanas [760,800,000 miles] at the speed of 2,000 yojanas and two krośas [16,004 miles] in a moment.

## Part XVIII: The Orbits of the Planets

5.22 Summary: In this chapter the orbits of the planets are described. According to the movements of the moon and other planets, all the inhabitants of the universe are prone to auspicious and inauspicious situations. This is referred to as the influence of the stars.

The sun-god, who controls the affairs of the entire universe, especially in regard to heat, light, seasonal changes and so on, is considered an expansion of Nārāyaṇa. He represents the three Vedas—Ṛg, Yajur and Sāma—and therefore he is known as Trayīmaya, the form of Lord Nārāyaṇa. Sometimes the sun-god is also called Sūrya Nārāyaṇa. The sun-god has expanded himself in twelve divisions, and thus he controls the six seasonal changes and causes winter, summer, rain and so on. Yogīs and karmīs following the varṇāśrama institution, who practice haṭha or aṣṭāṅga-yoga or who perform agnihotra sacrifices, worship Sūrya Nārāyaṇa for their own benefit. The demigod Sūrya is always in touch with the Supreme Personality of Godhead, Nārāyaṇa. Residing in outer space, which is in the middle of the universe, between Bhūloka and Bhuvarloka, the sun rotates through the time circle of the zodiac, represented by twelve rāśis, or signs, and assumes different names according to the sign he is in. For the moon, every month is divided into two fortnights. Similarly, according to solar calculations, a month is equal to the time the sun spends in one constellation; two months constitute one season, and there are twelve months in a year. The entire area of the sky is divided into two halves, each representing an ayana, the course traversed by the sun within a period of six months. The sun travels sometimes slowly, sometimes swiftly and sometimes at a moderate speed. In this way it travels within the three worlds, consisting of the heavenly planets, the earthly planets and outer space. These orbits are referred to by great learned scholars by the names Saṁvatsara, Parivatsara, Iḍāvatsara, Anuvatsara and Vatsara.

The moon is situated 100,000 yojanas above the rays of the sunshine. Day and night on the heavenly planets and Pitṛloka are calculated according to its waning and waxing. Above the moon by a distance of 200,000 yojanas are some stars, and above these stars is Śukra-graha (Venus), whose influence is always auspicious for the inhabitants of the entire universe. Above Śukra-graha by 200,000 yojanas is Budha-graha (Mercury), whose influence is sometimes auspicious and sometimes inauspicious. Next, above Budha-graha by 200,000 yojanas, is Aṅgāraka (Mars), which almost always has an unfavorable influence. Above Aṅgāraka by another 200,000 yojanas is the planet called Bṛhaspati-graha (Jupiter), which is always very favorable for qualified brāhmaṇas. Above Bṛhaspati-graha is the planet Śanaiścara (Saturn), which is very inauspicious, and above Saturn is a group of seven stars occupied by great saintly persons who are always thinking of the welfare of the entire universe. These seven stars circumambulate Dhruvaloka, which is the residence of Lord Viṣṇu within this universe.

5.22.1 rājovāca/ yad etad bhagavata ādityasya meruṁ dhruvaṁ ca pradakṣiṇena parikrāmato rāśīnām abhimukhaṁ pracalitaṁ cāpradakṣiṇaṁ bhagavatopavarṇitam amuṣya vayaṁ katham anumimīmahīti.

SYNONYMS: rājā uvāca—the King (Mahārāja Parīkṣit) inquired; yat—which; etat—this; bhagavataḥ—of the most powerful; ādityasya—of the sun (Sūrya Nārāyaṇa); merum—the mountain known as Sumeru; dhruvam ca—as well as the planet known as Dhruvaloka; pradakṣiṇena—by placing on the right; parikrāmataḥ—which is going around; rāśīnām—the different signs of the zodiac; abhimukham—facing toward; pracalitam—moving; ca—and; apradakṣiṇam—placing on the left; bhagavatā—by Your Lordship; upavarṇitam—described; amuṣya—of that; vayam—we (the hearer); katham—how; anumimīmahi—can accept it by argument and inference; iti—thus.

TRANSLATION: King Parīkṣit inquired from Śukadeva Gosvāmī: My dear lord, you have already affirmed the truth that the supremely powerful sun-god travels around Dhruvaloka with both Dhruvaloka and Mount Sumeru on his right. Yet at the same time the sun-god faces the signs of the zodiac and keeps Sumeru and Dhruvaloka on his left. How can we reasonably accept that the sun-god proceeds with Sumeru and Dhruvaloka on both his left and right simultaneously?

5.22.2: sa hovāca/ yathā kulāla-cakreṇa bhramatā saha bhramatāṁ tad-āśrayāṇāṁ pipīlikādīnāṁ gatir anyaiva pradeśāntareṣv apy upalabhyamānatvād evaṁ nakṣatra-rāśibhir upalakṣitena kāla-cakreṇa dhruvaṁ meruṁ ca pradakṣiṇena paridhāvatā saha paridhāvamānānāṁ tad-āśrayāṇāṁ sūryādīnāṁ grahāṇāṁ gatir anyaiva nakṣatrāntare rāśy-antare copalabhyamānatvāt.

SYNONYMS: saḥ—Śukadeva Gosvāmī; ha—very clearly; uvāca—answered; yathā—just as; kulāla-cakreṇa—a potter's wheel; bhramatā—turning around; saha—with; bhramatām—of those turning around; tat-āśrayāṇām—being located on that (wheel); pipīlika-ādīnām—of small ants; gatiḥ—the motion; anyā—other; eva—certainly; pradeśa-antareṣu—in different locations; api—also; upalabhyamānatvāt—due to being experienced; evam—similarly; nakṣatra-rāśibhiḥ—by the stars and signs; upalakṣitena—being seen; kāla-cakreṇa—with the great wheel of time; dhruvam—the star known as Dhruvaloka; merum—the mountain known as Sumeru; ca—and; pradakṣiṇena—on the right; paridhāvatā—

going around; saha—with; paridhāvamānānām—of those going around; tat-āśrayāṇām—whose shelter is that wheel of time; sūrya-ādīnām—headed by the sun; grahāṇām—of the planets; gatiḥ—the motion; anyā—other; eva—certainly; nakṣatra-antare—in different stars; rāśi-antare—in different signs; ca—and; upalabhyamānatvāt—due to being observed.

TRANSLATION: Śrī Śukadeva Gosvāmī clearly answered: When a potter's wheel is moving and small ants located on that big wheel are moving with it, one can see that their motion is different from that of the wheel because they appear sometimes on one part of the wheel and sometimes on another. Similarly, the signs and constellations, with Sumeru and Dhruvaloka on their right, move with the wheel of time, and the antlike sun and other planets move with them. The sun and planets, however, are seen in different signs and constellations at different times. This indicates that their motion is different from that of the zodiac and the wheel of time itself.

5.22.3: sa eṣa bhagavān ādi-puruṣa eva sākṣān nārāyaṇo lokānāṁ svastaya ātmānaṁ trayīmayaṁ karma-viśuddhi-nimittaṁ kavibhir api ca vedena vijijñāsyamāno dvādaśadhā vibhajya ṣaṭsu vasantādiṣv ṛtuṣu yathopa-joṣam ṛtu-guṇān vidadhāti.

SYNONYMS: saḥ—that; eṣaḥ—this; bhagavān—the supremely powerful; ādi-puruṣaḥ—the original person; eva—certainly; sākṣāt—directly; nārāyaṇaḥ—the Supreme Personality of Godhead, Nārāyaṇa; lokānām—of all the planets; svastaye—for the benefit; ātmānam—Himself; trayī-mayam—consisting of the three Vedas (Sāma, Yajur and Ṛg); karma-viśuddhi—of purification of fruitive activities; nimittam—the cause; kavibhiḥ—by great saintly persons; api—also; ca—and; vedena—by Vedic knowledge; vijijñāsyamānaḥ—being inquired about; dvādaśa-dhā—in twelve divisions; vibhajya—dividing; ṣaṭsu—in six; vasanta-ādiṣu—headed by spring; ṛtuṣu—seasons; yathā-upajoṣam—according to the enjoyment of their past activities; ṛtu-guṇān—the qualities of the different seasons; vidadhāti—he arranges.

TRANSLATION: The original cause of the cosmic manifestation is the Supreme Personality of Godhead, Nārāyaṇa. When great saintly persons, fully aware of the Vedic knowledge, offered prayers to the Supreme Person, He descended to this material world in the form of the sun to benefit all the planets and purify fruitive activities. He divided Himself into twelve parts and created seasonal forms, beginning with spring. In this way He created the seasonal qualities, such as heat, cold and so on.

5.22.4: tam etam iha puruṣās trayyā vidyayā varṇāśramācārānupathā uccāvacaiḥ karmabhir āmnātair yoga-vitānaiś ca śraddhayā yajanto 'ñjasā śreyaḥ samadhigacchanti.

SYNONYMS: tam—Him (the Supreme Personality of Godhead); etam—this; iha—in this world of mortality; puruṣāḥ—all people; trayyā—having three divisions; vidyayā—by the Vedic knowledge; varṇa-āśrama-ācāra—the practices of the varṇāśrama system; anupathāḥ—following; ucca-avacaiḥ—higher or lower according to the different positions in the varṇāśrama-dharma (brāhmaṇa, kṣatriya, vaiśya and śūdra); karmabhiḥ—by their respective activities; āmnātaiḥ—handed down; yoga-vitānaiḥ—by meditation and other yogic processes; ca—and; śraddhayā—with great faith; yajantaḥ—worshiping; añjasā—without difficulty; śreyaḥ—the ultimate benefit of life; samadhigacchanti—they attain.

TRANSLATION: According to the system of four varṇas and four āśramas, people generally worship the Supreme Personality of Godhead, Nārāyaṇa, who is situated as the sun-god. With great faith they worship the Supreme Personality as the Supersoul according to ritualistic ceremonies handed down in the three Vedas, such as agnihotra and similar higher and lower fruitive acts, and according to the process of mystic yoga. In this way they very easily attain the ultimate goal of life.

5.22.5: atha sa eṣa ātmā lokānāṁ dyāv-āpṛthivyor antareṇa nabho-valayasya kālacakra-gato dvādaśa māsān bhuṅkte rāśi-saṁjñān saṁvatsarāvayavān māsaḥ pakṣa-dvayaṁ divā naktaṁ ceti sapādarkṣa-dvayam upadiśanti yāvatā ṣaṣṭham aṁśaṁ bhuñjīta sa vai ṛtur ity upadiśyate saṁvatsarāvayavaḥ.

SYNONYMS: atha—therefore; saḥ—He; eṣaḥ—this; ātmā—the living force; lokānām—of all the three worlds; dyāv-ā-pṛthivyoḥ antareṇa—between the upper and lower portions of the universe; nabhaḥ-valayasya—of outer space; kāla-cakra-gataḥ—positioned in the wheel of time; dvādaśa māsān—twelve months; bhuṅkte—passes; rāśi-saṁjñān—named after the zodiac signs; saṁvatsara-avayavān—the parts of the whole year; māsaḥ—one month; pakṣa-dvayam—two fortnights; divā—a day; naktam ca—and a night; iti—thus; sapāda-ṛkṣa-dvayam—by stellar calculations, two and a quarter constellations; upadiśanti—they instruct; yāvatā—by as much time; ṣaṣṭham aṁśam—one sixth of his orbit; bhuñjīta—pass; saḥ—that portion; vai—indeed; ṛtuḥ—a season; iti—thus; upadiśyate—is instructed; saṁvatsara-avayavaḥ—a part of a year.

TRANSLATION: The sun-god, who is Nārāyaṇa, or Viṣṇu, the soul of all the worlds, is situated in outer space between the upper and lower portions of the universe. Passing through twelve months on the wheel of time, the sun comes in touch with twelve different signs of the zodiac and assumes twelve different names according to those signs. The aggregate of those twelve months is called a saṁvatsara, or an entire year. According to lunar calculations, two fortnights—one of the waxing moon and the other of the waning—form one month. That same period is one day and night for the planet Pitṛloka. According to stellar calculations, a month equals two and one quarter constellations. When the sun travels for two months, a season passes, and therefore the seasonal changes are considered parts of the body of the year.

5.22.6: atha ca yāvatārdhena nabho-vīthyāṁ pracarati taṁ kālam ayanam ācakṣate.

SYNONYMS: atha—now; ca—also; yāvatā—by as long as; ardhena—half; nabhaḥ-vīthyām—in outer space; pracarati—the sun moves; tam—that; kālam—time; ayanam—ayana; ācakṣate—is said.

TRANSLATION: Thus the time the sun takes to rotate through half of outer space is called an ayana, or its period of movement [in the north or in the south].

5.22.7: atha ca yāvan nabho-maṇḍalaṁ saha dyāv-āpṛthivyor maṇḍalābhyāṁ kārtsnyena sa ha bhuñjīta taṁ kālaṁ saṁvatsaraṁ parivatsaram iḍāvatsaram anuvatsaraṁ vatsaram iti bhānor māndya-śaighrya-sama-gatibhiḥ samāmananti.

SYNONYMS: atha—now; ca—also; yāvat—as long as; nabhaḥ-maṇḍalam—outer space, between the upper and lower world; saha—along with; dyau—of the upper world; āpṛthivyoḥ—of the lower world; maṇḍalābhyām—the spheres; kārtsnyena—entirely; saḥ—he; ha—indeed; bhuñjīta—may pass through; tam—that; kālam—time; saṁvatsaram—Saṁvatsara; parivatsaram—Parivatsara; iḍāvatsaram—Iḍāvatsara; anuvatsaram—Anuvatsara; vatsaram—Vatsara; iti—thus; bhānoḥ—of the sun; māndya—slow; śaighrya—speedy; sama—and equal; gatibhiḥ—by the speeds; samāmananti—the experienced scholars describe.

TRANSLATION: The sun-god has three speeds—slow, fast and moderate. The time he takes to travel entirely around the spheres of heaven, earth and space at these three speeds is referred to, by learned scholars, by the five names Saṁvatsara, Parivatsara, Iḍāvatsara, Anuvatsara and Vatsara.

PURPORT: According to solar astronomical calculations, each year extends six days beyond the calendar year, and according to lunar calculations, each year is six days shorter. Therefore, because of the movements of the sun and moon, there is a difference of twelve days between the solar and lunar years. As the Saṁvatsara, Parivatsara, Iḍāvatsara, Anuvatsara and Vatsara pass by, two extra months are added within each five years. This makes a sixth saṁvatsara, but because that saṁvatsara is extra, the solar system is calculated according to the above five names.

5.22.8: evaṁ candramā arka-gabhastibhya upariṣṭāl lakṣa-yojanata upalabhyamāno 'rkasya saṁvatsara-bhuktiṁ pakṣābhyāṁ māsabhuktiṁ sapādarkṣābhyāṁ dinenaiva pakṣa-bhuktim agracārī drutatara-gamano bhuṅkte.

SYNONYMS: evam—thus; candramā—the moon; arka-gabhastibhyaḥ—from the rays of the sunshine; upariṣṭāt—above; lakṣa-yojanataḥ—by a measurement of 100,000 yojanas; upalabhyamānaḥ—being situated; arkasya—of the sun globe; saṁvatsara-bhuktim—the passage of one year of enjoyment; pakṣābhyām—by two fortnights; māsa-bhuktim—the passage of one month; sapāda-ṛkṣābhyām—by two and a quarter days; dinena—by a day; eva—only; pakṣa-bhuktim—the passage of a fortnight; agracārī—moving impetuously; druta-tara-gamanaḥ—passing more speedily; bhuṅkte—passes through.

TRANSLATION: Above the rays of the sunshine by a distance of 100,000 yojanas [800,000 miles] is the moon, which travels at a speed faster than that of the sun. In two lunar fortnights the moon travels through the equivalent of a saṁvatsara of the sun, in two and a quarter days it passes through a month of the sun, and in one day it passes through a fortnight of the sun.

PURPORT: When we take into account that the moon is 100,000 yojanas, or 800,000 miles, above the rays of the sunshine, it is very surprising that the modern excursions to the moon could be possible. Since the moon is so distant, how space vehicles could go there is a doubtful mystery. Modern scientific calculations are subject to one change after another, and therefore they are uncertain. We have to accept the calculations of the Vedic literature. These Vedic calculations are steady; the astronomical calculations made long ago and recorded in the Vedic literature are correct even now. Whether the Vedic calculations or modern ones are better may remain a mystery for others, but as far as we are concerned, we accept the Vedic calculations to be correct.

5.22.9: atha cāpūryamāṇābhiś ca kalābhir amarāṇāṁ kṣīyamāṇābhiś ca kalābhiḥ pitṝṇām aho-rātrāṇi pūrva-pakṣāpara-pakṣābhyāṁ vitanvānaḥ sarva-jīva-nivaha-prāṇo jīvaś caikam ekaṁ nakṣatraṁ triṁśatā muhūrtair bhuṅkte.

SYNONYMS: atha—thus; ca—also; āpūryamāṇābhiḥ—gradually increasing; ca—and; kalābhiḥ—by the parts of the moon; amarāṇām—of the demigods; kṣīyamāṇābhiḥ—by gradually decreasing; ca—and; kalābhiḥ—by parts of the moon; pitṝṇām—of those on the planet known as Pitṛloka; ahaḥ-rātrāṇi—the days and nights; pūrva-pakṣa-apara-pakṣābhyām—by the period of waxing and waning; vitanvānaḥ—distributing; sarva-jīva-nivaha—of the total living entities; prāṇaḥ—the life; jīvaḥ—the chief living being; ca—also; ekam ekam—one after another; nakṣatram—a constellation of stars; triṁśatā—by thirty; muhūrtaiḥ—muhūrtas; bhuṅkte—passes through.

TRANSLATION: When the moon is waxing, the illuminating portions of it increase daily, thus creating day for the demigods and night for the pitās. When the moon is waning, however, it causes night for the demigods and day for the pitās. In this way the moon passes through each constellation of stars in thirty muhūrtas [an entire day]. The moon is the source of nectarean coolness that influences the growth of food grains, and therefore the moon-god is considered the life of all living entities. He is consequently called Jīva, the chief living being within the universe.

5.22.10: ya eṣa ṣoḍaśa-kalaḥ puruṣo bhagavān manomayo 'nnamayo 'mṛtamayo deva-pitṛ-manuṣya-bhūta-paśu-pakṣi-sarīsṛpa-vīrudhāṁ prāṇāpy āyana-śīlatvāt sarvamaya iti varṇayanti.

SYNONYMS: yaḥ—that; eṣaḥ—this; ṣoḍaśa-kalaḥ—having all sixteen parts (the full moon); puruṣaḥ—the person; bhagavān—having great power received from the Supreme Personality of Godhead; manaḥ-mayaḥ—the predominating deity of the mind; anna-mayaḥ—the source of potency for food grains; amṛta-mayaḥ—the source of the substance of life; deva—of all the demigods; pitṛ—of all the inhabitants of Pitṛloka; manuṣya—all human beings; bhūta—all living entities; paśu—of the animals; pakṣi—of the birds; sarīsṛpa—of the reptiles; vīrudhām—of all kinds of herbs and plants; prāṇa—life air; api—certainly; āyana-śīlatvāt—due to refreshing; sarva-mayaḥ—all-pervading; iti—thus; varṇayanti—the learned scholars describe.

TRANSLATION: Because the moon is full of all potentialities, it represents the influence of the Supreme Personality of Godhead. The moon is the predominating deity of everyone's mind, and therefore the moon-god is called Manomaya. He is also called Annamaya because he gives potency to all herbs and plants, and he is called Amṛtamaya because he is the source of life for all living entities. The moon pleases the demigods, pitās, human beings, animals, birds, reptiles, trees, plants and all other living entities. Everyone is satisfied by the presence of the moon. Therefore the moon is also called Sarvamaya [all-pervading].

5.22.11: tata upariṣṭād dvi-lakṣa-yojanato nakṣatrāṇi meruṁ dakṣiṇenaiva kālāyana īśvara-yojitāni sahābhijitāṣṭā-viṁśatiḥ.

SYNONYMS: tataḥ—from that region of the moon; upariṣṭāt—above; dvi-lakṣa-yojanataḥ—200,000 yojanas; nakṣatrāṇi—many stars; merum—Sumeru Mountain; dakṣiṇena eva—to the right side; kāla-ayane—in the wheel of time; īśvara-yojitāni—attached by the Supreme Personality of Godhead; saha—with; abhijitā—the star known as Abhijit; aṣṭā-viṁśatiḥ—twenty-eight.

TRANSLATION: There are many stars located 200,000 yojanas [1,600,000 miles] above the moon. By the supreme will of the Supreme Personality of Godhead, they are fixed to the wheel of time, and thus they rotate with Mount Sumeru on their right, their motion being different from that of the sun. There are twenty-eight important stars, headed by Abhijit.

PURPORT: The stars referred to herein are 1,600,000 miles above the sun, and thus they are 4,000,000 miles above the earth.

5.22.12: tata upariṣṭād uśanā dvi-lakṣa-yojanata upalabhyate purataḥ paścāt sahaiva vārkasya śaighrya-māndya-sāmyābhir gatibhir arkavac carati lokānāṁ nityadānukūla eva prāyeṇa varṣayaṁś cāreṇānumīyate sa vṛṣṭi-viṣṭambha-grahopaśamanaḥ.

SYNONYMS: tataḥ—from that bunch of stars; upariṣṭāt—above; uśanā—Venus; dvi-lakṣa-yojanataḥ—200,000 yojanas (1,600,000 miles); upalabhyate—is experienced; purataḥ—in front; paścāt—behind; saha—along with; eva—indeed; vā—and; arkasya—of the sun; śaighrya—speedy; māndya—slow; sāmyābhiḥ—equal; gatibhiḥ—the movements; arkavat—exactly like the sun; carati—rotates; lokānām—of all the planets within the universe; nityadā—constantly; anukūlaḥ—offering favorable conditions; eva—indeed; prāyeṇa—almost always; varṣayan—causing rainfall; cāreṇa—by infusing the clouds; anumīyate—is perceived; saḥ—he (Venus); vṛṣṭi-viṣṭambha—obstacles to rainfall; graha-upaśamanaḥ—nullifying planets.

TRANSLATION: Some 1,600,000 miles above this group of stars is the planet Venus, which moves at almost exactly the same pace as the sun according to swift, slow and moderate movements. Sometimes Venus moves behind the sun, sometimes in front of the sun and sometimes along with it. Venus nullifies the influence of planets that are obstacles to rainfall. Consequently its presence causes rainfall, and it is therefore considered very favorable for all living beings within this universe. This has been accepted by learned scholars.

5.22.13: uśanasā budho vyākhyātas tata upariṣṭād dvi-lakṣa-yojanato budhaḥ soma-suta upalabhyamānaḥ prāyeṇa śubha-kṛd yadārkād vyatiricyeta tadātivātābhra-prāyānāvṛṣṭy-ādi-bhayam āśaṁsate.

SYNONYMS: uśanasā—with Venus; budhaḥ—Mercury; vyākhyātaḥ—explained; tataḥ—from that (Venus); upariṣṭāt—above; dvi-lakṣa-yojanataḥ—1,600,000 miles; budhaḥ—Mercury; soma-sutaḥ—the son of the moon; upalabhyamānaḥ—is situated; prāyeṇa—almost always; śubha-kṛt—very auspicious to the inhabitants of the universe; yadā—when; arkāt—from the sun; vyatiricyeta—is separated; tadā—at that time; ativāta—of cyclones and other bad effects; abhra—clouds; prāya—almost always; anāvṛṣṭi-ādi—such as scarcity of rain; bhayam—fearful conditions; āśaṁsate—expands.

TRANSLATION: Mercury is described to be similar to Venus, in that it moves sometimes behind the sun, sometimes in front of the sun and sometimes along with it. It is 1,600,000 miles above Venus, or 7,200,000 miles above earth. Mercury, which is the son of the moon, is almost always very auspicious for the inhabitants of the universe, but when it does not move along with the sun, it forbodes cyclones, dust, irregular rainfall, and waterless clouds. In this way it creates fearful conditions due to inadequate or excessive rainfall.

5.22.14: ata ūrdhvam aṅgārako 'pi yojana-lakṣa-dvitaya upalabhyamānas tribhis tribhiḥ pakṣair ekaikaśo rāśīn dvādaśānubhuṅkte yadi na vakreṇābhivartate prāyeṇāśubha-graho 'gha-śaṁsaḥ.

SYNONYMS: ataḥ—from this; ūrdhvam—above; aṅgārakaḥ—Mars; api—also; yojana-lakṣa-dvitaye—at a distance of 1,600,000 miles; upalabhyamānaḥ—is situated; tribhiḥ tribhiḥ—with each three and three; pakṣaiḥ—fortnights; eka-ekaśaḥ—one after another; rāśīn—the signs; dvādaśa—twelve; anubhuṅkte—passes through; yadi—if; na—not; vakreṇa—with a curve; abhivartate—approaches; prāyeṇa—almost always; aśubha-grahaḥ—an unfavorable, inauspicious planet; agha-śaṁsaḥ—creating trouble.

TRANSLATION: Situated 1,600,000 miles above Mercury, or 8,800,000 miles above earth, is the planet Mars. If this planet does not travel in a crooked way, it crosses through each sign of the zodiac in three fortnights and in this way travels through all twelve, one after another. It almost always creates unfavorable conditions in respect to rainfall and other influences.

5.22.15: tata upariṣṭād dvi-lakṣa-yojanāntara-gatā bhagavān bṛhaspatir ekaikasmin rāśau parivatsaraṁ parivatsaraṁ carati yadi na vakraḥ syāt prāyeṇānukūlo brāhmaṇa-kulasya.

SYNONYMS: tataḥ—that (Mars); upariṣṭāt—above; dvi-lakṣa-yojana-antara-gatāḥ—situated at a distance of 1,600,000 miles; bhagavān—the most powerful planet; bṛhaspatiḥ—Jupiter; eka-ekasmin—in one after another; rāśau—sign; parivatsaram parivatsaram—for the period of Parivatsara; carati—moves; yadi—if; na—not; vakraḥ—curved; syāt—becomes; prāyeṇa—almost always; anukūlaḥ—very favorable; brāhmaṇa-kulasya—to the brāhmaṇas of the universe.

TRANSLATION: Situated 1,600,000 miles above Mars, or 10,400,000 miles above earth, is the planet Jupiter, which travels through one sign of the zodiac within the period of a Parivatsara. If its movement is not curved, the planet Jupiter is very favorable to the brāhmaṇas of the universe.

5.22.16: tata upariṣṭād yojana-lakṣa-dvayāt pratīyamānaḥ śanaiścara ekaikasmin rāśau triṁśan māsān vilambamānaḥ sarvān evānuparyeti tāvadbhir anuvatsaraiḥ prāyeṇa hi sarveṣām aśāntikaraḥ.

SYNONYMS: tataḥ—that (Jupiter); upariṣṭāt—above; yojana-lakṣa-dvayāt—by a distance of 1,600,000 miles; pratīyamānaḥ—is situated; śanaiścaraḥ—the planet Saturn; eka-ekasmin—in one after another; rāśau—zodiac signs; triṁśat māsān—for a period of thirty months in each; vilam-bamānaḥ—lingering; sarvān—all twelve signs of the zodiac; eva—certainly; anuparyeti—passes through; tāvadbhiḥ—by so many; anuvatsaraiḥ—Anuvatsaras; prāyeṇa—almost always; hi—indeed; sarveṣām—to all the inhabitants; aśāntikaraḥ—very troublesome.

TRANSLATION: Situated 1,600,000 miles above Jupiter, or 12,000,000 miles above earth, is the planet Saturn, which passes through one sign of the zodiac in thirty months and covers the entire zodiac circle in thirty Anuvatsaras. This planet is always very inauspicious for the universal situation.

5.22.17: tata uttarasmād ṛṣaya ekādaśa-lakṣa-yojanāntara upalabhyante ya eva lokānāṁ śam anubhāvayanto bhagavato viṣṇor yat paramaṁ padaṁ pradakṣiṇaṁ prakramanti.

SYNONYMS: tataḥ—the planet Saturn; uttarasmāt—above; ṛṣayaḥ—great saintly sages; ekādaśa-lakṣa-yojana-antare—at a distance of 1,100,000 yojanas; upalabhyante—are situated; ye—all of them; eva—indeed; lokānām—for all the inhabitants of the universe; śam—the good fortune; anubhāvayantaḥ—always thinking of; bhagavataḥ—of the Supreme Personality of Godhead; viṣṇoḥ—Lord Viṣṇu; yat—which; paramam padam—the supreme abode; pradakṣiṇam—placing on the right; prakramanti—circumambulate.

TRANSLATION: Situated 8,800,000 miles above Saturn, or 20,800,000 miles above earth, are the seven saintly sages, who are always thinking of the well-being of the inhabitants of the universe. They circumambulate the supreme abode of Lord Viṣṇu, known as Dhruvaloka, the polestar.

PURPORT: Śrīla Madhvācārya quotes the following verse from the Brahmāṇḍa Purāṇa: jñānānandātmano viṣṇuḥ/śiśumāra-vapuṣy atha/ūrdhva-lokeṣu sa vyāpta/ ādityādyās tad-āśritā// Lord Viṣṇu, who is the source of knowledge and transcendental bliss, has assumed the form of Śiśumāra in the seventh heaven, which is situated in the topmost level of the universe. All the other planets, beginning with the sun, exist under the shelter of this Śiśumāra planetary system.

## Part XIX: The Polestar (Dhruvaloka)

5.23 Summary: This chapter describes how all the planetary systems take shelter of the polestar, Dhruvaloka. It also describes the totality of these planetary systems to be Śiśumāra, another expansion of the external body of the Supreme Personality of Godhead. Dhruvaloka, the abode of Lord Viṣṇu within this universe, is situated 1,300,000 yojanas from the seven stars. In the planetary system of Dhruvaloka are the planets of the fire-god, Indra, Prajāpati, Kaśyapa and Dharma, all of whom are very respectful to the great devotee Dhruva, who lives on the polestar. Like bulls yoked to a central pivot, all the planetary systems revolve around Dhruvaloka, impelled by eternal time. Those who worship the virāṭ-puruṣa, the universal form of the Lord, conceive of this entire rotating system of planets as an animal known as śiśumāra. This imaginary śiśumāra is another form of the Lord. The head of the śiśumāra form is downward, and its body appears like that of a coiled snake. On the end of its tail is Dhruvaloka, on the body of the tail are Prajāpati, Agni, Indra and Dharma, and on the root of the tail are Dhātā and Vidhātā. On its waist are the seven great sages. The entire body of the śiśumāra faces toward its right and appears like a coil of stars. On the right side of this coil are the fourteen prominent stars from Abhijit to Punarvasu, and on the left side are the fourteen prominent stars from Puṣyā to Uttarāṣāḍhā. The stars known as Punarvasu and Puṣyā are on the right and left hips of the śiśumāra, and the stars known as Ārdrā and Aśleṣā are on the right and left feet of the śiśumāra. Other stars are also fixed on different sides of the Śiśumāra planetary system according to the calculations of Vedic astronomers. To concentrate their minds, yogīs worship the Śiśumāra planetary system, which is technically known as the kuṇḍalini-cakra.

5.23.1: śrī-śuka uvāca/ atha tasmāt paratas trayodaśa-lakṣa-yojanāntarato yat tad viṣṇoḥ paramaṁ padam abhivadanti yatra ha mahā-bhāgavato dhruva auttānapādir agninendreṇa prajāpatinā kaśyapena dharmeṇa ca samakāla-yugbhiḥ sabahu-mānaṁ dakṣiṇataḥ kriyamāṇa idānīm api kalpa-jīvinām ājīvya upāste tasyehānubhāva upavarṇitaḥ.

SYNONYMS: śrī-śukaḥ uvāca—Śrī Śukadeva Gosvāmī said; atha—thereupon; tasmāt—the sphere of the seven stars; parataḥ—beyond that; trayodaśa-lakṣa-yojana-antarataḥ—another 1,300,000 yojanas; yat—which; tat—that; viṣṇoḥ paramam padam—the supreme abode of Lord Viṣṇu, or the lotus feet of Lord Viṣṇu; abhivadanti—the Ṛg Veda mantras praise; yatra—on which; ha—indeed; mahā-bhāgavataḥ—the great devotee; dhruvaḥ—Mahārāja Dhruva; auttānapādiḥ—the son of Mahārāja Uttānapāda; agninā—by the fire-god; indreṇa—by the heavenly King, Indra; prajāpatinā—by the Prajāpati; kaśyapena—by Kaśyapa; dharmeṇa—by Dharmarāja; ca—also; samakāla-yugbhiḥ—who are engaged at the time; sa-bahu-mānam—always respectfully; dakṣiṇataḥ—on the right side; kriyamāṇaḥ—being circumambulated; idānīm—now; api—even; kalpa-jīvinām—of the living entities who exist at the end of the creation; ājīvyaḥ—the source of life; upāste—remains; tasya—his; iha—here; anubhāvaḥ—greatness in discharging devotional service; upavarṇitaḥ—already described (in the Fourth Canto of Śrīmad-Bhāgavatam).

TRANSLATION: Śukadeva Gosvāmī continued: My dear King, 1,300,000 yojanas [10,400,000 miles] above the planets of the seven sages is the place that learned scholars describe as the abode of Lord Viṣṇu. There the son of Mahārāja Uttānapāda, the great devotee Mahārāja Dhruva, still resides as the life source of all the living entities who live until the end of the creation. Agni, Indra, Prajāpati, Kaśyapa and Dharma all assemble there to offer him honor and respectful obeisances. They circumambulate him with their right sides toward him. I have already described the glorious activities of Mahārāja Dhruva [in the Fourth Canto of Śrīmad-Bhāgavatam].

5.23.2: sa hi sarveṣāṁ jyotir-gaṇānāṁ graha-nakṣatrādīnām animiṣeṇāvyakta-raṁhasā bhagavatā kālena bhrāmyamāṇānāṁ sthāṇur ivāvaṣṭambha īśvareṇa vihitaḥ śaśvad avabhāsate.

SYNONYMS: saḥ—that planet of Dhruva Mahārāja; hi—indeed; sarveṣām—of all; jyotiḥ-gaṇānām—the luminaries; graha-nakṣatra-ādīnām—such as the planets and stars; animiṣeṇa—who does not rest; avyakta—inconceivable; raṁhasā—whose force; bhagavatā—the most powerful; kālena—by the time factor; bhrāmyamāṇānām—being caused to revolve; sthāṇuḥ iva—like a post; avaṣṭambhaḥ—the pivot; īśvareṇa—by the will of the Supreme Personality of Godhead; vihitaḥ—established; śaśvat—constantly; avabhāsate—shines.

TRANSLATION: Established by the supreme will of the Supreme Personality of Godhead, the polestar, which is the planet of Mahārāja Dhruva, constantly shines as the central pivot for all the stars and planets. The unsleeping, invisible, most powerful time factor causes these luminaries to revolve around the polestar without cessation.

PURPORT: It is distinctly stated herein that all the luminaries, the planets and stars, revolve by the influence of the supreme time factor. The time factor is another feature of the Supreme Personality of Godhead. Everyone is under the influence of the time factor, but the Supreme Personality of Godhead is so kind and loves His devotee Mahārāja Dhruva so much that He has placed all the luminaries under the control of Dhruva's planet and has arranged for the time factor to work under him or with his cooperation. Everything is actually done according to the will and direction of the Supreme Personality of Godhead, but to make His devotee Dhruva the most important individual within the universe, the Lord has placed the activities of the time factor under his control.

5.23.3: yathā meḍhīstambha ākramaṇa-paśavaḥ saṁyojitās tribhis tribhiḥ savanair yathā-sthānaṁ maṇḍalāni caranty evaṁ bhagaṇā grahādaya etasminn antar-bahir-yogena kāla-cakra āyojitā dhruvam evāvalambya vāyunodīryamāṇā ākalpāntaṁ paricaṅ kramanti nabhasi yathā meghāḥ śyenādayo vāyu-vaśāḥ karma-sārathayaḥ parivartante evaṁ jyotirgaṇāḥ prakṛti-puruṣa-saṁyogānugṛhītāḥ karma-nirmita-gatayo bhuvi na patanti.

SYNONYMS: yathā—exactly like; meḍhīstambhe—to the pivot post; ākramaṇa-paśavaḥ—bulls for threshing rice; saṁyojitāḥ—being yoked; tribhiḥ tribhiḥ—by three; savanaiḥ—movements; yathā-sthānam—in their proper places; maṇḍalāni—orbits; caranti—traverse; evam—in the same way; bha-gaṇāḥ—the luminaries, like the sun, the moon, Venus, Mercury, Mars and Jupiter; graha-ādayaḥ—the different planets; etasmin—in this; antaḥ-bahiḥ-yogena—by connection with the inner or outer circles; kāla-cakre—in the wheel of eternal time; āyojitāḥ—fixed; dhruvam—Dhruvaloka; eva—certainly; avalambya—taking support of; vāyunā—by the wind; udīryamāṇāḥ—being propelled; ā-kalpa-antam—until the end of the creation; paricaṅ kramanti—revolve all around; nabhasi—in the sky; yathā—exactly like; meghāḥ—heavy clouds; śyena-ādayaḥ—birds such as the big eagle; vāyu-vaśāḥ—controlled by the air; karma-sārathayaḥ—whose chariot drivers are the results of their own past activities; parivartante—move around; evam—in this way; jyotiḥ-gaṇāḥ—the luminaries, the planets and stars in the sky; prakṛti—of material nature; puruṣa—and of the Supreme Personality, Kṛṣṇa; saṁyoga-anugṛhītāḥ—supported by the combined efforts; karma-nirmita—caused by their own past fruitive activities; gatayaḥ—whose movements; bhuvi—on the ground; na—not; patanti—fall down.

TRANSLATION: When bulls are yoked together and tied to a central post to thresh rice, they tread around that pivot without deviating from their proper positions—one bull being closest to the post, another in the middle, and a third on the outside. Similarly, all the planets and all the hundreds and thousands of stars revolve around the polestar, the planet of Mahārāja Dhruva, in their respective orbits, some higher and some lower. Fastened by the Supreme Personality of Godhead to the machine of material nature according to the results of their fruitive acts, they are driven around the polestar by the wind and will continue to be so until the end of creation. These planets float in the air within the vast sky, just as clouds with hundreds of tons of water float in the air or as the great śyena eagles, due to the results of past activities, fly high in the sky and have no chance of falling to the ground.

PURPORT: According to the description of this verse, the hundreds and thousands of stars and the great planets such as the sun, the moon, Venus, Mercury, Mars and Jupiter are not clustered together because of the law of gravity or any similar idea of the modern scientists. These planets and stars are all servants of the

Supreme Personality of Godhead, Govinda or Kṛṣṇa, and according to His order they sit in their chariots and travel in their respective orbits. The orbits in which they move are compared to machines given by material nature to the operating deities of the stars and planets, who carry out the orders of the Supreme Personality of Godhead by revolving around Dhruvaloka, which is occupied by the great devotee Mahārāja Dhruva. This is confirmed in the Brahma-saṁhitā (5.52) as follows: yac-cakṣur eṣa savitā sakala-grahāṇāṁ/rājā samasta-sura-mūrtir aśeṣa-tejāḥ/yasyājñayā bhramati sambhṛta-kāla-cakro/govindam ādi-puruṣaṁ tam ahaṁ bhajāmi// "I worship Govinda, the primeval Lord, the Supreme Personality of Godhead, under whose control even the sun, which is considered to be the eye of the Lord, rotates within the fixed orbit of eternal time. The sun is the king of all planetary systems and has unlimited potency in heat and light." This verse from Brahma-saṁhitā confirms that even the largest and most powerful planet, the sun, rotates within a fixed orbit, or kāla-cakra, in obedience to the order of the Supreme Personality of Godhead. This has nothing to do with gravity or any other imaginary laws created by the material scientists.

Material scientists want to avoid the ruling government of the Supreme Personality of Godhead, and therefore they imagine different conditions under which they suppose the planets move. The only condition, however, is the order of the Supreme Personality of Godhead. All the various predominating deities of the planets are persons, and the Supreme Personality of Godhead is also a person. The Supreme Personality orders the subordinate persons, the demigods of various names, to carry out His supreme will. This fact is also confirmed in Bhagavad-gītā (9.10), wherein Kṛṣṇa says: mayādhyakṣeṇa prakṛtiḥ/sūyate sa-carācaram/ hetunānena kaunteya/jagad viparivartate// "This material nature is working under My direction, O son of Kuntī, and it is producing all moving and unmoving beings. By its rule this manifestation is created and annihilated again and again."

The orbits of the planets resemble the bodies in which all living entities are seated because they are both machines controlled by the Supreme Personality of Godhead. As Kṛṣṇa says in Bhagavad-gītā (18.61): īśvaraḥ sarva-bhūtānāṁ/hṛd-deśe 'rjuna tiṣṭhati/bhrāmayan sarva-bhūtāni/yantrārūḍhāni māyayā// "The Supreme Lord is situated in everyone's heart, O Arjuna, and is directing the wanderings of all living entities, who are seated as on a machine, made of the material energy." The machine given by material nature—whether the machine of the body or the machine of the orbit, or kāla-cakra—works according to the orders given by the Supreme Personality of Godhead. The Supreme Personality of Godhead and material nature work together to maintain this great universe, and not only this universe but also the millions of other universes beyond this one.

The question of how the planets and stars are floating is also answered in this verse. It is not because of the laws of gravity. Rather, the planets and stars are enabled to float by manipulations of the air. It is due to such manipulations that big, heavy clouds float and big eagles fly in the sky. Modern airplanes like the 747 jet aircraft work in a similar way: by controlling the air, they float high in the sky, resisting the tendency to fall to earth. Such adjustments of the air are all made possible by the cooperation of the principles of puruṣa (male) and prakṛti (female). By the cooperation of material nature, which is considered to be prakṛti, and the Supreme Personality of Godhead, who is considered the puruṣa, all the affairs of the universe are going on nicely in their proper order. prakṛti, material nature, is also described in the Brahma-saṁhitā (5.44) as follows: sṛṣṭi-sthiti-pralaya-sādhana-śaktir ekā/chāyeva yasya bhuvanāni bibharti durgā/icchānurūpam api yasya ca ceṣṭate sā/govindam ādi-puruṣaṁ tam ahaṁ bhajāmi// "The external potency, māyā, who is of the nature of the shadow of the cit [spiritual] potency, is worshiped by all people as Durgā, the creating, preserving and destroying agency of this mundane world. I adore the primeval Lord Govinda, in accordance with whose will Durgā conducts herself." Material nature, the external energy of the Supreme Lord, is also known as Durgā, or the female energy that protects the great fort of this universe. The word Durgā also means fort. This universe is just like a great fort in which all the conditioned souls are kept, and they cannot leave it unless they are liberated by the mercy of the Supreme Personality of Godhead...

For example, as we have previously discussed, the moon is called jīva, which means that he is a living entity like us, but because of his pious activities he has been appointed to his post as the moon-god. Similarly, all the demigods are living entities who have been appointed to their various posts as the masters of the moon, the earth, Venus and so on because of their great service and pious acts. Only the predominating deity of the sun, Sūrya Nārāyaṇa, is an incarnation of the Supreme Personality of Godhead. Mahārāja Dhruva, the predominating deity of Dhruvaloka, is also a living entity...

Regarding the great eagles mentioned in this verse, it is understood that there are eagles so big that they can prey on big elephants. They fly so high that they can travel from one planet to another. They start flying in one planet and land in another, and while in flight they lay eggs that hatch into other birds while falling through the air. In Sanskrit such eagles are called śyena. Under the present circumstances, of course, we cannot see such huge birds, but at least we know of eagles that can capture monkeys and then throw them down to kill and eat them. Similarly, it is understood that there are gigantic birds that can carry off elephants, kill them and eat them.

The two examples of the eagle and the cloud are sufficient to prove that flying and floating can be made possible through adjustments of the air. The planets, in a similar way, are floating because material nature adjusts the air according to the orders of the Supreme Lord. It could be said that these adjustments constitute the law of gravity, but in any case, one must accept that these laws are made by the

Supreme Personality of Godhead. The so-called scientists have no control over them. The scientists can falsely, improperly declare that there is no God, but this is not a fact.

## Part XX: The Śiśumāra Planetary System

5.23.4: kecanaitaj jyotir-anīkaṁ śiśumāra-saṁsthānena bhagavato vāsudevasya yoga-dhāraṇāyām anuvarṇayanti.

SYNONYMS: kecana—some yogīs or learned scholars of astronomy; etat—this; jyotiḥ-anīkam—great wheel of planets and stars; śiśumāra-saṁsthānena—imagine this wheel to be a śiśumāra (dolphin); bhagavataḥ—of the Supreme Personality of Godhead; vāsudevasya—Lord Vāsudeva (the son of Vasudeva), Kṛṣṇa; yoga-dhāraṇāyām—in absorption in worship; anuvarṇayanti—describe.

TRANSLATION: This great machine, consisting of the stars and planets, resembles the form of a śiśumāra [dolphin] in the water. It is sometimes considered an incarnation of Kṛṣṇa, Vāsudeva. Great yogīs meditate upon Vāsudeva in this form because it is actually visible.
PURPORT: Transcendentalists such as yogīs whose minds cannot accommodate the form of the Lord prefer to visualize something very great, such as the virāṭ-puruṣa. Therefore some yogīs contemplate this imaginary śiśumāra to be swimming in the sky the way a dolphin swims in water. They meditate upon it as the virāṭ-rūpa, the gigantic form of the Supreme Personality of Godhead.

5.23.5: yasya pucchāgre 'vākśirasaḥ kuṇḍalī-bhūta-dehasya dhruva upakalpitas tasya lāṅgūle prajāpatir agnir indro dharma iti puccha-mūle dhātā vidhātā ca kaṭyāṁ saptarṣayaḥ; tasya dakṣiṇāvarta-kuṇḍalī-bhūta-śarīrasya yāny udagayanāni dakṣiṇa-pārśve tu nakṣatrāṇy upakalpayanti dakṣiṇāyanāni tu savye; yathā śiśumārasya kuṇḍalā-bhoga-sanniveśasya pārśvayor ubhayor apy avayavāḥ samasaṅkhyā bhavanti; pṛṣṭhe tv ajavīthī ākāśa-gaṅgā codarataḥ.

SYNONYMS: yasya—of which; puccha-agre—at the end of the tail; avākśirasaḥ—whose head is downward; kuṇḍalī-bhūta-dehasya—whose body, which is coiled; dhruvaḥ—Mahārāja Dhruva on his planet, the polestar; upakalpitaḥ—is situated; tasya—of that; lāṅgūle—on the tail; prajāpatiḥ—of the name Prajāpati; agniḥ—Agni; indraḥ—Indra; dharmaḥ—Dharma; iti—thus; puccha-mūle—at the base of the tail; dhātā vidhātā—the demigods known as Dhātā and Vidhātā; ca—also; kaṭyām—on the hip; sapta-ṛṣayaḥ—the seven saintly sages; tasya—of that; dakṣiṇa-āvarta-kuṇḍalī-bhūta-śarīrasya—whose body is like a coil turning toward the right side; yāni—which; udagayanāni—marking the northern courses; dakṣiṇa-pārśve—on the right side; tu—but; nakṣatrāṇi—constellations; upakalpayanti—are situated; dakṣiṇa-āyanāni—the fourteen stars, from Puṣyā to Uttarāṣāḍhā, marking the northern course; tu—but; savye—on the left side; yathā—just like; śiśumārasya—of the dolphin; kuṇḍalā-bhoga-sanniveśasya—whose body appears like a coil; pārśvayoḥ—on the sides; ubhayoḥ—both; api—certainly; avayavāḥ—the limbs; samasaṅkhyāḥ—of equal number (fourteen); bhavanti—are; pṛṣṭhe—on the back; tu—of course; ajavīthī—the first three stars marking the southern route (Mūlā, Pūrvaṣāḍhā and Uttarāṣāḍhā); ākāśa-gaṅgā—the Ganges in the sky (the Milky Way); ca—also; udarataḥ—on the abdomen.

TRANSLATION: This form of the śiśumāra has its head downward and its body coiled. On the end of its tail is the planet of Dhruva, on the body of its tail are the planets of the demigods Prajāpati, Agni, Indra and Dharma, and at the base of its tail are the planets of the demigods Dhātā and Vidhātā. Where the hips might be on the śiśumāra are the seven saintly sages like Vasiṣṭha and Aṅgirā. The coiled body of the Śiśumāra-cakra turns toward its right side, on which the fourteen constellations from Abhijit to Punarvasu are located. On its left side are the fourteen stars from Puṣyā to Uttarāṣāḍhā. Thus its body is balanced because its sides are occupied by an equal number of stars. On the back of the śiśumāra is the group of stars known as Ajavīthī, and on its abdomen is the Ganges that flows in the sky [the Milky Way].

5.23.6: punarvasu-puṣyau dakṣiṇa-vāmayoḥ śroṇyor ārdrāśleṣe ca dakṣiṇa-vāmayoḥ paścimayoḥ pādayor abhijid-uttarāṣāḍhe dakṣiṇa-vāmayor nāsikayor yathā-saṅkhyaṁ śravaṇa-pūrvāṣāḍhe dakṣiṇa-vāmayor locanayor dhaniṣṭhā mūlaṁ ca dakṣiṇa-vāmayoḥ karṇayor maghādīny aṣṭa nakṣatrāṇi dakṣiṇāyanāni vāma-pārśva-vaṅkriṣu yuñjīta tathaiva mṛga-śīrṣādīny udagayanāni dakṣiṇa-pārśva-vaṅkriṣu prātilomyena prayuñjīta śatabhiṣā-jyeṣṭhe skandhayor dakṣiṇa-vāmayor nyaset.

SYNONYMS: punarvasu—the star named Punarvasu; puṣyau—and the star named Puṣyā; dakṣiṇa-vāmayoḥ—on the right and left; śroṇyoḥ—loins; ārdrā—the star named Ārdrā; aśleṣe—the star named Aśleṣā; ca—also; dakṣiṇa-vāmayoḥ—at the right and left; paścimayoḥ—behind; pādayoḥ—feet; abhijit-uttarāṣāḍhe—the stars named Abhijit and Uttarāṣāḍhā; dakṣiṇa-vāmayoḥ—on the right and left; nāsikayoḥ—nostrils; yathā-saṅkhyam—according to numerical order; śravaṇa-pūrvāṣāḍhe—the stars named Śravaṇā and Pūrvāṣāḍhā; dakṣiṇa-vāmayoḥ—at the right and left; locanayoḥ—eyes; dhaniṣṭhā mūlam ca—and the stars named Dhaniṣṭhā and Mūlā; dakṣiṇa-vāmayoḥ—at the right and left; karṇayoḥ—ears; maghā-ādīni—the stars such as Maghā; aṣṭa nakṣatrāṇi—eight stars; dakṣiṇa-āyanāni—which mark the southern course; vāma-pārśva—of the left side; vaṅkriṣu—at the ribs; yuñjīta—may place; tathā eva—similarly; mṛga-śīrṣā-ādīni—such as Mṛgaśīrṣā; udagayanāni—marking the northern course; dakṣiṇa-pārśva-vaṅkriṣu—on the right side; prātilomyena—in the reverse order; prayuñjīta—may place; śatabhiṣā—Śatabhiṣā; jyeṣṭhe—Jyeṣṭhā; skandhayoḥ—on the two shoulders; dakṣiṇa-vāmayoḥ—right and left; nyaset—should place.

TRANSLATION: On the right and left sides of where the loins might be on the Śiśumāra-cakra are the stars named Punarvasu and Puṣyā. Ārdrā and Aśleṣā are on its right and left feet, Abhijit and Uttarāṣāḍhā are on its right and left nostrils, Śravaṇā and Pūrvāṣāḍhā are at its right and left eyes, and Dhaniṣṭhā and Mūlā are on its right and left ears. The eight stars from Maghā to Anurādhā, which mark the southern course, are on the ribs of the left of its body, and the eight stars from Mṛgaśīrṣā to Pūrvabhādra, which mark the northern course, are on the ribs on the right side. Śatabhiṣā and Jyeṣṭhā are on the right and left shoulders.

5.23.7: uttarā-hanāv agastir adharā-hanau yamo mukheṣu cāṅgārakaḥ śanaiścara upasthe bṛhaspatiḥ kakudi vakṣasy ādityo hṛdaye nārāyaṇo manasi candro nābhyām uśanā stanayor aśvinau budhaḥ prāṇāpānayo rāhur gale ketavaḥ sarvāṅgeṣu romasu sarve tārā-gaṇāḥ.

SYNONYMS: uttarā-hanau—on the upper jaw; agastiḥ—the star named Agasti; adharā-hanau—on the lower jaw; yamaḥ—Yamarāja; mukheṣu—on the mouth; ca—also; aṅgārakaḥ—Mars; śanaiścaraḥ—Saturn; upasthe—on the genitals; bṛhaspatiḥ—Jupiter; kakudi—on the back of the neck; vakṣasi—on the chest; ādityaḥ—the sun; hṛdaye—within the heart; nārāyaṇaḥ—Lord Nārāyaṇa; manasi—in the mind; candraḥ—the moon; nābhyām—on the navel; uśanā—Venus; stanayoḥ—on the two breasts; aśvinau—the two stars named Aśvin; budhaḥ—Mercury; prāṇāpānayoḥ—in the inner airs known as prāṇa and apāna; rāhuḥ—the planet Rāhu; gale—on the neck; ketavaḥ—comets; sarva-aṅgeṣu—all over the body; romasu—in the pores of the body; sarve—all; tārā-gaṇāḥ—the numerous stars.

TRANSLATION: On the upper chin of the śiśumāra is Agasti; on its lower chin, Yamarāja; on its mouth, Mars; on its genitals, Saturn; on the back of its neck, Jupiter; on its chest, the sun; and within the core of its heart, Nārāyaṇa. Within its mind is the moon; on its navel, Venus; and on its breasts, the Aśvinī-kumāras. Within its life air, which is known as prāṇāpāna, is Mercury, on its neck is Rāhu, all over its body are comets, and in its pores are the numerous stars.

5.23.8: etad u haiva bhagavato viṣṇoḥ sarva-devatāmayaṁ rūpam aharahaḥ sandhyāyāṁ prayato vāgyato nirīkṣamāṇa upatiṣṭheta namo jyotir-lokāya kālāyanāyānimiṣāṁ pataye mahā-puruṣāyābhidhīmahīti.

SYNONYMS: etat—this; u ha—indeed; eva—certainly; bhagavataḥ—of the Supreme Personality of Godhead; viṣṇoḥ—of Lord Viṣṇu; sarva-devatā-mayam—consisting of all the demigods; rūpam—form; ahaḥ-ahaḥ—always; sandhyāyām—in the morning, noon and evening; prayataḥ—meditating upon; vāgyataḥ—controlling the words; nirīkṣamāṇaḥ—observing; upatiṣṭheta—one should worship; namaḥ—respectful obeisances; jyotiḥ-lokāya—unto the resting place of all the planetary systems; kālāyanāya—in the form of supreme time; animiṣām—

of the demigods; pataye—unto the master; mahā-puruṣāya—unto the Supreme person; abhidhīmahi—let us meditate; iti—thus.

TRANSLATION: My dear King, the body of the śiśumāra, as thus described, should be considered the external form of Lord Viṣṇu, the Supreme Personality of Godhead. Morning, noon and evening, one should silently observe the form of the Lord as the Śiśumāra-cakra and worship Him with this mantra: "O Lord who has assumed the form of time! O resting place of all the planets moving in different orbits! O master of all demigods, O Supreme Person, I offer my respectful obeisances unto You and meditate upon You."

5.23.9: grahar̥kṣatārāmayam ādhidaivikaṁ/pāpāpahaṁ mantra-kṛtāṁ tri-kālam/
namasyataḥ smarato vā tri-kālaṁ/naśyeta tat-kālajam āśu pāpam

SYNONYMS: graha-ṛkṣa-tārā-mayam—consisting of all the planets and stars; ādhidaivikam—the leader of all the demigods; pāpa-apaham—the killer of sinful reactions; mantra-kṛtām—of those who chant the mantra mentioned above; tri-kālam—three times; namasyataḥ—offering obeisances; smarataḥ—meditating; vā—or; tri-kālam—three times; naśyeta—destroys; tat-kāla-jam—born at that time; āśu—very quickly; pāpam—all sinful reactions.

TRANSLATION: The body of the Supreme Lord, Viṣṇu, which forms the Śiśumāra-cakra, is the resting place of all the demigods and all the stars and planets. One who chants this mantra to worship that Supreme Person three times a day—morning, noon and evening—will surely be freed from all sinful reactions. If one simply offers his obeisances to this form or remembers this form three times a day, all his recent sinful activities will be destroyed.

PURPORT: Summarizing the entire description of the planetary systems of the universe, Śrīla Viśvanātha Cakravartī Ṭhākura says that one who is able to meditate upon this arrangement as the virāṭ-rūpa, or viśva-rūpa, the external body of the Supreme Personality of Godhead, and worship Him three times a day by meditation will always be free from all sinful reactions. Viśvanātha Cakravartī Ṭhākura estimates that Dhruvaloka, the polestar, is 3,800,000 yojanas above the sun. Above Dhruvaloka by 10,000,000 yojanas is Maharloka, above Maharloka by 20,000,000 yojanas is Janaloka, above Janaloka by 80,000,000 yojanas is Tapoloka, and above Tapoloka by 120,000,000 yojanas is Satyaloka. Thus the distance from the sun to Satyaloka is 233,800,000 yojanas, or 1,870,400,000 miles. The Vaikuṇṭha planets begin 26,200,000 yojanas (209,600,000 miles) above Satyaloka. Thus the Viṣṇu Purāṇa describes that the covering of the universe is 260,000,000 yojanas (2,080,000,000 miles) away from the sun. The distance from the sun to the earth is lower planetary systems called Atala, Vitala, Sutala, Talātala, Mahātala, Rasātala and Pātāla. Below these lower planets by 30,000 yojanas, Śeṣa Nāga is lying on the Garbhodaka Ocean. That ocean is 249,800,000 yojanas deep.

Thus the total diameter of the universe is approximately 500,000,000 yojanas, or 4,000,000,000 miles.

5.24 Summary: This chapter describes the planet Rāhu, which is 10,000 yojanas (80,000 miles) below the sun, and it also describes Atala and the other lower planetary systems. Rāhu is situated below the sun and moon. It is between these two planets and the earth. When Rāhu conceals the sun and moon, eclipses occur, either total or partial, depending on whether Rāhu moves in a straight or curving way.

Below Rāhu by another 1,000,000 yojanas are the planets of the Siddhas, Cāraṇas and Vidyādharas, and below these are planets such as Yakṣaloka and Rakṣaloka. Below these planets is the earth, and 70,000 yojanas below the earth are the lower planetary systems—Atala, Vitala, Sutala, Talātala, Mahātala, Rasātala and Pātāla. Demons and Rakṣasas live in these lower planetary systems with their wives and children, always engaged in sense gratification and not fearing their next births. The sunshine does not reach these planets, but they are illuminated by jewels fixed upon the hoods of snakes. Because of these shining gems there is practically no darkness. Those living in these planets do not become old or diseased, and they are not afraid of death from any cause but the time factor, the Supreme Personality of Godhead.

In the planet Atala, the yawning of a demon has produced three kinds of women, called svairiṇī (independent), kāmiṇī (lusty) and puṁścalī (very easily subdued by men). Below Atala is the planet Vitala, wherein Lord Śiva and his wife Gaurī reside. Because of their presence, a kind of gold is produced called hāṭaka. Below Vitala is the planet Sutala, the abode of Bali Mahārāja, the most fortunate king. Bali Mahārāja was favored by the Supreme Personality of Godhead, Vāmanadeva, because of his intense devotional service. The Lord went to the sacrificial arena of Bali Mahārāja and begged him for three paces of land, and on this plea the Lord took from him all his possessions. When Bali Mahārāja agreed to all this, the Lord was very pleased, and therefore the Lord serves as his doorkeeper. The description of Bali Mahārāja appears in the Eighth Canto of Śrīmad-Bhāgavatam.

When the Supreme Personality of Godhead offers a devotee material happiness, this is not His real favor. The demigods, who are very puffed up by their material opulence, pray to the Lord only for material happiness, not knowing anything better. Devotees like Prahlāda Mahārāja, however, do not want material happiness. Not to speak of material happiness, they do not want even liberation from material bondage, although one can achieve this liberation simply by chanting the holy name of the Lord, even with improper pronunciation.

Below Sutala is the planet Talātala, the abode of the demon Maya. This demon is always materially happy because he is favored by Lord Śiva, but he cannot achieve spiritual happiness at any time. Below Talātala is the planet Mahātala, where there are many snakes with hundreds and thousands of hoods. Below Mahātala is Rasātala, and below that is Pātāla, where the serpent Vasukī lives with his associates.

5.24.1: śrī-śuka uvāca/ adhastāt savitur yojanāyute svarbhānur nakṣatravac caratīty eke yo 'sāv amaratvaṁ grahatvaṁ cālabhata bhagavad-anukampayā svayam asurāpasadaḥ saiṁhikeyo hy atad-arhas tasya tāta janma karmāṇi copariṣṭād vakṣyāmaḥ.

SYNONYMS: śrī-śukaḥ uvāca—Śrī Śukadeva Gosvāmī said; adhastāt—below; savituḥ—the sun globe; yojana—a measurement equal to eight miles; ayute—ten thousand; svarbhānuḥ—the planet known as Rāhu; nakṣatra-vat—like one of the stars; carati—is rotating; iti—thus; eke—some who are learned in the Purāṇas; yaḥ—which; asau—that; amaratvam—a lifetime like those of the demigods; grahatvam—a position as one of the chief planets; ca—and; alabhata—obtained; bhagavat-anukampayā—by the compassion of the Supreme Personality of Godhead; svayam—personally; asura-apasadaḥ—the lowest of the asuras; saiṁhikeyaḥ—being the son of Siṁhikā; hi—indeed; a-tat-arhaḥ—not qualified for that position; tasya—his; tāta—O my dear King; janma—birth; karmāṇi—activities; ca—also; upariṣṭāt—later; vakṣyāmaḥ—I shall explain.

TRANSLATION: Śrī Śukadeva Gosvāmī said: My dear King, some historians, the speakers of the Purāṇas, say that 10,000 yojanas [80,000 miles] below the sun is the planet known as Rāhu, which moves like one of the stars. The presiding deity of that planet, who is the son of Siṁhikā, is the most abominable of all asuras, but although he is completely unfit to assume the position of a demigod or planetary deity, he has achieved that position by the grace of the Supreme Personality of Godhead. Later I shall speak further about him.

5.24.2: yad adas taraṇer maṇḍalaṁ pratapatas tad vistarato yojanāyutam ācakṣate dvādaśa-sahasraṁ somasya trayodaśa-sahasraṁ rāhor yaḥ parvaṇi tad-vyavadhana-kṛd vairanubandhaḥ surya-candramasāv abhidhāvati.

SYNONYMS: yat—which; adaḥ—that; taraṇeḥ—of the sun; maṇḍalam—globe; pratapataḥ—which is always distributing heat; tat—that; vistarataḥ—in terms of width; yojana—a distance of eight miles; ayutam—ten thousand; ācakṣate—they estimate; dvādaśa-sahasram—20,000 yojanas (160,000 miles); somasya—of the moon; trayodaśa—thirty; sahasram—one thousand; rāhoḥ—of the planet Rāhu; yaḥ—which; parvaṇi—on occasion; tat-vyavadhāna-kṛt—who created an obstruction to the sun and moon at the time of the distribution of nectar; vaira-anubandhaḥ—whose intentions are inimical; sūryā—the sun; candramasau—and the moon; abhidhāvati—runs after them on the full-moon night and the dark-moon day.

TRANSLATION: The sun globe, which is a source of heat, extends for 10,000 yojanas [80,000 miles]. The moon extends for 20,000 yojanas [160,000 miles], and Rāhu extends for 30,000 yojanas [240,000 miles]. Formerly, when nectar was being distributed, Rāhu tried to create dissension between the sun and moon by interposing himself between them. Rāhu is inimical toward both the sun and the moon, and therefore he always tries to cover the sunshine and moonshine on the dark-moon day and full-moon night.

PURPORT: As stated herein, the sun extends for 10,000 yojanas, and the moon extends for twice that, or 20,000 yojanas. The word dvādaśa should be understood to mean twice as much as ten, or twenty. In the opinion of Vijayadhvaja, the extent of Rāhu should be twice that of the moon, or text of the Bhāgavatam, Vijayadhvaja cites the following quotation concerning Rāhu; rāhu-soma-ravīṇāṁ tu maṇḍalā dvi-guṇoktitām. This means that Rāhu is twice as large as the moon, which is twice as large as the sun. This is the conclusion of the commentator Vijayadhvaja.

5.24.3: tan niśamyobhayatrāpi bhagavatā rakṣaṇāya prayuktaṁ sudarśanaṁ nāma bhāgavataṁ dayitam astraṁ tat tejasā durviṣahaṁ muhuḥ parivartamānam abhyavasthito muhūrtam udvijamānaś cakita-hṛdaya ārād eva nivartate tad uparāgam iti vadanti lokāḥ.

SYNONYMS: tat—that situation; niśamya—hearing; ubhayatra—around both the sun and moon; api—indeed; bhagavatā—by the Supreme Personality of Godhead; rakṣaṇāya—for their protection; prayuktam—engaged; sudarśanam—the wheel of Kṛṣṇa; nāma—named; bhāgavatam—the most confidential devotee; dayitam—the most favorite; astram—weapon; tat—that; tejasā—by its effulgence; durviṣaham—unbearable heat; muhuḥ—repeatedly; parivartamānam—moving around the sun and moon; abhyavasthitaḥ—situated; muhūrtam—for a muhūrta (forty-eight minutes); udvijamānaḥ—whose mind was full of anxieties; cakita—frightened; hṛdayaḥ—the core of whose heart; ārāt—to a distant place; eva—certainly; nivartate—flees; tat—that situation; uparāgam—an eclipse; iti—thus; vadanti—they say; lokāḥ—the people.

TRANSLATION: After hearing from the sun and moon demigods about Rāhu's attack, the Supreme Personality of Godhead, Viṣṇu, engages His disc, known as the Sudarśana cakra, to protect them. The Sudarśana cakra is the Lord's most beloved devotee and is favored by the Lord. The intense heat of its effulgence, meant for killing non-Vaiṣṇavas, is unbearable to Rāhu, and he therefore flees in fear of it. During the time Rāhu disturbs the sun or moon, there occurs what people commonly know as an eclipse.

PURPORT: Although Rāhu attempts to attack both the sun and the moon, they are protected by Lord Viṣṇu. Being very afraid of Lord Viṣṇu's cakra, Rāhu cannot stay in front of the sun or moon for more than a muhūrta (forty-eight minutes). The phenomenon that occurs when Rāhu blocks the light of the sun or moon is called an eclipse. The attempt of the scientists of this earth to go to the moon is as demoniac as Rāhu's attack. Of course. their attempts will be failures because no one can enter the moon or sun so easily. Like the attack of Rāhu, such attempts will certainly be failures.

5.24.4: tato 'dhastāt siddha-cāraṇa-vidyādharāṇāṁ sadanāni tāvan mātra eva.

SYNONYMS: tataḥ—the planet Rāhu; adhastāt—below; siddha-cāraṇa—of the planets known as Siddhaloka and Cāraṇaloka; vidyādharāṇām—and the planets of the Vidyādharas; sadanāni—the residential places; tāvat mātra—only that much distance (eighty thousand miles); eva—indeed.

TRANSLATION: Below Rāhu by 10,000 yojanas [80,000 miles] are the planets known as Siddhaloka, Cāraṇaloka and Vidyādhara-loka.

PURPORT: It is said that the residents of Siddhaloka, being naturally endowed with the powers of yogīs, can go from one planet to another by their natural mystic powers without using airplanes or similar machines.

5.24.5: tato 'dhastād yakṣa-rakṣaḥ-piśāca-preta-bhūta-gaṇānāṁ vihārājiram antarikṣaṁ yāvad vāyuḥ pravāti yāvan meghā upalabhyante.

SYNONYMS: tataḥ adhastāt—beneath the planets occupied by the Siddhas, Cāraṇas and Vidyādharas; yakṣa-rakṣaḥ-piśāca-preta-bhūta-gaṇānām—of Yakṣas, Rākṣasas, Piśācas, ghosts and so on; vihāra-ajiram—the place of sense gratification; antarikṣam—in the sky or outer space; yāvat—as far as; vāyuḥ—the wind; pravāti—blows; yāvat—as far as; meghāḥ—the clouds; upalabhyante—are seen.

TRANSLATION: Beneath Vidyādhara-loka, Cāraṇaloka and Siddhaloka, in the sky called antarikṣa, are the places of enjoyment for the Yakṣas, Rākṣasas, Piśācas, ghosts and so on. Antarikṣa extends as far as the wind blows and the clouds float in the sky. Above this there is no more air.

5.24.6: tato 'dhastāc chata-yojanāntara iyaṁ pṛthivī yāvad dhaṁsa-bhāsa-śyena-suparṇādayaḥ patattri-pravarā utpatantīti.

SYNONYMS: tataḥ adhastāt—beneath that; śata-yojana—of one hundred yojanas; antare—by an interval; iyam—this; pṛthivī—planet earth; yāvat—as high as; haṁsa—swans; bhāsa—vultures; śyena—eagles; suparṇa-ādayaḥ—and other birds; patattri-pravarāḥ—the chief among birds; utpatanti—can fly; iti—thus.

TRANSLATION: Below the abodes of the Yakṣas and Rākṣasas by a distance of 100 yojanas [800 miles] is the planet earth. Its upper limits extend as high as swans, hawks, eagles and similar large birds can fly.

## Part XXI: Subterranean Heavenly Planets

5.24.7: upavarṇitaṁ bhūmer yathā-sanniveśāvasthānam avaner apy adhastāt sapta bhū-vivarā ekaikaśo yojanāyutāntareṇāyāma-vistāreṇopakḷptā atalaṁ vitalaṁ sutalaṁ talātalaṁ mahātalaṁ rasātalaṁ pātālam iti.

SYNONYMS: upavarṇitam—stated previously; bhūmeḥ—of the planet earth; yathā-sanniveśa-avasthānam—according to the arrangement of the different places; avaneḥ—the earth; api—certainly; adhastāt—beneath; sapta—seven; bhū-vivarāḥ—other planets; eka-ekaśaḥ—in succession, up to the outer limit of the universe; yojana-ayuta-antareṇa—with an interval of ten thousand yojanas (eighty thousand miles); āyāma-vistāreṇa—by width and length; upakḷptāḥ—situated; atalam—named Atala; vitalam—Vitala; sutalam—Sutala; talātalam—Talātala; mahātalam—Mahātala; rasātalam—Rasātala; pātālam—Pātāla; iti—thus.

TRANSLATION: My dear King, beneath this earth are seven other planets, known as Atala, Vitala, Sutala, Talātala, Mahātala, Rasātala and Pātāla. I have already explained the situation of the planetary systems of earth. The width and length of the seven lower planetary systems are calculated to be exactly the same as those of earth.

5.24.8: eteṣu hi bila-svargeṣu svargād apy adhika-kāma-bhogaiśvaryānanda-bhūti-vibhūtibhiḥ susamṛddha-bhavanodyānākrīḍa-vihāreṣu daitya-dānava-kādraveyā nitya-pramuditānurakta-kalatrāpatya-bandhu-suhṛd-anucarā gṛha-pataya īśvarād apy apratihata-kāmā māyā-vinodā nivasanti.

SYNONYMS: eteṣu—in these; hi—certainly; bila-svargeṣu—known as the heavenly subterranean worlds; svargāt—than the heavenly planets; api—even; adhika—a greater quantity; kāma-bhoga—enjoyment of sense gratification; aiśvarya-ānanda—bliss due to opulence; bhūti—influence; vibhūtibhiḥ—by those things and wealth; su-samṛddha—improved; bhavana—houses; udyāna—gardens; ākrīḍa-vihāreṣu—in places for different types of sense gratification; daitya—the demons; dānava—ghosts; kādraveyāḥ—snakes; nitya—who are always; pramudita—overjoyed; anurakta—because of attachment; kalatra—to wife; apatya—children; bandhu—family relations; suhṛt—friends; anucarāḥ—followers; gṛha-patayaḥ—the heads of the households; īśvarāt—than those more capable, like the demigods; api—even; apratihata-kāmāḥ—whose fulfillment of lusty desires is unimpeded; māyā—illusory; vinodāḥ—who feel happiness; nivasanti—live.

TRANSLATION: In these seven planetary systems, which are also known as the subterranean heavens [bila-svarga], there are very beautiful houses, gardens and places of sense enjoyment, which are even more opulent than those in the higher planets because the demons have a very high standard of sensual pleasure, wealth and influence. Most of the residents of these planets, who are known as Daityas, Dānavas and Nāgas, live as householders. Their wives, children, friends and society are all fully engaged in illusory, material happiness. The sense enjoyment of the demigods is sometimes disturbed, but the residents of these planets enjoy life without disturbances. Thus they are understood to be very attached to illusory happiness.

5.24.9: yeṣu mahārāja mayena māyāvinā vinirmitāḥ puro nānā-maṇi-pravara-praveka-viracita-vicitra-bhavana-prākāra-gopura-sabhā-caitya-catvarāyatanādibhir nāgāsura-mithuna-pārāvata-śuka-sārikākīrṇa-kṛtrima-bhūmibhir vivareśvara-gṛhottamaiḥ samalaṅkṛtāś cakāsati.

SYNONYMS: yeṣu—in those lower planetary systems; mahā-rāja—O my dear King; mayena—by the demon named Maya; māyā-vinā—possessing advanced knowledge in the construction of material comforts; vinirmitāḥ—constructed; puraḥ—cities; nānā-maṇi-pravara—of valuable gems; praveka—with excellent; viracita—constructed; vicitra—wonderful; bhavana—houses; prākāra—walls; gopura—gates; sabhā—legislative meeting rooms; caitya—temples; catvara—schools; āyatana-ādibhiḥ—with hotels or recreation halls and so on; nāga—of living entities with snakelike bodies; asura—of demons, or godless persons; mithuna—by couples; pārāvata—pigeons; śuka—parrots; sārikā—mynas; ākīrṇa—crowded; kṛtrima—artificial; bhūmibhiḥ—possessing areas; vivara-īśvara—of the leaders of the planets; gṛha-uttamaiḥ—with first-class houses; samalaṅkṛtāḥ—decorated; cakāsati—shine magnificently.

TRANSLATION: My dear King, in the imitation heavens known as bila-svarga there is a great demon named Maya Dānava, who is an expert artist and architect. He has constructed many brilliantly decorated cities. There are many wonderful houses, walls, gates, assembly houses, temples, yards and temple compounds, as well as many hotels serving as residential quarters for foreigners. The houses for the leaders of these planets are constructed with the most valuable jewels, and they are always crowded with living entities known as Nāgas and Asuras, as well as many pigeons, parrots and similar birds. All in all, these imitation heavenly cities are most beautifully situated and attractively decorated.

5.24.10: udyānāni cātitarāṁ mana-indriyānandibhiḥ kusuma-phala-stabaka-subhaga-kisalayāvanata-rucira-viṭapa-viṭapināṁ latāṅgāliṅgitānāṁ śrībhiḥ samithuna-vividha-vihaṅgama-jalāśayānām amala-jala-pūrṇānāṁ jhaṣakulollaṅghana-kṣubhita-nīra-nīraja-kumuda-kuva-laya-kahlāra-nīlotpala-lohita-śatapatrādi-vaneṣu kṛta-niketanānām eka-vihārākula-madhura-vividha-svanādibhir indriyotsavair amara-loka-śriyam atiśayitāni.

SYNONYMS: udyānāni—the gardens and parks; ca—also; atitarām—greatly; manaḥ—to the mind; indriya—and to the senses; ānandibhiḥ—which cause pleasure; kusuma—by flowers; phala—of fruits; stabaka—bunches; subhaga—very beautiful; kisalaya—new twigs; avanata—bent low; rucira—attractive; viṭapa—possessing branches; viṭapinām—of trees; latā-aṅga-āliṅgitānām—which are embraced by the limbs of creepers; śrībhiḥ—by the beauty; sa-mithuna—in pairs; vividha—varieties; vihaṅgama—frequented by birds; jala-āśayānām—of reservoirs of water; amala-jala-pūrṇānām—full of clear and transparent water; jhaṣa-kula-ullaṅghana—by the jumping of different fish; kṣubhita—agitated; nīra—in the water; nīraja—of lotus flowers; kumuda—lilies; kuvalaya—flowers named kuvalaya; kahlāra—kahlāra flowers; nīla-utpala—blue lotus flowers; lohita—red; śata-patra-ādi—lotus flowers with a hundred petals and so on; vaneṣu—in forests; kṛta-niketanānām—of birds that have made their nests; eka-vihāra-ākula—full of uninterrupted enjoyment; madhura—very sweet; vividha—varieties; svana-ādibhiḥ—by vibrations; indriya-utsavaiḥ—invoking sense enjoyment; amara-loka-śriyam—the beauty of the residential places of the demigods; atiśayitāni—surpassing.

TRANSLATION: The parks and gardens in the artificial heavens surpass in beauty those of the upper heavenly planets. The trees in those gardens, embraced by creepers, bend with a heavy burden of twigs with fruits and flowers, and therefore they appear extraordinarily beautiful. That beauty could attract anyone and make his mind fully blossom in the pleasure of sense gratification. There are many lakes and reservoirs with clear, transparent water, agitated by jumping fish and decorated with many flowers such as lilies, kuvalayas, kahlāras and blue and red lotuses. Pairs of cakravākas and many other water birds nest in the lakes and always enjoy in a happy mood, making sweet, pleasing vibrations that are very satisfying and conducive to enjoyment of the senses.

5.24.11: yatra ha vāva na bhayam aho-rātrādibhiḥ kāla-vibhāgair upalakṣyate.

SYNONYMS: yatra—where; ha vāva—certainly; na—not; bhayam—fearfulness; ahaḥ-rātra-ādibhiḥ—because of days and nights; kāla-vibhāgaiḥ—the divisions of time; upalakṣyate—is experienced.

TRANSLATION: Since there is no sunshine in those subterranean planets, time is not divided into days and nights, and consequently fear produced by time does not exist.

5.24.12: yatra hi mahāhi-pravara-śiro-maṇayaḥ sarvaṁ tamaḥ prabādhante.

SYNONYMS: yatra—where; hi—indeed; mahā-ahi—of great serpents; pravara—of the best; śiraḥ-maṇayaḥ—the gems on the hoods; sarvam—all; tamaḥ—darkness; prabādhante—drive away.

TRANSLATION: Many great serpents reside there with gems on their hoods, and the effulgence of these gems dissipates the darkness in all directions.

5.24.13: na vā eteṣu vasatāṁ divyauṣadhi-rasa-rasāyanānna-pāna-snānādibhir ādhayo vyādhayo valī-palita-jarādayaś ca deha-vaivarṇya-daurgandhya-sveda-klama-glānir iti vayo 'vasthāś ca bhavanti.

SYNONYMS: na—not; vā—either; eteṣu—in these planets; vasatām—of those residing; divya—wonderful; auṣadhi—of herbs; rasa—the juices; rasāyana—and elixirs; anna—by eating; pāna—drinking; snāna-ādibhiḥ—by bathing in and so on; ādhayaḥ—mental troubles; vyādhayaḥ—diseases; valī—wrinkles; palita—grey hair; jarā—old age; ādayaḥ—and so on; ca—and; deha-vaivarṇya—the fading of bodily luster; daurgandhya—bad odor; sveda—perspiration; klama—fatigue; glāniḥ—lack of energy; iti—thus; vayaḥ avasthāḥ—miserable conditions due to increasing age; ca—and; bhavanti—are.

TRANSLATION: Since the residents of these planets drink and bathe in juices and elixirs made from wonderful herbs, they are freed from all anxieties and physical diseases. They have no experience of grey hair, wrinkles or invalidity, their bodily lusters do not fade, their perspiration does not cause a bad smell, and they are not troubled by fatigue or by lack of energy or enthusiasm due to old age.

5.24.14: na hi teṣāṁ kalyāṇānāṁ prabhavati kutaścana mṛtyur vinā bhagavat-tejasaś cakrāpadeśāt.

SYNONYMS: na hi—not; teṣām—of them; kalyāṇānām—who are by nature auspicious; prabhavati—able to influence; kutaścana—from anywhere; mṛtyuḥ—death; vinā—except; bhagavat-tejasaḥ—of the energy of the Supreme Personality of Godhead; cakra-apadeśāt—from that weapon named the Sudarśana cakra.

TRANSLATION: They live very auspiciously and do not fear death from anything but death's established time, which is the effulgence of the Sudarśana cakra of the Supreme Personality of Godhead.

PURPORT: This is the defect of material existence. Everything in the subterranean heavens is very nicely arranged. There are well situated residential quarters, there is a pleasing atmosphere, and there are no bodily inconveniences or mental anxieties, but nevertheless those who live there have to take another birth according to karma. Persons whose minds are dull cannot understand this defect of a materialistic civilization aiming at material comforts. One may make his living conditions very pleasing for the senses, but despite all favorable conditions, one must in due course of time meet death. The members of a demoniac civilization endeavor to make their living conditions very comfortable, but they cannot check death. The influence of the Sudarśana cakra will not allow their so-called material happiness to endure.

5.24.15: yasmin praviṣṭe 'sura-vadhūnāṁ prāyaḥ puṁsavanāni bhayād eva sravanti patanti ca.

SYNONYMS: yasmin—where; praviṣṭe—when entered; asura-vadhūnām—of the wives of those demons; prāyaḥ—almost always; puṁsavanāni—fetuses; bhayāt—because of fear; eva—certainly; sravanti—slip out; patanti—fall down; ca—and.

TRANSLATION: When the Sudarśana disc enters those provinces, the pregnant wives of the demons all have miscarriages due to fear of its effulgence.

5.24.16: athātale maya-putro 'suro balo nivasati yena ha vā iha sṛṣṭāḥ ṣaṇ-ṇavatir māyāḥ kāścanādyāpi māyāvino dhārayanti yasya ca jṛmbhamāṇasya mukhatas trayaḥ strī-gaṇā udapadyanta svairiṇyaḥ kāminyaḥ puṁścalya iti yā vai bilāyanaṁ praviṣṭaṁ puruṣaṁ rasena hāṭakākhyena sādhayitvā sva-vilāsāvalokanānurāga-smita-saṁlāpopagūhanādibhiḥ svairaṁ kila ramayanti yasminn upayukte puruṣa īśvaro 'haṁ siddho 'ham ity ayuta-mahā-gaja-balam ātmānam abhimanyamānaḥ katthate madāndha iva.

SYNONYMS: atha—now; atale—on the planet named Atala; maya-putraḥ asuraḥ—the demon son of Maya; balaḥ—Bala; nivasati—resides; yena—by whom; ha vā—indeed; iha—in this; sṛṣṭāḥ—propagated; ṣaṭ-ṇavatiḥ—ninety-six; māyāḥ—varieties of illusion; kāścana—some; adya api—even today; māyā-vinaḥ—those who know the art of magical feats (like manufacturing gold); dhārayanti—utilize; yasya—of whom; ca—also; jṛmbhamāṇasya—while yawning; mukhataḥ—from the mouth; trayaḥ—three; strī-gaṇāḥ—varieties of women; udapadyanta—were generated; svairiṇyaḥ—svairiṇī (one who only marries in her same class); kāminyaḥ—kāmiṇī (one who, being lusty, marries men from any group); puṁścalyaḥ—puṁścalī (one who wants to go from one husband to another); iti—thus; yāḥ—who; vai—certainly; bila-ayanam—the subterranean planets; praviṣṭam—entering; puruṣam—a male; rasena—by a juice; hāṭaka-ākhyena—made from an intoxicating herb known as hāṭaka; sādhayitvā—making sexually fit; sva-vilāsa—for their personal sense gratification; avalokana—by glances; anurāga—lustful; smita—by smiling; saṁlāpa—by talking; upagūhana-ādibhiḥ—and by embracing; svairam—according to their own desire; kila—indeed;

ramayanti—enjoy sex pleasure; yasmin—which; upayukte—when used; puruṣaḥ—a man; īśvaraḥ aham—I am the most powerful person; siddhaḥ aham—I am the greatest and most elevated person; iti—thus; ayuta—ten thousand; mahā-gaja—of big elephants; balam—the strength; ātmānam—himself; abhimanyamānaḥ—being full of pride; katthate—they say; mada-andhaḥ—blinded by false prestige; iva—like.

TRANSLATION: My dear King, now I shall describe to you the lower planetary systems, one by one, beginning from Atala. In Atala there is a demon, the son of Maya Dānava named Bala, who created ninety-six kinds of mystic power. Some so-called yogīs and svāmīs take advantage of this mystic power to cheat people even today. Simply by yawning, the demon Bala created three kinds of women, known as svairiṇī, kāmiṇī and puṁścalī. The svairiṇīs like to marry men from their own group, the kāmiṇīs marry men from any group, and the puṁścalīs change husbands one after another. If a man enters the planet of Atala, these women immediately capture him and induce him to drink an intoxicating beverage made with a drug known as hāṭaka [cannabis indica]. This intoxicant endows the man with great sexual prowess, of which the women take advantage for enjoyment. A woman will enchant him with attractive glances, intimate words, smiles of love and then embraces. In this way she induces him to enjoy sex with her to her full satisfaction. Because of his increased sexual power, the man thinks himself stronger than ten thousand elephants and considers himself most perfect. Indeed, illusioned and intoxicated by false pride, he thinks himself God, ignoring impending death.

5.24.17: tato 'dhastād vitale haro bhagavān hāṭakeśvaraḥ sva-pārṣada-bhūta-gaṇāvṛtaḥ prajāpati-sargopabṛṁhaṇāya bhavo bhavānyā saha mithunī-bhūta āste yataḥ pravṛttā sarit-pravarā hāṭakī nāma bhavayor vīryeṇa yatra citrabhānur mātariśvanā samidhyamāna ojasā pibati tan niṣṭhyūtaṁ hāṭakākhyaṁ suvarṇaṁ bhūṣaṇenāsurendrāvarodheṣu puruṣāḥ saha puruṣībhir dhārayanti.

SYNONYMS: tataḥ—the planet Atala; adhastāt—beneath; vitale—on the planet; haraḥ—Lord Śiva; bhagavān—the most powerful personality; hāṭakeśvaraḥ—the master of gold; sva-pārṣada—by his own associates; bhūta-gaṇa—who are ghostly living beings; āvṛtaḥ—surrounded; prajāpati-sarga—of the creation of Lord Brahmā; upabṛṁhaṇāya—to increase the population; bhavaḥ—Lord Śiva; bhavānyā saha—with his wife, Bhavānī; mithunī-bhūtaḥ—being united in sex; āste—remains; yataḥ—from that planet (Vitala); pravṛttā—being emanated; sarit-pravarā—the great river; hāṭakī—Hāṭakī; nāma—named; bhavayoḥ vīryeṇa—due to the semina and ovum of Lord Śiva and Bhavānī; yatra—where; citra-bhānuḥ—the fire-god; mātariśvanā—by the wind; samidhyamānaḥ—being brightly inflamed; ojasā—with great strength; pibati—drinks; tat—that; niṣṭhyūtam—spit out with a hissing sound; hāṭaka-ākhyam—named Hāṭaka; suvarṇam—gold; bhūṣaṇena—by different types of ornaments; asura-indra—of the great asuras;

avarodheṣu—in the homes; puruṣāḥ—the males; saha—with; puruṣībhiḥ—their wives and women; dhārayanti—wear.

TRANSLATION: The next planet below Atala is Vitala, wherein Lord Śiva, who is known as the master of gold mines, lives with his personal associates, the ghosts and similar living entities. Lord Śiva, as the progenitor, engages in sex with Bhavānī, the progenitress, to produce living entities, and from the mixture of their vital fluid the river named Hāṭakī is generated. When fire, being made to blaze by the wind, drinks of this river and then sizzles and spits it out, it produces gold called Hāṭaka. The demons who live on that planet with their wives decorate themselves with various ornaments made from that gold, and thus they live there very happily.

PURPORT: It appears that when Bhava and Bhavānī, Lord Śiva and his wife, unite sexually, the emulsification of their secretions creates a chemical which when heated by fire can produce gold. It is said that the alchemists of the medieval age tried to prepare gold from base metal, and Śrīla Sanātana Gosvāmī also states that when bell metal is treated with mercury, it can produce gold.

5.24.18: tato 'dhastāt sutale udāra-śravāḥ puṇya-śloko virocanātmajo balir bhagavatā mahendrasya priyaṁ cikīrṣamāṇenāditer labdha-kāyo bhūtvā vaṭu-vāmana-rūpeṇa parākṣipta-loka-trayo bhagavad-anukampayaiva punaḥ praveśita indrādiṣv avidyamānayā susamṛddhayā śriyābhijuṣṭaḥ sva-dharmeṇārādhayaṁs tam eva bhagavantam ārādhanīyam apagata-sādhvasa āste 'dhunāpi.

SYNONYMS: tataḥ adhastāt—beneath the planet known as Vitala; sutale—on the planet known as Sutala; udāra-śravāḥ—very greatly celebrated; puṇya-ślokaḥ—very pious and advanced in spiritual consciousness; virocana-ātmajaḥ—the son of Virocana; baliḥ—Bali Mahārāja; bhagavatā—by the Supreme Personality of Godhead; mahā-indrasya—of the King of heaven, Indra; priyam—the welfare; cikīrṣamāṇena—desiring to perform; āditeḥ—from Aditi; labdha-kāyaḥ—having obtained His body; bhūtvā—appearing; vaṭu—brahmacāri; vāmana-rūpeṇa—in the form of a dwarf; parākṣipta—wrested away; loka-trayaḥ—the three worlds; bhagavat-anukampayā—by the causeless mercy of the Supreme Personality of Godhead; eva—certainly; punaḥ—again; praveśitaḥ—caused to enter; indra-ādiṣu—even among the demigods like the King of heaven; avidyamānayā—not existing; susamṛddhayā—much enriched by such exalted opulence; śriyā—by good fortune; abhijuṣṭaḥ—being blessed; sva-dharmeṇa—by discharging devotional service; ārādhayan—worshiping; tam—Him; eva—certainly; bhagavantam—the Supreme Personality of Godhead; ārādhanīyam—who is most worshipable; apagata-sādhvasaḥ—without fear; āste—remains; adhunā api—even today.

TRANSLATION: Below the planet Vitala is another planet, known as Sutala, where the great son of Mahārāja Virocana, Bali Mahārāja, who is celebrated as the most pious king, resides even now. For the welfare of Indra, the King of heaven, Lord Viṣṇu appeared in the form of a dwarf brahmacārī as the son of Aditi and tricked Bali Mahārāja by begging for only three paces of land but taking all the three worlds. Being very pleased with Bali Mahārāja for giving all his possessions, the Lord returned his kingdom and made him richer than the opulent King Indra. Even now, Bali Mahārāja engages in devotional service by worshiping the Supreme Personality of Godhead in the planet of Sutala.

5.24.28: tato 'dhastāt talātale mayo nāma dānavendras tri-purādhipatir bhagavatā purāriṇā tri-lokī-śaṁ cikīrṣuṇā nirdagdha-sva-pura-trayas tat-prasādāl labdha-pado māyāvinām ācāryo mahādevena parirakṣito vigata-sudarśana-bhayo mahīyate.

SYNONYMS: tataḥ—the planet known as Sutala; adhastāt—below; talātale—in the planet known as Talātala; mayaḥ—Maya; nāma—named; dānava-indraḥ—the king of the Dānava demons; tri-pura-adhipatiḥ—the Lord of the three cities; bhagavatā—by the most powerful; purāriṇā—Lord Śiva, who is known as Tripurāri; tri-lokī—of the three worlds; śam—the good fortune; cikīrṣuṇā—who was desiring; nirdagdha—burned; sva-pura-trayaḥ—whose three cities; tat-prasādāt—by Lord Śiva's mercy; labdha—obtained; padaḥ—a kingdom; māyāvinām ācāryaḥ—who is the ācārya, or master, of all the conjurers; mahā-devena—by Lord Śiva; parirakṣitaḥ—protected; vigata-sudarśana-bhayaḥ—who is not afraid of the Supreme Personality of Godhead and His Sudarśana cakra; mahīyate—is worshiped.

TRANSLATION: Beneath the planet known as Sutala is another planet, called Talātala, which is ruled by the Dānava demon named Maya. Maya is known as the ācārya [master] of all the māyāvīs, who can invoke the powers of sorcery. For the benefit of the three worlds, Lord Śiva, who is known as Tripurāri, once set fire to the three kingdoms of Maya, but later, being pleased with him, he returned his kingdom. Since that time, Maya Dānava has been protected by Lord Śiva, and therefore he falsely thinks that he need not fear the Sudarśana cakra of the Supreme Personality of Godhead.

5.24.29: tato 'dhastān mahātale kādraveyāṇāṁ sarpāṇāṁ naika-śirasāṁ krodhavaśo nāma gaṇaḥ kuhaka-takṣaka-kāliya-suṣeṇādi-pradhānā mahā-bhogavantaḥ patattri-rājādhipateḥ puruṣa-vāhād anavaratam udvijamānāḥ sva-kalatrāpatya-suhṛt-kuṭumba-saṅgena kvacit pramattā viharanti.

SYNONYMS: tataḥ—the planet Talātala; adhastāt—beneath; mahātale—in the planet known as Mahātala; kādraveyāṇām—of the descendants of Kadrū;

sarpāṇām—who are big snakes; na eka-śirasām—who have many hoods; krodha-vaśaḥ—always subject to anger; nāma—named; gaṇaḥ—the group; kuhaka—Kuhaka; takṣaka—Takṣaka; kāliya—Kāliya; suṣeṇa—Suṣeṇa; ādi—and so on; pradhānāḥ—who are the prominent ones; mahā-bhogavantaḥ—addicted to all kinds of material enjoyment; patattri-rāja-adhipateḥ—from the king of all birds, Garuḍa; puruṣa-vāhāt—who carries the Supreme Personality of Godhead; anavaratam—constantly; udvijamānāḥ—afraid; sva—of their own; kalatra-apatya—wives and children; suhṛt—friends; kuṭumba—relatives; saṅgena—in the association; kvacit—sometimes; pramattāḥ—infuriated; viharanti—they sport.

TRANSLATION: The planetary system below Talātala is known as Mahātala. It is the abode of many-hooded snakes, descendants of Kadrū, who are always very angry. The great snakes who are prominent are Kuhaka, Takṣaka, Kāliya and Suṣeṇa. The snakes in Mahātala are always disturbed by fear of Garuḍa, the carrier of Lord Viṣṇu, but although they are full of anxiety, some of them nevertheless sport with their wives, children, friends and relatives.

PURPORT: It is stated here that the snakes who live in the planetary system known as Mahātala are very powerful and have many hoods. They live with their wives and children and consider themselves very happy, although they are always full of anxiety because of Garuḍa, who comes there to destroy them. This is the way of material life. Even if one lives in the most abominable condition, he still thinks himself happy with his wife, children, friends and relatives.

5.24.30: tato 'dhastād rasātale daiteyā dānavāḥ paṇayo nāma nivāta-kavacāḥ kāleyā hiraṇya-puravāsina iti vibudha-pratyanīkā utpattyā mahaujaso mahā-sāhasino bhagavataḥ sakala-lokānubhāvasya harer eva tejasā pratihata-balāvalepā bileśayā iva vasanti ye vai saramayendra-dūtyā vāgbhir mantra-varṇābhir indrād bibhyati.

SYNONYMS: tataḥ adhastāt—below the planetary system Mahātala; rasātale—on the planet called Rasātala; daiteyāḥ—the sons of Diti; dānavāḥ—the sons of Danu; paṇayaḥ nāma—named Paṇis; nivāta-kavacāḥ—Nivāta-kavacas; kāleyāḥ—Kāleyas; hiraṇya-puravāsinaḥ—Hiraṇya-puravāsīs; iti—thus; vibudha-pratyanīkāḥ—enemies of the demigods; utpattyāḥ—from birth; mahā-ojasaḥ—very powerful; mahā-sāhasinaḥ—very cruel; bhagavataḥ—of the Personality of Godhead; sakala-loka-anubhāvasya—who is auspicious for all planetary systems; hareḥ—of the Supreme Personality of Godhead; eva—certainly; tejasā—by the Sudarśana cakra; pratihata—defeated; bala—strength; avalepāḥ—and pride (because of bodily strength); bila-īśayāḥ—the snakes; iva—like; vasanti—they live; ye—who; vai—indeed; saramayā—by Saramā; indra-dūtyā—the messenger of Indra; vāgbhiḥ—by the words; mantra-varṇābhiḥ—in the form of a mantra; indrāt—from King Indra; bibhyati—are afraid.

TRANSLATION: Beneath Mahātala is the planetary system known as Rasātala, which is the abode of the demoniac sons of Diti and Danu. They are called Paṇis, Nivāta-kavacas, Kāleyas and Hiraṇya-puravāsīs [those living in Hiraṇya-pura]. They are all enemies of the demigods, and they reside in holes like snakes. From birth they are extremely powerful and cruel, and although they are proud of their strength, they are always defeated by the Sudarśana cakra of the Supreme Personality of Godhead, who rules all the planetary systems. When a female messenger from Indra named Saramā chants a particular curse, the serpentine demons of Mahātala become very afraid of Indra.

PURPORT: It is said that there was a great fight between these serpentine demons and Indra, the King of heaven. When the defeated demons met the female messenger Saramā, who was chanting a mantra, they became afraid, and therefore they are living in the planet called Rasātala.

5.24.31: tato 'dhastāt pātāle nāga-loka-patayo vāsuki-pramukhāḥ śaṅkha-kulika-mahāśaṅkha-śveta-dhanañjaya-dhṛtarāṣṭra-śaṅkhacūḍa-kambalāśvatara-devadattādayo mahā-bhogino mahāmarṣā nivasanti yeṣām u ha vai pañca-sapta-daśa-śata-sahasra-śīrṣāṇāṁ phaṇāsu viracitā mahā-maṇayo rociṣṇavaḥ pātāla-vivara-timira-nikaraṁ sva-rociṣā vidhamanti.

SYNONYMS: tataḥ adhastāt—beneath that planet Rasātala; pātāle—on the planet known as Pātāla; nāga-loka-patayaḥ—the masters of the Nāgalokas; vāsuki—by Vāsuki; pramukhāḥ—headed; śaṅkha—Śaṅkha; kulika—Kulika; mahā-śaṅkha—Mahāśaṅkha; śveta—Śveta; dhanañjaya—Dhanañjaya; dhṛtarāṣṭra—Dhṛtarāṣṭra; śaṅkha-cūḍa—Śaṅkhacūḍa; kambala—Kambala; aśvatara—Aśvatara; deva-datta—Devadatta; ādayaḥ—and so on; mahā-bhoginaḥ—very addicted to material happiness; mahā-amarṣāḥ—greatly envious by nature; nivasanti—live; yeṣām—of all of them; u ha—certainly; vai—indeed; pañca—five; sapta—seven; daśa—ten; śata—one hundred; sahasra—one thousand; śīrṣāṇām—of those possessing hoods; phaṇāsu—on those hoods; viracitāḥ—fixed; mahā-maṇayaḥ—very valuable gems; rociṣṇavaḥ—full of effulgence; pātāla-vivara—the caves of the Pātāla planetary system; timira-nikaram—the mass of darkness; sva-rociṣā—by the effulgence of their hoods; vidhamanti—disperse.

TRANSLATION: Beneath Rasātala is another planetary system, known as Pātāla or Nāgaloka, where there are many demoniac serpents, the masters of Nāgaloka, such as Śaṅkha, Kulika, Mahāśaṅkha, Śveta, Dhanañjaya, Dhṛtarāṣṭra, Śaṅkhacūḍa, Kambala, Aśvatara and Devadatta. The chief among them is Vāsuki. They are all extremely angry, and they have many, many hoods—some snakes five hoods, some seven, some ten, others a hundred and others a thousand. These hoods are bedecked with valuable gems, and the light emanating from the gems illuminates the entire planetary system of bila-svarga.

## Part XXII: Anantadeva

5.25.1: śrī-śuka uvāca/ tasya mūla-deśe triṁśad-yojana-sahasrāntara āste yā vai kalā bhagavatas tāmasī samākhyātānanta iti sātvatīyā draṣṭṛ-dṛśyayoḥ saṅkarṣaṇam aham ity abhimāna-lakṣaṇaṁ yaṁ saṅkarṣaṇam ity ācakṣate.

SYNONYMS: śrī-śukaḥ uvāca—Śrī Śukadeva Gosvāmī said; tasya—of the planet Pātāla; mūla-deśe—in the region beneath the base; triṁśat—thirty; yojana—eight-mile units of measurement; sahasra-antare—at an interval of one thousand; āste—remains; yā—which; vai—indeed; kalā—an expansion of an expansion; bhagavataḥ—of the Supreme Personality of Godhead; tāmasī—related to darkness; samākhyātā—called; anantaḥ—Ananta; iti—thus; sātvatīyāḥ—the devotees; draṣṭṛ-dṛśyayoḥ—of matter and spirit; saṅkarṣaṇam—the drawing together; aham—I; iti—thus; abhimāna—by self-conception; lakṣaṇam—symptomized; yam—whom; saṅkarṣaṇam—Saṅkarṣaṇa; iti—thus; ācakṣate—learned scholars describe.

TRANSLATION: Śrī Śukadeva Gosvāmī said to Mahārāja Parīkṣit: My dear King, approximately 240,000 miles beneath the planet Pātāla lives another incarnation of the Supreme Personality of Godhead. He is the expansion of Lord Viṣṇu known as Lord Ananta or Lord Saṅkarṣaṇa. He is always in the transcendental position, but because He is worshiped by Lord Śiva, the deity of tamo-guṇa or darkness, He is sometimes called tāmasī. Lord Ananta is the predominating Deity of the material mode of ignorance as well as the false ego of all conditioned souls. When a conditioned living being thinks, "I am the enjoyer, and this world is meant to be enjoyed by me," this conception of life is dictated to him by Saṅkarṣaṇa. Thus the mundane conditioned soul thinks himself the Supreme Lord.

5.25.2: yasyedaṁ kṣiti-maṇḍalaṁ bhagavato 'nanta-mūrteḥ sahasra-śirasa ekasminn eva śīrṣaṇi dhriyamāṇaṁ siddhārtha iva lakṣyate.

SYNONYMS: yasya—of whom; idam—this; kṣiti-maṇḍalam—universe; bhagavataḥ—of the Supreme Personality of Godhead; ananta-mūrteḥ—in the form of Anantadeva; sahasra-śirasaḥ—who has thousands of hoods; ekasmin—on one; eva—only; śīrṣaṇi—hood; dhriyamāṇam—is being sustained; siddhārthaḥ iva—and like a white mustard seed; lakṣyate—is seen.

TRANSLATION: Śukadeva Gosvāmī continued: This great universe, situated on one of Lord Anantadeva's thousands of hoods, appears just like a white mustard seed. It is infinitesimal compared to the hood of Lord Ananta.

5.25.3: yasya ha vā idaṁ kālenopasañjihīrṣato 'marṣa-viracita-rucira-bhramad-bhruvor antareṇa sāṅkarṣaṇo nāma rudra ekādaśa-vyūhas try-akṣas tri-śikhaṁ śūlam uttambhayann udatiṣṭhat.

SYNONYMS: yasya—of whom; ha vā—indeed; idam—this (material world); kālena—in due course of time; upasañjihīrṣataḥ—desiring to destroy; amarṣa—by anger; viracita—formed; rucira—very beautiful; bhramat—moving; bhruvoḥ—the two eyebrows; antareṇa—from between; sāṅkarṣaṇaḥ nāma—named Sāṅkarṣaṇa; rudraḥ—an incarnation of Lord Śiva; ekādaśa-vyūhaḥ—who has eleven expansions; tri-akṣaḥ—three eyes; tri-śikham—having three points; śūlam—a trident; uttambhayan—raising; udatiṣṭhat—arose.

TRANSLATION: At the time of devastation, when Lord Anantadeva desires to destroy the entire creation, He becomes slightly angry. Then from between His two eyebrows appears three-eyed Rudra, carrying a trident. This Rudra, who is known as Sāṅkarṣaṇa, is the embodiment of the eleven Rudras, or incarnations of Lord Śiva. He appears in order to devastate the entire creation.

## Part XXIII: Hellish Planets

5.26.4: rājovāca/ narakā nāma bhagavan kiṁ deśa-viśeṣā athavā bahis tri-lokyā āhosvid antarāla iti.

SYNONYMS: rājā uvāca—the King said; narakāḥ—the hellish regions; nāma—named; bhagavan—O my Lord; kim—whether; deśa-viśeṣāḥ—a particular country; athavā—or; bahiḥ—outside; tri-lokyāḥ—the three worlds (the universe); āhosvit—or; antarāle—in the intermediate spaces within the universe; iti—thus.

TRANSLATION: King Parīkṣit inquired from Śukadeva Gosvāmī: My dear lord, are the hellish regions outside the universe, within the covering of the universe, or in different places on this planet?

5.26.5: ṛṣir uvāca/ antarāla eva tri-jagatyās tu diśi dakṣiṇasyām adhastād bhūmer upariṣṭāc ca jalād yasyām agniṣvāttādayaḥ pitṛ-gaṇā diśi svānāṁ gotrāṇāṁ paramena samādhinā satyā evāśiṣa āśāsānā nivasanti.

SYNONYMS: ṛṣiḥ uvāca—the great sage replied; antarāle—in the intermediate space; eva—certainly; tri-jagatyāḥ—of the three worlds; tu—but; diśi—in the direction; dakṣiṇasyām—southern; adhastāt—beneath; bhūmeḥ—on the earth; upariṣṭāt—a little above; ca—and; jalāt—the Garbhodaka Ocean; yasyām—in which; agniṣvāttā-ādayaḥ—headed by Agniṣvāttā; pitṛ-gaṇāḥ—the persons known as pitās; diśi—direction; svānām—their own; gotrāṇām—of the families; paramena—with great; samādhinā—absorption in thoughts of the Lord; satyāḥ—

in truth; eva—certainly; āśiṣaḥ—blessings; āśāsānāḥ—desiring; nivasanti—they live.

TRANSLATION: The great sage Śukadeva Gosvāmī answered: All the hellish planets are situated in the intermediate space between the three worlds and the Garbhodaka Ocean. They lie on the southern side of the universe, beneath Bhū-maṇḍala, and slightly above the water of the Garbhodaka Ocean. Pitṛloka is also located in this region between the Garbhodaka Ocean and the lower planetary systems. All the residents of Pitṛloka, headed by Agniṣvāttā, meditate in great samādhi on the Supreme Personality of Godhead and always wish their families well.

PURPORT: As previously explained, below our planetary system are seven lower planetary systems, the lowest of which is called Pātālaloka. Beneath Pātālaloka are other planets, known as Narakaloka, or the hellish planets. At the bottom of the universe lies the Garbhodaka Ocean. Therefore the hellish planets lie between Pātālaloka and the Garbhodaka Ocean.

5.26.6: yatra ha vāva bhagavān pitṛ-rājo vaivasvataḥ sva-viṣayaṁ prāpiteṣu sva-puruṣair jantuṣu sampareteṣu yathā-karmāvadyaṁ doṣam evānullaṅghita-bhagavac-chāsanaḥ sagaṇo damaṁ dhārayati.

SYNONYMS: yatra—where; ha vāva—indeed; bhagavān—the most powerful; pitṛ-rājaḥ—Yamarāja, the king of the pitās; vaivasvataḥ—the son of the sun-god; sva-viṣayam—his own kingdom; prāpiteṣu—when caused to reach; sva-puruṣaiḥ—by his own messengers; jantuṣu—the human beings; sampareteṣu—dead; yathā-karma-avadyam—according to how much they have violated the rules and regulations of conditional life; doṣam—the fault; eva—certainly; anullaṅghita-bhagavat-śāsanaḥ—who never oversteps the Supreme Personality of Godhead's order; sagaṇaḥ—along with his followers; damam—punishment; dhārayati—executes.

TRANSLATION: The King of the pitās is Yamarāja, the very powerful son of the sun-god. He resides in Pitṛloka with his personal assistants and, while abiding by the rules and regulations set down by the Supreme Lord, has his agents, the Yamadūtas, bring all the sinful men to him immediately upon their death. After bringing them within his jurisdiction, he properly judges them according to their specific sinful activities and sends them to one of the many hellish planets for suitable punishments.

PURPORT: Yamarāja is not a fictitious or mythological character; he has his own abode, Pitṛloka, of which he is king. Agnostics may not believe in hell, but Śukadeva Gosvāmī affirms the existence of the Naraka planets, which lie between the Garbhodaka Ocean and Pātālaloka. Yamarāja is appointed by the Supreme

Personality of Godhead to see that the human beings do not violate His rules and regulations...The conditioned souls, who have come to this material world for sense gratification, are allowed to enjoy their senses under certain regulative principles. If they violate these regulations, they are judged and punished by Yamarāja. He brings them to the hellish planets and properly chastises them to bring them back to Kṛṣṇa consciousness. By the influence of māyā, however, the conditioned souls remain infatuated with the mode of ignorance. Thus in spite of repeated punishment by Yamarāja, they do not come to their senses, but continue to live within the material condition, committing sinful activities again and again.

5.26.7: tatra haike narakān eka-viṁśatiṁ gaṇayanti atha tāṁs te rājan nāma-rūpa-lakṣaṇato 'nukramiṣyāmas tāmisro 'ndhatāmisro rauravo mahārauravaḥ kumbhīpākaḥ kālasūtram asipatravanaṁ sūkaramukham andhakūpaḥ kṛmibhojanaḥ sandaṁśas taptasūrmir vajrakaṇṭaka-śālmalī vaitaraṇī pūyodaḥ prāṇarodho viśasanaṁ lālābhakṣaḥ sārameyādanam avīcir ayaḥpānam iti; kiñca kṣārakardamo rakṣogaṇa-bhojanaḥ śūlaproto dandaśūko 'vaṭa-nirodhanaḥ paryāvartanaḥ sūcīmukham ity aṣṭā-viṁśatir narakā vividha-yātanā-bhūmayaḥ.

SYNONYMS: tatra—there; ha—certainly; eke—some; narakān—the hellish planets; eka-viṁśatim—twenty-one; gaṇayanti—count; atha—therefore; tān—them; te—unto you; rājan—O King; nāma-rūpa-lakṣaṇataḥ—according to their names, forms and symptoms; anukramiṣyāmaḥ—we shall outline one after another; tāmisraḥ—Tāmisra; andha-tāmisraḥ—Andhatāmisra; rauravaḥ—Raurava; mahā-rauravaḥ—Mahāraurava; kumbhī-pākaḥ—Kumbhīpāka; kāla-sūtram—Kālasūtra; asi-patravanam—Asi-patravana; sūkara-mukham—Sūkaramukha; andha-kūpaḥ—Andhakūpa; kṛmi-bhojanaḥ—Kṛmibhojana; sandaṁśaḥ—Sandaṁśa; tapta-sūrmiḥ—Taptasūrmi; vajra-kaṇṭaka-śālmalī—Vajrakaṇṭaka-śālmalī; vaitaraṇī—Vaitaraṇī; pūyodaḥ—Pūyoda; prāṇa-rodhaḥ—Prāṇarodha; viśasanam—Viśasana; lālā-bhakṣaḥ—Lālābhakṣa; sārameyādanam—Sārameyādana; avīciḥ—Avīci; ayaḥ-pānam—Ayaḥpāna; iti—thus; kiñca—some more; kṣāra-kardamaḥ—Kṣārakardama; rakṣaḥ-gaṇa-bhojanaḥ—Rakṣogaṇa-bhojana; śūla-protaḥ—Śūlaprota; danda-śūkaḥ—Dandaśūka; avaṭa-nirodhanaḥ—Avaṭa-nirodhana; paryāvartanaḥ—Paryāvartana; sūcī-mukham—Sūcīmukha; iti—in this way; aṣṭā-viṁśatiḥ—twenty-eight; narakāḥ—hellish planets; vividha—various; yātanā-bhūmayaḥ—lands of suffering in hellish conditions.

TRANSLATION: Some authorities say that there is a total of twenty-one hellish planets, and some say twenty-eight. My dear King, I shall outline all of them according to their names, forms and symptoms. The names of the different hells are as follows: Tāmisra, Andhatāmisra, Raurava, Mahāraurava, Kumbhīpāka, Kālasūtra, Asi-patravana, Sūkaramukha, Andhakūpa, Kṛmibhojana, Sandaṁśa, Taptasūrmi, Vajrakaṇṭaka-śālmalī, Vaitaraṇī, Pūyoda, Prāṇarodha, Viśasana,

Lālābhakṣa, Sārameyādana, Avīci, Ayaḥpāna, Kṣārakardama, Rakṣogaṇa-bhojana, Śūlaprota, Dandaśūka, Avaṭa-nirodhana, Paryāvartana and Sūcīmukha. All these planets are meant for punishing the living entities.

5.26.8: tatra yas tu para-vittāpatya-kalatrāṇy apaharati sa hi kāla-pāśa-baddho yama-puruṣair ati-bhayānakais tāmisre narake balān nipātyate anaśanānudapāna-daṇḍa-tāḍana-santarjanādibhir yātanābhir yātyamāno jantur yatra kaśmalam āsādita ekadaiva mūrcchām upayāti tāmisra-prāye.

SYNONYMS: tatra—in those hellish planets; yaḥ—a person who; tu—but; para-vitta-apatya-kalatrāṇi—the money, wife and children of another; apaharati—takes away; saḥ—that person; hi—certainly; kāla-pāśa-baddhaḥ—being bound by the ropes of time or Yamarāja; yama-puruṣaiḥ—by the assistants of Yamarāja; ati-bhayānakaiḥ—who are very fearful; tāmisre narake—into the hell known as Tāmisra; balāt—by force; nipātyate—is thrown; anaśana—starvation; anudapāna—without water; daṇḍa-tāḍana—beaten with rods; santarjana-ādibhiḥ—by scolding and so on; yātanābhiḥ—by severe punishments; yātyamānaḥ—being punished; jantuḥ—the living entity; yatra—where; kaśmalam—misery; āsāditaḥ—obtained; ekadā—sometimes; eva—certainly; mūrcchām—fainting; upayāti—obtains; tāmisra-prāye—in that condition, which is almost entirely dark.

TRANSLATION: My dear King, a person who appropriates another's legitimate wife, children or money is arrested at the time of death by the fierce Yamadūtas, who bind him with the rope of time and forcibly throw him into the hellish planet known as Tāmisra. On this very dark planet, the sinful man is chastised by the Yamadūtas, who beat and rebuke him. He is starved, and he is given no water to drink. Thus the wrathful assistants of Yamarāja cause him severe suffering, and sometimes he faints from their chastisement.

5.26.9: evam evāndhatāmisre yas tu vañcayitvā puruṣaṁ dārādīn upayuṅkte yatra śarīrī nipātyamāno yātanā-stho vedanayā naṣṭa-matir naṣṭa-dṛṣṭiś ca bhavati yathā vanaspatir vṛścyamāna-mūlas tasmād andhatāmisraṁ tam upadiśanti.

SYNONYMS: evam—in this way; eva—certainly; andhatāmisre—in the hellish planet known as Andhatāmisra; yaḥ—the person who; tu—but; vañcayitvā—cheating; puruṣam—another person; dāra-ādīn—the wife and children; upayuṅkte—enjoys; yatra—where; śarīrī—the embodied person; nipātyamānaḥ—being forcibly thrown; yātanā-sthaḥ—always situated in extremely miserable conditions; vedanayā—by such suffering; naṣṭa—lost; matiḥ—whose consciousness; naṣṭa—lost; dṛṣṭiḥ—whose sight; ca—also; bhavati—becomes; yathā—as much as; vanaspatiḥ—the trees; vṛścyamāna—being cut; mūlaḥ—whose root; tasmāt—because of this; andhatāmisram—Andhatāmisra; tam—that; upadiśanti—they call.

TRANSLATION: The destination of a person who slyly cheats another man and enjoys his wife and children is the hell known as Andhatāmisra. There his condition is exactly like that of a tree being chopped at its roots. Even before reaching Andhatāmisra, the sinful living being is subjected to various extreme miseries. These afflictions are so severe that he loses his intelligence and sight. It is for this reason that learned sages call this hell Andhatāmisra.

5.26.10: yas tv iha vā etad aham iti mamedam iti bhūta-droheṇa kevalaṁ sva-kuṭumbam evānudinaṁ prapuṣṇāti sa tad iha vihāya svayam eva tad-aśubhena raurave nipatati.

SYNONYMS: yaḥ—one who; tu—but; iha—in this life; vā—or; etat—this body; aham—I; iti—thus; mama—mine; idam—this; iti—thus; bhūta-droheṇa—by envy of other living entities; kevalam—alone; sva-kuṭumbam—his family members; eva—only; anudinam—day to day; prapuṣṇāti—supports; saḥ—such a person; tat—that; iha—here; vihāya—giving up; svayam—personally; eva—certainly; tat—of that; aśubhena—by the sin; raurave—in Raurava; nipatati—he falls down.

TRANSLATION: A person who accepts his body as his self works very hard day and night for money to maintain his own body and the bodies of his wife and children. While working to maintain himself and his family, he may commit violence against other living entities. Such a person is forced to give up his body and his family at the time of death, when he suffers the reaction for his envy of other creatures by being thrown into the hell called Raurava.

PURPORT: If one simply considers his body to be his self, as do the animals, he is not very sinful. However, if one needlessly commits sins to maintain his body, he is put into the hell known as Raurava. This is the opinion of Śrīla Viśvanātha Cakravartī Ṭhākura. Although animals are certainly in the bodily concept of life, they do not commit any sins to maintain their bodies, mates or offspring. Therefore animals do not go to hell. However, when a human being acts enviously and cheats others to maintain his body, he is put into a hellish condition.

5.26.11: ye tv iha yathaivāmunā vihiṁsitā jantavaḥ paratra yama-yātanām upagataṁ ta eva ruravo bhūtvā tathā tam eva vihiṁsanti tasmād rauravam ity āhū rurur iti sarpād ati-krūra-sattvasyāpadeśaḥ.

SYNONYMS: ye—those who; tu—but; iha—in this life; yathā—as much as; eva—certainly; amunā—by him; vihiṁsitāḥ—who were hurt; jantavaḥ—the living entities; paratra—in the next life; yama-yātanām upagatam—being subjected to miserable conditions by Yamarāja; te—those living entities; eva—indeed;

ruravaḥ—rurus (a kind of envious animal); bhūtvā—becoming; tathā—that much; tam—him; eva—certainly; vihiṁsanti—they hurt; tasmāt—because of this; rauravam—Raurava; iti—thus; āhuḥ—learned scholars say; ruruḥ—the animal known as ruru; iti—thus; sarpāt—than the snake; ati-krūra—much more cruel and envious; sattvasya—of the entity; apadeśaḥ—the name.

TRANSLATION: In this life, an envious person commits violent acts against many living entities. Therefore after his death, when he is taken to hell by Yamarāja, those living entities who were hurt by him appear as animals called rurus to inflict very severe pain upon him. Learned scholars call this hell Raurava. Not generally seen in this world, the ruru is more envious than a snake.

PURPORT: According to Śrīdhara Svāmī, the ruru is also known as the bhāra-śṛṅga (ati-krūrasya bhāra-śṛṅgākhya-sattvasya apadeśaḥ saṁjñā). Śrīla Jīva Gosvāmī confirms this in his Sandarbha: ruru-śabdasya svayaṁ muninaiva ṭīkā-vidhānāl lokeṣv aprasiddha evāyaṁ jantu-viśeṣaḥ. Thus although rurus are not seen in this world, their existence is confirmed in the śāstras.

5.26.12: evam eva mahārauravo yatra nipatitaṁ puruṣaṁ kravyādā nāma ruravas taṁ kravyeṇa ghātayanti yaḥ kevalaṁ dehambharaḥ.

SYNONYMS: evam—thus; eva—certainly; mahā-rauravaḥ—the hell known as Mahāraurava; yatra—where; nipatitam—being thrown; puruṣam—a person; kravyādāḥ nāma—named kravyāda; ruravaḥ—the ruru animals; tam—him (the condemned person); kravyeṇa—for eating his flesh; ghātayanti—kill; yaḥ—who; kevalam—only; dehambharaḥ—intent upon maintaining his own body.

TRANSLATION: Punishment in the hell called Mahāraurava is compulsory for a person who maintains his own body by hurting others. In this hell, ruru animals known as kravyāda torment him and eat his flesh.

5.26.13: yas tv iha vā ugraḥ paśūn pakṣiṇo vā prāṇata uparandhayati tam apakaruṇaṁ puruṣādair api vigarhitam amutra yamānucarāḥ kumbhīpāke tapta-taile uparandhayanti.

SYNONYMS: yaḥ—a person who; tu—but; iha—in this life; vā—or; ugraḥ—very cruel; paśūn—animals; pakṣiṇaḥ—birds; vā—or; prāṇataḥ—in a live condition; uparandhayati—cooks; tam—him; apakaruṇam—very cruel-hearted; puruṣa-ādaiḥ—by those who eat human flesh; api—even; vigarhitam—condemned; amutra—in the next life; yama-anucarāḥ—the servants of Yamarāja; kumbhīpāke—in the hell known as Kumbhīpāka; tapta-taile—in boiling oil; uparandhayanti—cook.

TRANSLATION: For the maintenance of their bodies and the satisfaction of their tongues, cruel persons cook poor animals and birds alive. Such persons are condemned even by man-eaters. In their next lives they are carried by the Yamadūtas to the hell known as Kumbhīpāka, where they are cooked in boiling oil.

5.26.14: yas tv iha brahma-dhruk sa kālasūtra-saṁjñake narake ayuta-yojana-parimaṇḍale tāmramaye tapta-khale upary-adhastād agny-arkābhyām ati-tapyamāne 'bhiniveśitaḥ kṣut-pipāsābhyāṁ ca dahyamānāntar-bahiḥ-śarīra āste śete ceṣṭate 'vatiṣṭhati paridhāvati ca yāvanti paśu-romāṇi tāvad varṣa-sahasrāṇi.

SYNONYMS: yaḥ—anyone who; tu—but; iha—in this life; brahma-dhruk—the killer of a brāhmaṇa; saḥ—such a person; kālasūtra-saṁjñake—named Kālasūtra; narake—in the hell; ayuta-yojana-parimaṇḍale—having a circumference of eighty thousand miles; tāmra-maye—made of copper; tapta—heated; khale—in a level place; upari-adhastāt—above and beneath; agni—by fire; arkābhyām—and by the sun; ati-tapyamāne—which is being heated; abhiniveśitaḥ—being made to enter; kṣut-pipāsābhyām—by hunger and thirst; ca—and; dahyamāna—being burned; antaḥ—internally; bahiḥ—externally; śarīraḥ—whose body; āste—remains; śete—sometimes lies; ceṣṭate—sometimes moves his limbs; avatiṣṭhati—sometimes stands; paridhāvati—sometimes runs here and there; ca—also; yāvanti—as many; paśu-romāṇi—hairs on the body of an animal; tāvat—that long; varṣa-sahasrāṇi—thousands of years.

TRANSLATION: The killer of a brāhmaṇa is put into the hell known as Kālasūtra, which has a circumference of eighty thousand miles and which is made entirely of copper. Heated from below by fire and from above by the scorching sun, the copper surface of this planet is extremely hot. Thus the murderer of a brāhmaṇa suffers from being burned both internally and externally. Internally he is burning with hunger and thirst, and externally he is burning from the scorching heat of the sun and the fire beneath the copper surface. Therefore he sometimes lies down, sometimes sits, sometimes stands up and sometimes runs here and there. He must suffer in this way for as many thousands of years as there are hairs on the body of an animal.

5.26.15: yas tv iha vai nija-veda-pathād anāpady apagataḥ pākhaṇḍaṁ copagatas tam asi-patravanaṁ praveśya kaśayā praharanti tatra hāsāv itas tato dhāvamāna ubhayato dhārais tāla-vanāsi-patraiś chidyamāna-sarvāṅgo hā hato 'smīti paramayā vedanayā mūrcchitaḥ pade pade nipatati sva-dharmahā pākhaṇḍānugataṁ phalaṁ bhuṅkte.

SYNONYMS: yaḥ—anyone who; tu—but; iha—in this life; vai—indeed; nija-veda-pathāt—from his own path, recommended by the Vedas; anāpadi—even without an emergency; apagataḥ—deviated; pākhaṇḍam—a concocted, atheistic system; ca—and; upagataḥ—gone to; tam—him; asi-patravanam—the hell known as Asi-patravana; praveśya—making enter; kaśayā—with a whip; praharanti—they beat; tatra—there; ha—certainly; asau—that; itaḥ tataḥ—here and there; dhāvamānaḥ—running; ubhayataḥ—on both sides; dhāraiḥ—by the edges; tāla-vana-asi-patraiḥ—by the swordlike leaves of palm trees; chidyamāna—being cut; sarva-aṅgaḥ—whose entire body; hā—alas; hataḥ—killed; asmi—I am; iti—thus; paramayā—with severe; vedanayā—pain; mūrcchitaḥ—fainted; pade pade—at every step; nipatati—falls down; sva-dharma-hā—the killer of his own principles of religion; pakhaṇḍa-anugatam phalam—the result of accepting an atheistic path; bhuṅkte—he suffers.

TRANSLATION: If a person deviates from the path of the Vedas in the absence of an emergency, the servants of Yamarāja put him into the hell called Asi-patravana, where they beat him with whips. When he runs hither and thither, fleeing from the extreme pain, on all sides he runs into palm trees with leaves like sharpened swords. Thus injured all over his body and fainting at every step, he cries out, "Oh, what shall I do now! How shall I be saved!" This is how one suffers who deviates from the accepted religious principles.

PURPORT: As Kali-yuga advances, people are becoming godless and taking up so-called secularism. They do not know the punishment awaiting them in Asi-patravana, as described in this verse.

5.26.16: yas tv iha vai rājā rāja-puruṣo vā adaṇḍye daṇḍaṁ praṇayati brāhmaṇe vā śarīra-daṇḍaṁ sa pāpīyān narake 'mutra sūkaramukhe nipatati tatrātibalair viniṣpiṣyamāṇāvayavo yathaiveheksukhaṇḍa ārta-svareṇa svanayan kvacin mūrcchitaḥ kaśmalam upagato yathaivehā-dṛṣṭa-doṣā uparuddhāḥ.

SYNONYMS: yaḥ—anyone who; tu—but; iha—in this life; vai—indeed; rājā—a king; rāja-puruṣaḥ—a king's man; vā—or; adaṇḍye—unto one not punishable; daṇḍam—punishment; praṇayati—inflicts; brāhmaṇe—unto a brāhmaṇa; vā—or; śarīra-daṇḍam—corporal punishment; saḥ—that person, king or government officer; pāpīyān—the most sinful; narake—in the hell; amutra—in the next life; sūkaramukhe—named Sūkaramukha; nipatati—falls down; tatra—there; ati-balaiḥ—by very strong assistants of Yamarāja; viniṣpiṣyamāṇa—being crushed; avayavaḥ—the different parts of whose body; yathā—like; eva—certainly; iha—here; ikṣu-khaṇḍaḥ—sugarcane; ārta-svareṇa—with a pitiable sound; svanayan—crying; kvacit—sometimes; mūrcchitaḥ—fainted; kaśmalam upagataḥ—becoming illusioned; yathā—just like; eva—indeed; iha—here; adṛṣṭa-doṣāḥ—who is not at fault; uparuddhāḥ—arrested for punishment.

TRANSLATION: In his next life, a sinful king or governmental representative who punishes an innocent person, or who inflicts corporal punishment upon a brāhmaṇa, is taken by the Yamadūtas to the hell named Sūkaramukha, where the most powerful assistants of Yamarāja crush him exactly as one crushes sugarcane to squeeze out the juice. The sinful living entity cries very pitiably and faints, just like an innocent man undergoing punishments. This is the result of punishing a faultless person.

5.26.17: yas tv iha vai bhūtānām īśvaropakalpita-vṛttīnām avivikta-para-vyathānāṁ svayaṁ puruṣopakalpita-vṛttir vivikta-para-vyatho vyathām ācarati sa paratrāndhakūpe tad-abhidroheṇa nipatati tatra hāsau tair jantubhiḥ paśu-mṛga-pakṣi-sarīsṛpair maśaka-yūkā-matkuṇa-makṣikādibhir ye ke cābhidrugdhās taiḥ sarvato 'bhidruhyamāṇas tamasi vihata-nidrā-nirvṛtir alabdhāvasthānaḥ parikrāmati yathā kuśarīre jīvaḥ.

SYNONYMS: yaḥ—any person who; tu—but; iha—in this life; vai—indeed; bhūtānām—to some living entities; īśvara—by the supreme controller; upakalpita—designed; vṛttīnām—whose means of livelihood; avivikta—not understanding; para-vyathānām—the pain of others; svayam—himself; puruṣa-upakalpita—designed by the Supreme Personality of Godhead; vṛttiḥ—whose livelihood; vivikta—understanding; para-vyathaḥ—the painful conditions of others; vyathām ācarati—but still causes pain; saḥ—such a person; paratra—in his next life; andhakūpe—to the hell named Andhakūpa; tat—to them; abhidroheṇa—by the sin of malice; nipatati—falls down; tatra—there; ha—indeed; asau—that person; taiḥ jantubhiḥ—by those respective living entities; paśu—animals; mṛga—wild bcasts; pakṣi—birds; sarīsṛpaiḥ—snakcs; maśaka—mosquitoes; yūkā—lice; matkuṇa—worms; makṣika-ādibhiḥ—flies and so on; ye ke—whoever else; ca—and; abhidrugdhāḥ—persecuted; taiḥ—by them; sarvataḥ—everywhere; abhidruhyamāṇaḥ—being injured; tamasi—in the darkness; vihata—disturbed; nidrā-nirvṛtiḥ—whose resting place; alabdha—not being able to obtain; avasthānaḥ—a resting place; parikrāmati—wanders; yathā—just as; ku-śarıre—in a low-grade body; jıvaḥ—a living entity.

TRANSLATION: By the arrangement of the Supreme Lord, low-grade living beings like bugs and mosquitoes suck the blood of human beings and other animals. Such insignificant creatures are unaware that their bites are painful to the human being. However, first-class human beings—brāhmaṇas, kṣatriyas and vaiśyas—are developed in consciousness, and therefore they know how painful it is to be killed. A human being endowed with knowledge certainly commits sin if he kills or torments insignificant creatures, who have no discrimination. The Supreme Lord punishes such a man by putting him into the hell known as Andhakūpa, where he is attacked by all the birds and beasts, reptiles, mosquitoes, lice, worms, flies, and any other creatures he tormented during his life. They attack him from all sides, robbing him of the pleasure of sleep. Unable to rest, he constantly wanders about in the darkness. Thus in Andhakūpa his suffering is just like that of a creature in the lower species.

5.26.18: yas tv iha vā asaṁvibhajyāśnāti yat kiñcanopanatam anirmita-pañca-yajño vāyasa-saṁstutaḥ sa paratra kṛmibhojane narakādhame nipatati tatra śata-sahasra-yojane kṛmi-kuṇḍe kṛmi-bhūtaḥ svayaṁ kṛmibhir eva bhakṣyamāṇaḥ kṛmi-bhojano yāvat tad aprattāprahūtādo 'nirveśam ātmānaṁ yātayate.

SYNONYMS: yaḥ—any person who; tu—but; iha—in this life; vā—or; asaṁvibhajya—without dividing; aśnāti—eats; yat kiñcana—whatever; upanatam—obtained by Kṛṣṇa's grace; anirmita—not performing; pañca-yajñaḥ—the five kinds of sacrifice; vāyasa—with the crows; saṁstutaḥ—who is described as equal; saḥ—such a person; paratra—in the next life; kṛmibhojane—named Kṛmibhojana; naraka-adhame—into the most abominable of all hells; nipatati—falls down; tatra—there; śata-sahasra-yojane—measuring 100,000 yojanas (800,000 miles); kṛmi-kuṇḍe—in a lake of worms; kṛmi-bhūtaḥ—becoming one of the worms; svayam—he himself; kṛmibhiḥ—by the other worms; eva—certainly; bhakṣyamāṇaḥ—being eaten; kṛmi-bhojanaḥ—eating worms; yāvat—as long as; tat—that lake is wide; apratta-aprahūta—unshared and unoffered food; adaḥ—one who eats; anirveśam—who has not performed atonement; ātmānam—to himself; yātayate—gives pain.

TRANSLATION: A person is considered no better than a crow if after receiving some food, he does not divide it among guests, old men and children, but simply eats it himself, or if he eats it without performing the five kinds of sacrifice. After death he is put into the most abominable hell, known as Kṛmibhojana. In that hell is a lake 100,000 yojanas [800,000 miles] wide and filled with worms. He becomes a worm in that lake and feeds on the other worms there, who also feed on him. Unless he atones for his actions before his death, such a sinful man remains in the hellish lake of Kṛmibhojana for as many years as there are yojanas in the width of the lake.

PURPORT: If we do not perform yajña and distribute prasāda to others, our lives are condemned. Only after performing yajña and distributing the prasāda to all dependents—children, brāhmaṇas and old men—should one eat. However, one who cooks only for himself or his family is condemned, along with everyone he feeds. After death he is put into the hell known as Kṛmibhojana.

5.26.19: yas tv iha vai steyena balād vā hiraṇya-ratnādīni brāhmaṇasya vāpaharaty anyasya vānāpadi puruṣas tam amutra rājan yama-puruṣā ayasmayair agni-piṇḍaiḥ sandaṁśais tvaci niṣkuṣanti.

SYNONYMS: yaḥ—any person who; tu—but; iha—in this life; vai—indeed; steyena—by thievery; balāt—by force; vā—or; hiraṇya—gold; ratna—gems; ādīni—and so on; brāhmaṇasya—of a brāhmaṇa; vā—or; apaharati—steals; anyasya—of others; vā—or; anāpadi—not in a calamity; puruṣaḥ—a person; tam—him; amutra—in the next life; rājan—O King; yama-puruṣāḥ—the agents of Yamarāja; ayaḥ-mayaiḥ—made of iron; agni-piṇḍaiḥ—balls heated in fire; sandaṁśaiḥ—with tongs; tvaci—on the skin; niṣkuṣanti—tear to pieces.

TRANSLATION: My dear King, a person who in the absence of an emergency robs a brāhmaṇa—or, indeed, anyone else—of his gems and gold is put into a hell known as Sandaṁśa. There his skin is torn and separated by red-hot iron balls and tongs. In this way, his entire body is cut to pieces.

5.26.20: yas tv iha vā agamyāṁ striyam agamyaṁ vā puruṣaṁ yoṣid abhigacchati
tāv amutra kaśayā tāḍayantas tigmayā sūrmyā lohamayyā puruṣam
āliṅgayanti striyaṁ ca puruṣa-rūpayā sūrmyā.

SYNONYMS: yaḥ—any person who; tu—but; iha—in this life; vā—or; agamyām—unsuitable; striyam—a woman; agamyam—unsuitable; vā—or; puruṣam—a man; yoṣit—a woman; abhigacchati—approaches for sexual intercourse; tau—both of them; amutra—in the next life; kaśayā—by whips; tāḍayantaḥ—beating; tigmayā—very hot; sūrmyā—by an image; loha-mayyā—made of iron; puruṣam—the man; āliṅgayanti—they embrace; striyam—the woman; ca—also; puruṣa-rūpayā—in the form of a man; sūrmyā—by an image.

TRANSLATION: A man or woman who indulges in sexual intercourse with an unworthy member of the opposite sex is punished after death by the assistants of Yamarāja in the hell known as Taptasūrmi. There such men and women are beaten with whips. The man is forced to embrace a red-hot iron form of a woman, and the woman is forced to embrace a similar form of a man. Such is the punishment for illicit sex.

5.26.21: yas tv iha vai sarvābhigamas tam amutra niraye vartamānaṁ
vajrakaṇṭaka-śālmalīm āropya niṣkarṣanti.

SYNONYMS: yaḥ—anyone who; tu—but; iha—in this life; vai—indeed; sarva-abhigamaḥ—indulges in sex life indiscriminately, with both men and animals; tam—him; amutra—in the next life; niraye—in the hell; vartamānam—existing; vajrakaṇṭaka-śālmalīm—a silk-cotton tree with thorns like thunderbolts; āropya—mounting him on; niṣkarṣanti—they pull him out.

TRANSLATION: A person who indulges in sex indiscriminately—even with animals—is taken after death to the hell known as Vajrakaṇṭaka-śālmalī. In this hell there is a silk-cotton tree full of thorns as strong as thunderbolts. The agents of Yamarāja hang the sinful man on that tree and pull him down forcibly so that the thorns very severely tear his body.

PURPORT: Sometimes people disbelieve these descriptions of hell, but whether one believes or not, everything must be carried out by the laws of nature, which no one can avoid.

5.26.22: ye tv iha vai rājanyā rāja-puruṣā vā apākhaṇḍā dharma-setūn bhindanti te samparetya vaitaraṇyāṁ nipatanti bhinna-maryādās tasyāṁ niraya-parikhā-bhūtāyāṁ nadyāṁ yādo-gaṇair itas tato bhakṣyamāṇā ātmanā na viyujyamānāś cāsubhir uhyamānāḥ svāghena karma-pākam anusmaranto viṇ-mūtra-pūya-śoṇita-keśa-nakhāsthi-medo-māṁsa-vasā-vāhinyām upatapyante.

SYNONYMS: ye—persons who; tu—but; iha—in this life; vai—indeed; rājanyāḥ—members of the royal family, or kṣatriyas; rāja-puruṣāḥ—government servants; vā—or; apākhaṇḍāḥ—although born in responsible families; dharma-setūn—the bounds of prescribed religious principles; bhindanti—transgress; te—they; samparetya—after dying; vaitaraṇyām—named Vaitaraṇī; nipatanti—fall down; bhinna-maryādāḥ—who have broken the regulative principles; tasyām—in that; niraya-parikhā-bhūtāyām—the moat surrounding hell; nadyām—in the river; yādaḥ-gaṇaiḥ—by ferocious aquatic animals; itaḥ tataḥ—here and there; bhakṣyamāṇāḥ—being eaten; ātmanā—with the body; na—not; viyujyamānāḥ—being separated; ca—and; asubhiḥ—the life airs; uhyamānāḥ—being carried; sva-aghena—by his own sinful activities; karma-pākam—the result of his impious activities; anusmarantaḥ—remembering; viṭ—of stool; mūtra—urine; pūya—pus; śoṇita—blood; keśa—hair; nakha—nails; asthi—bones; medaḥ—marrow; māṁsa—flesh; vasā—fat; vāhinyām—in the river; upatapyante—are afflicted with pain.

TRANSLATION: A person who is born into a responsible family—such as a kṣatriya, a member of royalty or a government servant—but who neglects to execute his prescribed duties according to religious principles, and who thus becomes degraded, falls down at the time of death into the river of hell known as Vaitaraṇī. This river, which is a moat surrounding hell, is full of ferocious aquatic animals. When a sinful man is thrown into the River Vaitaraṇī, the aquatic animals there immediately begin to eat him, but because of his extremely sinful life, he does not leave his body. He constantly remembers his sinful activities and suffers terribly in that river, which is full of stool, urine, pus, blood, hair, nails, bones, marrow, flesh and fat.

5.26.23: ye tv iha vai vṛṣalī-patayo naṣṭa-śaucācāra-niyamās tyakta-lajjāḥ paśu-caryāṁ caranti te cāpi pretya pūya-viṇ-mūtra-śleṣma-malā-pūrṇārṇave nipatanti tad evātibībhatsitam aśnanti.

SYNONYMS: ye—persons who; tu—but; iha—in this life; vai—indeed; vṛṣalī-patayaḥ—the husbands of the śūdras; naṣṭa—lost; śauca-ācāra-niyamāḥ—whose cleanliness, good behavior and regulated life; tyakta-lajjāḥ—without shame; paśu-caryām—the behavior of animals; caranti—they execute; te—they; ca—also; api—indeed; pretya—dying; pūya—of pus; viṭ—stool; mūtra—urine; śleṣma—mucus; malā—saliva; pūrṇa—full; arṇave—in an ocean; nipatanti—fall; tat—that; eva—only; atibībhatsitam—extremely disgusting; aśnanti—they eat.

TRANSLATION: The shameless husbands of lowborn śūdra women live exactly like animals, and therefore they have no good behavior, cleanliness or regulated life. After death, such persons are thrown into the hell called Pūyoda, where they are put into an ocean filled with pus, stool, urine, mucus, saliva and similar things. Śūdras who could not improve themselves fall into that ocean and are forced to eat those disgusting things.

PURPORT: Śrīla Narottama dāsa Ṭhākura has sung, karma-kāṇḍa, jñāna-kāṇḍa,/ kevala viṣera bāṇḍa,/amṛta baliyā yebā khāya/nānā yoni sadā phire, /kadarya bhakṣaṇa kare,/tāra janma adaḥ-pate yāya// He says that persons following the paths of karma-kāṇḍa and jñāna-kāṇḍa (fruitive activities and speculative thinking) are missing the opportunities for human birth and gliding down into the cycle of birth and death. Thus there is always the chance that he may be put into the Pūyoda Naraka, the hell named Pūyoda, where one is forced to eat stool, urine, pus, mucus, saliva and other abominable things. It is significant that this verse is spoken especially about śūdras. If one is born a śūdra, he must continually return to the ocean of Pūyoda to eat horrible things. Thus even a born śūdra is expected to become a brāhmaṇa; that is the meaning of human life.

5.26.24: ye tv iha vai śva-gardabha-patayo brāhmaṇādayo mṛgayā vihārā atīrthe ca mṛgān nighnanti tān api samparetāû lakṣya-bhūtān yama-puruṣā iṣubhir vidhyanti.

SYNONYMS: ye—those who; tu—but; iha—in this life; vai—or; śva—of dogs; gardabha—and asses; patayaḥ—maintainers; brāhmaṇa-ādayaḥ—brāhmaṇas, kṣatriyas and vaiśyas; mṛgayā vihārāḥ—taking pleasure in hunting animals in the forest; atīrthe—other than prescribed; ca—also; mṛgān—animals; nighnanti—kill; tān—them; api—indeed; samparetān—having died; lakṣya-bhūtān—becoming the targets; yama-puruṣāḥ—the assistants of Yamarāja; iṣubhiḥ—by arrows; vidhyanti—pierce.

TRANSLATION: If in this life a man of the higher classes [brāhmaṇa, kṣatriya and vaiśya] is very fond of taking his pet dogs, mules or asses into the forest to hunt and kill animals unnecessarily, he is placed after death into the hell known as

Prāṇarodha. There the assistants of Yamarāja make him their targets and pierce him with arrows.

PURPORT: In the Western countries especially, aristocrats keep dogs and horses to hunt animals in the forest. Whether in the West or the East, aristocratic men in the Kali-yuga adopt the fashion of going to the forest and unnecessarily killing animals. Men of the higher classes (the brāhmaṇas, kṣatriyas and vaiśyas) should cultivate knowledge of Brahman, and they should also give the śūdras a chance to come to that platform. If instead they indulge in hunting, they are punished as described in this verse. Not only are they pierced with arrows by the agents of Yamarāja, but they are also put into the ocean of pus, urine and stool described in the previous verse.

5.26.25: ye tv iha vai dāmbhikā dambha-yajñeṣu paśūn viśasanti tān amuṣmiû loke vaiśase narake patitān niraya-patayo yātayitvā viśasanti.

SYNONYMS: ye—persons who; tu—but; iha—in this life; vai—indeed; dāmbhikāḥ—very proud of wealth and a prestigious position; dambha-yajñeṣu—in a sacrifice performed to increase prestige; paśūn—animals; viśasanti—kill; tān—them; amuṣmin loke—in the next world; vaiśase—Vaiśasa or Viśasana; narake—into the hell; patitān—fallen; niraya-patayaḥ—assistants of Yamarāja; yātayitvā—causing sufficient pain; viśasanti—kill.

TRANSLATION: A person who in this life is proud of his eminent position, and who heedlessly sacrifices animals simply for material prestige, is put into the hell called Viśasana after death. There the assistants of Yamarāja kill him after giving him unlimited pain.

5.26.26: yas tv iha vai savarṇāṁ bhāryāṁ dvijo retaḥ pāyayati kāma-mohitas taṁ pāpa-kṛtam amutra retaḥ-kulyāyāṁ pātayitvā retaḥ sampāyayanti.

SYNONYMS: yaḥ—any person who; tu—but; iha—in this life; vai—indeed; savarṇām—of the same caste; bhāryām—his wife; dvijaḥ—a person of a higher caste (such as a brāhmaṇa, kṣatriya or vaiśya); retaḥ—the semen; pāyayati—causes to drink; kāma-mohitaḥ—being deluded by lusty desires; tam—him; pāpa-kṛtam—performing sin; amutra—in the next life; retaḥ-kulyāyām—in a river of semen; pātayitvā—throwing; retaḥ—semen; sampāyayanti—force to drink.

TRANSLATION: If a foolish member of the twice-born classes [brāhmaṇa, kṣatriya and vaiśya] forces his wife to drink his semen out of a lusty desire to keep her under control, he is put after death into the hell known as Lālābhakṣa. There he is thrown into a flowing river of semen, which he is forced to drink.

PURPORT: Generally only low-class men engage in this black art, but if a man born in a higher class does so, after death he is put into the hell known as Lālābhakṣa. There he is immersed in the river known as Śukra-nadī and forced to drink semen.

5.26.27: ye tv iha vai dasyavo 'gnidā garadā grāmān sārthān vā vilumpanti rājāno rāja-bhaṭā vā tāṁś cāpi hi paretya yamadūtā vajra-daṁṣṭrāḥ śvānaḥ sapta-śatāni viṁśatiś ca sarabhasaṁ khādanti.

SYNONYMS: ye—persons who; tu—but; iha—in this life; vai—indeed; dasyavaḥ—thieves and plunderers; agni-dāḥ—who set fire; garadāḥ—who administer poison; grāmān—villages; sārthān—the mercantile class of men; vā—or; vilumpanti—plunder; rājānaḥ—kings; rāja-bhaṭāḥ—government officials; vā—or; tān—them; ca—also; api—indeed; hi—certainly; paretya—having died; yamadūtāḥ—the assistants of Yamarāja; vajra-daṁṣṭrāḥ—having mighty teeth; śvānaḥ—dogs; sapta-śatāni—seven hundred; viṁśatiḥ—twenty; ca—and; sarabhasam—voraciously; khādanti—devour.

TRANSLATION: In this world, some persons are professional plunderers who set fire to others' houses or administer poison to them. Also, members of the royalty or government officials sometimes plunder mercantile men by forcing them to pay income tax and by other methods. After death such demons are put into the hell known as Sārameyādana. On that planet there are 720 dogs with teeth as strong as thunderbolts. Under the orders of the agents of Yamarāja, these dogs voraciously devour such sinful people.

PURPORT: Whether a highway robber or a government thief, such a man will be punished in his next life by being thrown into the hell known as Sārameyādana, where he will suffer greatly from the bites of ferocious dogs.

5.26.28: yas tv iha vā anṛtaṁ vadati sākṣye dravya-vinimaye dāne vā kathañcit sa vai pretya narake 'vīcimaty adhaḥ-śirā niravakāśe yojana-śatocchrāyād giri-mūrdhnaḥ sampātyate yatra jalam iva sthalam aśma-pṛṣṭham avabhāsate tad avīcimat tilaśo viśīryamāṇa-śarīro na mriyamāṇaḥ punar āropito nipatati.

SYNONYMS: yaḥ—anyone who; tu—but; iha—in this life; vā—or; anṛtam—a lie; vadati—speaks; sākṣye—giving witness; dravya-vinimaye—in exchange for goods; dāne—in giving charity; vā—or; kathañcit—somehow; saḥ—that person; vai—indeed; pretya—after dying; narake—in the hell; avīcimati—named Avīcimat (having no water); adhaḥ-śirāḥ—with his head downward; niravakāśe—without support; yojana-śata—of eight hundred miles; ucchrāyāt—having a height; giri—of a mountain; mūrdhnaḥ—from the top; sampātyate—is thrown; yatra—where; jalam iva—like water; sthalam—land; aśma-pṛṣṭham—having a surface of stone;

avabhāsate—appears; tat—that; avīcimat—having no water or waves; tilaśaḥ—in pieces as small as seeds; viśīryamāṇa—being broken; śarīraḥ—the body; na mriyamāṇaḥ—not dying; punaḥ—again; āropitaḥ—raised to the top; nipatati—falls down.

TRANSLATION: A person who in this life bears false witness or lies while transacting business or giving charity is severely punished after death by the agents of Yamarāja. Such a sinful man is taken to the top of a mountain eight hundred miles high and thrown headfirst into the hell known as Avīcimat. This hell has no shelter and is made of strong stone resembling the waves of water. There is no water there, however, and thus it is called Avīcimat [waterless]. Although the sinful man is repeatedly thrown from the mountain and his body broken to tiny pieces, he still does not die but continuously suffers chastisement.

5.26.29: yas tv iha vai vipro rājanyo vaiśyo vā soma-pīthas tat-kalatraṁ vā surāṁ vrata-stho 'pi vā pibati pramādatas teṣāṁ nirayaṁ nītānām urasi padākramyāsye vahninā dravamāṇaṁ kārṣṇāyasaṁ niṣiñcanti.

SYNONYMS: yaḥ—anyone who; tu—but; iha—in this life; vai—indeed; vipraḥ—a learned brāhmaṇa; rājanyaḥ—a kṣatriya; vaiśyaḥ—a vaiśya; vā—or; soma-pīthaḥ—drink soma-rasa; tat—his; kalatram—wife; vā—or; surām—liquor; vrata-sthaḥ—being situated in a vow; api—certainly; vā—or; pibati—drinks; pramādataḥ—out of illusion; teṣām—of all of them; nirayam—to hell; nītānām—being brought; urasi—on the chest; padā—with the foot; ākramya—stepping; asye—in the mouth; vahninā—by fire; dravamāṇam—melted; kārṣṇāyasam—iron; niṣiñcanti—they pour into.

TRANSLATION: Any brāhmaṇa or brāhmaṇa's wife who drinks liquor is taken by the agents of Yamarāja to the hell known as Ayaḥpāna. This hell also awaits any kṣatriya, vaiśya, or person under a vow who in illusion drinks soma-rasa. In Ayaḥpāna the agents of Yamarāja stand on their chests and pour hot melted iron into their mouths.

5.26.30: atha ca yas tv iha vā ātma-sambhāvanena svayam adhamo janma-tapo-vidyācāra-varṇāśramavato varīyaso na bahu manyeta sa mṛtaka eva mṛtvā kṣārakardame niraye 'vāk-śirā nipātito durantā yātanā hy aśnute.

SYNONYMS: atha—furthermore; ca—also; yaḥ—anyone who; tu—but; iha—in this life; vā—or; ātma-sambhāvanena—by false prestige; svayam—himself; adhamaḥ—very degraded; janma—good birth; tapaḥ—austerities; vidyā—knowledge; ācāra—good behavior; varṇa-āśrama-vataḥ—in terms of strictly following the principles of varṇāśrama; varīyasaḥ—of one who is more honorable; na—not; bahu—much; manyeta—respects; saḥ—he; mṛtakaḥ—a dead body; eva—only; mṛtvā—after dying; kṣārakardame—named Kṣārakardama; niraye—in the hell; avāk-śirā—with his head downward; nipātitaḥ—thrown; durantāḥ yātanāḥ—severe painful conditions; hi—indeed; aśnute—suffers.

TRANSLATION: A lowborn and abominable person who in this life becomes falsely proud, thinking "I am great," and who thus fails to show proper respect to one more elevated than he by birth, austerity, education, behavior, caste or spiritual order, is like a dead man even in this lifetime, and after death he is thrown headfirst into the hell known as Kṣārakardama. There he must great suffer great tribulation at the hands of the agents of Yamarāja.

PURPORT: One should not become falsely proud. One must be respectful toward a person more elevated than he by birth, education, behavior, caste or spiritual order. If one does not show respect to such highly elevated persons but indulges in false pride, he receives punishment in Kṣārakardama.

5.26.31: ye tv iha vai puruṣāḥ puruṣa-medhena yajante yāś ca striyo nṛ-paśūn khādanti tāṁś ca te paśava iva nihatā yama-sadane yātayanto rakṣo-gaṇāḥ saunikā iva svadhitināvadāyāsṛk pibanti nṛtyanti ca gāyanti ca hṛṣyamāṇā yatheha puruṣādāḥ.

SYNONYMS: ye—persons who; tu—but; iha—in this life; vai—indeed; puruṣāḥ—men; puruṣa-medhena—by sacrifice of a man; yajante—worship (the goddess Kālī or Bhadra Kālī); yāḥ—those who; ca—and; striyaḥ—women; nṛ-paśūn—the men used as sacrifice; khādanti—eat; tān—them; ca—and; te—they; paśavaḥ iva—like the animals; nihatāḥ—being slain; yama-sadane—in the abode of Yamarāja; yātayantaḥ—punishing; rakṣaḥ-gaṇāḥ—being Rākṣasas; saunikāḥ—the killers; iva—like; svadhitinā—by a sword; avadāya—cutting to pieces; asṛk—the blood; pibanti—drink; nṛtyanti—dance; ca—and; gāyanti—sing; ca—also; hṛṣyamāṇāḥ—being delighted; yathā—just like; iha—in this world; puruṣa-adāḥ—the man-eaters.

TRANSLATION: There are men and women in this world who sacrifice human beings to Bhairava or Bhadra Kalı and then eat their victims' flesh. Those who perform such sacrifices are taken after death to the abode of Yamarāja, where their victims, having taken the form of Rākṣasas, cut them to pieces with sharpened swords. Just as in this world the man-eaters drank their victims' blood, dancing and singing in jubilation, their victims now enjoy drinking the blood of the sacrificers and celebrating in the same way.

5.26.32: ye tv iha vā anāgaso 'raṇye grāme vā vaiśrambhakair upasṛtān upaviśrambhayya jijīviṣūn śūla-sūtrādiṣūpaprotān krīḍanakatayā yātayanti te 'pi ca pretya yama-yātanāsu śūlādiṣu protātmānaḥ kṣut-tṛḍbhyāṁ cābhihatāḥ kaṅka-vaṭādibhiś cetas tatas tigma-tuṇḍair āhanyamānā ātma-śamalaṁ smaranti.

SYNONYMS: ye—persons who; tu—but; iha—in this life; vā—or; anāgasaḥ—who are faultless; araṇye—in the forest; grāme—in the village; vā—or; vaiśrambhakaiḥ—by means of good faith; upasṛtān—brought near; upaviśrambhayya—inspiring with confidence; jijīviṣūn—who want to be protected; śūla-sūtra-ādiṣu—on a lance, thread, and so on; upaprotān—fixed; krīḍanakatayā—like a plaything; yātayanti—cause pain; te—those persons; api—certainly; ca—and; pretya—after dying; yama-yātanāsu—the persecutions of Yamarāja; śūla-ādiṣu—on lances and so on; prota-ātmānaḥ—whose bodies are fixed; kṣut-tṛḍbhyām—by hunger and thirst; ca—also; abhihatāḥ—overwhelmed; kaṅka-vaṭa-ādibhiḥ—by birds such as herons and vultures; ca—and; itaḥ tataḥ—here and there; tigma-tuṇḍaiḥ—having pointed beaks; āhanyamānāḥ—being tortured; ātma-śamalam—own sinful activities; smaranti—they remember.

TRANSLATION: In this life some people give shelter to animals and birds that come to them for protection in the village or forest, and after making them believe that they will be protected, such people pierce them with lances or threads and play with them like toys, giving them great pain. After death such people are brought by the assistants of Yamarāja to the hell known as Śūlaprota, where their bodies are pierced with sharp, needlelike lances. They suffer from hunger and thirst, and sharp-beaked birds such as vultures and herons come at them from all sides to tear at their bodies. Tortured and suffering, they can then remember the sinful activities they committed in the past.

5.26.33: ye tv iha vai bhūtāny udvejayanti narā ulbaṇa-svabhāvā yathā dandaśūkās te 'pi pretya narake dandaśūkākhye nipatanti yatra nṛpa dandaśūkāḥ pañca-mukhāḥ sapta-mukhā upasṛtya grasanti yathā bileśayān.

SYNONYMS: ye—persons who; tu—but; iha—in this life; vai—indeed; bhūtāni—to living entities; udvejayanti—cause unnecessary pain; narāḥ—men; ulbaṇa-svabhāvāḥ—angry by nature; yathā—just like; dandaśūkāḥ—snakes; te—they; api—also; pretya—after dying; narake—in the hell; dandaśūka-ākhye—named Dandaśūka; nipatanti—fall down; yatra—where; nṛpa—O King; dandaśūkāḥ—serpents; pañca-mukhāḥ—having five hoods; sapta-mukhāḥ—having seven hoods; upasṛtya—reaching up; grasanti—eat; yathā—just like; bileśayān—mice.

TRANSLATION: Those who in this life are like envious serpents, always angry and giving pain to other living entities, fall after death into the hell known as Dandaśūka. My dear King, in this hell there are serpents with five or seven hoods. These serpents eat such sinful persons just as snakes eat mice.

5.26.34: ye tv iha vā andhāvaṭa-kusūla-guhādiṣu bhūtāni nirundhanti tathāmutra teṣv evopaveśya sagareṇa vahninā dhūmena nirundhanti.

SYNONYMS: ye—persons who; tu—but; iha—in this life; vā—or; andha-avaṭa—a blind well; kusūla—granaries; guha-ādiṣu—and in caves; bhūtāni—the living entities; nirundhanti—confine; tathā—similarly; amutra—in the next life; teṣu—in those same places; eva—certainly; upaveśya—causing to enter; sagareṇa—with poisonous fumes; vahninā—with fire; dhūmena—with smoke; nirundhanti—confine.

TRANSLATION: Those who in this life confine other living entities in dark wells, granaries or mountain caves are put after death into the hell known as Avaṭa-nirodhana. There they themselves are pushed into dark wells, where poisonous fumes and smoke suffocate them and they suffer very severely.

5.26.35: yas tv iha vā atithīn abhyāgatān vā gṛha-patir asakṛd upagata-manyur didhakṣur iva pāpena cakṣuṣā nirīkṣate tasya cāpi niraye pāpa-dṛṣṭer akṣiṇī vajra-tuṇḍā gṛdhrāḥ kaṅka-kāka-vaṭādayaḥ prasahyoru-balād utpāṭayanti.

SYNONYMS: yaḥ—a person who; tu—but; iha—in this life; vā—or; atithīn—guests; abhyāgatān—visitors; vā—or; gṛha-patiḥ—a householder; asakṛt—many times; upagata—obtaining; manyuḥ—anger; didhakṣuḥ—one desiring to burn; iva—like; pāpena—sinful; cakṣuṣā—with eyes; nirīkṣate—looks at; tasya—of him; ca—and; api—certainly; niraye—in hell; pāpa-dṛṣṭeḥ—of he whose vision has become sinful; akṣiṇī—the eyes; vajra-tuṇḍāḥ—those who have powerful beaks; gṛdhrāḥ—vultures; kaṅka—herons; kāka—crows; vaṭa-ādayaḥ—and other birds; prasahya—violently; uru-balāt—with great force; utpāṭayanti—pluck out.

TRANSLATION: A householder who receives guests or visitors with cruel glances, as if to burn them to ashes, is put into the hell called Paryāvartana, where he is gazed at by hard-eyed vultures, herons, crows and similar birds, which suddenly swoop down and pluck out his eyes with great force.

PURPORT: A guest who comes to one's home should be received very politely. If he is unwanted, the householder should not stare at him with blinking eyes, for one who does so will be put into the hell known as Paryāvartana after death, and there many ferocious birds like vultures, crows, and coknis will suddenly come upon him and pluck out his eyes.

5.26.36: yas tv iha vā āḍhyābhimatir ahaṅkṛtis tiryak-prekṣaṇaḥ sarvato 'bhiviśaṅkī artha-vyaya-nāśa-cintayā pariśuṣyamāṇa-hṛdaya-vadano nirvṛtim anavagato graha ivārtham abhirakṣati sa cāpi pretya tad-utpādanotkarṣaṇa-saṁrakṣaṇa-śamala-grahaḥ sūcīmukhe narake nipatati yatra ha vitta-grahaṁ pāpa-puruṣaṁ dharmarāja-puruṣā vāyakā iva sarvato 'ṅgeṣu sūtraiḥ parivayanti.

SYNONYMS: yaḥ—any person who; tu—but; iha—in this world; vā—or; āḍhya-abhimatiḥ—proud because of wealth; ahaṅkṛtiḥ—egotistic; tiryak-prekṣaṇaḥ—whose vision is crooked; sarvataḥ abhiviśaṅkī—always fearful of being cheated by others, even by superiors; artha-vyaya-nāśa-cintayā—by the thought of expenditure and loss; pariśuṣyamāṇa—dried up; hṛdaya-vadanaḥ—his heart and face; nirvṛtim—happiness; anavagataḥ—not obtaining; grahaḥ—a ghost; iva—like; artham—wealth; abhirakṣati—protects; saḥ—he; ca—also; api—indeed; pretya—after dying; tat—of those riches; utpādana—of the earning; utkarṣaṇa—increasing; saṁrakṣaṇa—protecting; śamala-grahaḥ—accepting the sinful activities; sūcīmukhe—named Sūcīmukha; narake—in the hell; nipatati—falls down; yatra—where; ha—indeed; vitta-graham—as a money-grabbing ghost; pāpa-puruṣam—very sinful man; dharmarāja-puruṣāḥ—the commanding men of Yamarāja; vāyakāḥ iva—like expert weavers; sarvataḥ—all over; aṅgeṣu—on the limbs of the body; sūtraiḥ—by threads; parivayanti—stitch.

TRANSLATION: One who in this world or this life is very proud of his wealth always thinks, "I am so rich. Who can equal me?" His vision is twisted, and he is always afraid that someone will take his wealth. Indeed, he even suspects his superiors. His face and heart dry up at the thought of losing his wealth, and therefore he always looks like a wretched fiend. He is not in any way able to obtain actual happiness, and he does not know what it is to be free from anxiety. Because of the sinful things he does to earn money, augment his wealth and protect it, he is put into the hell called Sūcīmukha, where the officials of Yamarāja punish him by stitching thread through his entire body like weavers manufacturing cloth.

5.26.37: evaṁ-vidhā narakā yamālaye santi śataśaḥ sahasraśas teṣu sarveṣu ca sarva evādharma-vartino ye kecid ihoditā anuditāś cāvani-pate paryāyeṇa viśanti tathaiva dharmānuvartina itaratra iha tu punar-bhave ta ubhaya-śeṣābhyāṁ niviśanti.

SYNONYMS: evam-vidhāḥ—of this sort; narakāḥ—the many hells; yama-ālaye—in the province of Yamarāja; santi—are; śataśaḥ—hundreds; sahasraśaḥ—thousands; teṣu—in those hellish planets; sarveṣu—all; ca—also; sarve—all; eva—indeed; adharma-vartinaḥ—persons not following the Vedic principles or regulative principles; ye kecit—whosoever; iha—here; uditāḥ—mentioned; anuditāḥ—not mentioned; ca—and; avani-pate—O King; paryāyeṇa—according to the degree of different kinds of sinful activity; viśanti—they enter; tathā eva—similarly; dharma-anuvartinaḥ—those who are pious and act according to the regulative principles or Vedic injunctions; itaratra—elsewhere; iha—on this planet; tu—but; punaḥ-bhave—into another birth; te—all of them; ubhaya-śeṣābhyām—by the remainder of the results of piety or vice; niviśanti—they enter.

TRANSLATION: My dear King Parīkṣit, in the province of Yamarāja there are hundreds and thousands of hellish planets. The impious people I have mentioned—and also those I have not mentioned—must all enter these various planets according to the degree of their impiety. Those who are pious, however, enter other planetary systems, namely the planets of the demigods. Nevertheless, both the pious and impious are again brought to earth after the results of their pious or impious acts are exhausted.

5.26.38: nivṛtti-lakṣaṇa-mārga ādāv eva vyākhyātaḥ; etāvān evāṇḍa-kośo yaś caturdaśadhā purāṇeṣu vikalpita upagīyate yat tad bhagavato nārāyaṇasya sākṣān mahā-puruṣasya sthaviṣṭhaṁ rūpam ātmamāyā-guṇamayam anuvarṇitam ādṛtaḥ paṭhati śṛṇoti śrāvayati sa upageyaṁ bhagavataḥ paramātmano 'grāhyam api śraddhā-bhakti-viśuddha-buddhir veda.

SYNONYMS: nivṛtti-lakṣaṇa-mārgaḥ—the path symptomized by renunciation, or the path of liberation; ādau—in the beginning (the Second and Third Cantos); eva—indeed; vyākhyātaḥ—described; etāvān—this much; eva—certainly; aṇḍa-kośaḥ—the universe, which resembles a big egg; yaḥ—which; caturdaśa-dhā—in fourteen parts; purāṇeṣu—in the purāṇas; vikalpitaḥ—divided; upagīyate—is described; yat—which; tat—that; bhagavataḥ—of the Supreme Personality of Godhead; nārāyaṇasya—of Lord Nārāyaṇa; sākṣāt—directly; mahā-puruṣasya—of the Supreme Person; sthaviṣṭham—the gross; rūpam—form; ātma-māyā—of His own energy; guṇa—of the qualities; mayam—consisting; anuvarṇitam—described; ādṛtaḥ—venerating; paṭhati—one reads; śṛṇoti—or hears; śrāvayati—or explains; saḥ—that person; upageyam—song; bhagavataḥ—of the Supreme Personality of Godhead; paramātmanaḥ—of the Supersoul; agrāhyam—difficult to understand; api—although; śraddhā—by faith; bhakti—and devotion; viśuddha—purified; buddhiḥ—whose intelligence; veda—understands.

TRANSLATION: In the beginning [the Second and Third Cantos of Śrīmad-Bhāgavatam] I have already described how one can progress on the path of liberation. In the Purāṇas the vast universal existence, which is like an egg divided into fourteen parts, is described. This vast form is considered the external body of the Lord, created by His energy and qualities. It is generally called the virāṭ-rūpa. If one reads the description of this external form of the Lord with great faith, or if one hears about it or explains it to others to propagate bhāgavata-dharma, or Kṛṣṇa consciousness, his faith and devotion in spiritual consciousness, Kṛṣṇa consciousness, will gradually increase. Although developing this consciousness is very difficult, by this process one can purify himself and gradually come to an awareness of the Supreme Absolute Truth.

5.26.39: śrutvā sthūlaṁ tathā sūkṣmaṁ/rūpaṁ bhagavato yatiḥ/sthūle nirjitam ātmānaṁ/śanaiḥ sūkṣmaṁ dhiyā nayed iti

SYNONYMS: śrutvā—after hearing of (from the disciplic succession); sthūlam—gross; tathā—as well as; sūkṣmam—subtle; rūpam—form; bhagavataḥ—of the Supreme Personality of Godhead; yatiḥ—a sannyāsī or devotee; sthūle—the gross form; nirjitam—conquered; ātmānam—the mind; śanaiḥ—gradually; sūkṣmam—the subtle. spiritual form of the Lord; dhiyā—by intelligence; nayet—one should lead it to; iti—thus.

TRANSLATION: One who is interested in liberation, who accepts the path of liberation and is not attracted to the path of conditional life, is called yati, or a devotee. Such a person should first control his mind by thinking of the virāṭ-rūpa, the gigantic universal form of the Lord, and then gradually think of the spiritual form of Kṛṣṇa [sac-cid-ānanda-vigraha] after hearing of both forms. Thus one's mind is fixed in samādhi. By devotional service one can then realize the spiritual form of the Lord, which is the destination of devotees. Thus his life becomes successful.

5.26.40 bhū-dvīpa-varṣa-sarid-adri-nabhaḥ-samudra-/pātāla-diṅ-naraka-bhāgaṇa-loka-saṁsthā/ gītā mayā tava nṛpādbhutam īśvarasya /sthūlaṁ vapuḥ sakala-jīva-nikāya-dhāma

SYNONYMS: bhū—of this planet earth; dvīpa—and other different planetary systems; varṣa—of tracts of land; sarit—rivers; adri—mountains; nabhaḥ—the sky; samudra—oceans; pātāla—lower planets; dik—directions; naraka—the hellish planets; bhāgaṇa-loka—the luminaries and higher planets; saṁsthā—the situation; gītā—described; mayā—by me; tava—for you; nṛpa—O King; adbhutam—wonderful; īśvarasya—of the Supreme Personality of Godhead; sthūlam—gross; vapuḥ—body; sakala-jīva-nikāya—of all the masses of living entities; dhāma—which is the place of repose.

TRANSLATION: My dear King, I have now described for you this planet earth, other planetary systems, and their lands [varṣas], rivers and mountains. I have also described the sky, the oceans, the lower planetary systems, the directions, the hellish planetary systems and the stars. These constitute the virāṭ-rūpa, the gigantic material form of the Lord, on which all living entities repose. Thus I have explained the wonderful expanse of the external body of the Lord.

# 12

# Surya-Siddhanta

## CHAPTER 1.
## OF THE MEAN MOTIONS OF THE PLANETS.

CONTENTS:—1, homage to the Deity; 2-9, revelation of the present treatise; 10-11, modes of dividing time; 11-12, subdivisions of a day; 12-14, of a year; 14-17, of the Ages; 18-19, of an AEon; 20-21, of Brahma's life; 21-23, part of it already elapsed; 24, time occupied in the work of creation; 25-27, general account of the movements of the planets; 28, subdivisions of the circle; 29-33, number of revolutions of the planets, and of the moon's apsis and node, in an Age; 34-39, number of days and months, of different kinds, in an Age; 40, in an AEon; 41-44, number of revolutions, in an AEon, of the apsides and nodes of the planets; 45-47, time elapsed from the end of creation to that of the Golden Age; 48-51, rule for the reduction to civil days of the whole time since the creation; 51-52, method of finding the lords of the day, the month, and the year; 53-54, rule for finding the mean place of a planet, and of its apsis and node; 55, to find the current year of the cycle of Jupiter; 56, simplification of the above calculations; 57-58, situation of the planets, and of the moon's apsis and node, at the end of the Golden Age; 59-60, dimensions of the earth; 60-61, correction, for difference of longitude, of the mean place of a planet as found; 62, situation of the principal meridian; 63-65, ascertainment of difference of longitude by difference between observed and computed time of a lunar eclipse; 66, difference of time owing to difference of longitude; 67, to find the mean place of a planet for any required hour of the day; 68-70, inclination of the orbits of the planets.

1. To him whose shape is inconceivable and unmanifested, who is unaffected by the qualities, whose nature is quality, whose form is the support of the entire creation—to Brahma be homage!
2. When but little of the Golden Age (*kṛta yuga*) was left, a great demon *(asura),* named Maya, being desirous to know that mysterious, supreme, pure and exalted science.
3. That chief auxiliary of the scripture *(vedānga),* in its entirety-the cause, namely of the motion of the heavenly bodies (*jyotis*), performed, in propitiation of the Sun, very severe religious austerities.
4. Gratified by these austerities, and rendered propitious, the Sun himself delivered unto that Maya, who besought a boon, the system of the planets.

**The blessed Sun spoke:**

5. Thine intent is known to me; I am gratified by thine austerities; I will give thee the science upon which time is founded, the grand system of the planets.
6. No one is able to endure my brilliancy; for communication I have no leisure; this person, who is a part of me, shall relate to thee the whole.
6a. [additional verse in some versions] "Go therefore to Romaka-city, thine own residence; there, undergoing incarnation as a barbarian, owning to a curse of Brahma, I will impart to thee this science."
7. Thus having spoken, the god disappeared, having given directions unto the part of himself. This latter person thus addressed Maya, as he stood bowed forward, his hands suppliantly joined before him:
8. Listen with concentrated attention to the ancient and exalted science, which has been spoken, in each successive Age, to the Great Sages *(maharshi),* by the Sun himself.
9. This is that very same original text-book which the Sun of old promulgated: only, by reason of the revolution of the Ages, there is here a difference of times.
10. Time is the destroyer of the worlds; another Time has for its nature to bring to pass. This latter, according as it is gross or minute, is called by two names, real *(mūrta)* and unreal *(amūrta).*
11. That which begins with respirations *(prāna)* is called real; that which begins with atoms *(truṭi)* is called unreal. Six respirations make a *vinādī,* sixty of these a *nādī;*
12. And sixty nādīs make a sidereal day and night. Of thirty of these sidereal days is composed a month; a civil *(sāvana)* month consists of as many sunrises;
13. A lunar month, of as many lunar days *(tithi);* a solar *(saura)* month is determined by the entrance of the sun into a sign of the zodiac: twelve months make a year. This is called a day of the gods.
*14. The day and night of the gods and of the demons are mutually opposed to one another. Six times sixty of them are a year of the gods, and likewise of the demons.*
15. Twelve thousand of these divine years are denominated a Quadruple Age *(caturyuga);* of ten thousand times four hundred and thirty-two solar years

16. Is composed that Quadruple Age, with its dawn and twilight. The difference of the Golden and the other Ages, as measured by the difference in the number of the feet of Virtue in each, is as follows:
17. The tenth part of an Age, multiplied successively by four, three, two, and one, gives the length of the Golden and the other Ages, in order: the sixth part of each belongs to its dawn and twilight.
18. One and seventy Ages are styled here a Patriarchate *(manvantara);* at its end is said to be a twilight which has the number of years of a Golden Age, and which is a deluge.
19. In an AEon *(kalpa)* are reckoned fourteen such Patriarchs *(manu)* with their respective twilights; at the commencement of the AEon is a fifteenth dawn, having the length of a Golden Age.

**The AEon is accordingly thus composed:**

20. The AEon, thus composed of a thousand Ages, and which brings about the destruction of all that exists, is styled a day of Brahma; his night is of the same length.
21. His extreme age is a hundred, according to this valuation of a day and a night. The half of his life is past; of the remainder, this is the first AEon.
22. And of this AEon, six Patriarchs *(manu)* are past, with their respective twilights; and of the Patriarch Manu son of Vivasvant, twenty-seven Ages are past;
23. Of the present the twenty-eighth, Age, this Golden Age is past: from this point, reckoning up the time, one should compute together the whole number.
24. One hundred times four hundred and seventy-four divine years passed while the All-wise was employed in creating the animate and inanimate creation, plants, stars, gods, demons, and the rest.
25. The planets moving westward with exceeding velocity, but constantly beaten by the asterisms, fall behind, at a rate precisely equal, proceeding each in its own path.
26. Hence they have an eastward motion. From the number of their revolutions is derived their daily motion, which is different according to the size of their orbits; in proportion to this daily motion they pass through the asterisms.
27. One which moves swiftly passes through them in a short time; one which moves slowly, in a long time. By their movement, the revolution is accounted complete at the end of the asterism Revatī.
28. Sixty seconds (*vikalā*) make a minute (*kalā*): sixty of these, a degree (*bhāga*); of thirty of the latter is composed a sign (*rāśi*); twelve of these are a revolution (*bhagaṇa*).
29. In an Age (yuga), the revolutions of the sun, Mercury, and Venus, and of the conjunctions (*śīghra*) of Mars, Saturn, and Jupiter, moving eastward, are four million, three hundred and twenty thousand;

*30. Of the moon, fifty-seven million, seven hundred and fifty-three thousand, three hundred and thirty-six; of Mars, two million, two hundred and ninety-six thousand, eight hundred and thirty-two.*
31. Of Mercury's conjunction (*śīghra*), seventeen million, nine hundred and thirty-seven thousand, and sixty; of Jupiter, three hundred and sixty-four thousand, two hundred and twenty;
32. Of Venus' conjunction (*śīghra*), seven million, twenty-two thousand, three hundred and seventy-six; of Saturn, one hundred and forty-six thousand five hundred and sixty-eight;
33. Of the moon's apsis (*ucca*), in an Age, four hundred and eight-eight thousand, two hundred and three; of its node (*pāta*), in the contrary direction two hundred and thirty-two thousand, two hundred and eight-eight;
34. Of the asterisms, on billion, five hundred and eighty-two million, two hundred and thirty-seven thousand, eight hundred and twenty-eight. The number of risings of the asterisms, diminished by
the number of the revolutions of each planet respectively, gives the number of risings of the planets in an Age.
35. The number of lunar months is the difference between the number of revolutions of the sun and of the moon. If from it the number of solar months be subtracted, the remainder is the number of intercalary months.
36. Take the civil days from the lunar, the remainder is the number of omitted lunar days *(tithikshaya,)*. From rising to rising of the sun are reckoned terrestrial civil days;
37. Of these there are, in an Age, one billion, five hundred and seventy-seven million, nine hundred and seventeen thousand, eight hundred and twenty-eight; of lunar days, one billion, six hundred and three million, and eighty;
38. Of intercalary months, one million, five hundred and ninety-three thousand, three hundred and thirty-six; of omitted lunar days, twenty-five million, eighty-two thousand, two hundred and fifty-two;
39. Of solar months, fifty-one million, eight hundred and forty thousand. The number of risings of the asterisms, diminished by that of the revolutions of the sun, gives the number of terrestrial days. I
40. The intercalary months, the omitted lunar days, the sidereal, lunar, and civil days-these, multiplied by a thousand, are the number of revolutions, etc., in an AEon.
41. The revolutions of the sun's apsis (*mānda*), moving eastward, in an AEon, are three hundred and eighty-seven; of that of Mars, two hundred and four; of that of Mercury, three hundred and sixty-eight;
42. Of that of Jupiter, nine hundred; of that of Venus, five hundred and thirty-five; of the apsis of Saturn, thirty-nine. Farther, the revolutions of, the nodes, retrograde, are:
43. Of that of Mars, two hundred and fourteen; of that of Mercury, four hundred and eighty-eight; of that of Jupiter, one hundred and seventy-four; of that of Venus, nine hundred and three;

44. Of the node of Saturn, the revolutions in an AEon are six hundred and sixty-two: the revolutions of the moon's apsis and node have been given here already.
45. Now add together the time of the six Patriarchs *(manu),* with their respective twilights, and with the dawn at the commencement of the AEon *(kalpa);* farther, of the Patriarch Manu, son of Vivasvant,
46. The twenty-seven Ages *(yuga)* that are past, and likewise the present Golden Age *(kṛta yuga);* from their sum subtract the time of creation, already stated in terms of divine years,
47. In solar years: the result is the time elapsed at the end of the Golden Age; namely, one billion, nine hundred and fifty-three million, seven hundred and twenty thousand solar years.
48. To this add the number of years of the time since past. Reduce the sum to months, and add the months expired of the current year, beginning with the light half of Cāitra.
49. Set the result down in two places; multiply it by the number of intercalary months, and divide by that of solar months, and add to the last result the number of intercalary months thus found; reduce the sum to days, and add the days expired of the current month;
50. Set the result down in two places; multiply it by the number of omitted lunar days, and divide by that of lunar days; subtract from the last result the number of omitted lunar days thus obtained: the remainder is, at midnight, on the meridian of Lankā,
51. The sum of days, in civil reckoning. From this may be found the lords of the day, the month, and the year, counting from the sun. If the number be divided by seven, the remainder marks the lord of the day, beginning with the sun.
52. Divide the same number by the number of days in a month and in a year, multiply the one quotient by two and the other by three, add one to each product, and divide by seven; the remainders indicate the lords of the month and of the year.
53. Multiply the sum of days *(dinarāśi)* by the number of revolutions of any planet, and divide by the number of civil days; the result is the position of that planet, in virtue of its mean motion, in revolutions and parts of a revolution.
54. Thus also are ascertained the places of the conjunction *(śīghra)* and apsis *(mandocca)* of each planet, which have been mentioned as moving eastward; and in like manner of the nodes, which have a retrograde motion, subtracting the result from a whole circle.
55. Multiply by twelve the past revolutions of Jupiter, add the signs of the current revolution, and divine by sixty; the remainder marks the year of Jupiter's cycle, counting from Vijaya.
56. The processes which have thus been stated in full detail, are practically applied in an abridged form. The calculation of the mean place of the planets may be made from any epoch *(yuga)* that may be fixed upon.

57. Now, at the end of the Golden Age *(kṛta yuga),* all the planets, by their mean motion— excepting, however, their nodes and apsides *(mandocca)*—are in conjunction in the first of Aries.
58. The moon's apsis *(ucca)* is in the first of Capricorn, and its node is in the first of Libra; and the rest, which have been stated above to have a slow motion— their position cannot be expressed in whole signs.
59. Twice eight hundred *yojanas* are the diameter of the earth: the square root of ten times the square of that is the earth's circumference.
60. This, multiplied by the sine of the co-latitude *(lambajyā)* of any place, and divided by radius *(trijīvā),* is the corrected *(sphuṭa)* circumference of the earth at that place. Multiply the daily motion of a planet by the distance in longitude *(deśantarā)* of any place, and divide by its corrected circumference;
61. The quotient, in minutes, subtract from the mean position of the planet as found, if the place be east of the prime meridian *(rekhā)* ; add, if it be west; the result is the planet's mean position at the given place.
62. Situated upon the line which passes through the haunt of the demons *(rākshasa)* and the mountain which is the seat of the gods, are Rohītaka and Avantī, as also the adjacent lake.
63. When, in a total eclipse of the moon, the emergence *(unmīlana)* takes place after the calculated time for its occurrence, then the place of the observer is to the east of the central meridian.
64. When it takes place before the calculated time, his place is to the west: the same thing may be ascertained likewise from the immersion *(nimīlana).* Multiply by the difference of the two times in nādīs the corrected circumference of the earth at the place of observation.
65. And divide by sixty, the result, in yojanas, indicates the distance of the observer from the meridian, to the east or to the west, upon his own parallel; and by means of that is made the correction for difference of longitude.
66. The succession of the week-day *(vāra)* takes place, to the east of the meridian, at a time after midnight equal to the difference of longitude in nādīs; to the west of the meridian, at a corresponding time before midnight.
67. Multiply the mean daily motion of a planet by the number of nādīs of the time fixed upon, and divide by sixty: subtract the quotient from the place of the planet, if the time be before midnight; add, if it be after: the result is its place at the given time.
68. The moon is, by its node, caused to deviate from the limit of its declination *(krānti),* northward and southward, to a distance, when greatest, of an eightieth part of the minutes of a circle;
69. Jupiter, to the ninth part of that multiplied by two; Mars, to the same amount multiplied by three; Mercury, Venus, and Saturn are by their nodes caused to deviate to the same amount multiplied by four.
70. So also, twenty-seven, nine, twelve, six, twelve, and twelve, multiplied respectively by ten, give the number of minutes of mean latitude *(vikshepa)* of the moon and the rest, in, their order.

# CHAPTER II.
# OF THE TRUE PLACES OF THE PLANETS.

CONTENTS:—1-3, causes of the irregularities of the planetary motions; 4-5, disturbing influence of the apsis and conjunction; 6-8, of the node; 9-11, different degree of irregularity of the motion of the different planets; 12-13, different kinds of planetary motion; 14, purpose of this chapter; 15-16, rule for constructing the table of sines; 17-22, table of sines; 22-27, table of versed sines; 28, inclination of the ecliptic, and rule for finding the declination of any point in it; 29-30, to find the sine and cosine of the anomaly; 31-32, to find, by interpolation, the sine or versed sine corresponding to any given arc; 33, to find, in like manner, the arc corresponding to a given sine or versed sine; 34-37, dimensions of the epicycles of the planets; 38, to find the true dimensions of the epicycle at any point in the orbit; 39, to find the equation of the apsis, or of the centre; 40-42, to find the equation of the conjunction, or the annual equation; 43-45, application of these equations in finding the true places of the different planets; 46, correction of the place of a planet for difference between mean and apparent solar time; 47-49, how to correct the daily motion of the planets for the effect of the apsis; 50-51, the same for that of the conjunction; 51-55, retrogradation of the lesser planets; 56, correction of the place of the node; 57-58, to find the celestial latitude of a planet, and its declination as affected by latitude; 59, to find the length of the day of any planet; 60, to find the radius of the diurnal circle; 61-63, to find the day-sine, and the respective length of the day and night; 64, to find the number of asterisms traversed by a planet, and of days elapsed, since the commencement of the current revolution; 65, to find the *yoga;* 66, to find the current lunar day, and the time in it of a given instant; 67-69, of the divisions of the lunar month called *karaṇa.*

1. Forms of Time, of invisible shape, stationed in the zodiac *(bhagaṇa)*, called the conjunction *(śīghrocca),* apsis*(mandocca),* and node *(pāta)*, are causes of the motion of the planets.
2. The planets, attached to these beings by cords of air, are drawn away by them, with the right and left hand, forward or backward, according to nearness, toward their own place.
3. A wind, moreover, called *provector (pravaha)* impels them toward their own apices (*ucca*); being drawn away forward and backward, they proceed by a varying motion.
4. The so-called apex *(ucca),* when in the half-orbit in front of the planet, draws the planet forward; in like manner, when in the half-orbit behind the planet, it draws it backward.

5. When the planets, drawn away by their apices *(ucca),* move forward in their orbits, the amount of the motion so caused is called their excess (*dhana*); when they move backward, it is called their deficiency (*ṛṇa*).
6. In like manner, also, the node, Rāhu, by its proper force, causes the deviation in latitude *(vikshepa)* of the moon and the other planets, northward and southward, from their point of declination *(apakrama).*
7. When in the half-orbit behind the planet, the node causes it to deviate northward; when in the half-orbit in front, it draws it away southward.
8. In the case of Mercury and Venus, however, when the node is thus situated with regard to the conjunction *(śīghra),* these two planets are caused to deviate in latitude, in the manner stated, by the attraction exercised by the node upon the conjunction.
9. Owing to the greatness of its orb, the sun is drawn away only a very little; the moon, by reason of the smallness of its orb, is drawn away much more;
10. Mars and the rest, on account of their small size, are, by the supernatural beings *(dāivata)* called conjunction *(śīghrocca)* and apsis *(mandocca),* drawn away very far, being caused to vacillate exceedingly.
11. Hence the excess *(dhana)* and deficiency *(ṛṇa)* of these latter is very great, according to their rate of motion. Thus do the planets, attracted by those beings, move in the firmament, carried on by the wind.
12. The motion of the planets is of eight kinds: retrograde *(vakra),* somewhat retrograde *(anuvakra),* transverse *(kuṭila)*, slow (*mānda*) very slow (*mandatara*), even (*sama*); also, very swift (*śīghratara*), and swift (*śīghra*).
13. Of these, the very swift (*atiśīghra*), that called swift, the slow , the very slow, the even—all these five are forms of the motion called direct (*ṛju*); the somewhat retrograde is retrograde.
14. By reason of this and that rate of motion, from day to day, the planets thus come to an accordance with their observed places (*dṛś*)—this, their correction (*sphuṭīkaraṇa*), I shall carefully explain.
15. The eighth part of the minutes of a sign is called the first sine (*jyārdha*); that, increased by the remainder left after subtracting from it the quotient arising from dividing it by itself, is the second sine.
16. Thus, dividing the tabular sines in succession by the first, and adding to them, in each case, what is left after subtracting the quotients from the first, the result is twenty-four tabular sines (*jyārdhapiṇḍa*), in order, as follows;
17. Two hundred and twenty-five; four hundred and forty-nine; six hundred and seventy-one; eight hundred and ninety eleven hundred and five; thirteen hundred and fifteen;
18. Fifteen hundred and twenty; seventeen hundred and nineteen, nineteen hundred and ten; two thousand and ninety-three;
19. Two thousand two hundred and sixty-seven; two thousand four hundred and one-one; two thousand five hundred and eighty-five; two thousand seven hundred and twenty-eight;

20. Two thousand eight hundred and fifty-nine; two thousand nine hundred and seventy-eight; three thousand three hundred and seventy-two; three thousand four hundred and nine;
21. Three thousand two hundred and fifty-six; three thousand three hundred and twenty one; three thousand three hundred and seventy-two; three thousand four hundred and nine;
22. Three thousand four hundred and thirty-one; three thousand four hundred and thirty-eight. Subtracting these, in reversed order, from the half-diameter, gives the tabular versed-sines (*utkramajyārdhapiṇḍaka*):
23. Seven; twenty-nine; sixty-six; one hundred and seventeen; one hundred and eighty-two; two hundred and sixty-one; three hundred and fifty-four;
24. Four hundred and sixty; five hundred and seventy-nine; seven hundred and ten; eight hundred and fifty-three; one thousand and seven; eleven hundred and seventy-one;
25. Thirteen hundred and forty-five; fifteen hundred and twenty-eight; seventeen hundred and nineteen; nineteen hundred and eighteen;
26. Two thousand one hundred and twenty-three; two thousand two hundred and thirteen; three thousand five hundred and forty-eight; two thousand seven hundred and sixty-seven;
27. Two thousand nine hundred and eighty-nine; three thousand two hundred and thirteen; three thousand four hundred and thirty-eight; these are the versed sines.
28. The sine of greatest declination is thirteen hundred ninety-seven; by this multiply any sine, and divide by radius; the arc corresponding to the result is said to be the declination.
29. Subtract the longitude of a planet from that of its apsis *(mandocca);* so also, subtract it from that of its conjunction *(śīghra);* the remainder is its anomaly *(kendra);* from that is found the quadrant *(pada);* from this, the base-sine *(bhujajyā),* and likewise that of the perpendicular *(koṭi).*
30. In an odd, *(vishama)* quadrant, the base-sine is taken from the part past, the perpendicular from that to come; but in an even *(yugma)* quadrant, the base-sine *(bāhujyā)* is taken from the part to come, and the perpendicular-sine from that past.
31. Divide the minutes contained in any arc by two hundred and twenty-five; the quotient is the number of the preceding tabular sine *(jyāpiṇḍaka).* Multiply the remainder by the difference of the preceding and following tabular sines, and divide by two hundred and twenty-five;
32. The quotient thus obtained add to the tabular sine called the preceding; the result is the required sine. The same method is prescribed also with respect to the versed sines.
33. Subtract from any given sine the next less tabular sine; multiply the remainder by two hundred and twenty-five, and divide by the difference between the next less and next greater tabular sines; add the quotient to the product of the serial

number of the next less sine into two hundred and twenty-five the result is the required arc.

34. The degrees of the sun's epicycle of the apsis (*mandaparidhi*) are fourteen, of that of the moon, thirty-two, at the end of the even quadrants; and at the end of the odd quadrants, they are twenty minutes less for both.

35. At the end of the even quadrants, they are seventy-five, thirty, thirty-three, twelve, forty-nine; at the odd (*oja*) they are seventy-two, twenty-eight, thirty-two, eleven, forty-eight,

36. For Mars and the rest; farther, the degrees of the epicycle of the conjunction (*śīghra*) are, at the end of the even quadrants, two hundred and thirty-five, one hundred and thirty-three, seventy, two hundred and sixty-two, thirty-nine;

37. At the end of the odd quadrants, they are stated to be two hundred and thirty-two, one hundred and thirty-two, seventy-two, two hundred and sixty, and forty, as made use of in the calculation for the conjunction (*śīghrakarman*).

38. Multiply the base-sine (*bhujajyā*) by the difference of the epicycles at the odd and even quadrants, and divide by radius (*trijyā*); the result, applied to the even epicycle (*vṛtta*), and additive (*dhana*) or subtractive (*ṛṇa*), according as this is less or greater than the odd, gives the corrected (*sphuṭa*) epicycle.

39. By the corrected epicycle multiply the base-sine (*bhujajyā*) and perpendicular-sine (*koṭijyā*) respectively, and divide by the number of degrees in a circle: then, the arc corresponding to the result from the base-sine *(bhujajyāphala)* is the equation of the apsis *(mānda phala),* in minutes, etc.

40. The result from the perpendicular-sine *(koṭiphala)* of the distance from the conjunction is to be added to radius, when the distance *(kendra)* is in the half-orbit beginning with Capricorn; but when in that beginning with Cancer, the result from the perpendicular-sine is to be subtracted.

41. To the square of this sum or difference add the square of the result from the base-sine *(bāhuphala);* the square root of their sum is the hypothenuse *(karṇa)* called variable *(cala).* Multiply the result from the base-sine by radius, and divide by the variable hypothenuse:

42. The arc corresponding to the quotient is, in etc., the equation of the conjunction *(śāighrya phala);* it is employed in the first and in the fourth process of correction (*karman*) for Mars and the other planets.

43. The process of correction for the apsis *(mānda karman)* is the only one required for the sun and moon: for Mars and the other planets are prescribed that for the conjunction (*śāighrya),* that for the apsis *(mānda)*, again that for the apsis, and that for the conjunction—four, in succession.

44. To the mean place of the planet apply half the equation of the conjunction (*śīghraphala)*, likewise half the equation of the apsis; to the mean place of the planet apply the whole equation of the apsis *(mandaphala),* and also that of the conjunction.

45. In the case of all the planets, and both in the process of correction for the conjunction and in that for the apsis, the equation is additive *(dhana)* when the

distance *(kendra)* is in the half-orbit beginning with Aries; subtractive *(ṛṇa),* when in the half-orbit beginning with Libra.
46. Multiply the daily motion *(bhukti)* of a planet by the sun's result from the base-sine *(bāhuphala),* and divide by the number of minutes in a circle *(bhacakra)*; the result, in minutes, apply to the planet's true place, in the same direction as the equation was applied to the sun.
47. From the mean daily motion of the moon subtract the daily motion of its apsis *(mānda),* and, having treated the difference in the manner prescribed by the next rule, apply the result, as an additive or subtractive equation, to the daily motion.
48. The equation of a planet's daily motion is to be calculated like the place of the planet in the process for the apsis: multiply the daily motion by the difference of tabular sines corresponding to the base-sine *(dorjyā)* of anomaly, and then divide by two hundred and twenty-five;
49. Multiply the result by the corresponding epicycle of the apsis *(mandaparidhi),* and divide by the number of degrees in a circle (*bhagaṇa*); the result, in minutes, is additive when in the half-orbit beginning with Cancer, and subtractive when in that beginning with Capricorn.
50. Subtract the daily motion of a planet, thus corrected for the apsis *(mānda),* from the daily motion of its conjunction *(śīghra)*; then multiply the remainder by the difference between the last hypothenuse and radius,
51. And divide by the variable hypothenuse *(cala karṇa)*: the result is additive to the daily motion when the hypothenuse is greater than radius, and subtractive when this is less; if, when subtractive, the equation is greater than the daily motion, deduct the latter from it, and the remainder is the daily motion in a retrograde *(vakra)* direction.
52. When at a great distance from its conjunction *(śīghrocca)*, a planet, having its substance drawn to the left and right by slack cords, comes. then to have a retrograde motion.
53. Mars and the rest, when their degrees of commutation *(kendra)*, in the fourth process, are, respectively, one hundred and sixty-four, one hundred and forty-four, one hundred and thirty, one hundred and sixty-three, one hundred and fifteen,
54. Become retrograde *(vakrin)*: and when their respective commutations are equal to the number of degrees remaining after subtracting those numbers, in each several case, from a whole circle, they cease retrogradation.
55. In accordance with the greatness of their epicycles of the conjunction *(śīghraparidhi)* , Venus and Mars cease retrograding in the seventh sign, Jupiter and Mercury in the eighth, Saturn in the ninth.
56. To the nodes of Mars, Saturn, and Jupiter, the equation of the conjunction is to be applied, as to the planets themselves respectively; to those of Mercury and Venus, the equation of the apsis, as found by the third process, in the contrary direction.
57. The sine of the arc found by subtracting the place of the node from that of the planet—or, in the case of Venus and Mercury, from that of the conjunction—

being multiplied by the extreme latitude, and divided by the last hypothenuse—or, in the case of the moon, by radius—given the latitude *(vikshepa).*
58. When latitude and declination *(apakrama)* are of like direction, the declination *(krānti)* is increased by the latitude; when of different direction, it is diminished by it, to find the true *(spashṭa)* declination: that of the, sun remains as already determined.
59. Multiply the daily motion of a planet by the time of rising of the sign in which it is, and divide by eighteen hundred; the quotient add to, or subtract from, the number of respirations in a revolution: the result is the number of respirations in the day and night of that planet.
60. Calculate the sine and versed sine of declination: then radius, diminished by the versed-sine, is the day-radius: it is either south or north.
61. Multiply the sine of the declination by the equinoctial shadow, and divide by twelve; the result is the earth-sine (*kshitijyā*); this, multiplied by radius and divided by the day-radius, gives the sine of the ascensional difference (*cara*): the number of respirations due to the ascensional difference
62. Is shown by the corresponding arc. Add these to, and subtract them from, the fourth part of the corresponding day and night, and the sum and remainder are, when declination is north, the half-day and half-night;
63. When declination is south, the reverse; these, multiplied by two, are the day and the night. The day and the night of the asterisms (*bha*) may be found in like manner, by means of their declination, increased or diminished by their latitude.
64. The portion *(bhoga)* of an asterism (*bha*) is eight hundred minutes; of a lunar day (*tithi*), in like manner, seven hundred and twenty. If the longitude of a planet, in minutes, be divided by the portion of an asterism, the result is its position in asterisms: by means of the daily motion are found the days, etc.
65. From the number of minutes in the sum of the longitudes of the sun and moon are found the *yogas*, by dividing that sum by the portion (*bhoga*) of an asterism. Multiply the minutes past and to come of the current yoga by sixty, and divide by the sum of the daily motions of the two planets: the result is the time in nādīs.
66. From the number of minutes in the longitude of the moon diminished by that of the sun are found the lunar days *(tithi),* by dividing the difference by the portion *(bhoga)* of a lunar day. Multiply the minutes past and to come of the current lunar day by sixty, and divide by the difference of the daily motions of the two planets: the result is the time in nādīs.
67. The fixed *(dhruva) karaṇas,* namely *śakuni, nāga, catushpada* the third, and *kinstughna,* are counted from the latter half of the fourteenth day of the dark half-month.
68. After these, the karaṇas called movable *(cara)* , namely *bava,* etc., seven of them: each of these karaṇas occurs eight times in a month.
69. Half the portion *(bhoga)* of a lunar day is established as that of the karaṇas. Thus has been declared the corrected *(sphuṭa)* motion of the sun and the other planets.

# CHAPTER III.
# OF DIRECTION, PLACE, AND TIME.

CONTENTS :—1-6, construction of the dial, and description of its parts; 7, the measure of amplitude; 8, of the gnomon, hypothenuse, and shadow, any two being given, to find the third; 9-12, precession of the equinoxes; 12-13, the equinoctial shadow; 13-14, to find, from the equinoctial shadow, the latitude and co-latitude; 14-17, the sun's declination being known, to find, from a given shadow at noon, his zenith-distance, the latitude, and its sine and cosine; 17, latitude being given, to find the equinoctial shadow; 17-20, to find, from the latitude and the sun's zenith-distance at noon, his declination and his true and mean longitude; 20-22, latitude and declination being given, to find the noon-shadow and hypothenuse; 22-23, from the sun's declination and the equinoctial shadow and the measure of amplitude; 23-25, to find, from the equinoctial shadow and the measure of amplitude at any given time, the base of the shadow; 25-27, to find the hypothenuse of the shadow when the sun is upon the prime vertical; 27-28, the sun's declination and the latitude being given, to find the sine and the measure of amplitude; 28-33, to find the sines of the altitude and zenith-distance of the sun, when upon the south-east and south-west vertical circles; 33-34, to find the corresponding shadow and hypothenuse; 34-36, the sun's ascensional difference and the hour-angle being given, to find the sines of his altitude and zenith-distance, and the corresponding shadow and hypothenuse; 37-39, to find, by a contrary process, from the shadow of a given time, the sun's altitude and zenith-distance, and the hour-angle; 40-41, the latitude and the sun's amplitude being known, to find his declination and true longitude; 41-42, to draw the path described by the extremity of the shadow; 42-45, to find the arcs of right and oblique ascension corresponding to the several signs of the ecliptic; 46-48, the sun's longitude and the time being known, to find the point of the ecliptic which is upon the horizon; 49, the sun's longitude and the hour-angle being known, to find the point of the ecliptic which is upon the meridian; 50-51, determination of time by means of these data.

1. On a stony surface, made water-level, or upon hard plaster, made level, there draw an even circle, of a radius equal to any required number of the digits *(angula)* of the gnomon *(śanku)*.
2. At its centre set up the gnomon, of twelve digits of the measure fixed upon; and where the extremity of its shadow touches the circle in the former and after parts of the day,

3. There fixing two points upon the circle and calling them the forenoon and afternoon points, draw midway between them, by means of a fish-figure *(timi)*, a north and south line.
4. Midway between the north and south directions draw, by a fish-figure, an east and west line: and in like manner also, by fish-figures (*matsya*) between the four cardinal directions, draw the intermediate directions.
5. Draw a circumscribing square, by means of the lines going out from the centre; by the digits of its base-line *(bhujasūtra)* projected upon that is any given shadow reckoned.
6. The east and west line is called the prime vertical *(samamaṇḍala);* it is likewise denominated the east and west hour circle *(unmaṇḍala)* and the equinoctial circle *(vishuvanmaṇḍala)*.
7. Draw likewise an east and west line through the extremity of the equinoctial shadow *(vishuvadbhā);* the interval between any given shadow and the line of the equinoctial shadow is denominated the measure of amplitude *(agrā)*.
8. The square root of the sum of the squares of the gnomon and shadow is the hypothenuse: if from the square of the latter the square of the gnomon be subtracted, the square root of the remainder is the shadow: the gnomon is found by the converse process.
9. In an Age *(yuga),* the circle of the asterisms *(bha)* falls back eastward thirty score of revolutions. Of the result obtained after multiplying the sum of days *(dyugaṇa)* by this number, and dividing by the number of natural days in an Age,
10. Take the part which determines the sine, multiply it by three, and divide by ten; thus are found the degrees called those of the precession *(ayana)*. From the longitude of a planet as corrected by these are to be calculated the declination, shadow, ascensional difference *(caradala),* etc.
11. The circle, as thus corrected, accords with its observed place at the solstice *(ayana)* and at either equinox; it has moved eastward, when the longitude of the sun, as obtained by calculation, is less than that derived from the shadow,
12. By the number of degrees of the difference; then, turning back, it has moved westward by the amount of difference, when the calculated longitude is greater. In like manner, the equatorial shadow which is cast at mid-day at one's place of observation.
13. Upon the north and south line of the dial—that is the equinoctial shadow *(vishuvatprabhā)* of that place. Radius, multiplied respectively by gnomon and shadow, and divided by the equinoctial hypothenuse,
14. Gives the sines of co-latitude *(lamba)* and of latitude *(aksha)* the corresponding arcs are co-latitude and latitude, always south. The mid-day shadow is the base *(bhuja)*; if radius be multiplied by that,
15. And the product divided by the corresponding hypothenuse, the result, converted to arc, is the sun's zenith-distance *(nata),* in minutes: this, when the base is south, is north, and then the base is north, is south. Of the sun's zenith-distance and his declination, in minutes,

16. Take the sum, when their direction is different—the difference, when it is the same; the result is the latitude, in minutes. From this find the sine of latitude; subtract its square from the square of radius, and the square-root of the remainder.
17. The sine of co-latitude. The sine of latitude, multiplied by twelve, and divided by the sine of co-latitude, gives the equinoctial shadow. The difference of the latitude of the place of observation and the sun's meridian zenith-distance in degrees *(natabhāgas),* if their direction be the same, or their sum,
18. If their direction be different is the sun's declination: if the sine of this latter be multiplied by radius and divided by the sine of greatest declination, the result, converted to arc, will be the sun's longitude, if he is in the quadrant commencing with Aries;
19. If in that commencing with Cancer, subtract from a half-circle; if in that commencing with Libra, add a half-circle; if in that commencing with Capricorn, subtract from a circle: the result, in each case, is the true *(sphuṭa)* longitude of the sun at mid-day.
20. To this if the equation of the apsis (*mānda phala*) be repeatedly applied, with a contrary sign, the sun's mean longitude will be found. The sum of the latitude of the place of observation and the sun's declination; if their direction is the same, or, in the contrary case, their difference,
21. Is the sun's meridian zenith-distance *(natānśās);* of that find the base-sine *(bāhujyā)* and the perpendicular-sine *(koṭijyā).* If, then, the base-sine and radius be multiplied respectively by the measure of the gnomon in digits,
22. And divided by the perpendicular-sine, the results are the shadow and hypothenuse at mid-day. The sine of declination, multiplied by the equinoctial hypothenuse, and divided by the gnomon-sine *(śankujivā),*
23. Gives, when farther multiplied by the hypothenuse of any given shadow, and divided by radius *(madhyakarṇa),* the sun's measure of amplitude *(arkāgrā)* corresponding to that shadow. The sum of the equinoctial shadow and the sun's measure of amplitude *(arkāgrā),* when the sun is in the southern hemisphere, is the base, north;
*24. When the sun is in the northern hemisphere, the base is found, if north, by subtracting the measure of amplitude from the equinoctial shadow; if south, by a contrary process-according to the direction of the interval between the end of the shadow and the east and west axis.*
25. The mid-day base is invariably the midday shadow. Multiply the sines of co-latitude and of latitude respectively by the equinoctial shadow and by twelve,
26. And divide by the sine of declination: the results are the hypothenuse when the sun is on the prime vertical *(samamaṇḍala).* When north declination is less than the latitude, then the mid-day
hypothenuse *(śrava),*
27. Multiplied by the equinoctial shadow, and divided by the mid-day measure of amplitude *(agrā),* is the hypothenuse. If the sine of declination of a given time be multiplied by radius and divided by the sine of co-latitude, the result is the sine of amplitude *(agramāurvikā).*

28. And this, being farther multiplied by the hypothenuse of a given shadow at that time, and divided by radius, gives the measure of amplitude *(agrā),* in digits *(angula),* etc. If from half the square of radius the square of the sine of amplitude *(agrajyā)* be subtracted, and the remainder multiplied by twelve,
29. And again multiplied by twelve, and then farther divided by the square of the equinoctial shadow increased by half the square of the gnomon—the result obtained by the wise
30. Is called the "surd"*(karaṇī)*: this let the wise man set down in two places. Then multiply the equinoctial shadow by twelve, and again by the sine of amplitude,
31. And divide as before: the result is styled the "fruit" *(phala).* Add its square to the "surd," and take the square root of their sum; this, diminished and increased by the "fruit," for the southern and northern hemispheres,
32. Is the sine of altitude *(śanku)* of the southern intermediate directions *(vidiś)*; and equally, whether the sun's revolution take place to the south or to the north of the gnomon *(śanku)*— only, in the latter case, the sine of altitude is that of the northern intermediate directions.
33. The square root of the difference of the squares of that and of radius is styled the sine of zenith-distance *(dṛś).* If, then, the sine of zenith-distance and radius be multiplied respectively by twelve, and divided by the sine of altitude,
34. The results are the shadow and hypothenuse at the angles *(koṇa),* under the given circumstances of time and place. If radius be increased by the sine of ascensional difference *(cara)* when declination is north, or diminished by the same, when declination is south,
35. The result is the day-measure *(antyā)* ; this, diminished by the versed sine *(utkramajyā)* of the hour-angle *(nata),* then multiplied by the day-radius and divided by radius, is the "divisor" *(cheda);* the latter, again, being multiplied by the sine of co-latitude *(lamba),* and divided
36. By radius, gives the sine of altitude *(śanku):* subtract its sine from that of radius, and the square root of the remainder is the sine of zenith-distance *(dṛś)* : the shadow and its hypothenuse are found as in the preceding process.
37. If radius be multiplied by a given shadow, and divided by the corresponding hypothenuse, the result is the sine of zenith-distance *(dṛś)*: the square root of the difference between the square of that and the square of radius
38. Is the sine of altitude *(śanku);* which, multiplied by radius and divided by the sine of co-latitude *(lamba),* gives the "divisor" *(cheda);* multiply the latter by radius, and divide by the radius of the diurnal circle,
39. And the quotient is the sine of the sun's distance from the horizon *(unnata)*; this, then, being subtracted from the day-measure *(antyā),* and the remainder turned into arc by means of the table of versed sines, the final result is the hour-angle *(nata),* in respirations (*asu),* east or west.
40. Multiply the sine of co-latitude by any given measure of amplitude *(agrā),* and divide by the corresponding hypothenuse in digits; the result is the sine of declination. This, again, is to be multiplied by radius, and divided by the sine of greatest declination;

41. The quotient, converted into arc, is, in signs, etc., the sun's place in the quadrant; by means of the quadrant is then found the actual longitude of the sun at that point. Upon a given day, the distances of three bases, at noon, in the forenoon, and in the afternoon, being laid off,
42. From the point of intersection of the lines drawn between them by means of two fish-figures *(matsya),* and with a radius touching the three points, is described the path of the shadow. Multiply by the day-radius of three signs, and divide by their own respective day-radii,
43. In succession, the sines of one, of two, and of three signs; the quotients, converted into arc, being subtracted, each from the one following, give, beginning with Aries, the times of rising *(udayāsavas)* at Lankā;
44. Namely sixteen hundred and seventy, seventeen hundred and ninety-five, and nineteen hundred and thirty-five respirations. And these, diminished each by its portion of ascensional difference *(carakhaṇḍa),* as calculated for a given place, are the times of rising at that place.
45. Invert them, and add their own portions of ascensional difference inverted, and the sums are the three signs beginning with Cancer: and these same six, in inverse order, are the other six; commencing with Libra.
46. From the longitude of the sun at a given time are to be calculated the ascensional equivalents of the parts past and to come of the sign in which he is: they are equal to the number of degrees traversed and to be traversed, multiplied by the ascensional equivalent *(udayāsavas)* of the sign, and divided by thirty;
47. Then, from the given time, reduced to respirations, subtract the equivalent, in respirations, of the part of the sign to come, and also the ascensional equivalents *(lagnāsavas)* of the following signs, in succession—so likewise, subtract the equivalents of the part past, and of the signs past, in inverse order;
48. If there be a remainder, multiply it by thirty and divide by the equivalent of the un subtracted sign; add the quotient, in degrees, to the whole signs, or subtract it from them: the result is the point of the ecliptic *(lagna)* which is at that time upon the horizon *(kshitija).*
49. So, from the east or west hour-angle *(nata)* of the sun, in nādīs, having made a similar calculation, by means of the equivalents in right ascension *(lankodayāsavas),* apply the result as an additive or subtractive equation to the sun's longitude: the result is the point of the ecliptic then upon the meridian *(madhyalagna).*
50. Add together the ascensional equivalents, in respirations, of the part of the sign to be traversed by the point having less longitude, of the part traversed by that having greater longitude, and of the intervening signs—thus is made the ascertainment of time *(kālasādhana).*
51. When the longitude of the point of the ecliptic upon the horizon *(lagna)* is less than that of the sun, the time is in the latter part of the night; when greater, it is in

the day-time; when greater than the longitude of the sun increased by half a revolution, it is after sunset.

# CHAPTER IV. OF ECLIPSES, AND ESPECIALLY OF LUNAR ECLIPSES.

CONTENTS:—1, dimensions of the sun and moon; 2-3, measurement of their apparent dimensions; 4-5, measurement of the earth's shadow; 6, conditions of the occurrence of an eclipse; 7-8, ascertainment of longitude at the time of conjunction or of opposition; 9, causes of eclipses; 10-11, to determine whether there will be an eclipse, and the amount of obscuration; 12-15, to find half the time of duration of the eclipse, and half that of total obscuration; 16-17, to ascertain the times of contact and of separation, and, in a total eclipse, of immersion and emergence; 18-21, to determine the amount of obscuration at a given time; 22-23, to find the time corresponding to a given amount of obscuration; 24-25, measurement of the deflection of the ecliptic, at the point occupied by the eclipsed body, from an east and west line; 26, correction of the scale of projection for difference of altitude.

1. The diameter of the sun's disk is six thousand five hundred yojanas; of the moon's, four hundred and eighty.

2. These diameters, each multiplied by the true motion; and divided by the mean motion, of its own planet, give the corrected *(sphuṭa)* diameters. If that of the sun be multiplied by the number of the sun's revolutions in an Age, and divided by that of the moons,

3. Or if it be multiplied by the moon's orbit *(kakshā),* and divided by the sun's orbit, the result will be its diameter upon the moon's orbit: all these, divided by fifteen, give the measures of the diameters in minutes.

4. Multiply the 'earth's diameter by the true daily motion of the moon, and divide by her mean motion: the result is the earth's corrected diameter *(śucī)*. The difference between the earth's diameter and the corrected diameter of the sun

5. Is to be multiplied by the moon' s mean diameter, and, divided by the sun's mean diameter: subtract the result from the earth's corrected diameter *(śucī),* and the remainder is the diameter of the shadow; which is reduced to minutes as before.

6. The earth's shadow is distant half the signs from the sun: when the longitude of the moon's node is the same with that of the shadow, or with that of the sun, or when it is a few degrees greater or less, there will be an eclipse.

7. The longitudes of the sun and moon, at the moment of the end of the day of new moon *(amāvāsyā)*, are equal, in signs, etc.; at the end of the day of full moon *(paurnamāsī)* they are equal in degrees, etc., at a distance of half the signs.

8. When diminished or increased by the proper equation of motion for the time, past or to come, of opposition or conjunction, they are made to agree, to minutes: the place of the node at the same time is treated ill the contrary manner.
9. The moon is the eclipser of the sun, coming to stand underneath it, like a cloud: the moon, moving eastward, enters the earth's shadow, and the latter becomes its eclipser.
10. Subtract the moon's latitude at the time of opposition or conjunction from half the sum of the measures of the eclipsed and eclipsing bodies: whatever the remainder is, that is said to be the amount obscured.
11. When that remainder is greater than the eclipsed body, the eclipse is total; when the contrary, it is partial; when the latitude is greater than the half sum, there takes place no obscuration (*grāsa*).
12. Divide by two the sum and difference respectively of the eclipsed and eclipsing bodies: from the square of each of the resulting quantities subtract the square of the latitude, and take the square roots of the two remainders.
13. These, multiplied by sixty and divided by the difference of the daily motions of the sun and moon, give, in nādīs, etc., half the duration *(sthiti)* of the eclipse, and half the time of total obscuration.
14. Multiply the daily motions by the half-duration, in nādīs, and divide by sixty: the result, in minutes, subtract for the time of contact *(pragraha),* and add for that of separation *(moksha),* respectively;
15. By the latitudes thence derived, the half-duration, and likewise the half-time of total obscuration, are to be calculated anew, and the process repeated. In the case of the node, the proper correction, in minutes, etc., is to be applied in the contrary manner.
16. The middle of the eclipse is to be regarded as occurring at the true close of the lunar day: if from that time the time of half-duration be subtracted, the moment of contact *(grāsa)* is found; if the same be added, the moment of separation.
17. In like manner also, if from and to it there be subtracted and added, in the case of a total eclipse, the half-time of total obscuration, the results will be the moments called those of immersion and emergence.
18. If from half the duration of the eclipse any given interval be subtracted, and the remainder multiplied by the difference of the daily motions of the sun and moon, and divided by sixty, the result will be the perpendicular *(koṭi)* in minutes.
19. In the case of an eclipse *(graha)* of the sun, the perpendicular in minutes is to be multiplied by the mean half-duration, and divided by the true *(sphuṭa)* half-duration, to give the true perpendicular in minutes.
20. The latitude is the base *(bhuja)*: the square root of the sum of their squares is the hypothenuse *(śrava):* subtract this from half the sum of the measures, and the remainder is the amount of obscuration *(grāsa)* at the given time.
21. If that time be after the middle of the eclipse, subtract the interval from the half-duration on the side of separation, and treat the remainder as before: the result is the amount remaining obscured on the side of separation.

22. From half the sum of the eclipsed and eclipsing bodies subtract any given amount of obscuration, in minutes: from the square of the remainder subtract the square of the latitude at the time, and take the square root of their difference.
23. The result is the perpendicular *(koṭi)* in minutes which, in an eclipse of the sun, is to be multiplied by the true, and divided by the mean, half-duration—and this, converted into time by the same manner as when finding the duration of the eclipse, gives the time of the given amount of obscuration *(grāsa)*.
24. Multiply the sine of the hour-angle *(nata)* by the sine of the latitude *(aksha)*, and divide by radius: the arc corresponding to the result is the degrees of deflection *(valanānśās)* , which are north and south in the eastern and western hemispheres *(kapāla)* respectively.
25. From the position of the eclipsed body increased by three signs calculate the degrees of declination: add them to the degrees of deflection, if of like direction; take their difference, if of different direction: the corresponding sine is the deflection *(valana)*—in digits, when divided by seventy.
26. To the altitude in time *(unnata)* add a day and a half, and divide by a half-day; by the quotient divide the latitudes and the disks; the results are the measures of those quantities in digits *(angula)*.

# CHAPTER V.
## Of Parallax in a Solar Eclipse.

CONTENTS:—1, when there is no parallax in longitude, or no parallax in latitude; 2, causes of parallax; 3, to find the orient-sine; 4-5, the meridian-sine; 5-7, and the sines of ecliptic zenith-distance and altitude; 7-8, to find the amount, in time, of the parallax in longitude; 9, its application in determining the moment of apparent conjunction; 10-11, to find the amount, in arc, of the parallax in latitude; 12-13, its application in calculating an eclipse; 14-17, application of the parallax in longitude in determining the moments of contact, of separation, etc.

1. When the sun's place is coincident with the meridian ecliptic-point *(madhyalagna),* there takes place no parallax in longitude (*harija*)*;* farther, when terrestrial latitude *(aksha)* and north declination of the meridian ecliptic-point *(madhyabha)* are the same, there takes place no parallax in latitude *(avanati).*
2. How parallax in latitude arises by reason of the difference of place *(deśa)* and time *(kāla),* and also parallax in longitude *(lambana)* from direction *(diś)* eastward or the contrary—that is now to be explained.
3. Calculate, by the equivalents in oblique ascension *(udayāsavas)* of the observer's place, the orient ecliptic-point *(lagna)* for the moment of conjunction *(parvavinādyas)*: multiply the sine of its longitude by the sine of greatest declination, and divide by the sine of co-latitude *(lamba):* the result is the quantity known as the orient-sine *(udaya).*

4. Then, by means of the equivalents in right ascension *(lankodayāsavas),* find the ecliptic-point *(lagna)* called that of the meridian *(madhya):* of the declination of that point and the latitude of the observer take the sum, when their direction is the same; otherwise, take their difference.
5. The result is the meridian zenith-distance, in degrees *(natānśās):* its sine is denominated the meridian-sine *madhyajyā.* Multiply the meridian-sine by the orient-sine, and divide by radius: square the result,
6. And subtract it from the square of the meridian-sine: the square root of the remainder is the sine of ecliptic zenith-distance. *(dṛkkshepa);* the square root of the difference of the squares of that and radius is the sine of ecliptic-altitude *(dṛggati).*
7. The sine and cosine of meridian zenith-distance *(natānśās)* are the approximate *(asphuṭa)* sines of ecliptic zenith-distance and altitude *(dṛkkshepa, dṛggati).* Divide the square of the sine of one sign by the sine called that of ecliptic-altitude *(dṛggatijīvā);* the quotient is the "divisor" *(cheda).*
8. By this "divisor" divide the sine of the interval between the meridian ecliptic-point *(madhyalagna)* and the sun's place: the quotient is to be regarded as the parallax in longitude *(lambana)* of the sun and moon, eastward or westward, in nādīs, etc.
9. When the sun's longitude is greater than that of the meridian ecliptic-point *(madhyalagna),* subtract the parallax in longitude from the end of the lunar day; when less, add the same: repeat the process until all is fixed.
10. If the sine of ecliptic zenith-distance *(dṛkkshepa)* be multiplied by the difference of the mean motions of the sun and moon, and divided by fifteen time's radius, the result will be the parallax in latitude *(avanati).*
11. Or the parallax in latitude is the quotient arising form dividing the sine of the ecliptic zenith-distance (*dṛkkshepa*) by seventy, or, from multiplying it by forty-nine, and dividing it by radius.
12. The parallax in latitude is to be regarded as south or north according to the direction of the meridian-sine (*madhyajyā*). When it and the moon's latitude are like direction, take their sum; otherwise, their difference:
13. With this calculate the half-duration (*sthiti*), half total obscuration (*vimarda*), amount of obscuration (*grāsa*), etc., in the manner already taught; likewise the scale of projection (*pramāna*), the deflection (*valana*), the required amount of obscuration, etc., as in the case of a lunar eclipse.
14. For the end of the lunar day, diminished and increased by the half-duration, as formerly, calculate again the parallax in longitude for the times of contact *(grāsa)* and of separation *(moksha),* and find the difference between these and the parallax in longitude *(harija)* for the middle of the eclipse.
15. If, in the eastern hemisphere, the parallax in longitude for the contact is greater than that for the middle, and that for the separation less; and if, in the western hemisphere, the contrary is the case—

16. Then the difference of parallax in longitude is to be added to the half-duration on the side of separation, and likewise on that of contact *(pragrahaṇa);* when the contrary is true, it is to be subtracted.
17. These rules are given for cases where the two parallaxes are in the same hemisphere: where they are in different hemispheres, the sum of the parallaxes in longitude is to be added to the corresponding half-duration. The principles here stated apply also to the half-time of total obscuration.

# CHAPTER VI.
# Of the Projection of Eclipses.

Contents: —1, value of a projection; 2-4, general directions; 5-6, how to lay off the deflection and latitude for the beginning and end of the eclipse; 7, to exhibit the points of contact and separation; 8-10, how to lay off the deflection for the middle of the eclipse; 11, to show the amount of greatest obscuration; 12, reversal of directions in the western hemisphere; 13, least amount of obscuration observable; 14-16, to draw the path of the eclipsing body; 17-19, to show the amount of obscuration at a given time; 20-22, to exhibit the points of immersion and emergence in a total eclipse; 23, color of the part of the moon obscured; 24, caution as to communicating a knowledge of these matters.

1. Since, without a projection *(chedyaka)* , the precise *(sphuṭa)* differences of the two eclipses are not understood, I shall proceed to explain the exalted doctrine of the projection.
2. Having fixed, upon a well prepared surface, a point, describe from it, in the first place, with a radius of forty-nine digits *(angula),* a circle for the deflection *(valana)*:
3. Then a second circle, with a radius equal to half the sum of the eclipsed and eclipsing bodies; this is called the aggregate circle *(samāsa);* then a third, with a radius equal to half the eclipsed body.
4. The determination of the directions, north, south, east, and west is as formerly. In a lunar eclipse, contact *(grahaṇa)* takes place on the east, and separation *(moksha)* on the west; in a solar eclipse, the contrary.
5. In a lunar eclipse, the deflection *(valana)* for the contact is to be laid off in its own proper direction; but that for the separation in reverse; in an eclipse of the sun, the contrary is the case.
6. From the extremity of either deflection draw a line to the centre: from the point where that cuts the aggregate-circle *(samāsa)* are to be laid off the latitudes of contact and of separation.
7. From the extremity of the latitude, again draw a line to the central point: where that, in either case, touches the eclipsed body, there point out the contact and separation.

8. Always, in a solar eclipse, the latitudes are to be drawn in the figure *(parilekha)* in their proper direction; in a lunar eclipse, in the opposite direction. In accordance with this, then, for the middle of the eclipse,

9. The deflection is to be laid off-eastward, when it and the latitude are of the same direction; when they are of different directions, it is to be laid off westward: this is for a lunar eclipse; in a solar, the contrary is the case.

10. From the end of the deflection, again, draw a line to the central point, and upon this line of the middle layoff the latitude, in the direction of the deflection.

11. From the extremity of the latitude describe a circle with a radius equal to half the measure of the eclipsing body: whatever of the disk of the eclipsed body is enclosed within that circle, so much is swallowed up by the darkness *(tamas).*

12. By the wise man who draws the projection *(chedyaka),* upon the ground or upon a board, a reversal of directions is to be made in the eastern and western hemispheres.

13. Owing to her clearness, even the twelfth part of the moon, when eclipsed *(grasta),* is observable; but, owing to his piercing brilliancy, even three minutes of the sun, when eclipsed, are not observable.

14. At the extremities of the latitudes make three points of corresponding names; then, between that of the contact and that of the middle, and likewise between that of the separation and that of the middle,

15. Describe two fish-figures *(matsya):* from the middle of these having drawn out two lines projecting through the mouth and tail, wherever their intersection takes place,

16. There, with a line touching the three points, describe an arc: that is called the path of the eclipsing body, upon which the latter will move forward.

17. From half the sum of the eclipsed and eclipsing bodies subtract the amount of obscuration, as calculated for any given time: take a little stick equal to the remainder, in digits, and, from the central point,

18. Lay it off toward the path upon either side—when the time is before that of greatest obscuration, toward the side of contact; when the obscuration is decreasing, in the direction of separation—and where the stick and the path of the eclipsing body.

19. Meet one another, from that point describe a circle with a radius equal to half the eclipsing body: whatever of the eclipsed body is included within it, that points out as swallowed up by the darkness *(tamas).*

20. Take a little stick equal to half the difference of the measures *(māna),* and lay it off in the direction of contact, calling it the stick of immersion *(nimīlana)*: where it touches the path,

21. From that point, with a radius equal to half the eclipsing body, draw a circle, as in the former case: where this meets the circle of the eclipsed body, there immersion takes place.

22. So also for the emergence *(unmīlana)*, lay it off in the direction of separation, and describe a circle, as before: it will show the point of emergence in the manner explained.

23. The part obscured, when less than half, will be dusky *(sadhūmra)* ; when more than half it will be black; when emerging, it is dark copper-color *(kṛshnatāmra)*; when the obscuration if total, it is tawny *(kapila)*.
24. This mystery of the gods is not to be imparted indiscriminately: it is to be made known to the well-tried pupil, who remains a year under instruction.

# CHAPTER VII.
# OF PLANETARY CONJUNCTIONS.

CONTENTS :—1, general classification of planetary conjunctions; 2-6, method of determining at what point on the ecliptic, and at what time, two planets will come to have the same longitude; 7-10, how to find the point on the ecliptic to which a planet, having latitude, will be referred by a circle passing through the north and south points of the horizon; 11, when a planet must be so referred; 12, how to ascertain the interval between two planets when in conjunction upon such a north and south line; 18-14, dimensions of the lesser planets; 15-18, modes of exhibiting the coincidence between the calculated and actual places of the planets; 18-20, definition of different kinds of conjunction; 20-21, when a planet, in conjunction, is vanquished or victor; 22, farther definition of different kinds of conjunction; 28, usual prevalence of Venus in a conjunction; 28, planetary conjunctions with the moon; 24, conjunctions apparent only; why calculated.

1. Of the star-planets there take place, with one another, encounter *(yuddha)* and conjunction *(samāgama)*; with the moon, conjunction *(samāgama);* with the sun, heliacal setting *(astamana)*.
2. When the longitude of the swift-moving planet is greater than that of the slow one, the conjunction *(samyoga)* is past; otherwise, it is to come: this is the case when the two are moving eastward; if, however, they are retrograding (*vakrin*), the contrary is true.
3. When the longitude of the one moving eastward is greater, the conjunction (*samāgama*) is past; but when that of the one that is retrograding is greater, it is to come. Multiply the distance in longitude of the planets, in minutes, by the minutes of daily motion of each,
4. And divide the products by the difference of daily motion, if both are moving with direct, or both with retrograde, motion: if one is retrograding, divide by the sum of daily motions.
5. Te quotient, in minutes, etc., is to be subtracted when the conjunction is past, and added when it is to come: if the two are retrograding, the contrary: if one is retrograding, the quotients are additive and subtractive respectively.
6. Thus the two planets, situated in the zodiac, are made to be of equal longitude, to minutes. Divide in like manner the distance in longitude, and a quotient is obtained which is the time, in days, etc.

7. Having calculated the measure of the day and night, and likewise the latitude *(vikshepa),* in minutes; having determined the meridian-distance *(nata)* and altitude *(unnata),* in time, according to the corresponding orient ecliptic-point *(lagna)*—
8. Multiply the latitude by the equinoctial shadow, and divide by twelve; the quotient multiply by the meridian-distance in nādīs, and divide by the corresponding half-day:
9. The result, when latitude is north, is subtractive in the eastern hemisphere, and additive in the western; when latitude is south, on the other hand, it is additive in the eastern hemisphere, and likewise subtractive in the western.
10. Multiply the minutes of latitude by the degrees of declination of the position of the planet increased by three signs: the result, in seconds *(vikalā),* is additive or subtractive, according as declination and latitude are of unlike or like direction.
11. In calculating the conjunction *(yoga)* of a planet and an asterism *(nakshatra),* in determining the setting and rising of a planet, and in finding the elevation of the moon's cusps, this operation for apparent longitude *(dṛkkarman)* is first prescribed.
12. Calculate again the longitudes of the two planets for the determined time, and from these their latitudes: when the latter are of the same direction, take their difference; otherwise, their sum: the result is the interval of the planets.
13. The diameters upon the moon's orbit of Mars, Saturn, Mercury, and Jupiter, are declared to be thirty, increased successively by half the half ; that of Venus is sixty.
14. These, divided by the sum of radius and the fourth hypothenuse, multiplied by two, and again multiplied by radius, are the respective corrected *(sphuṭa)* diameters: divided by fifteen. they are the measures *(māna)* in minutes.
15. Exhibit, upon the shadow-ground, the planet at the extremity of its shadow reversed: it is viewed at the apex of the gnomon in its mirror.
16. Take two gnomons, five cubits *(hasta)* in height, stationed according to the variation of direction, separated by the interval of the two planets, and buried at the base one cubit.
17. Then fix the two hypothenuses of the shadow, passing from the extremity of the shadow through the apex of each gnomon: and, to a person situated at the point of union of the extremities of the shadow and hypothenuse, exhibit.
18. The two planets in the sky, situated at the apex each of its own gnomon, and arrived at a coincidence of observed place (*dṛś*). When there is contact of the stars, it is styled "depiction" (*ullekha*); when there is separation, "division" (*bheda*);
19. An encounter (*yuddha*) is called "ray-obliteration" (*anśuvimarda*) when there is a mutual mingling of rays: when the interval is less than a degree, the encounter is named "dexter" (*apasavya*)—if, in this case, one be faint (*aṇu*).
20. If the interval be more than a degree, it is "conjunction" (*samāgama*), if both are endued with power (*bala*). One that is vanquished (*jita*) in a dexter encounter (*apasavya yuddha*), one that is covered, faint (*aṇu*), destitute of brilliancy,

21. One that is rough, colorless, struck down (*vidhvasta*), situated to the south, is utterly vanquished (*vijita*). One situated to the north, having brilliancy, large, is victor (*jayin*)—and even in the south if powerful (*balin*).
22. Even when closely approached, if both are brilliant, it is "conjunction" (*samāgama*): if the two are very small, and struck down, it is "front" *(kuṭā)* and "conflict" *(vigraha),* respectively.
23. Venus is generally victor, whether situated to the north or to the south. Perform in like manner the calculation of the conjunction *(samyoga)* of the planets with the moon.
24. Unto the good and evil fortune of men is this system set forth: the planets move on upon their own paths, approaching one another at a distance.

# CHAPTER VIII.
# OF THE ASTERISMS.

*CONTENTS:—1-9, positions of the asterisms; 10-19, of certain fixed stars; 12, direction to test by observation the accuracy of these positions; 13, splitting of Rohiṇī's wain; 14-10, how to determine the conjunction of a planet with an asterism; 16-19, which is the junction-star in each asterism; 20-21, positions of other fixed stars.*

1. Now are set forth the positions of the asterisms *(bha),* in minutes. If the share of each one, then, be multiplied by ten, and increased by the minutes in the portions *(bhoga)* of the past asterisms *(dhishṇya),* the result will be the polar longitudes *(dhruva).*
2. Forty-eight, forty, sixty-five, fifty-seven, fifty-eight, four, seventy-eight, seventy-six, fourteen,
3. Fifty-four, sixty-four, fifty, sixty, forty, seventy-four, seventy-eight, sixty-four.
4. Fourteen, six, four: Uttara-Ashāḍhā, *(vaisva)* is at the middle of the portion *(bhoga)* of Pūrva-Ashāḍhā *(apya)* ; Abhijit, likewise, is at the end of Pūrva-Ashāḍhā; the position of Śravaṇa is at the end of Uttara-Ashāḍhā;
5. Śravishthā, on the other hand, is at the point of connection of the third and fourth quarters *(pada)* of Śravaṇa: then, in their own portions, eighty, thirty-six, twenty-two,
6. Seventy-nine. Now their respective latitudes, reckoned from the point of declination *(apakrama)* of each: ten, twelve, five, north; south, five, ten, nine;
7. North, six; nothing; south, seven; north, nothing, twelve, thirteen; south, eleven, two; then thirty-seven, north;
8. South, one and a half, three, four, nine, five and a half, five; north, also, sixty, thirty, and also thirty-six;
9. South, half a degree; twenty-four, north, twenty-six degrees; nothing—for Aśvinī *(dasra),* etc., in succession.
10. Agastya is at the end of Gemini, and eighty degrees south; and Mṛgavyādha is situated in the twentieth degree of Gemini;

11. His latitude *(vikshepa)*, reckoned from his point of declination *(apakrama)*, is forty degrees south: Agni *(hutabhuj)* and Brahmahṛdaya are in Taurus, the twenty-second degree;
12. And they are removed in latitude *(vikshipta),* northward, eight and thirty degrees respectively. Having constructed a sphere, one may examine the corrected *(sphuṭa)* latitude and polar longitude *(dhruvaka).*
13. In Taurus, the seventeenth degree, a planet of which the latitude is a little more than two degrees, south, will split the wain of Rohiṇī.
14. Calculate, as in the case of the planets, the day and night of the asterisms, and perform the operation for apparent longitude *(dṛkkarman),* as before: the rest is by the rules for the conjunction *(melaka)* of planets, using the daily motion of the planet as a divisor: the same is the case as regards the time.
15. When the longitude of the planet is less than the polar longitude *(dhruvaka)* of the asterism, the conjunction *(yoga)* is to come; when greater, it is past: when the planet is retrograding *(vakragati),* the contrary is to be recognized as true of the conjunction *(samāgama).*
16. Of the two Phalgunīs, the two Bhādrapadās, and like wise the two Ashāḍhās, of Viśākha, Aśvinī, and Mṛgaśīrsha *(sāumya),* the junction-star *(yogatārā)* is stated to be the northern *(uttara)* :
17. That which is the western northern star, being the second situated westward, that is the junction-star of Hasta; of Śravishṭhā it is the western :
18. Of Jyeshṭhā, Śravaṇa, Anurādhā *(maitra),* and Pushya *(bārhaspatya),* it is the middle star: of Bharaṇī, Kṛttikā *(āgneya),* and Maghā *(pitrya)* , and likewise of Revatī. it is the southern:
19. Of Rohiṇī, Punarvasu *(āditya),* and Mūla, it is the eastern, and so also of Āśleshā *(sārpa)*: in the case of each of the others, the junction-star *(yogatārakā)* is the great *(sthūla)* one,
20. Situated five degrees eastward from Brahmahṛdaya is Prajāpati: it is at the end of Taurus, and thirty-eight degrees north.
21. Apāmvatsa is five degrees north from Citrā: somewhat greater than it, as also six degrees to the north of it, is Āpas.

# CHAPTER IX.
# OF HELIACAL RISINGS AND SETTINGS.

CONTENTS:—1, subject of the chapter; 2-3, under what circumstances, and at which horizon, the planets rise and set heliacally; 4-5, method of calculating their distances in oblique ascension from the sun; 6-9, distances from the sun at which they disappear and re-appear; 10-11, how to find the time of heliacal setting or rising, past or to come; 12-15, distances from the sun at which the asterisms and fixed stars disappear and re-appear; 16-17, mode of determining their times of rising and setting; 18, what asterisms and stars never set heliacally.

1. Now is set forth the knowledge of the risings *(udaya)* and settings *(astamaya)* of the heavenly bodies of inferior brilliancy, whose orbs are overwhelmed by the rays of the sun.
2. Jupiter, Mars, and Saturn, when their longitude is greater than that of the sun, go to their setting in the west; when it is less, to their rising in the east: so likewise Venus and Mercury, when retrograding.
3. The moon, Mercury, and Venus, having a swifter motion, go to their setting in the east when of less longitude than the sun; when of greater, to their rising in the west.
4. Calculate the longitudes of the sun and of the planet—in the west, for the time of sunset; in the east, for that of sunrise— and then make also the calculation of apparent longitude *(dṛkkarman)* of the planet.
5. Then the ascensional equivalent, in respirations, of the interval between the two *(lagnāntaraprāṇās)* will give, when divided by sixty, the degrees of time *(kālāṇśās);* or, in the west, the ascensional equivalent, in respirations, of the interval between the two when increased each by six signs.
6. The degrees of setting *(astāṇśās)* are, for Jupiter eleven; for Saturn, fifteen; for Mars, moreover, they are seventeen:
7. Of Venus, the setting in the west and the rising in the east take place, by reason of her greatness, at eight degrees; the setting in the east and the rising in the west occur, owing to her inferior size, at ten degrees:
8. So also Mercury makes his setting and rising at a distance from the sun of twelve or fourteen degrees, according as be is retrograding or rapidly advancing.
9. At distances, in degrees of time *(kālabhāgās),* greater , than these, the planets become visible to men; at less distances they become invisible, their forms being swallowed up *(grasta)* by the brightness of the sun.
10. The difference, in minutes, between the numbers thus stated and the planet's degrees of time *(kālāṇśās),* when divided by the difference of daily motions—or, if the planet be retrograding, by the sum of daily motions—gives a result which is the time, in days, etc.
11. The daily motions, multiplied by the corresponding ascensional equivalents *(tallagnāsavas),* and divided by eighteen hundred, give the daily motions in time *(kālagati);* by means of these is found the distance, in days etc., of the time past or to come.
12. Svātī , Agastya, Mṛgavyādha, Citrā, Jyeshṭhā, Punarvasu, Abhijit, and Brahmahṛdaya rise and set at thirteen degrees.
13. Hasta, Śravaṇa, the Phalgunīs. Śravishṭhā, Rohinī, and Maghā become visible at fourteen degrees; also Viśākhā and Aśvinī.
14. Kṛttikā, Anurādhā *(maitra),* and Mūla, and likewise Āśleshā and Ārdrā *(rāudrarksha),* are seen at fifteen degrees; so, too, the pair of Ashāḍhās.
*15. Bharaṇī , Pushya, and Mṛgaśirsha, owing to their faintness, are seen at twenty-one degrees; the rest of the asterisms become visible and invisible at seventeen degrees.*

16. The degrees of visibility *(dṛśyānśās),* if multiplied by eighteen hundred and divided by the corresponding ascensional equivalent *(udayāsavas)*, give, as a result, the corresponding degrees on the ecliptic *(kshetrānśās)*; by means of them, likewise, the time of visibility and of invisibility may be ascertained.
17. Their rising takes place in the east, and their setting in the west; the calculation of their apparent longitude *(dṛkkarman)* is to be made according to previous rules; the ascertainment of the time, in days etc., is always by the daily motion of the sun alone.
18. Abhijit, Brahmahṛdaya, Svātī, Śravaṇa *(vāishṇava)*, Śravishṭhā *(vāsava)*, and Uttara-Bhādrapada *(ahirbudhnya)*, owing to their northern situation, are not extinguished by the sun's rays.

# CHAPTER X.
# OF THE MOON'S RISING AND SETTING, AND OF THE ELEVATION OF HER CUSPS.

CONTENTS—: 1, of the heliacal rising and setting of the moon; 2-5, how to find the interval from sunset to the setting or rising of the moon; 6-8, method of determining the moon's relative altitude and distance from the sun at sunset; 9, to ascertain the measure of the illuminated part of her disk; 10-14, method of delineating the moon's appearance at sunset; 15, how to make the same calculation and delineation for sunrise.

1. The calculation of the heliacal rising *(udaya)* and setting *(asta)* of the moon, too, is to be made by the rules already given. At twelve degrees distance from the sun she becomes visible in the west, or invisible in the east.
2. Add six signs to the longitudes of the sun and moon respectively, and find, as in former processes, the ascensional equivalent, in respirations, of their interval *(lagnāntarāsavas)* : if the sun and moon be in the same sign, ascertain their interval in minutes.
*3. Multiply the daily motions of the sun and moon by the result, in* nāḍīs, *and divide by sixty; add to the longitude of each the correction for its motion, thus found, and find anew their interval, in respirations*
4. And so on, until the interval, in respirations, of the sun and moon is fixed: by so many respirations does the moon, in the light half *(śukla),* go to her setting after the sun.
5. Add half a revolution to the sun's longitude, and calculate the corresponding interval, in respirations: by so many respirations does the moon, in the dark half-month *(kṛshṇapaksha),* come to her rising after sunset.
6. Of the declinations of the sun and moon, if their direction be the same, take the difference; in the contrary case, take the sum: the corresponding sine is to be regarded as south or north, according to the direction of the moon from the sun.

7. Multiply this by the hypothenuse of the moon's mid-day shadow, and, when it is north, subtract it from the sine of latitude *(aksha)* multiplied by twelve; when it is south, add it to the same.
8. The result, divided by the sine of co-latitude *(lamba),* gives the base *(bhuja),* in its own direction; the gnomon is the perpendicular *(koṭi);* the square root of the sum of their square is the hypothenuse.
9. The number of minutes in the longitude of the moon diminished by that of the sun gives, when divided by nine hundred, her illuminated part *(śukla)* : this, multiplied by the number of digits *(angula)* of the moon's disk, and divided by twelve, gives the same corrected *(sphuṭa).*
10. Fix a point, calling it the sun: from that lay off the base, in its own proper direction; then the perpendicular, towards the west; and also the hypothenuse, passing through the extremity of the perpendicular and the central point.
11. From the point of intersection of the perpendicular and the hypothenuse describe the moon's disk, according to its dimensions at the given time. Then, by means of the hypothenuse, first make a determination of directions;
12. And layoff upon the hypothenuse, from the point of its intersection with the disk, in an inward direction, the measure of the illuminated part: between the limit of the illuminated part and the north and south points draw two fish-figures *(matsya)*;
13. From the point of intersection of the lines passing through their midst describe an arc touching the three points: as the disk already drawn appears, such is the moon upon that day.
14. After making a determination of directions by means of the perpendicular, point out the elevated *(unnata)* cusp at the extremity of the cross-line: having made the perpendicular *(koṭi)* to be erect *(unnata),* that is the appearance of the moon.
15. In the dark half-month subtract the longitude of the sun increased by six signs from that of the moon, and calculate, in the same manner as before, her dark part. In this case lay off the base in a reverse direction, and the circle of the moon on the west.

# CHAPTER XI.
# OF CERTAIN MALIGNANT ASPECTS OF THE SUN AND MOON.

CONTENTS:—1-5, definition and description of the malignant aspects of the sun and moon, when of equal declination; 6-11, to find the longitude of the sun and moon, when their declinations are equal; 12-13, to ascertain the corresponding time; 14-15, to determine the duration of the aspect, and the moment of its beginning and end; 16-18, its continuance and its influences; 19, when such an aspect may occur more than once, or not at all; 20, occurrence of the yoga of like name and character; 21, of unlucky points in the circle of asterisms; 22, caution as to these unlucky aspects and points; 23, introductory to the following chapters.

1. When the sun and moon are upon the same side of either solstice, and when, the sum of their longitudes being a circle, they are of equal declination, it is styled *vāidhrta.*
2. When the moon and sun are upon opposite sides of either solstice, and their minutes of declination are the same, it is *vyatīpāta,* the sum of their longitudes being a half-circle.
3. Owing to the mingling of the nets of their equal rays, the fire arising from the wrathfulness of their gaze, being driven on by the provector *(pravaha),* is originated unto the calamity of mortals.
4. Since a fault *(pāta)* at this time often causes the destruction of mortals, it is known as *vyatipata,* or, by a difference of title, *vāidhṛti.*
5. Being black, of frightful shape, bloody-eyed, big-bellied, the source of misfortune to all, it is produced again and again.
6. When the longitudes of the sun and moon, being increased by the degrees, etc., found for the coincidence of the solstice with its observed place, are together nearly a circle or nearly a half-circle, calculate the corresponding declinations.
7. Then, if the declination of the moon, she being in an odd quadrant, is, when corrected by her latitude *(vikshepa),* greater than the declination of the sun, the aspect *(pāta)* is already past;
8. If less, it is still to come: in an even quadrant, the contrary is the case. If the moon's declination is to be subtracted from her latitude, the rules as to the quadrant are to be reversed.
9. Multiply the sines of the two declinations by radius, and divide by the sine of greatest declination: the difference of the arcs corresponding to the results, or half that difference, is to be added to the moon's longitude when the aspect *(pāta)* is to come;
10. And is to be subtracted from the moon's longitude when the aspect is past. If the same quantity be multiplied by the sun's motion and divided by the moon's motion, the result is an equation, in minutes, which is to be applied to the sun's place, in the same direction as the other to the moon's.
11. So also is to be applied, in the contrary direction, a like equation to the place of the moon's node. This operation is to be repeated, until the declinations of the two bodies come to be the same.
12. The aspect *(pāta)* is at the time of equality of declinations; if, then, the moon's longitude, as thus increased or diminished, be less than her longitude at midnight, the aspect is past; if greater, it is to come.
13. The minutes of interval between the moon's longitude as finally established and that at midnight give, when multiplied by sixty and divided by the moon's daily motion, the time of the aspect, in nādīs.

14. Multiply the half-sum of the dimensions *(māna)* of the sun and moon by sixty, and divide by the difference of their daily motions: the result is half the duration *(sthiti),* in *nādī*, etc.
15. The corrected *(sphuṭa)* time of the aspect *(pāta)* is the middle: if that be diminished by the half-duration, the result is the time of the commencement; if increased by the same, it is the time of the end.
16. The time intervening between the moments of the beginning and end is to be looked upon as exceedingly terrible, having the likeness of a consuming fire, forbidden for all works.
17. While any parts of the disks of the sun and moon have the same declination, so long is there a continuance of this aspect, causing the destruction of all works.
18. So, from a knowledge of the time of its occurrence, very great advantage is obtained, by means of bathing, giving, prayer, ancestral offerings, vows, oblations, and other like acts.
19. When the equality of declinations of the sun and moon takes place in the neighbourhood of the equator, the aspect may then again occur a second time: in the contrary case, it may fail to occur.
20. If the sum of the longitudes of the sun and moon, in minutes, on being divided by the portion *(bhoga)* of an asterism *(bha),* yields a quotient between sixteen and seventeen, there is another, a third, *vyatīpāta.*
21. Of the asterisms *(dhishṇya)* Āśleshā *(sārpa),* Jyeshṭhā *(āindra),* and Revatī *(paushṇya),* the last quarters are junctions of the asterisms *(bhasandhi)*; the first quarter in the asterisms following these respectively is styled *gaṇḍānta.*
22. In all works, one must avoid the terrible trio of *vyatīpātas,* as also the trio of *gaṇḍāntas,* and this trio of junctions of asterisms.
23. Thus hath been related that supreme, pure, excellent, mysterious, and grand system of the heavenly bodies: what else dost thou desire to know?

# CHAPTER XII.
# COSMOGONY, GEOGRAPHY, DIMENSIONS OF THE CREATION.

CONTENTS:—1-9, inquiries; 10-28, development of the creative agencies, of the elements and of the existing creation; 29-31, form and disposition of the stellar and planetary systems; 32-44, situation, form, structure, and divisions of the earth; 45-72, varying phenomena of night and day in different latitudes and zones; 73-77, revolutions of the stars and planets; 78-79, regents of the different divisions of time; 80-90, dimensions of the planetary, stellar, and ethereal orbits.

1. Then the demon Maya, prostrating himself with hands suppliantly joined before him who derived his being from the part of the Sun, and revering him with exceeding devotion, inquired as follows:

2. O blessed one! of what measure is the earth? of what form? how supported? how divided? and how are there in it seven interterranean *(pātāla)* earths?
3. And how does the sun cause the varying distinction of day and night? how does he revolve about the earth, enlightening all creatures?
4. For what reason are the day and night of the gods and of the demons opposed to one another? or how does that take place by means of the sun's completion of his revolution?
5. Why does the day of the Fathers consist of a month, but that of mortals of sixty nādīs? for what reason is not this latter everywhere the case?
6. Whence is it that the regents of the days, years, months, and hours *(horā)* are not the same? How does the circle of asterisms *(bhagaṇa)* revolve? what is the support of it with the planets?
7. The orbits of the planets and stars, uplifted from the earth one above another—what are their heights? what their intervals? what their dimensions? and what the order in which they are fixed?
8. Why are the rays of the sun hot in the summer, and not so n the winter? how far do his rays penetrate? How many modes of measuring time *(māna)* are there? and how are they employed?
9. Resolve these my difficulties, O blessed one, creator of creatures! for there is not found besides thee another resolver, who beholdeth all things.
10. Having heard the words thus uttered with, devotion by Maya, he then again promulgated this mysterious and supreme Book *(adhyāya):*
11. Listen with concentrated attention: I will proclaim the secret doctrine called the transcendental *(adhyātma):* there is nothing which may not be bestowed on those who are exceedingly devoted to me.
12. Vāsudeva, the supreme principle of divinity *(brahman)* whose form is all that is *(tat),* the supreme Person *(purusha),* unmanifested, free from qualities, superior to the twenty-five principles, imperishable,
13. Contained within matter *(prakṛti),* divine, pervading everything, without and within, the attractor—he, having in the first place created the waters, deposited in them energy.
14. That became a golden egg, on all sides enveloped in darkness: in it first became manifested the unrestrained, the everlasting one.
15. He in the scripture *(chandas)* is denominated the golden-wombed *(hiraṇyagarbha),* the blessed; as being the first *(ādi)* existence, he is called Āditya ; as being generator, the sun.
16. This sun, likewise named Savitar, the supreme source of light *(jyotis)* upon the border of darkness—he revolves, bringing beings into being, the creator of creatures.
17. He is extolled as natural illuminator, destroyer of darkness, great. The Hymns *(ṛcas)* are his disk, the Songs *(sāmāni)* his beams, the Liturgy *(yajūnshi)* his form.
18. He, the blessed one, is composed of the trio of sacred scriptures, the soul of time, the producer of time, mighty, the soul of the universe, all-penetrating, subtle: in him is the universe established.

19. Having made for his chariot, which is composed of the universe, a wheel consisting of the year, and having yoked the seven metres as his steeds, he revolves continually.
20. Three quarters are immortal, secret; this one quarter hath become manifest. In order to the production of the animated creation, he, the mighty one, produced Brahma, the principle of consciousness *(ahankāra)*.
21. Bestowing upon him the Scriptures *(veda)* as gifts, and establishing him within the egg as grandfather of all worlds, he himself then revolves, causing existence.
22. Then Brahma, wearing the form of the principle of consciousness *(ahankāra)*, produced mind in the creation: from mind was born the moon; from the eyes, the sun, the repository of light;
23. From mind, the ether; thence, in succession, wind, fire, waters, earth—these five elements *(mahābhūta)* were produced by the successive addition of one quality.
24. Agni and Soma, the sun and moon: then Mars etc. were produced, in succession, from light, earth, ether, water, wind.
25. Again, dividing himself twelve-fold, he, the mighty one, produced what is known as the signs; and yet farther, what has the form of the asterisms *(nakshatra)*, twenty-seven-fold.
26. Then he wrought out the whole animate and inanimate creation, from the gods downward, producing forms of matter *(prakṛti)* from the upper, middle, and lower currents *(srotas)*.
27. Having produced them in succession, as stated, by a difference of quality and function, he fashioned the distinctive character of each, according to the showing of the Scripture *(veda)*—
28. That is, of the planets, asterisms, and stars, of the earth, and of the universe, he the mighty one; of gods, demons, and mortals, and of the Perfected *(siddha)*, in their order.
29. This Brahma-egg is hollow; within it is the universe, consisting of earth, sky, etc.; it has the form of a sphere, like a receptacle made of a pair of caldrons.
30. A circle within the Brahma-egg is styled the orbit of the ether *(vyoman)*: within that is the revolution of the asterisms *(bha)*; and likewise, in order, one below the other,

31. Revolve Saturn, Jupiter, Mars, the sun, Venus, Mercury, and the moon; below, in succession, the Perfected *(siddha)*, the Possessors of Knowledge *(vidyādhara)*, and the clouds.
32. Quite in the middle of the egg, the earth-globe *(bhūgola)* stands in the ether, bearing the supreme might of Brahma, which is of the nature of self-supporting force.
33. Seven cavities within it, the abodes of serpents *(nāga)* and demons *(asura)*, endowed with the savor of heavenly plants, delightful, are the interterranean *(pātāla)* earths.
34. A collection of manifold jewels, a mountain of gold, is Meru, passing through the middle of the earth-globe, and protruding on either side.

35. At its upper end are stationed, along with Indra, the gods, and the Great Sages *(maharshi);* at its lower end, in like manner, the demons *(asura)* have their place—each the enemy of the other.
36. Surrounding it on every side is fixed next this great ocean, like a girdle about the earth, dividing the two hemispheres of the gods and of the demons.
37. And on all sides of the midst of Meru, in equal divisions of the ocean, upon islands *(dvīpa),* in the different directions, are the eastern and other cities, fashioned by the gods.
38. At a quadrant of the earth's circumference eastward, in the clime *(varsha)* Bhadrāśva, is the city famed as Yamakoṭi, having walls and gateways of gold.
39. To the southward, in the clime Bhārata, is in like manner, the great city Lankā: to the west, in the clime called Ketumāla, is declared to be the city named Romaka.
40. Northward, in the clime Kuru, is declared to be the city called that of the Perfected *(siddha)*; in it dwell the magnanimous Perfected, free from trouble.
41. These are situated also at a distance from one another of a quadrant of the earth's circumference; to the north of them, at the same distance, is Meru, the abode of the gods *(sura).*
42. Above them goes the sun when situated at the equinoxes; they have neither equinoctial shadow nor elevation of the pole *(akshonnati).*
43. In both directions from Meru are two pole-stars *(dhruvatārā),* fixed in the midst of the sky: to those who are situated in places of no latitude *(niraksha),* both these have their place in the horizon.
44. Hence there is in those cities no elevation of the pole, the two pole-stars being situated in their horizon; but their degrees of co-latitude *(lambaka)* are ninety: at Meru the degrees of latitude *(aksha)* are of the same number.
45. In the half-revolution beginning with Aries, the sun, being in the hemisphere of the gods, is visible to the gods: but while in that beginning with Libra, he is visible to the demons, moving in their hemisphere.
46. Hence, owing to his exceeding nearness, the rays of the sun are hot in the hemisphere of the gods in summer, but in that of the demons in winter: in the contrary season, they are sluggish.
47. At the equinox, both gods and demons see the sun in the horizon; their day and night are mutually opposed to each other.
48. The sun, rising at the first of Aries, while moving on northward for three signs, completes the former half-day of the dwellers upon Meru;
49. In like manner, while moving through the three signs beginning with Cancer, he completes the latter half of their day: he accomplishes the same for the enemies of the gods while moving through the three signs beginning with Libra and the three beginning with Capricorn, respectively.
50. Hence are their night and day mutually opposed to one another; and the measure of the day and night is by the completion of the sun's revolution.
51. Their mid-day and midnight, which are opposed to one another, are at the end of each half-revolution from solstice to solstice *(ayana).* The gods and demons each suppose themselves to be uppermost.

52. Others, too, who are situated upon the same diameter *(samasūtrastha),* think one another underneath—as the dwellers in Bhadrāśva and in Ketumāla, and the inhabitants of Lankā and of the city of the Perfected, respectively.
53. And everywhere upon the globe of the earth, men think their own place to be uppermost: but since it is a globe in the ether, where should there be an upper, or where an under side of it?
54. Owing to the littleness of their own bodies, men, looking in every direction from the position they occupy, behold this earth, although it is globular, as having the form of a wheel.
55. To the gods, this sphere of asterisms revolves toward the right; to the enemies of the gods, toward the left; in a situation of no latitude, directly overhead—always in a westerly direction.
56. Hence, in the latter situation, the day is of thirty nādīs, and the night likewise: in the two hemispheres of the gods and demons there take place a deficiency and an excess, always opposed to one another.
57. During the half-revolution beginning with Aries, there is always an excess of the day to the north, in the hemisphere of the gods—greater according to distance north—and a corresponding deficiency of the night; in the hemisphere of the demons, the reverse.
58. In the half-revolution beginning with Libra, both the deficiency and excess of day and night in the two hemispheres are the opposite of this: the method of determining them, which is always dependent upon situation *(deśa)* and declination, has been before explained.
59. Multiply the earth's circumference by the sun's declination in degrees, and divide by the number of degrees in a circle: the result, in yojanas, is the distance from the place of no latitude where the sun is passing overhead.
*60. Subtract from a quarter of the earth's circumference the number of yojanas thus derived from the greatest declination: at the distance of the remaining number of yojanas*
61. There occurs once, at the end of the sun's half-revolution from solstice to solstice, a day of sixty nādīs, and a night of the same length, mutually opposed to one another, in the two hemispheres of the gods and of the demons.
62. In the intermediate region, the deficiency and excess of day and night are within the limit of sixty nādīs; beyond, this sphere of asterisms *(bha)* revolves perversely.
63. Subtract from a quarter of the earth's circumference the number of yojanas derived from the declination found by the sine of two signs: at that distance from the equator the sun is not seen, in the hemisphere of the gods, when in Sagittarius and Capricorn;
64. So also, in the hemisphere of the demons, when in Gemini and Cancer: in the quarter of the earth's circumference where her shadow is lost, the sun may be shown to be visible.

65. Subtract from the fourth part of the earth's periphery *(kakshā)* the number of yojanas derived from the declination found by the sine of one sign: at the distance from the place of no latitude of the remaining number of yojanas,
66. The sun, when situated in Sagittarius, Capricorn, Scorpio, and Aquarius, is not seen in the hemisphere of the gods; in that of the demons, on the other hand, when in the four signs commencing with Taurus.
67. At Meru, the gods behold the sun, after but a single rising, during the half of his revolution beginning with Aries; the demons in like manner, during that beginning with Libra.
68. The sun, during his northern and southern progresses *(ayana)* revolves directly over a fifteenth part of the earth's circumference, on the side both of the gods and of the demons.
69. Between those limits, the shadow is cast both southward and northward; beyond them, it falls toward the Meru of either hemisphere respectively.
70. When passing overhead at Bhadrāśva, the sun is rising in Bhārata; it is moreover, at that time, midnight in Ketumāla, and sunset in Kuru.
*71. In like manner also he produces, by his revolution, in Bhārata and the other climes, noon, sunrise, midnight, and sunset, reckoning from east to west.*
72. To one going toward Meru, there take place an elevation of the pole *(dhruva)* and a depression of the circle of asterisms; to one going toward the place of no latitude, on the contrary, a depression of the former and an elevation of the latter.
73. The circle of asterisms, bound at the two poles, impelled by the provector *(pravaha)* winds, revolves eternally: attached to that are the orbits of the planets, in their order.
74. The gods and demons behold the sun, after it is once risen, for half a year; the Fathers *(pitaras),* who have their station in the moon, for a half-month *(paksha)*; and men upon the earth, during their own day.
75. The orbit *(kakshā)* of one that is situated higher up is large; that of one situated lower down is small. Upon a great orbit the degrees are great; so also, upon a small one, they are small.
76. A planet situated upon a small circuit *(bhramaṇa)* traverses the circle of constellations *(bhagaṇa)* in a little time; one revolving on a large circle *(maṇḍala),* in a long time.
77. The moon, upon a very small orbit, makes many revolutions: Saturn, moving upon a great orbit, makes, as compared with her, a much less number of revolutions.
78. Counting downward from Saturn, the fourth successively is regent of the day; and the third, in like manner, is declared to be the regent of the year;
79. Reckoning upward from the moon are found, in succession, the regents of the months; the regents of the hours *(horā),* also, occur in downward order from Saturn.
80. The orbit *(kakshā)* of the asterisms *(bha)* is the circuit *(bhramaṇa)* of the sun multiplied by sixty: by so many yojanas does the circle of the asterisms revolve above all.

81. If the stated number of revolutions of the moon in an AEon *(kalpa)* be multiplied by the moon's orbit, the result is to be known as the orbit of the ether: so far do the rays of the sun penetrate.
82. If this be divided by the number of revolutions of any planet in an AEon *(kalpa),* the result will be the orbit of that planet: divide this by the number of terrestrial days, and the result is the daily eastward motion of them all.
83. Multiply this number of yojanas of daily motion by the orbit of the moon, and divide by a planet's own orbit; the result is, when divided by fifteen, its daily motion in minutes.
84. Any orbit, multiplied by the earth's diameter and divided by the earth's circumference, gives the diameter of that orbit; and this, being diminished by the earth's diameter and halved, gives the distance of the planet.
85. The orbit of the moon is three hundred and twenty-four thousand yojanas: that of Mercury's conjunction *(śīghra)* is one million and forty-three thousand, two hundred and nine:
86. That of Venus's conjunction *(śīghra)* is two million, six hundred and sixty-four thousand, six hundred and thirty-seven: next, that of the sun, Mercury, and Venus is four million, three hundred and thirty-one thousand, five hundred:
87. That of Mars, too, is eight million, one hundred and forty-six thousand, nine hundred and nine; that of the moon's apsis *(ucca)* is thirty-eight million, three hundred and twenty-eight thousand, four hundred and eighty-four:
88. That of Jupiter, fifty-one million, three hundred and seventy-five thousand, seven hundred and sixty-four: of the moon's node, eighty million, five hundred and seventy-two thousand, eight hundred and sixty-four:
89. Next, of Saturn, one hundred and twenty-seven million, six hundred and sixty-eight thousand, two hundred and fifty-five: of the asterisms, two hundred and fifty-nine million, eight hundred and ninety thousand, and twelve:
90. The entire circumference of the sphere of the Brahma-egg is eighteen quadrillion, seven hundred and twelve trillion, eighty billion, eight hundred and sixty-four million: within this is the pervasion of the sun's rays.

# CHAPTER XIII. OF THE ARMILIARY SPHERE, AND OTHER INSTRUMENTS.

CONTENTS:—1-13, construction and equipment of the armillary sphere; 13-15, position of certain points and sines upon it; 15-16, its adjustment and revolution; 17-25, other instruments, especially for the determination of time.

1. Then, having bathed in a secret and pure place, being pure, adorned, having worshipped with devotion the sun, the planets, the asterisms *(bha),* and the elves *(guhyaka),*
2. Let the teacher, in order to the instruction of the pupil himself beholding everything clearly, in accordance with the knowledge handed down by successive communication, and learned from the mouth of the master *(guru)*—
3. Prepare the wonder-working fabric of the terrestrial and stellar sphere *(bhūbhagola).* Having fashioned an earth-globe of wood, of the desired size,
4. Fix a staff, passing through the midst of it and protruding at either side, for Meru; and likewise a couple of sustaining hoops (*kakshā*), and the equinoctial hoop;
5. These are to be made with graduated divisions (angula) of degrees of the circle (*bhagaṇa*). Farther—by means of the several day-radii, as adapted to the scale established for those other circles,
6. And by means of the degrees of declination and latitude (vikshepa) marked off upon the latter—at their own respective distances in declination, according to the declination of Aries etc., three
7. Hoops are to be prepared and fastened: these answer also inversely for Cancer, etc. In the same manner, three for Libra etc., answering also inversely for Capricorn, etc.,
8. And situated in the southern hemisphere, are to be made and fastened to the two hoop-supporters. Those likewise of the asterisms *(bha)* situated in the southern and northern hemispheres, of Abhijit,
9. Of the Seven Sages *(saptarshayas),* of Agastya, of Brahma etc., are to be fixed. Just in the midst of all, the equinoctial *(vāishuvatī)* hoop is fixed.
10. Above the points of intersection of that and the supporting hoops are the two solstices *(ayana)* and the two equinoxes *(vishuvat).* From the place of the equinox, with the exact number of degrees, as proportioned to the whole circle,
11. Fix, by oblique chords, the spaces *(kshetra)* of Aries and the rest; and so likewise another hoop, running obliquely from solstice *(ayana)* to solstice,
12. And called the circle of declination *(krānti)*: upon that the sun constantly revolves, giving light: the moon and the other planets also, by their own nodes, which are situated in the ecliptic *(apamaṇḍala),*
13. Being drawn away from it, are beheld at the limit of their removal in latitude *(vikshepa)* from the corresponding point of declination. The orient ecliptic-point (*lagna*) is that at the orient horizon; the occident point (*astamgachat*) is similarly determined.
14. The meridian ecliptic-point (*madhyama*) is as calculated by the equivalents in right ascension (*lankodayās*), for mid-heaven (*khamadhya*) above. The sine which is between the meridian (*madhya*) and the horizon (*kshitija*) is styled the day-measure (*antyā*).
15. And the sine of the sun's ascensional difference (caradala)is to be recognized as the interval between the equator (vishuvat) and the horizon. Having turned upward one's own place, the circle

of the horizon is midway of the sphere.
16. As covered with a casing *(vastra)* and as left uncovered, it is the sphere surrounded by Lokāloka. By the application of water is made ascertainment of the revolution of time.
17. One may construct a sphere-instrument combined with quicksilver: this is a mystery; if plainly described, it would be generally intelligible in the world.
18. Therefore let the supreme sphere be constructed according to the instruction of the preceptor *(guru).* In each successive age *(yuga),* this construction, having become lost, is, by the Sun's
19. Favor, again revealed to some one or other, at his pleasure. So also, one should construct instruments *(yantra)* in order to the ascertainment of time.
20. When quite alone, one should apply quicksilver to the wonder-causing instrument. By the gnomon *(śanku),* staff *(yashṭi),* arc *(dhanus),* wheel *(cakra),* instruments for taking the shadow, of various kinds,
21. According to the instruction of the preceptor *(guru),* is to be gained a knowledge of time by the diligent. By water-instruments, the vessel *(kapāla),* etc., by the peacock, man, monkey, and by stringed sand-receptacles, one may determine time accurately.
22. Quicksilver-holes, water, and cords, ropes *(śulba),* and oil and water, mercury, and sand are used in these: these applications too, are difficult
23. A copper vessel, with a hole in the bottom, set in a basin of pure water, sinks sixty times in a day and night, and is an accurate hemispherical instrument.
24. So also, the man-instrument *(narayantra)* is good in the day-time, and when the sun is clear. The best determination of time by means of determinations of the shadow has been explained.
25. He who thoroughly knows the system of the planets and asterisms, and the sphere, attains the world of the planets in the succession of births, his own possessor.

# CHAPTER XIV. OF THE DIFFERENT MODES OF RECKONING TIME.

CONTENTS:—1-2, enumeration of the modes of .measuring time, and general explanation of their uses; 3, solar time; 4-6, of the periods of eighty-six days; 7-11, of points and divisions in the sun's revolution; 12-13, lunar time; 14, time of the Fathers; 15, sidereal time; 15-16, of the months and their asterisms; 17, of the twelve-year cycle of Jupiter; 18-19, civil, or mean solar, time; 20-21, time of the gods, Prajāpati, and Brahma; 22-26, conclusion of the work.

1. The modes of measuring time *(māna)* are nine, namely those of Brahma, of the gods, of the Fathers, of Prajāpati, of Jupiter, and solar *(saura)* , civil *(sāvana),* lunar, and sidereal time.

2. Of four modes, namely solar, lunar, sidereal, and civil time, practical use is made among men; by that of Jupiter is to be determined the year of the cycle of sixty years; of the rest, no use is ever made,
3. By solar *(sāura)* time are determined the measure of the day and night, the *shaḍaśītikmukhas,* the solstice *(ayana)*, the equinox *(vishuvat),* and the propitious period of the sun's entrance into a sign *(sankrānti).*
4. Beginning with Libra, the *shaḍaśītikmukha* is at the end of the periods of eighty-six *(shaḍaśīti)* days, in succession: there are four of them, occurring in the signs of double character *(dvisvabhāva)*
5. Namely, at the twenty-sixth degree of Sagittarius, at the twenty-second of Pisces, at the eighteenth degree of Gemini, and at the fourteenth of Virgo.
6. From the latter point, the sixteen days of Virgo which remain are suitable for sacrifices: anything given to the Fathers *(pitaras)* in them is inexhaustible.
7. In the midst of the zodiac *(bhasakra)* are the two equinoxes *(vishuvat),* situated upon the same diameter *(samasūtraga),* and likewise the two solstices (*ayana*); these four are well known.
8. Between these are, in each case, two entrances (*sankrānti*); from the immediateness of the entrance are to be known the two feet of Vishnu.
*9. From the sun's entrance (*sankrānti*) into Capricorn, six months are his northern progress (*uttarāyana*); so likewise, form the beginning of Cancer, six months are his southern progress (*dakshiṇāyana*).*
*10. Thence also are reckoned the season (*ṛtu*), the cool season (*śiśira*) and the rest, each prevailing through two signs. These twelve, commencing with Aries, are the month; of them is made up the year.*
11. Multiply the number of minutes in the sun's measure *(māna)* by sixty, and divide by his daily motion: a time equal to half the result, in nādīs, is propitious before the sun's entrance into a sign *(sankrānti),* and likewise after it.
12. As the moon, setting out from the sun, moves from day today eastward, that is the lunar method of reckoning time *(māna)*: a lunar day *(tithi)* is to be regarded as corresponding to twelve degrees of motion.
13. The lunar day *(tithi),* the karaṇa, the general ceremonies, marriage, shaving, and the performance of vows, fastings, and pilgrimages, are determined by lunar time.
14. Of thirty lunar days is composed the lunar month, which is declared to be a day and a night of the Fathers: the end of the month and of the half-month *(paksha)* are at their mid-day and midnight respectively.
15. The constant revolution of the circle of asterisms *(bhacakra)* is called a sidereal day. The months are to be known by the names of the asterisms *(nakshatra),* according to the conjunction *(yoga)* at the end of a lunar period *(parvan).*
16. To the months Kārttika etc., belong, as concerns the conjunction *(samayoga),* the asterisms Kṛttikā etc., two by two: but three months, namely the last, the next to the last and the fifth, have triple asterisms.

17. In Vaiśākha etc., a conjunction *(yoga)* in the dark half-month *(kṛshṇa),* on the fifteenth lunar day *(tithi),* determines in like manner the years Kārttika etc. of Jupiter, from his heliacal setting *(asta)* and rising *(udaya).*
18. From rising to rising of the sun, that is called civil (*sāvana*) reckoning. By that are determined the civil days (*sāvana*), and by these is the regulation of the time of sacrifice;
19. Likewise the removal of uncleanness form child-bearing etc., and the regents of days, months, and years: the mean motion of the planets, too, is computed by civil time.
20. The mutually opposed day and night of the gods (*sura*) and demons (*asura*), which has been already explained, is time of the gods, being measured by the completion of the sun's revolution.
21. The space of a Patriarchate (*manvantara*) is styled time of Prajāpati: in it is no distinction of day from night. An AEon (*kalpa*) is called time of Brahma.
22. Thus hath been told thee that supreme mystery, lofty and wonderful, that sacred knowledge *(brahman),* most exalted; pure, all guilt destroying;
23. And the highest knowledge of the heaven, the stars, and the planets hath been exhibited: he who knoweth it thoroughly obtaineth in the worlds of the sun etc. an everlasting place.
24. With these words, taking leave of Maya, and being suitably worshipped by him, the part of the sun ascended to heaven, and entered his own disk.
25. So then Maya, having personally learned from the sun that divine knowledge, regarded himself as having attained his desire, and as purified from sin.
26. Then, too, the sages *(ṛshi),* learning that Maya had received from the sun this gift, drew near and surrounded him, and, reverently asked the knowledge.
27. And he graciously bestowed upon them the grand system of the planets, of mysteries in the world the most wonderful, and equal to the Scripture *(brahman).*
(The end)

*(Translation of Sūrya-siddhānta by Rev. Ebenezer Burgess; Please see Chapter 8 for further information)

# 13

# Siddhanta Shiromani

*TRANSLATION OF THE GOLĀDHYĀYA OF THE SIDDHĀNTA - ŚIROMAṆI*

## CHAPTER 1

*In praise of the advantages of the study of the Spheric.*

Salutation to *Gaṇesa*!

Invocation

1. Having saluted that god, who when called upon brings all undertakings to a successful issue, and also that goddess, through whose benign favour the tongues of poets, gifted with a flow of words ever new and with elegance, sweetness and playfulness, sport in their mouths as in a place of recreation, as dancing girls adorned with beauty disport themselves in the dance with elegance and with every variety of step, I proceed to indite this work on the Sphere. It has been freed from all error, and rendered intelligible to the lowest capacity.

*Object of the work.*

2. Inasmuch as no calculator can hope to acquire in the assemblage of the learned a distinguished reputation as an Astronomer, without a clear understanding of the principles upon which all the calculations of the mean and other places of the planets are founded, and to remove the doubts which my arise in his own mind, I therefore proceed to treat of the sphere, in such a manner as to make the reasons of all my calculations manifest. On inspecting the Globe they become clear and manifest as if submitted to the eye, and are as completely at command, as the wild apple (*ānwlā*) held in the palm of the hand.

*Ridicule of an ignorance of the Spheric*

3. As a feast with abundance of all things but without clarified butter, and as a kingdom without a king, and an assemblage without eloquent speakers have little to recommend them; so the Astronomer who has no knowledge of the spheric, commands no consideration.
4. As a foolish impudent disputant, who ignorant of grammar (rudely) enters into the company of the learned and vainly prates, is brought to ridicule, and put to shame by frowns and ironical remarks of even children of any smartness, so he, who is ignorant of the spheric, is exposed in an assemblage of the Astronomers, by the various questions of really accomplished Astronomers.

*Object of the Armillary sphere.*

5. The Armillary sphere is said, by the wise, to be a representation of the celestial sphere, for the purpose of ascertaining the proofs of the positions of the Earth, the stars, and the planets: this is a species of figure, and hence it is deemed by the wise to be an object of mathematical calculation.

*In praise of mathematics.*

6. It is said by ancient astronomers that the purpose of the science is judicial astrology, and this indeed depends upon the influence of the horoscope, and this on the true places of the planets: these (true places) can be found only by a perfect knowledge of the spheric. A knowledge of the spheric is not to be attained without mathematical calculation. How then can a man, ignorant of mathematics, comprehend the doctrine of the sphere etc.?

*Who is likely to undertake the study with effect.*

7. Mathematical calculations are of two kinds, Arithmetical and Algebraical: he who has mastered both forms, is qualified if he have previously acquired (a perfect knowledge of) the Grammar (of the Sanskrit Language,) to undertake the study of the various branches of Astronomy. Otherwise he may acquire the name (but never the substantial knowledge) of an Astronomer.

*In praise of Grammar.*

8. He who has acquired a perfect knowledge of grammar, which has been termed *Vedavadana* i. e. the mouth of the *Vedas* and domicile of *Saraswati*, may acquire a knowledge of every other science— nay of the *Vedas* themselves. For this reason it is that none, but he who has acquired a thorough knowledge of Grammar, is qualified to undertake the study of other sciences.

*The opinion of others on this work, quoted with a view of extending the study of it.*

9. O learned man; if you intend the to study the spheric, study the Treatise of *Bhaśkara*, it is neither too concise nor idly diffuse: it contains every essential principle of the science, and is of easy comprehension; it is moreover written in an eloquent style, is made interesting with questions; it imparts to all who study it that manner of correct expression in learned assemblages, approved of by accomplished scholars.

End of Chapter 1.

# CHAPTER II.

*Questions on the General view of the Sphere.*

*Questions regarding the earth.*

1. This Earth being encircled by the revolving planets, remains stationary in the heavens, within the orbits of all the revolving fixed stars; tell me by whom or by what is it supported, that it falls not downwards (in space)?
2. Tell me also, after a full examination of all the various opinions on the subject, its figure and magnitude, how its principal islands mountains and seas are situated in it?

*Questions regarding those calculations used in ascertaining planets' true places and their causes.*

3. Tell me, O my father, why the place of a planet found out from well calculated *Ahargaṇa* (or enumeration of mean terrestrial days, elapsed from the commencement of the *Kalpa*) by applying the rule of proportion to the revolutions in the *Yuga* etc. is not a true one? (i. e. why is it only a mean and not the true place) and why the rules for finding the true places of the different planets are not of the same kind? What are the *Desāntara*, *Udayāntara*, *Bhujāntara*, and *Chara* corrections? What is the *Mandochcha* (slow or 1st Apogee) and *Śīghraochcha* (quick or 2nd Apogee)? What is the node?
4. What is the *Kendra* and that which arises from it (i. e. the sine, cosine, etc. of it)? What is the *Mandaphala* (the first equation) and *Śīghraphala* (the 2nd equation) which depend on the sine of the *Kendra*? Why does the place of a planet become true, when the *Mandaphala* or *Śīghraphala* are (at one time) added to and (at another) subtracted from it? What is the twofold correction called *Dṛikkarma* which learned astronomers have applied (to the true place of a planet) at the rising and setting of the planet? Answer me all these questions plainly, if you have a thorough knowledge of the sphere.

*Questions regarding the length of the day and night.*

Tell me, O you acute Astronomer, why, when the Sun is on the northern hemisphere, is the day long and the night short, and the day short and the night long when the Sun is on the southern hemisphere?

*Questions regarding the length of the day and night of the, gods,* Daityas, Pitṛis *and* Brahmā.

How is it that the day and night of the gods and their enemies *Daityas* correspond in length with the solar years? How is it that the night and day of the *Pitṛis* is equal in length to a (synodic) lunar month, and how is it that the day and night of *Brahmā* is 2000 *Yugas* in length?

*Questions regarding the periods of risings of the signs of the Zodiac.*

Why, O Astronomer, is it that the 12 signs of the Zodiac which are all of equal length, rise in unequal times (even at the Equator,) and why are not those periods or rising the same in all countries?

*Questions as to the places of* Dyujyā, *the* Kujyā, *etc.*

8. Shew me, O learned one, the places of the *Dyujyā* (the radius or the diurnal circle), the *Kujyā* (the sine of that part of the arc of the diurnal circle intercepted between the horizon and the six o'clock line, i. e. of the ascensional difference in terms of a small circle), and show me also the places of the declination, *Sama-śanku*, *Agra* (the sine of amplitude), latitude and co-latitude etc., in this Armillary sphere as these places are in the heavens.

*Questions regarding certain differences in the times and places of solar and lunar Eclipses.*

If the middle of a lunar Eclipse takes place at the end of the *Tithi* (at the full moon), why does not the middle of the solar Eclipse take place in like manner at the change? Why is the Eastern limb of the Moon in a lunar Eclipse first involved in obscurity, and the western limb of the Sun first eclipsed in a solar Eclipse?

*Questions regarding parallaxes.*

9. What, O most intelligent one, is the *Lambana* and what is the *Nati*? why is the *Lambana* applied to the *Tithi* and the *Nati* applied to the latitude (of the Moon)? and why are these corrections settled by means (of the radius) of the Earth?

*Questions regarding the phases of the Moon.*

10. Ah! why, after being full, does the Moon, having lost her pure brightness, lose her circularity, as it were, by her too close association, caused by her diurnal revolution with the night: and why again after having arrived in the same sign as the Sun, does she thenceforth, by successive augmentation of her pure brightness, as from association with the Sun, attain her circular form ?

End of the second Chapter.

---

# CHAPTER III.

*Called Bhuvana-kośa or Cosmography.*

*The excellence of the Supreme Being.*

1. The Supreme Being *Para Brahma* the first principle, excels eternally. From the soul (*Purusha*) and nature (*Prakriti,*) when excited by the first principle, arose the first Great Intelligence called the *Mahatattwa* or *Buddhitattwa*: from it sprung self-consciousness (*Ahankāra*:) from it were produced the Ether, Air, Fire, Water, and Earth; and by the combination of these was made the universe *Brahmānda*, in the centre of which is the Earth: and from *Brahmā Chaturāna*, residing on the surface of the Earth, sprung all animate and inanimate things.

*Description of the Earth.*

2. This Globe of the Earth formed of (the five elementary principles)Earth, Air, Water, the Ether, and Fire, is perfectly round, and encompassed by the orbits of the Moon, Mercury, Venus, the Sun, Mars, Jupiter, and Saturn, and by the constellations. It has no (material) supporter; but stands firmly in the expanse of heaven by its own inherent force. On its surface throughout subsist (in security) all animate and inanimate objects, *Danujas* and human beings, gods and *Daityas*. 3. It is covered on all sides with multitudes of mountains, groves, towns and sacred edifices, as is the bulb of the Nauclea's globular flower with its multitude of anthers.

*Refutation of the supposition that the Earth has successive supporters.*

4. If the Earth were supported by any material substance or living creature, then that would require a second supporter, and for that second a third would be required. Here we have the absurdity of an interminable series. If the last of the series be supposed to remain firm by it's own inherent power, then why may not

the same power be supposed to exist in the first, that is in the Earth? For is not the Earth one of the forms of the eight-fold divinity i. e. of *Śiva*.

*Refutation of the objection as to how the Earth has its own inherent power.*

5. As heat is an inherent property of the Sun and of Fire, as cold of the Moon, fluidity of water, and hardness of stones, and as the Air is volatile, so the earth is naturally immovable. For oh! the properties existing in things are wonderful.
6. The property of attraction is inherent in the Earth. By this property the Earth attracts any unsupported heavy thing towards it: The thing appears to be falling. The ethereal expanse being equally outspread all around, where can the Earth fall?

*Opinion of the* Bauddhas.

7. Observing the revolution of constellations, the *Bauddhas* thought that the Earth had no support, and as no heavy body is seen stationary in the air, they asserted that, the earth goes eternally downwards in space.

*Opinion of the* Jainas.

8. The *Jainas* and others maintain that there are two Suns and two Moons, and also two sets of constellations, which rise in constant alternation. 'To them I give this appropriate answer.

*Refutation of the opinion of the* Bauddhas.

9. Observing as you do, O *Bauddha*, that every heavy body projected into the air, comes back again to, and overtakes the Earth, how then can you idly maintain that the Earth is falling down in space?

*Refutation of the opinion of the* Jainas.

10. But what shall I say to thy folly, O *Jaina*, who without object or use supposest a double set of constellations, two Suns and two Moons? Dost thou not see that the visible circumpolar constellations take a whole day to complete their revolutions?

*Refutation of the supposition that the Earth is level.*

11. If this blessed Earth were level, like a plane mirror, then why is not the sun, revolving above at a distance from the Earth, visible to men as well as to the gods? (on the *Paurāṇika* hypothesis, that it is always revolving about *Meru*, above and horizontally to the Earth.

12. If the Golden mountain (*Meru*) is the cause of night, then why is it not visible when it intervenes between us and the Sun? And *Meru* being admitted (by the *Paurāṇikas*) to lie to the North, how comes it to pass that the Sun rises (for half the year) to the South?

*Reason of the false appearance of he plane form of the Earth.*

13. As the one-hundredth part of the circumference of a circle is (scarcely different from) a plane, and as the Earth is an excessively large body, and a man exceedingly small (in comparison,) the whole visible portion of the Earth consequently appears to a man on its surface to be perfectly plane.

*Proof of the correctness of alleged circumference of the Earth.*

14. That the correct dimensions of circumference of the Earth have been state may be proved by the simple Rule of proportion in this mode: (ascertain the difference in *Yujanas* between two towns in an exact north and south line, and ascertain also the difference of the latitudes of those towns: then say) if the difference of latitude gives this distance in *Yujanas*, what will the whole circumference of 360 degrees give?

*To confirm the same circumference of the Earth.*

15. As it is ascertained by calculation that the city of *Ujjayinī* is situated at a distance from the equator equal to the one-sixteenth part of the whole circumference: this distance, therefore, multiplied by 16 will be the measure of the Earth's circumference. What reason then is there in attributing (as the *Paurāṇika*s do) such an immense magnitude to the earth?

16. For the position of the moon's cusps, the conjunction of the planets, eclipses, the time of the risings and settings of the planets, the lengths of the shadows of the gnomon, etc. are all consistent with this (estimate of the extent of the) circumference, and not with any other; therefore it is declared that the correctness of the aforesaid measurement of the earth is proved both directly and indirectly,—(directly, by its agreeing with the phenomena;—indirectly, by no other estimate agreeing with the phenomena).

17. *Lankā* is situated in the middle of the Earth: *Yamakoṭi* is situated to the East of *Lankā*, and *Romkapattana* to the west. The city of *Siddhapura* lies underneath *Lankā*. *Sumeru* is situated to the North (under the North Pole,) and *Vadavānala* to the South of *Lankā* (under the south Pole):

18. These six places are situated at a distance of one-fourth part of the Earth's circumference each from its adjoining one. So those who have a knowledge of Geography maintain. At *Meru* reside the gods and the *Siddhas*, whilst at *Vadavānala*, are situated all the hells and the *Daityas*.

19. A man on whatever part of the Globe he may be, thinks the Earth to be under his feet, and that he is standing up right upon it; but two individuals placed at 90° from each other, fancy each that the other is standing in a horizontal line, as it were at right angles to himself.
20. Those who are placed at the distance of half the Earth's circumference from each other are mutually antipodes, as a man on the bank of a river and his shadow reflected in the water; But as well those who are situated at the distance of 90° as those who are situated at that of 180° from you, maintain their positions without difficulty. They stand with the same ease as we do here in our position.

*Positions of Dwīpas and Seas.*

21. Most learned astronomers have stated that *Jambūdwīpa* embraces the whole northern hemisphere lying to the north of the salt sea: and that the other six *Dwīpa*s and the (seven) Seas viz. those of salt, milk, etc. are all situated in the southern hemisphere.
22. To the south of the equator lies the salt sea, and to the south of it the sea of milk, whence sprung the nectar, the Moon and the Goddess *Lakshmī* , and where the Omnipresent *Vāsudeva*, to whose Lotus-feet *Brahmā* and all the gods bow in reverence, holds His favorite residence.
23. Beyond the sea of milk lie in succession the seas of curds, clarified butter, sugar-cane-juice, and wine: and last of all, that of sweet Water, which surrounds *Vadavānala*. The *Pātāla Lokas* or infernal regions, form the concave strata of the Earth.
24. In those lower regions dwell the race of serpents (who live) in the light shed by the rays issuing from the multitude of the brilliant jewels of their crests, together with the multitude of *Asuras*; and there the *Siddhas* enjoy themselves with the pleasing persons of beautiful females resembling the finest gold in purity.
25. The *Śāka*, *Śāmala*, *Kauśa*, *Krāuncha*, *Gomedaka*, and *Pushkara Dwīpa*s are situated in regular alternation: each *Dwīpa* lying, it is said, between two of these seas.

*Positions of the Mountains in Jambū Dwīpa and its nine Khanḍas parts causes by the mountains.*

26. To the North of *Lankā* lies the *Himālaya* mountain and beyond that the *Hemakūṭa* mountain and beyond that again the *Nishadha* mountain. These three Mountains stretch from sea to sea. In like manner to the north of *Siddhapura* lie in succession the *Śṛingavan*, *Sukla* and *Nīla* mountains. To the valleys lying between these mountains the wise have given the name of *Varshas*.
27. This valley which we inhabit is called the Bharātavarsha; to the North of it lies the *Kinnaravarsha*, and beyond it again the *Harivarsha*, and know that the north of *Siddhapura* in like manner are situated the *Kuru*, *Hiraṇmaya* and *Ramyaka Varshas*.

28. To the north of *Yamakoṭi* lies the *Mālyavān* mountain, and to the north of *Romakapattana* the *Gandhamādana* mountain. These two mountains are terminated by the *Nila* and *Nishadha* mountains, and the space between these two is called the *Ilāvṛita Varsha*.
29. The country lying between the *Mālyavān* mountain and the sea, is called the *Bhadrāśwa-Varsha* by the learned; and geographers have denominated the country between the *Gandhamādana* and the sea, the *Ketumāla-Varsha*.
30. The *Ilāvṛita*-Varsha, which is bounded by the *Nishadha, Nila, Gandhamādana* and *Mālyavān* mountains, is distinguished by a peculiar splendour. It is a land rendered brilliant by its shining gold, and thickly covered with the bowers of the immortal gods.

*Position of the mountain* Meru *in* Ilāvṛita.

31. In the middle of the *Ilāvṛita Varsha* stands the mountain *Meru* which is composed of gold and of precious stones, the abode of the immortal gods. Expounders of the *Purāṇas* have further described this *Meru* to be the pericarp of the earthlotus whence *Brahmā* had his birth.
32. The four mountains *Mandara, Sudandha, Vipula* and *Supārśwa* serve as buttresses to support this *Meru*, and upon these four hills grow severally the *Kadamba*, *Jambū*, *Vaṭa* and *Pippala* trees which are as banners on those four hills.
33. From the clear juice which flows from the fruit of the *Jambū* springs the *Jambū-Nādī*; from contact with this juice earth becomes gold: and it is from this fact that gold is called *Jambūnada*: the multitude of the immortal gods and *Siddhas*, turning with distaste from nectar, delight to quaff this delicious beverage.
34. And it is well known that upon those four hills [the buttresses of *Meru*] are four gardens, (1st) *Chaitraratha* of varied brilliancy, (2nd) *Nandana* which is the delight of the *Apsaras*, *(*3rd*)* the *Dhṛiti* which gives refreshment to the gods, and (4th) the resplendent *Vaibhrāja*.
35. And in these gardens are beautified four reservoirs, viz. the *Aruṇa*, the *Mānasa*, the *Mahāhrada* and the *Śwetajala*, in due order: and these are the lakes in the waters of which the celestial spirits, when fatigued with their dalliance with the fair goddesses, love to disport themselves.
36. *Meru* divided itself into three peaks, upon which are situated the three cities sacred to *Vishnu*, *Brahmā* and *Śiva*, and beneath them are the eight cities sacred to *Indra*, *Agni*, *Yama*, *Nairṛita*, *Varuṇa*, *Vāyu*, *Śasī*, and *Īśa*,

*Some peculiarity*

37. The sacred Ganges, springing from the Foot of *Vishnu*, falls upon mount *Meru*, and thence separating itself into four streams descends through the heavens down upon the four *Vishkumbhas* or buttress hills, and thus falls into the four reservoirs.

38. The first called *Sītā*, went to *Bhadrāśwa-varsha*, the second, called *Alakanandā*, to *Bharāta-varsha*, the third, called *Chakshu*, to *Ketumāla-varsha*, and the fourth, called *Bhadrā* to *Uttara Kuru*.
39. And this sacred river has so rare an efficacy that if her name be listened to, if she be sought to be seen, if seen, touched or bathed in, if her waters be tasted, if her name be uttered, or brought to mind, and her virtues be celebrated, she purifies in many ways thousands of sinful men [from their sins].
40. And if a man make a pilgrimage to this sacred stream, the whole line of his progenitors, bursting the bands, bound away in liberty, and dance with joy; nay even, by a man's approach to its banks they repulse the slaves of *Yama*, and, escaping from *Naraka*, secure an abode in the happy regions of Heaven.

*The 9 Khaṇḍas and 7 Kulachālas of Bharāta-varsha*

41. Here in this *Bharāta-varsha* are embraced the following nine *Khaṇḍa*s viz. *Aindra*, *Kaśeru*, *Tāmraparna*, *Gahastimat*, *Kumarikā*, *Nāga*, *Saumya*, V*aruṇa*, and lastly *Gāndharva*.
42. In the *Kumārikā* alone is found the subdivision of men into castes; in the remaining *Khaṇḍa*s are found all the tribes of *Antyajas* or outcaste tribes of men. In this region are also seven *Kulāchalas*, viz. the *Māhendra*, *Śukti*, *Malaya*, *Ṛikshaka*, *Pāriyātra*, the *Sahya*, and *Vindhya* hills.

*Arrangement of the seven* Lokas *worlds.*

43. The country to the south of the equator is called the *Bhūrloka*, that to the north the *Bhuvaloka* and *Meru* is called the *Swarloka*, next is the *Maharloka* in the Heavens beyond this is the *Janaloka*, then the *Tapoloka* and last of all the *Satyaloka*. These Lokas are gradually attained by increasing religious merits.
44. When it is sunrise at *Lankā*, it is then midday at *Yamakoṭi* (90° east of *Lankā*), sunset at *Siddhapura* and midnight at *Romakapattana*.

*Points of the compass why* Meru *is due north all places.*

45. Assume the point of the horizon at which the sun rises as the east point, and that at which he sets as the west point, and then determine the other two points, i. e., the north and south through the *Matsya* effected by the east and west points. The line connecting the north and south points will be a meridian line and this line in whatever place it is drawn will fall upon the north point: hence *Meru* lies due north of all places.

*A curious fact is rehearsed. Geographical Anomaly.*

46. Only *Yamakoṭi* lies due east Geographical Anomaly from *Ujjayinī*, at the distance of 90° from it; but *Lankā*: and not *Ujjayinī* lies due west from *Yamakoṭi*.

47. The same is the case everywhere; no place can lie west of that which is to its east except on the equator, so that east and west are strangely related.

*Right sphere.*

48. A man situated on the equator sees both the north and south poles touching the horizon, and the celestial sphere resting (as it were) upon the two poles as centres of motion and revolving vertically over his head in the heavens, as the Persian water-wheel.

*Oblique sphere.*

49. As a man proceeds north from the equator, he observes the constellations to revolve obliquely, being deflected from his vertical point: and the north pole elevated above his horizon. The degrees between the pole and the horizon are the degrees of latitude. These degrees arc caused by the *Yojanas*.

*How the degrees of latitude are produced from the distance in* Yojanas *and vice versa.*

50. The number of *Yojanas* multiplied by 360 and divided by the circumference of the circle is the number of degrees in the earth or in the planetary orbit in the heavens. The *Yojanas* are found from the degrees by reversing the calculation.

*Parallel sphere.*

51. The gods who live in the mount *Meru* observe at their zenith the north pole, while the *Daityas* in *Vadavānala* the south pole. But while the gods behold the constellations revolving from left to right, to the *Daityas* they appear to revolve from right to left. But to both gods and *Daityas* the equatorial constellations appear to revolve on and correspond with the horizon.

*Dimensions of the Earth's circumference.*

52. The circumference of the earth has been pronounced to be 4967 *Yojanas* and the diameter of the same has been declared to be 1581 1/24 *Yojanas* in length. The superficial area of the Earth, like the net enclosing the hand ball, is 78,53,034 square *Yojanas*, and is found by multiplying the circumference by the diameter.

*The error of Lalla is exposed in regard to the superficial area of the Earth.*

53. The superficial area of the Earth, like the net enclosing the hand ball, is most erroneously stated by *Lalla*: the true area not amounting to one hundredth part of that so idly assumed by him. His dimensions are contrary to what is found by

factual inspection : my charge of error therefore cannot be pronounced to be rude and uncalled for. But if any doubt be entertained, I beg you, O learned mathematicians, to examine well and with the utmost impartiality whether the amount stated by me or that stated by him is the correct one.

*Shows the wrongness of the Rule given by Lalla.*

54. If a piece of cloth be cut in a circular form with a diameter equal to half the circumference of the sphere, then half of the sphere will be (entirely) covered by that circular cloth and there will still he some cloth to spare.
55. As the area of this piece of cloth is to be found nearly 2 1/2 times the area of a great circle of the sphere: and the area of the piece of cloth covering the other half of the sphere is also the same.
56. Therefore the area of the whole sphere cannot be more than 5 times the area of the great circle of the sphere. How then has he multiplied by the circumference?
57. As the area of a great circle multiplied by the circumference is without reason, the rule (therefore of *Lalla* for the superficial contents of the sphere) is wrong, and the superficial area of the Earth (given by him) is consequently wrong.

*Otherwise*

58, 59. Suppose the length of the [equatorial] circumference of the globe equal to 4 times the number of sines and such oblong sections equal to the number of the length of the said circumference and marked with the vertical lines, as there are seen formed by nature on the *Ānwlā* fruit marked off by the lines running from the top of it to its bottom.
60. If we determine the superficial area of one of these sections by means of its parts, we have it in this form. Sum or all the sines diminished by half of the radius and divided by the same.
61. As the superficial area of one section thus determined is equal to the diameter of the globe, the product found by multiplying the diameter by the circumference has therefore been asserted to be the superficial contents of a sphere.

*The grand deluges or dissolutions.*

62. The earth is said to swell the extent of one *Yojana* equally all around in a day of *Brahmā* by reason of the decay of the natural productions which grow upon it in the *Brahmā* deluge that increase is again lost. In the grand deluge the Earth itself is reduced to a state of nonentity.

*Arc four-fold.*

63. That extinction which is daily taking place amongst created beings called the Dainandina or daily extinction. The *Brahmā* extinction or deluge takes place at

the end of *Brahmā*'s day; for all created beings are then absorbed in *Brahmā*'s body.
64. As on the extinction of *Brahmā* himself all things are dissolved into nature, wise men therefore call that dissolution the Prakritika or resolution into nature. Things thus in a state of extinction having their destinies severally fixed are again produced in separate forms when nature is excited (by the Creator).
65. The devout men, who have destroyed all their virtues and sins by a knowledge of the soul, having abstracted the minds from worldly acts, concentrate their thoughts on the Supreme Being, and after their death, as they attain the state from which there is no return, the wise men therefore denominate this state the *Ātyantika* dissolution. Thus the dissolutions are four-fold.

*The universe.*

66. The earth and its mountains, the gods and *Dānavas*, men and others and also the orbits of the constellations and planets and the *Lokas* which, it is said, are arranged one above the other, are all included in what has been denominated the *Brahmānda* (universe).

*Dimensions of the Brahmānda*

67. Some astronomers have asserted the circumference of the circle of Heaven to be 18,712,069,200,000,000 *Yojanas* in length. Some say that this is the length of the zone which binds the two hemispheres of the *Brahmānda*. Some *Paurāṇika*s say that this is the length of the circumference of the *Lokāloka Parvata.*

68. Those, however, who have had a most perfect mastery of the clear doctrine of the sphere, have declared that this is the length of that circumference bounding the limits, to which the darkness dispelling rays of the Sun extend.
69. But let this be the length of the circumference of the *Brahmānda* or not: but it is my opinion that each planet traverses a distance corresponding to this number of *Yojanas* in the course of a *Kulpa* or a day of *Brahmā* and that it has been called the *Khakakshā* by the ancients.

End of third Chapter called the *Bhuvana-kośa* or cosmography.

---

# CHAPTER IV.

CALLED MADHYA-GATI-VĀSANĀ.

*On the principles of the Rules for finding the mean places of the Planets.*

*Places of the several winds*

1. The seven winds have thus been named: viz.

1st. The *Āvaha* or atmosphere.
2nd. The *Pravaha* beyond it.
3rd. The *Udvaha*.
4th. The *Samvaha*.
5th. The *Suvaha*.
6th. The *Parivaha*.
7th. The *Parāvaha*.

2. The atmosphere extends to the height of 12 *Yojanas* from the Earth: within this limit are the clouds, lightning, etc. The *Pravaha* wind which is above the atmosphere moves constantly to the westward with uniform motion.
3. As this sphere of the universe includes the fixed stars and planets, it therefore being impelled by the *Pravaha* wind, is carried round with the stars and planets in a constant revolution.

*An illustration of the motion of the planets.*

4. The Planets moving eastward in the heavens with a slow motion, appear as if fixed on account of the rapid motion of the sphere of the Heavens to the west, as insects moving reversely on a whirling potter's wheel appear to be stationary.

*Sidereal and terrestrial days and their lengths*

5. If a star and the Sun rise simultaneously, the star will rise again (on the following morning) in 60 sidereal *ghaṭikās*: the Sun, however, will rise later by the number of *Asus* (sixths of a sidereal minute), found by dividing the product of the Sun's daily motion and the *Asus* which the sign, in which the Sun is, takes in rising, by 1800.
6. The time thus found added to the 60 sidereal *Ghaṭikās*, forms a true terrestrial day or natural day. The length of this day is variable, as it depends on the Sun's daily motion and on the time in rising.

*Revolutions of the Sun in a year are less than the revolutions of stars by one.*

7. A sidereal day consists invariably of 60 sidereal *ghaṭikās* a mean *sāvana* day of the Sun or terrestrial day consists of that time with an addition of the number *asus* equal to the number of the Sun's daily mean motion. Thus the number of terrestrial days in a year is less by one than the number of revolutions made by the fixed stars.

*Length of solar year.*

8. The length of the (solar) year is 365 days, 15 *ghaṭikās*, 30 *palas*, 22 ½ *vipalas* reckoned in Bhumi *sāvana* or terrestrial days: The 1/ 12* of this is called a *saura* (solar) month, viz. 30 days, 26 *ghaṭikās*, 17 *palas*, 31 *vipalas*, 52 1/2 *pravipalas*. Thirty *sāvana* or terrestrial days make a *sāvana* month.

[*The text of this fraction is difficult to read but dividing the solar year by 12 gives the appropriate solar months. —DG]

*Length of lunar month or lunation.*

9. The time in which the Moon completing a revolution with the difference between the daily motion and that of the Sun, again overtakes the Sun, (which moves at a slower rate) is called a Lunar month. It 29 days, 31 *ghaṭikās*, 50 *palas* in length.

*The reason of additive months called Adhimāsas.*

10. An *Adhimāsa* or additive month which is lunar, occurs in the duration of 32 ½ *saura* (solar) months found by dividing the lunar month by the difference between this and the *saura* month. From this, the number of the additive months in a *Kalpa* may also be found by proportion.
11. As a mean lunar month is shorter in length than a mean *saura* month, the lunar months are therefore more in number than the *saura* in a *kalpa*. The difference between the number of lunar and *saura* months in a *kalpa* is called by astronomers the number of avamas in that period.

*The reason of subtractive day called* avama.

12. An avama or subtractive day which is *sāvana* occurs in 64 1/11 *tithis* (lunar days) found by dividing 30 by the difference between the lunar and *sāvana* month. From this, the number of avamas in a *yuga* may be found by proportion.
13. If the *Adhimasas* are found from *saura* days or months, then the result found is in the lunar months, If the *Adhimāsas* are found from lunar days or months, then the result is in *saura* months, and the remainder is of the like denomination.
14. The *avamas* or subtractive days if found from lunar days, are in *sāvana* time: if found from *sāvana* time they are lunar and the remainder is so likewise.

*A question.*

15. Why, O Astronomer, in finding the *Ahargaṇa* do you add *saura* months to the lunar months *Chaitra* etc.: and tell me also why the remainders of *Adhimāsas* and *Avama* are rejected: for you know that to give a true result in using the rule of proportion, remainders should be taken into account?

*Reason of omitting to include the* Adhimāsa *śesha in finding the* Ahargaṇa.

16. As the lunar month ends at the change of the Moon and the *saura* month terminates when the Sun enters a stellar sign, the accumulating portion of an *Adhimāsa* always lies after each new Moon and before the Sun enters the sign.
17. Now the number of *tithis* (lunar days) elapsed since the change of the Moon and supposed as if *saura*, is added to the number of *saura* days: but as this number exceeds the proper amount by the quantity of the *Adhimās-śesha* therefore the *Adhimās-śesha* is omitted.
18. There is always a portion of a *Avama-śesha* between the time of sun-rise and the end of the *tithi*. By omitting to subtract it, the *Ahargaṇa* is found at the time of sun-rise: if it were not omitted, the *Ahargaṇa* would represent the time of the end of the *tithi*.

*Reason of the correction called the* Udayāntara karma.

19, 20, 21 and 22. As the true, terrestrial day is of variable length, the *Ahargaṇa* has been found mean terrestrial days: the places of the planets found by this *Ahargaṇa* when rectified by the amount of the correction called the *Udayāntara* whether additive or subtractive will be found to be at the time of sun-rise at *Lankā*. The ancient Astronomers have not thus rectified the places of the planets by this correction, as it is of a variable and small amount. The difference between the number of *asus* of the right ascension of the mean Sun and the number of *asus* equal to the number of minutes of the mean longitude of the Sun is the difference between the true and mean *Ahargaṇas*. Multiply this difference by the daily motion of the planet and divide the product by the number of *asus* in a nycthemeron. The result in minutes is to be subtracted from the places of the planets, if the *asus* fall short of the *kalās* or minutes, otherwise the result is to be added to the places of the planets. Instead of the right ascension, if oblique ascension be taken this *Udayāntara* correction which is to be applied to the places of the planets, includes also the *chara* correction or the correction for the ascensional difference.

*Reason of the correction called the* Desāntara.

23. The places of the planets which are found being rectified by this *Udayāntara* correction at the time of sun-rise at *Lankā* may be found, being applied with the *Desāntara* correction, at the time of sun-rise at a given place. This *Desāntara* correction is two-fold, one is east and west and the other is north and south. This north and south correction is called *Chara*.
24. The line which passes from *Lankā*, *Ujjayinī*, *Kurukshetra* and other places to *Meru* (or the North Pole of the Earth) has been denominated the *Madhyarekhā* mid-line of the Earth, by the Astronomers. The sun rises at any place east of this line before it rises to that line: and after it has risen all the line at places to its west.

On this account, an amount of the correction which is produced from the difference between the time of sun-rise at the mid-line and that at a given place, is subtractive or additive to the places of the planets, as the given place be east or west of the mid-line.
25. As the circle which is described around *Meru* or North Pole of the Earth, at the distance in *Yojanas* reckoned from *Meru* to given place and produced from co-latitude of the place is called rectified circumference of the Earth (parallel of latitude) therefore, the circumference of the Earth is multiplied by the sine of co-latitude and divided by the radius.

End of 4th Chapter called *Madhya-Gati Vāsanā.*

---

# CHAPTER V.

*On the principles on which the Rules for finding the true places of the Planets are grounded.*

*On the canon of sines.*

1. The planes of a Sphere are intersected by sines of *bhuja* and *koṭi*, as a piece of cloth by upright and transverse threads. Before describing the spheric, I shall first explain the cannon of sines.
2. Take any radius, and suppose it the hypothenuse (of a right-angled triangle). The sine of *bhuja* is the base, and the sine of *koṭi* is the square root of the difference of the difference of the squares of the radius and the base. The sines of degrees of *bhuja* and *koṭi* subtracted separately from the radius will he the versed sines of *koṭi* and *bhuja* (respectively).
3. The versed sine is like the arrow intersecting the bow and the string, or the arc and the sine. The square root of half the square of the radius is the sine of an arc of 45°, The co-sine of an arc of 45° is of the same length as the sine of that arc.
4. Half the radius is the sine of an arc of 30°: The co-sine of an arc of 30° is the sine of an arc of 60°. Half the root of the sum of the squares of the sine and versed sine of an arc, is the sine of half that arc.
5. Or, the sine of half that arc is the square-root of half the product of the radius and the versed sine. The sines and co-sines of the halves of the arcs before found may thus be found to any extent.
6. Thus a Mathematician may find (in a quadrant of a circle) 3, 6, 12, 24 etc., sines to any required extent. or in a circle described with a given radius and divided into 360°, the required sines may be found by measuring their lengths in digits.

*Reason of correction which is required to find the true from the mean place of planet.*

7. As the centre of the circle of the constellation of the Zodiac coincides with the centre of the Earth: and the centre of the circle in which the planet revolves does not coincide with the centre of the Earth: the spectator, therefore, on the Earth does not find the planet in its mean place in the Zodiac. Hence Astronomers apply the correction called *bhuja phala* to the mean place of the planet

*Mode of illustration of the above fact.*

8. On the northern side of a wall running due east and west, let the teacher draw a diagram illustrative of the fact for the satisfaction of his pupils.

*A verse to encourage those who may be disposed to despond in consequence of the difficulties of the science.*

9. But this science is of divine origin, revealing facts not cognizable by the senses. Springing from the supreme *Brahmā* himself it was brought down to the Earth by *Vasishṭha* and other holy Sages in regular succession; though it was deemed of too secret a character to be divulged to men or to the vulgar. Hence, this is not to be communicated to those who revile its revelations, nor to ungrateful, evil-disposed and bad men: nor to men who take up their residence with its professors for but a short time. Those professors of this science who transgress these limitations imposed by holy Sages, will incur a loss of religious merit, and shorten their days on Earth.

*Construction of a diagram to illustrate the excentric theory.*

10. In the first place then, describe a circle with the compass opened to the length of the radius (3438). This is called the *kakshāvṛitta*, or concentric circle; at the centre of the circle draw a small sphere of the Earth with a radius equal to 1/13th of the mean daily motion of the planet.
11. In this concentric circle, having marked it with 360°, find the place of the higher apsis and that of the planet, counting from the 1st point of stellar Aries; then draw a (perpendicular) diameter passing through the centre of the Earth and the higher apsis (which is called *uchcha-rekhā*, the line of the apsides) and draw another transverse diameter also passing through the centre.
12. On this line which passes to the highest apsis from the centre of the Earth, take a point at a distance from the Earth's centre equal to the excentricity or the sine of the greatest equation of the centre, and with that point as centre and the radius, describe the *prativṛitta* or excentric circle; the *uchcha-rekhā* answers the like purpose also in this circle, but make the transverse diameter different in it.
13 and 14. Where the *uchcha-rekhā* perpendicular diameter (when produced) cuts the excentric circle, that is the place of the higher apsis in it also. From this mark the first stellar Aries, at the distance in degree of the higher apsis in antecedentia: the place of the planet must be then fixed counting the degree from

the mark of the 1st Aries in the usual order. The distance between the higher apsis and the planet is called the *kendra*. The right line let fall from the planet on the *uchcha-rekhā* is the sine of *bhuja* of the *kendra*. The right line falling from the planet on the transverse diameter is the cosine of the *kendra*, it is upright and the sine of *bhuja* is transverse line.

*The principle on which the rule for finding the amount of equation of centre it based.*

15. As the distance between the diameters of the two circles is equal to the excentricity and the co-sin of the *kendra* is above and below the excentricity when the *kendra* is *mrigādi* and *karkyādi* (respectively).
16 and 17. Therefore the sum or difference of the co-sine and excentricity (respectively) is here the *sphuṭa koṭi* (i.e. the upright side of a right-angled triangle from the place or the planet in the excentric to the transverse diameter in the concentric,) the sine of the *bhuja* is the *bhuja* (the base) and the square-root of the sum of the squares of the *sphuṭa koṭi* and *bhuja* is called *karṇa*, hypothenuse. This hypothenuse is the distance between the Earth's centre and the planet's place in the excentric circle. The planet will be observed in that point of the concentric cut by the hypothenuse. The equation of the centre is the distance between the mean and apparent places of the planet: when the mean place is more advanced than the apparent place then the equation thus found is subtractive; when it is behind the true place, the equation is additive.

*The reason for assuming the manda-spashṭa planet as a mean in finding the 2nd equation.*

18. The mean planet moves in its *manda-prativṛitta* (first excentric)the *manda-spashṭa* planet (i. e. whose mean place is rectified by the first equation) moves in its *śīghra-prativṛitta* (second excentric). The *manda-spashṭa* is therefore here assumed as the mean planet in the second process (i. e. in finding the second equation).

*The reason for the invention of the higher apsis.*

19. The place in the concentric in which the revolving planet in its own excentric is seen by observers is its true place. To find the distance between the true and mean places of the planet; the higher apsis has been inserted by former Astronomers.
20. That point of the excentric which is most distant from the Earth has been denominated the higher apsis (or *uchcha*): that point is not fixed but moves; a motion of the higher apsis has therefore been established by those conversant with the science.
21. The lower apsis is at a distance of six signs from the higher apsis: when the planet is in either its higher or lower apsis, then its true place coincides with its

mean place, because the line of the hypothenuse falls on the mean place of the planet in the concentric.

*The cause of variation of apparent size of planet's disc.*

22. As the planet when in the higher apsis is at its greatest distance from the Earth, and when in the lower apsis at its least distance, therefore its disc appears small and large accordingly. In like manner, its disc appears small and large accordingly as the planet is near to and remote from the Sun.
23. To prevent the student from becoming confused, I have separately explained the proof of finding the equation by the *Prativṛitta Bhangi* of the diagram of the excentric. I shall now proceed to explain the same proof in a different manner by the diagram of a *nīchochcha-vṛitta* (epicycle).

*Construction of Diagram to illustrate the theory of epicycle.*

24. Taking the mean place of the planet in the concentric as the centre, with a radius equal to the excentricity of the planet, draw a circle. This is called *nīchochcha-vṛitta* or epicycle. Then draw a line from the centre of the Earth passing through the mean place of the planet.
25. That place in the epicycle most distant from the centre of the Earth, cut by the line is supposed to be the place of the higher apsis: and the point in the epicycle nearest to the Earth's centre, the lower apsis. In the epicycle draw a transverse line passing through the centre of it.
26. As the mean planet revolves with its *kendra-gati* (the motion from its higher apsis) in the 1st and 2nd epicycle marked with the 12 signs and 360 degrees towards the reverse signs, and according to the order of the signs respectively from its higher apsis.
27. Mark off therefore the places of the first and second *kendras* or distances from their respective higher apsides in the manner directed in the last verse: the planet must be fixed at those points. The (perpendicular) line from the planet to the *uchcha-rekhā* is the sine of the *bhuja* of the *kendra*: and from the planet on the transverse line is the cosine.

*To find the hypothenuse and the equation of centre.*

28 and 29. The *bhuja phala* and *koṭi phala* of the *kendra* which are found are sine and cosine in the epicycle. As the *koṭi phala* is above the radius (of the concentric) in *mrigādi kendra* and within the radius in *karkyādi-kendra*, the sum and difference, therefore, of the *koṭi phala* and the radius is here the *sphuti-koṭi* (upright line), the *bhuja phala* is the *bhuja* (the base) and the *karṇa* hypothenuse (to complete the right-angled triangle) is the line intercepted between the centre of the Earth and the planet. The equation of the centre is here the arc intercepted

between the mean place of the planet and the point cut by the hypothenuse. The equation thus found is to be added or subtracted as was before explained.

*Construction of the mixed diagrams of the excentric and epicycle.*

30. The planet appears to move forward from *mandochcha*, or 1st higher apsis, in the excentric circle with its *kendra-gati* (the motion form its *mandochcha*) and in the order of the signs and to the
East: From its *śīghrachcha*, 2nd higher apsis, it moves in antecedentia or reversely, as it is thrown backwards.
31. When the epicycle however is used, the reverse of this takes place, the planet moving in antecedentia from its 1st higher apsis and in the order of the sign from its 2nd higher apsis. Now as the actual motion in both cases is the same, while the appearances are thus diametrically opposed, it must be admitted therefore that these expedients are the mere invention of wise astronomers to ascertain the amount of equation.
32. If the diagrams (of the excentric and epicycle) be drawn unitedly, and the place of the planet be marked off in the manner before explained, then the planet will necessarily be in the point of the intersection of the excentric by the epicycle.
33. As in the oil-man's press, the wooden press (moving in the direction in which the bullock fastened to it goes) moves (also itself) in the opposite direction to that in which the bullock goes, thus the motion of the planet, though it moves in the excentric circle, appears in antecedentia in the epicycle.

*Explains why the 5 minor planets require both the 1st and 2nd equations to their true places.*

34. As the centre of the 1st epicycle is in the concentric, let the planet therefore move in the concentric with its mean motion: In the concentric is the centre of the *śīghra nīchochcha, vṛitta* or of the 2nd epicycle: In the second or *śīghra* epicycle is found the true place of the planet.
35. The first process, or process of finding the 1st equation, is used in the first place, in order to ascertain the position of the centre of the *śīghra nīchochcha vṛitta* or of the 2nd epicycle, and the 2nd process, or the process of the 2nd equation, to ascertain the actual place of the planet. As these two processes are mutually dependent, it on this account becomes necessary to have recourse to the repetition of these two processes.

*Explains reason of omission of hypothenuse in the* manda *process.*

36 and 37. Some say that the hypothenuse is not used in the 1st process, because the difference (in the two modes of computation)is inconsiderable, but others maintain that since in this process the periphery of the first epicycle being multiplied by the hypothenuse and divided by the radius becomes true, and that, if the

hypothenuse then be used, the result is the same as it was before, therefore the hypothenuse is not employed. No objection is to be made why the case in the 2nd process, because the proofs of equation are different here.

*Reason of* Natakarma

*38.* As no observer on the surface of the Earth sees the planet moving in the excentric, deflected from his zenith, in that place of the concentric, where an observer situated at the centre of the Earth observes it in the eastern or western hemisphere, and at noon both observers see it in the same place, Therefore the correction called *Natakarma* is declared (by astronomers). The proof of this is the same as in finding the parallax.

*The reason of the invention of the appellation of* kendra.

41. The word *kendra* means the centre of a circle: it is on that account applied to the distance between the planet and higher apsis, for the centre of the *nīchochcha-vṛitta* or epicycle, is always at the distance of the planet from the place of the higher apsis.

*Sphuṭa-kakshā or corrected orbit.*

42. The circumference in *yojanas* of the planet's orbit by the *śīghra-karṇa* (or 2nd hypothenuse), and divided by the radius (3438) is *sphuṭa-kakshā* (corrected orbit). The planet is (that moment) being carried by the *pravaha* wind, and moves at a distance equal to half the diameter of the *śīghra-kakshā* from the earth's centre.

*Reason for bujantara correction.*

43. When the sun's *manda-phala* i. e. the equation of the centre is subtractive, the apparent or real time of sun-rise takes place before the time of mean sun-rise: when the equation of the centre is additive, the real is after the mean sun-rise, on that account the amount of that correction arising from the sun's *manda-phala* converted into *asus* of time has been properly declared to be subtractive or additive.

44. Those who have wits as sharp as the sharp point of the inmost blade of the *dorbha* or *darbha* grass, find the subject above explained by diagrams, a matter of no difficulty whatever: but men of weak and blunt understanding find this subject as heavy and immovable as the high mountain that has been shorn of its wings by the thunderbolt of *Indra*.

End of Chapter V. on the principles on which the rules for finding the true places of the planets are grounded.

# CHAPTER VI.

*Called* Golabandha, *on the construction of an Armillary Sphere.*

1. Let a mathematician, who is as skilful in mechanics as in his knowledge of the sphere, construct an armillary sphere with circles made of polished pieces of straight bamboo; and marked with the number of degrees in the circle.
2. In the first place, let him mark a straight and cylindrical *dhruva-yashṭi*, or polar axis, of any excellent wood he pleases: then let him place loosely in the middle of it a small sphere to represent the earth. Let him then firmly secure the spheres beyond it of the Moon, Mercury, Venus, the Sun, Mars, Jupiter, Saturn and the fixed stars: Beyond them let him place two spheres called *khagola* and *ḍriggola* unconnected with each other, and fastened to the hollow cylinders.

*The prime vertical, the meridian and the* koṇavṛittas.

3. Fix vertically the four circles and another circle called horizon transversely in the middle of
them, so that one of those vertical circles called *Samamanḍala*, prime vertical, may pass through the east and west points of the horizon, the other called *yāmyottara-vṛitta*, meridian the north and south points, and the remaining two called *koṇavṛittas* N. E. and S. W. and N. W. and S. E. points.

*The unmanḍala or six o'clock line.*

4. Then fix a circle passing through the points of the horizon intersected by the prime vertical, and passing also through the south and north poles at a distance below and above the horizon equal to the latitude of the place, This is called the *unmanḍala*, or six o'clock line, and is necessary to illustrate the increase and decrease in the length of the days and nights.

*Thc cquinoctial.*

5. The equinoctial (called *nādī-valaya*), marked with 60 *ghaṭis*, should be placed so as to pass through the east and west points of the horizon, and also to pass over the meridian at a distance south from the zenith equal to the latitude, and at a distance north of the nadir also equal to the latitude of the place.

*Azimuth or vertical circle.*

6. Let the azimuth or vertical circle be next attached within the other circles, fixed by a pair of nails at the zenith and nadir, so as to revolve freely on them: It should be capable of being placed so as to cover the planet, wherever it may happen to be.

7. Only one azimuth circle may be used for all the planets; or else eight azimuth circles may be made, viz. one for each of the 7 planets and the 8th for the nonagesimal point. The azimuth circle for the nonagesimal point is called the *dṛikshepa-vṛitta*.

*The Dṛiggola.*

8. Let two hollow cylinders project beyond the two poles north and south of the *khagola* celestial sphere, and on these cylinders let the skilful astronomer place the *dṛiggola* double sphere as follows.
9. When the system of the *khagola*, celestial sphere, is mixed with the ecliptic, and all the other circles forming the *bhagola* (which will be presently shown) it is then called *dṛiggola*, double sphere. As in this the figures formed by the circles of the two spheres *khagola* and *bhagola* are seen, it is therefore called *dṛiggola* double sphere.

*THE BHAGOLA.*

10. Let two circles be firmly fixed on the axis of the poles answering to the meridian and horizon (of the *khagola*); they are called the *ādhāra-vṛittas*, or circles of support: Let the equinoctial circle also be fixed on them marked with 60 *ghaṭis* like the prime vertical (of the *khagola*).

*The Ecliptic.*

11. Make the ecliptic (of the same size) and mark it with 12 signs; in this the Sun moves: and also in it revolves the Earth's shadow at a distance of 6 signs from the Sun. The *krānti-pāta* or vernal equinox, moves in it contrary to the order of the signs: The *spashṭa-pātas* have a like motion: the places of these should be marked in it.
12. Let the ecliptic be fixed on the equinoctial in the point of vernal equinox *krānti-pāta* and in a point (autumnal equinox) 6 signs from that: it should be so placed that the point of it, distant 3 signs eastward from the vernal equinox, shall be 240 north of the equinoctial, and the 3 signs westward shall be at the same distance south from the equinoctial.

*Planet's orbit.*

13. Divide a circle called *kshepa-vṛitta* representing the orbit of a planet into 12 signs and mark in it the places of the *spashṭa-pātas*, rectified nodes, as has been before prescribed. Then this circle should be so placed in connection with the ecliptic as it has been placed in connection with the equinoctial.
14. The ecliptic and the *kshepa-vṛitta* should be so placed that the latter may intersect the former at the ascending and descending nodes, and pass through

points distant 3 signs from the ascending node east and west at a distance from the ecliptic north and south equal to the rectified greatest latitude of the planet.
15. The greatest (mean) latitudes of the planets being multiplied by the radius and divided by the *śīghra-karṇa* second hypothenuse becomes *spashṭa*, rectified. The *kshepa-vṛitta*, or circles representing the orbits of the six planets, should be made separately. The Moon and the rest revolve in their own orbits.

*Declination and latitude.*

16. The declination is an arc of a great meridian circle: cutting the equinoctial at right angles, and continued till it touch the ecliptic celestial latitude is in like manner an arc of a great circle (which passes through the ecliptic poles) intercepted between the ecliptic and the *kshepa-vṛitta*. The corrected declination is the distance of the planet from the equinoctial in a circle of declination.

*Precession of the equinox.*

17. The point of intersection of the equinoctial and ecliptic circles is the *krāntī-pāta* or intersecting point for declination. The retrograde revolutions of that point in a *kalpa* amount to 30,000 according to the author of the *Sūrya-siddhānta*.
18. The motion of the solstitial points spoken of by *Munjāla* and others is the same with this motion of the equinox: according to these authors its revolutions are 199,669 in a *kalpa*.
19. The place of the *krāntī-pāta*, or the amount of the precession of the equinox determined through the revolutions of the *krāntī-pāta* must be added to the place of a planet; and the declination then ascertained. The ascensional difference and periods of rising of the signs depend on the declination: hence the precession must be added to ascertain the ascensional difference and horoscope.
20. Thus the points of intersection of the ecliptic and the orbits of the Moon and other planets are the *kshepa-pātas*, or intersecting points for the *kshepa* celestial latitude. The revolutions of the *kshepa-pātas* are also contrary to the order of the signs, hence to find their latitudes, the places of the *kshepa-pātas* must be added to the places of the planets (before found).
21. As the *manda-spashṭa* planet (or the mean planet corrected by the 1st equation) and its ascending node revolve in the *śīghra-prativṛitta* or 2nd excentric, hence the amount of the latitude is to be ascertained from (the place of) the *manda-spashṭa* planet added to the node found by calculation.
22. Or the amount of the latitude may be found from the *spashṭa* planet added to the node which the *śīghra-phala* 2nd equation is added to or subtracted from accordingly as it was subtractive or additive. As the Moon's node revolves in the concentric circle, the amount of the latitude, therefore, is to be found from the true place of the Moon added to the mean node.
23. The exact revolutions of the nodes of Mercury and Venus will be found by adding the revolutions of their *śīghra-kendras* to the revolutions of their nodes

which have been stated: if it be asked why these smaller amounts have been stated, I answer, it is for greater facility of calculation. Hence their nodes which are found from their stated revolutions are to be added to the places of their *śīghra-kendras*.
24. To find the *kendra* the place of the planet is subtracted from the *śīghrochcha*: then take the *kendra* with the *pāta* added and let the place of the planet be added thereto. Therefore from the *śīghrochchas* of these two planets with the *pātas* added, their latitudes are directed by the ancient astronomers to be found.
25 and 26. The *pātas* or nodes of these two planets added to the *śīghrochchas* from which the true places of the planets have been subtracted, become *spashṭa* or rectified. It is the *spashṭa-pāta* which is found in the *bhagola* (above described) In the sphere of a planet, take the ecliptic above described as the concentric circle, to this circle the second excentric circle should be attached, as was explained before, and a circle representing the orbit of a planet (and which consequently would represent the real second excentric) should be also attached to the latter circle with the amount of latitude detailed for it. In this latter circle mark off the mean places of the nodes of the (superior) planets, and also mark in it the mean place of the nodes of Mercury and Venus added to their respective *śīghra-kendras*.

*Diurnal circles called* ahoratra-vṛittas.

27. Next the *ahoratra-vṛittas* or diurnal circles, must be made on both sides of the equinoctial at every or any degree of declination that may be required:—and they must all be marked with 60 *ghaṭis*: The radius of the diurnal circle is called *dyujyā*.
28. From the vernal equinox mark the 12 signs in direct order, and then let diurnal circles be attached at the extremity of each sign.
29. On either side of the equinoctial, three diurnal circles should be attached in the order of the signs; these again will answer for the three following signs. The *bhagola* has thus been described. This is to be known also as the *khechara-gola*, the sphere of a planet.
30. Or in the plane of the ecliptic bind the orbits of Saturn and of the other planets with cross diameters to support them, but these must be bound below (within) the ecliptic in successive circles one within the other, like the circles woven one within the other by the spider.
31. Having thus secured the *bhagola* on the axis or *yashṭi*, after placing it within the hollow cylinders on which the *khagola* is to be fastened, make the *bhagola* revolve: —it will do so freely without reference to the *khagola* as its motion is on the solid axis. The *khagola* and *dṛiggola* remain stationary whilst the *bhagola* revolves.

End of Chapter VI. on the construction of an armillary sphere.

# CHAPTER VII.

*Called Tripraśna-vāsanā on the Principles of the Rules for resolving the question on time, space, and directions.*

*The ascensional difference and its place.*

1. The time called *chara-khunda*, or ascensional difference is found by that arc of a diurnal circle intercepted between the horizon and the six o'clock line. The sine of that arc is called the *kujyā* in the diurnal circle: but, when reduced to relative value in a great circle, it is called *charajyā* or sine of ascensional difference.
2. The horizon, as seen at the equator, or in a right sphere, is denominated in other places the *unmaṇḍala* six o'clock line; but as the Sun appears at any place to rise on its own horizon, the difference between the times of the Sun's rising is the ascensional difference.

*Determination of the question when the chara correction is additive and when subtractive.*

3. When the sun is in the northern hemisphere, it rises at any place (north of the equator) before it does to that on the equator; but it sets after it sets to that on the equator. Therefore the correction depending on the ascensional difference is to be subtracted at sunrise of a given place from the place of the planet and to be added at sunset to the place of the planet.
4. When the Sun is in the southern hemisphere the reverse of this takes place, as the part of the *unmaṇḍala* in that hemisphere lies below the horizon. The halves of the sphere north and south of the equinoctial are called the northern and southern hemispheres.

*Cause of increase and decrease in length of days and nights.*

5. The days are longer and the nights shorter (than they are on equator) when the Sun is in the northern hemisphere: and that the days are shorter and the nights longer when the Sun is in the southern hemisphere. For, the length of the night is represented by that arc of the diurnal circle below the horizon, and the length of the day by that arc above the horizon.
6. But at the equator the days and nights are always of the same length, as there is no *unmaṇḍala* there except the horizon.

A circumstance of peculiar curiosity, however, occurs in those places having a latitude greater than 66° N. viz. than the complement of the Sun's greatest declination.

*Determination of place and time of perpetual day and night.*

7. Whenever the northern declination of the Sun exceeds the complement of the latitude, then there will be perpetual day for such time as that excess continued; and when the southern declination of the Sun shall exceed the complement of the latitude, then there will be perpetual night during the continuance of that excess. On *Meru*, therefore, day and night are each of half a year's length.

*Place of Meru.*

8. To the Celestial Beings the equinoctial is horizon: so also is to the *daityas*. For, the northern and southern poles are situated respectively in their zeniths.
9. The Celestial Beings on *Meru* behold the Sun whilst he is in the northern hemisphere, always revolving above the horizon from left to right: but daityas the inhabitants of the southern polar regions behold him whilst he is in the southern hemisphere revolving above their horizon from the right to the left.

*Definition of the artificial day and night and the day and night of the pitṛis.*

10. Thus it is day whilst the Sun is visible, and night whilst he is invisible. As the determination of
night and day is made in regard to men residing on the surface of the Earth, so also is that of the *pitṛis* or deceased ancestors who dwell on the upper part of the Moon.

*The meaning of the fact stated by the astrological professors or sanhitikas.*

11. As for the doctrine of astrologers, that it was day with the gods at *Meru* whilst the Sun was in the *uttarāyaṇa* (or moving from the winter to the summer solstice) and night whilst the Sun was in the *dakshiṇāyana* (or moving from the summer to the winter solstice), it can only be said in defense of such an assertion, that it is day when the Sun is turned towards the day, and it is night when turned towards the night. Their doctrine has reference merely to judicial astrology and the fruits it foretells.
12. By the degrees by which the Sun proceeds in his northern course to the end of Gemini, he moves back from that sign: entering also the same diurnal circles in his descent as he did in his ascent. Is it not therefore that the Sun is visible in his descent to the gods in the place where he was first seen by them in his ascent?

*Length of the day of the* pitṛis.

13. The *pitṛis* reside on the upper part of the Moon and fancy the fountain of nectar to be beneath themselves. They behold the Sun on the day of our *amāvāsyā* or new Moon in their zenith. That therefore is the time of their midday.
14. They (i. e. the *pitṛis*) cannot see the Sun when he is opposite the lower part of the Moon: it is therefore, midnight with the *pitṛis* on the day of the *pūrnīmā* or

full Moon. The Sun rises to them in the middle of the *krishna paksha* or dark half of the Moon, and sets in the middle of the *sukla paksha* or light half of the Moon. This is clearly established from the context.

*The explanation of a day of* Brahmā.

15. As *Brahmā* being at an immense distance from the Earth, always sees the Sun till the time of the *pralaya* or general deluge, and sleeps for the same time, therefore the day and night of *Brahmā* are together of 2000 *mahāyugas* in length.

*The time taken by each sign in rising above the horizon.*

16. As the portion of the ecliptic which is more oblique than the other, rises and sets in a shorter time and that which is more upright takes a longer time in rising and setting, hence the times of rising of the several signs are various.
17. The (six) signs from Capricorn to Gemini or ascending signs which are inclined towards the south with their respective declinations whilst they rise even at the equator are still more inclined towards the south in the northern latitudes (on account of the obliquity of the starry sphere towards the south); hence they arise in still shorter times than they do at the equator.
18. At the equator, the signs from Cancer or descending signs incline whilst they rise to the northerly direction, but they will have upright direction in consequence of the northern latitude, hence they rise in longer times. The difference between the period of the rising of a sign in a given latitude, and at the equator under the same meridian, is equivalent to the *charakhanḍa* of that sign.
19. Each quarter of the ecliptic rises in 15 *ghaṭis* or 6 hours to those on the equator: and the 6 signs of the northern as well the 6 of the southern hemisphere appear to rise each in 12 hours or 30 *ghaṭis* in every or any latitude.
20. The three signs from the commencement of Aries to the end of Gemini, i. e. the first quarter of the ecliptic, pass the *unmanḍala* in 15 *ghaṭis*; but the horizon is below the *unmanḍala*, they therefore previously pass it in time less than 15 *ghaṭis* by the *charakhanḍa*s.
21. The three signs from the end of Virgo to the end of Sagittarius, i. e. the 3rd quarter of the ecliptic, pass the *unmanḍala* in 15 *ghaṭis*; but they pass the horizon of a place afterwards which is above the *unmanḍala* in 15 *ghaṭis* added to the *charakhanḍa*s.
22. The three signs from the end of Gemini to the end of Virgo, i. e. the 2nd quarter of the ecliptic or those from the end of Sagittarius to the end of Pisces i. e. the 4th quarter of the ecliptic, pass the horizon in the time equal to the remainder of 30 *ghaṭis* diminished by the time which the first or third quarter takes to pass the horizon respectively. For this reason, the times which the signs contained in the 1st and 4th quarters of the ecliptic, or ascending signs, and those contained in the 2nd and 3rd quarters, or descending signs take to pass the horizon at a given

place are found by subtracting the *charakhaṇḍa*s of the signs from and adding them to the times which those signs take in rising on the equator respectively.
23. Having placed the 1st Aries in the horizon and set the sphere in motion, the tutor should show the above facts to the pupils, that they may understand as well what has been explained as any other facts which have not been now mentioned.
24. In whatever time any sign rises above the horizon the sign which is the 7th from it, will take exactly the same time in setting: as one half of the ecliptic is always above the horizon.
25. When the complement of latitude is less than 24° (i.e. than the extreme amount of the Sun's declination taken to be 24° by Hindu astronomers) then neither the rising periods or the signs, nor the ascensional differences and other particulars will correspond with what has been here explained. The facts of those countries (having latitudes greater than 66°) which are different from what has been explained on account of their totally different circumstances, are not here mentioned, as those countries are not inhabited by men.

*Etymology of the word* lagna.

26. That point of the ecliptic which is (at any time) on the eastern horizon is called the *lagna* or horoscope. This is expressed in signs, degrees, etc. reckoned from the first point of stellar Aries. That point which is on the western horizon is called the *asta-lagna* or setting horoscope. The point of the ecliptic on the meridian is called the *madhya-lagna* or middle horoscope (culminating point of the ecliptic).

*The reason for finding the exact place of the Sun at the time of question in order to find lagna.*

27. If when you want to find the *lagna*, the given *ghaṭis* are *sāvana-ghaṭis*, then they will become sidereal by finding the Sun's instantaneous place i, e. the place of the Sun for the hour given. The times of rising of the signs which are sidereal must be subtracted from these *ghaṭis*; (of the question) reduced to a like denomination. When the hours of the question are already sidereal, there is no necessity for finding the sun's real place for that time.

*Determination of latitudes in which different signs are always above and below the horizon.*

28. In those countries having a north latitude of 69° 20' the signs sagittarius and capricornus are never visible: and the signs gemini and cancer remain always above the horizon.
29. In those places having a northern latitude of 78° 15', the four signs scorpio, sagittarius, capricornus, and aquarius are never seen, and the four signs taurus, gemini, cancer, and leo, always appear revolving above the horizon.

30. On that far-famed hill of gold *Meru* which has a latitude of 90° N. the six signs of the southern hemisphere never appear above the horizon and the six northern signs are always above the horizon.

*An error of Lalla exposed.*

31. *Lalla* has declared that when the *asus* of *charakhaṇḍa* are equal to the time which any sign takes to rise on the equator, then that sign will always remain visible above the horizon: but this assertion is without reason. Were it so, then in places having a latitude of 66°, the whole twelve signs (or the ecliptic would always be visible, and would all appear at once on all occasions, as the times of their rising on the equator are equal to the *asus* of their *chara-khaṇḍa*s: but this is not the fact.

*Another gross error of* Lalla

32. *Lalla* has also stated in his work on the sphere that where the north latitude is 66° 30', sagittarius and capricornus are not visible, and also that in north latitude 75°, scorpio and aquarius are never there visible: but this also an idle assertion. How, my learned friend, has he managed to make so gross and palpable an error of three degrees?
33. The altitude of the polar star and its zenith distance as found by observation, give respectively the latitude and the *lambānsā* or complement of the latitude. Or the zenith distance and altitude of the Sun at mid-day when on the equinoctial give the latitude and its complement.
34. The *unnata* the time found in that arc of the diurnal circle which is intercepted between the eastern or western horizon and the planet above it, is *sāvana*. This is used in finding the shadow of the planet. The sine of the *unnata* which is oblique, like the *āksha-karṇa*, by reason of the latitude, is called *chhedaka* not *śanku* because it is upright.
35. In order to find the shadow of the Moon, the *udita* (the time elapsed from the rising of a planet) which has been found by some astronomers by means of repeated calculation is erroneous, for the *udita*, (found by repeated calculation) is not *sāvana*. The labour of the astronomer that does not thoroughly understand mathematics as well the doctrine of the sphere, in writing a book of instruction on the science is futile and is utterly useless.

Determination of śanku and *dṛigjya*.

36. The degrees of altitude are found in the *dṛinmanḍala* or vertical circle, being the degrees elevation in it above the horizon; the degrees of zenith distance are (as their name imports) the degrees in the same circle by which the object is distant from the zenith or mid-heaven of the observer: the *śanku* is the sine of the degrees of altitude: and the *dṛigjya* is the sine of the zenith distance.

*Of* sama-*śanku,* koṇa-*śanku and* madhya-*śanku.*

37. When the Sun in his ascent arrives at the prime vertical, the *śanku* found at the moment is the *sama-śanku* found at the moments of his passing the *koṇa-vṛitta* and the meridian are respectively termed the *koṇa-śanku* and *madhya-śanku.*

*Reason of the correction of parallax to the sine of altitude.*

38. One-half of the vertical circle in which a planet is observed should be visible, but only one-half less the portion opposite the radius of the Earth is visible to observers on the surface of the Earth. Therefore 1/13th part of the daily motion of the planet observed is to be subtracted from the sine of altitude or from the *śanku* to find the shadow:

*The sine of amplitude and the* udayāsta-*sūtra.*

39. The *agra* (the sine of amplitude) is the sine of the arc of the horizon intercepted between the prime vertical and the planet's diurnal circle in the east or west i.e. between the east or west point of the horizon, and the point of the horizon at which the planet rises or sets. The line connecting the points of the extremities of the east and west *agra* it called the *udayāsta-sūtra*, the line of rising and setting.
40. The *śanku-tala* or base of the *śanku* stretches during the day to the south of the *udayāsta-sūtra*; because the diurnal circle have during the day a southern inclination (in northern latitude) above the horizon. But, below the horizon at night, the base lies to the north of the *udayāsta-sūtra* as then the diurnal circles incline to the north. The *śanku-tala*'s place has thus been rightly defined.
41. The *śanku-tala* lies to the south of the extreme point of *agra* when that *agra* is north and when the *agra* is south, the *śanku-tala* lies still to the south of it. The difference and sum of the sine of amplitude and *śanku-tala* has been denominated the *bahu* or *bhuja*; it is the sine of the degrees lying between the prime vertical and the planet on the place of the horizon.
42.The sine of the zenith distance being the hypothenuse then the third side or the *koṭi* being the square root or the difference of their squares will be found: it is an east and west portion of the diameter of the prime vertical. I now propose to explain the triangles which are created by reason of the Sun's varying declination: and shall then proceed to explain briefly also the latitudinal triangles or those created by different latitudes.
43. In the 1st triangle of declination. 1st The sine of declination = *bhuja* or base, the radius of diurnal circle corresponding with the declination above given = *Koṭi* or perpendicular, and radius of large circle = hypothenuse. 2nd. Or in a right sphere. The sine of 1, 2 or 3 signs = hypothenuse : The declination of 1, 2 or 3 signs in six o'clock line = *bhujas*.

44. Sines of arcs of diurnal circles corresponding with the declination above given = *koṭis*. These sines being converted into terms of a large circle and their arcs taken, they will then express the times in *asus* which each sign of the ecliptic takes in rising at the equator i.e. the right ascensions of those signs or the *lankodayas*, that is the 2nd will be found when the 1st is subtracted from two found conjointly, and the 3rd will be found when the sum of the 1st and 2nd is subtracted from three found conjointly.

*Triangles arise from latitude.*

45. In the right-angled triangle formed by the *śanku* or gnomon when the Sun is on the equinoctial. 1st. The *śanku* of 12 digits = the *koṭi*. The *palabhā* or the shadow of *śanku* or gnomon = the *bhuja*, And the *āksha-karṇa* = the *karṇa* or hypothenuse 2nd. The sine of latitude = *bhuja*. The sine of co-latitude = *koṭi* and radius = hypothenuse This triangle is found in the plane of the meridian.
46. Or the sine of declination reckoned on the *unmanḍala* from the east and westline = *koṭi*, *Kujyā*, the sine of ascensional difference in the diurnal circle of the given day = *bhuja* The sine of amplitude in the horizon = hypothenuse
This is a well known triangle.
47. Or the *sama śanku* in the prime vertical being = *koṭi*, The sine of amplitude = *bhuja*, The *taddhṛiti* in the diurnal circle = hypothenuse Or taking the sine of declination = *bhuja* and the *sama-śanku* hypothenuse *taddhṛiti minus kujyā* = *koṭi*
48.The *unmanḍala śanku* being = *bhuja* The sine of declination will then be = hypothenuse And *agradi khanḍa* or 1st portion of the sine of amplitude will be = *koṭi* Or Making the *unmanḍala śanku* = *koṭi* the *agrāgra-khanḍa* or 2nd portion of the sine of amplitude is= *bhujya* the *kujyā* then becomes = hypothenuse
49. The *śanku* being= *koṭi* and the *śanku-tala* = *bhuja* Then the *chhedaka* or *hṛiti* = hypothenuse Those who have a clear knowledge of the spherics having thus immediately formed thousands of triangles should explain the doctrine of the sphere to their pupils.

End of Chapter VII. on the principles of the rules for resolving the questions on time, space and directions.

---

# CHAPTER VIII.

*Called Grahana Vāsanā*
*In explanation of the cause of eclipses of the Sun and Moon.*

*The cause of the directions of the beginning and end of the solar eclipse.*

1. The Moon, moving like a cloud in a lower sphere, overtakes the Sun hence it arises that the western side of the Sun's disk is first obscured, and that the eastern side is the last part relieved from the Moon's dark body: and to some places the Sun is eclipsed and to others is not eclipsed (although he is above the horizon) on account of their different orbits.

*The cause of the parallax in longitude and that in latitude.*

2. At the change of the Moon it often so happens that an observer placed at the centre of the Earth, would find the Sun when far from the zenith obscured by the intervening body of the Moon, whilst another observer on the surface of the Earth will not at the same time find him to be so obscured, as the Moon will appear to him to be depressed from the line of vision extending from his eye to the Sun. Hence arises the necessity for the correction of parallax in celestial longitude and parallax in latitude in solar eclipses in consequence of the difference of the distances of the Sun and Moon.

*The reason of the correction of parallax not being necessary in lunar eclipses.*

3. When the Sun and Moon are in opposition, the Earth's shadow envelopes the Moon in darkness. As the Moon is actually enveloped in darkness, its eclipse is equally seen by every one on the Earth's surface: and as the Earth's shadow and the Moon which enters it, are at the same distance from the Earth there is therefore no call for the correction of the parallax in a lunar eclipse.

*The cause of the directions, of the beginning and end of the lunar eclipse.*

4. As the Moon moving eastward enters the dark shadow of the Earth: therefore its eastern side is first of all involved in obscurity and its western is the last portion of its disc which emerges from darkness as it advances in it course.
5. As the Sun is a body of vast size, and the Earth insignificantly small in comparison: the shadow made by the Sun from the Earth is therefore of a conical form terminating in a sharp point. It extends to a distance considerably beyond that of the Moon's orbit.
6. The length of the Earth's shadow, and its breadth at the part traversed by the Moon, may be easily found by proportion. In the lunar eclipse the Earth's shadow is northwards or southwards of the Moon when its latitude is south or north. Hence the latitude of the Moon is here to be supposed inverse (i.e. it is to be marked reversly in the projection to find the centre of the Earth's shadow from the Moon.)

*The determination of the coverer in the eclipse of the Sun and Moon.*

7. As the horns of the Moon, when it is half obscured form very obtuse angles: and the duration of a lunar eclipse is also very great, hence the coverer of the Moon is much larger than it.
8. The horns of the Sun on the contrary with half of it's disc is obscured form very acute angles: and the duration of a solar eclipse is short: hence it may be safely inferred that the dimensions of the body causing the obscuration in a solar eclipse are smaller than and different from the body causing an eclipse of the Moon.
9. Those learned astronomers, who, being too exclusively devoted to the doctrine of the sphere, believe and maintain that *Rāhu* cannot be the cause of the obscuration of the Sun and Moon, founding their assertions on the above mentioned contrarieties, and differences in the parts of the body first obscured, in the place, time, causes of obscuration etc. must be admitted to assert what is at variance with the *Sanhitā*, the *Vedas* and *Purāṇas*.
10. All discrepancy, however, between the assertions above referred to and the sacred scriptures may be reconciled by understanding that it is the dark *Rāhu* which entering the Earth's shadow obscures the Moon, and which again entering the Moon (in a solar eclipse) obscures the Sun by the power conferred upon it by the favour of *Brahmā*.

*What is the cause of parallax, and why it is calculated from the radius of the Earth.*

11. As the spectator is elevated above the centre of the earth by half its diameter, he therefore sees the Moon depressed from its place. Hence the parallax in longitude is calculated from the radius of the Earth, as is also the parallax in latitude.

*Construction of diagram to illustrate the cause of parallax.*

12. Draw upon a smooth wall, the sphere of the earth reduced to any convenient scale, and the orbits of the Moon and Sun at proportionate distances: next draw a transverse diameter and also a perpendicular diameter to both orbits.
13, 14 and 15. Those points of the orbits cut by this diameter are on the (rational) horizon. And the point above cut by the perpendicular diameter will represent the observer's zenith: Then placing the Sun and Moon with their respective zenith distances let the learned astronomer show the manner in which the parallax arises. Let him draw one line passing the centre of the earth to the Sun's disc: and another which is called the *dṛiksūtra* or line of vision, let him draw from the observer on the Earth's surface to the Sun's disc. The minutes contained in the arc, intercepted between these two lines give the Moon's parallax from the Sun.
16. (At the new Moon) the Sun and Moon will always appear by a line drawn from the centre of the earth to be in exactly the same place and to have the same longitude but when the Moon is observed from the surface of the Earth in the

*dṛiksūtra* or line of vision, it appears to be depressed and hence the name *lambana*, or depression, for parallax.
17. (When the now Moon happens in the zenith) then the line drawn from the Earth's centre will coincide with that drawn from its surface, hence a planet has no parallax when in the zenith. Now on a wall running due north and south draw a diagram as above prescribed;
18. The orbits now drawn, must be considered as *dṛikshepa-vṛittas* or the azimuth circles for the nonagesimal. The sine of the zenith distance of the nonagesimal or of the latitude of the zenith is the *dṛikshepa* of both the Sun and Moon.
19. Mark the nonagesimal points on the *dṛikshepa-vṛittas* at the distance from the zenith equal to the latitude of the points. From these two points (supposing them as the Sun and Moon) find as before the minutes of parallax in altitude. These minutes are here *Nati-kalās*, i. e. the minutes of the parallax in latitude of the Moon from the Sun.
20. The difference north and south between the two orbits i. e. the measure of their mutual inclination, is the same in every part of the orbit as it is in the nonagesimal point, hence this difference called *nati* is ascertained through the *dṛikshepa* or the sine of the zenith distance of the nonagesimal.
21. The amount by which the Moon is depressed below the Sun deflected from the zenith wherever it he, is the east and west difference between the Sun and Moon in a vertical circle.
22. For this reason, the difference is two-fold, being partly east and west, and partly north and south. And the ecliptic is here east and west, and the circle secondary to it is north and south. (It follows from this, that the east and west difference lies in the ecliptic, and the north and south difference in the secondary to it.)
23. The difference east and west has been denominated *lambana* or parallax in longitude, whilst that running north and south is parallax in latitude.
24. The parallax in minutes as observed in a vertical circle, forms the hypothenuse of a right angle triangle, of which the *nati-kalā* or the minutes of the parallax in latitude form one of the sides adjoining the right angle then the third side found by taking the square-root of the difference of the squares of the two preceding sides will be *sphuṭa-lambana-liptā* or the minutes of the parallax in longitude.
25. The amounts in minutes of parallax in a vertical circle may be found by multiplying the sine of the Sun's zenith distance of the minutes of the extreme or horizontal parallax and dividing the product by the radius. Thus the *nati* will be found from the *dṛikshepa* or the sine of the nonagesimal zenith distance.
26. The extreme or horizontal parallax of the Moon from the Sun amounts to 1\15 part of the difference of the Sun's and Moon's daily motion. For 1/13 part of the *yojanas*, the distance of which any planet traverses per diem (according to the *siddhāntas*) is equal to the Earth's radius.
27. The minutes of the parallax in longitude of the Moon from the Sun divided by the difference in degrees of the daily motions of the Sun and Moon will be

converted into *ghaṭis*. If the Moon be to the east, it is thrown forward from the Sun, if to the west it is thrown backward (by the parallax).
28. And if the Moon be advanced from the Sun, then it must be inferred that the conjunction has already taken place by reason of the Moon's quicker motion; if depressed behind the Sun, then it may be inferred that the conjunction is to come by the same reason. Hence the parallax in time, if the Moon be to the east is to be subtracted from the end of the *tithi* or the hour of ecliptic conjunction, and to be added when the Moon is to the west.
29. The latitude of the Moon is north and south distance between the Sun and Moon, and the *nati* also is north and south. Hence the *śara* or latitude applied with the nati or the parallax in latitude, becomes the apparent latitude (of the Moon from the Sun).

*āyana-valana*

30. In either the 1st Libra or the 1st Aries in the equinoctial point of intersection of the equinoctial and ecliptic, the north and south lines of the two circles i. e. their secondaries are different and are at a distance of the extreme declination (of the Sun i. e. 24°) from each other.
31. Hence, the *āyana-valana* will then be equal to the sine of 24°:—The north and south lines of these two circles however are coincident at the solstitial points.
32, 33 and 34. And the north and south lines being there coincident, it follows as a matter of course that the east of those two circles will be the same. Hence at the solstitial points there is no (*āyana*) *valana*. When the planet is in any point of the ecliptic between the equinoctial and solstitial points, *āyana-valana* is then found by proportion, or by multiplying the co-sine of the longitude of the planet by the sine of 24°, and dividing the product by the *dyujyā* or the co-sine of the declination of the planet. This *āyana-valana* is called north or south as the planet be in the ascending or descending signs respectively. Thus in like manner at the point of intersection of the prime vertical and equinoctial, the six o'clock line is the north and south line of the equinoctial, whilst the horizon (of the given place) is the north and south line of the prime vertical. The distance of these north and south lines is equal to the latitude (of the place).

*ĀKSHA-VALANA.*

35. Hence at (the east or west point of) the horizon, *āksha-valana* is equal to the sine of the latitude. At midday the north and south line of the equinoctial and prime vertical is the same. Hence at midday there is no *āksha-valana*.
36. For any intervening spot, the *āksha-valana* is to be found from the sine of the *nata* by proportion. First, the degrees of *nata* are (nearly) to be found by multiplying the time from noon by 90 and dividing product by the half length of day.

*SPASHṬA-VALANA.*

37. Then the sine of the *nata* degrees multiplied by the sine of latitude, and divided by the co-sine of the declination of the planet will be the *āksha-valana*. If the *nata* be to the east, the *āksha-valana* is called north. If west, then it is called south (in the north terrestrial latitude). The sum and difference of the *āyana* and *āksha-valana* must be taken for the *spashṭa-valana*, viz. their sum when the *āyana* and *āksha-valanas* are both of the same denomination, and their difference when of different denominations i. e. one north and the other south.
38. When the planet is at either the points of the intersection of the ecliptic and prime vertical, the *spashṭa-valana* found by adding or subtracting the *āyana* and *āksha-valanas* (as they happen to be of the same or different denominations) is for that time at its maximum.
39. But at a point of the ecliptic distant from the point or intersection three signs either forward or backward, there is no *spashṭa-valana*: for, at those points the north and the south lines of the two circles are coincident.
40. However, were you to attempt to shew by the use of the versed sine, that there was then no *spashṭa-valana* at those points, you could not succeed. The calculation must be worked by the right sine. I repeat this to impress the rule more strongly on your mind.

*Another way of refutation of using the versed line.*

41. As all the circles of declination meet at the poles; it is therefore evident that the north and south line perpendicular to the east and west line in the plane of the equinoctial, will fall in the poles.
42. But all the circles of celestial latitude meet in the pole of the ecliptic-called the *kadamba*, 24° distant from the equinoctial pole. And it is this ecliptic pole which causes and makes manifest the *valana*.
43. In the ecliptic poles always lies the north and south line which is perpendicular to the east and west line in the plane of the ecliptic. To illustrate this, a circle should be attached to the sphere, taking the equinoctial pole for a centre, and 24° for radius. This circle is called the *kadamba-bhrama-vṛitta* or the circle in which the *kadamba* revolves (round the pole). The sines in this circle correspond with the sines of the declination. All the secondary circles to the prime vertical meet in the point of intersection of the meridian and horizon, and this point of intersection is called *sama* i. e. north or south point of horizon. Now from the planet draw circles on the sphere so as to meet in the *sama*, in the equinoctial pole and also in the ecliptic pole. The three different kinds of *valana* will now clearly appear between these circles: viz. the *āksha valana* is the distance between the two circles just described passing through the *sama* and equinoctial pole. 2. The *āyana-valana* is the distance between the circles passing through the ecliptic and equinoctial poles. 3. The *spashṭa-valana* is the distance between the circles passing through the *sama* and *kadamba*. These three *valanas* are at the distance of a quadrant from the planet and are the same in an directions.

*Second mode of illustrating the* SPASHṬA-VALANA.

48 and 49. Or (to illustrate the subject further) making the planet as the pole of a sphere, draw a circle at 90° from it: then in that circle you will observe the *āksha valana* which, in it, is the distance of the point intersected by the equinoctial from the point cut by the prime vertical. The distance of the point cut by the equinoctial from that cut by the ecliptic is the *āyana*—and the distance between the points cut by the ecliptic and prime vertical the *spashṭa-valana*.
50. In this case the plane of the ecliptic is always east and west— celestial latitude forming its north and south line. Those therefore who (like *Śrīpati* or *Lalla*) would add the *śara* celestial latitude to find the *valana*, labour under a grievous delusion.
51. The 1st of Capricorn and the ecliptic pole reach the meridian at the same time (in any latitude): so also with regard to the 1st Cancer. Hence at the solstitial points there is no *āyana-valana*.
52. As the 1st Capricorn revolves in the sphere, so the ecliptic pole revolves in its own small circle (called the *kadamba-bhrama-vṛitta* round the pole).
53 and 54. When the 1st of Aquarius or the 1st of Pisces comes to the meridian, the distance in the form of a sine in the *kadamba-bhrama-vṛitta*, between the ecliptic pole and the meridian is the *āyana-valana*. This *valana* corresponds with the *krāntijyā* or the sine of declination found from the degrees corresponding to the time elapsed from the 1st Capricornus leaving the meridian.
55. As the versed sine is like the sagitta and the sine is the half chord (therefore the versed sine of the distance of the ecliptic pole from the meridian will not express the proper quantity of *valana* as has been asserted by *Lalla*, etc.: but the right sine of that distance does so precisely). The *āyana-valana* will be found from the declination of the longitude of the Sun added with three signs or 90°.
56. Those people who have directed that the versed sine of the declination of that point three signs in advance of the Sun should be used, have thereby vitiated the whole calculation. *Āksha-valana* may be in like manner ascertained and illustrated: but it is found by the right sine, (and not by the versed sine).
57. He who prescribes rules at variance with former texts and does not shew the error of their authors is much to be blamed. Hence I am acquitted of blame having thus clearly exposed the errors of my predecessors.

*Another way of refutation, using the versed sine.*

58. The inapplicability of the versed sine may be further illustrated as follows. Make the ecliptic pole the centre and draw the circle called the *jina-vṛitta* with a radius equal to 24°.
59. Then make a moveable secondary circle to the ecliptic to revolve on the two ecliptic poles. This circle will pass over the equinoctial poles, when it comes to the end of the sign of Gemini.

60. By whatever number of degrees this secondary circle is advanced beyond the end of Gemini, by precisely the same number of degrees, it is advanced beyond the equinoctial pole in this small *jina-vṛitta*. The sine of those degrees will be there found to correspond exactly with and increase as does the sine of the declination.
61. And this sine is the *āyana-valana*: *valana* is the *valana* at the end of the *dynjyā*. For the distance between the equinoctial pole and planet is always equal to the arc of which the *dynjya* is the sine i. e. the cosine of the declination.
62. But as the value of the result found is required terms of the radius, it is consequently to be converted in those terms As the *jina-vṛitta* was drawn from the ecliptic pole centre, with a radius equal to the greatest declination, so now, making the *sama* centre draw a circle round it with a radius equal to the degrees of the place's latitude. (This circle is called *āksha-vṛitta*)
63 and 64. To the two *samas* or north and south points of the horizon as poles, attach a moveable secondary circle to the prime vertical. Now, if this moveable circle be brought over the planet, then its distance counted in the *āksha-vṛitta* of small circle from the equinoctial pole will be exactly equal that of the planet from the zenith in the prime vertical. The sine of the planet's zenith distance in the prime vertical, when reduced to the value of the radius of *āksha-vṛitta* represent the *āksha-valana*.
65. As in the *āyana-valana* so also in this *āksha-valana*, the result at the end of the *dynjyā* is found; this therefore must be converted into terms of the radius. From this illustration it is evident that it may be accurately ascertained from the zenith distance in the prime vertical.

*See verse 41, Chap. VII.*

66. I will show now how the *āksha-valana* may be also ascertained from the time from the planets being on the meridian in its diurnal circle. Add or subtract the *śakutala* to and from the sine of amplitude according as they are of the same or of different denominations (for the *bāhu* or *bhuja*).
67. The sine of the latitude of the given place multiplied by the sine of the *asus* of the time from the planet's being on the meridian, and divided by the square-root of the difference between the squares of the *bhuja* (above found) and of the radius, will be exactly the *āksha-valana*.
68. Or the *āksha-valana* may be thus roughly found. Multiply the time from the planet's being on the meridian and divide the product by the half length of day, the result are the *nata* degrees. The sine of these *nata* degrees multiplied by the sine of the latitude and divided by the *dynjyā* or the cosine of the declination, will give the rough *āksha-valana*.

*Further illustration.*

69. Place the disc of the Sun at the point at which the diurnal circle intersects the ecliptic. The arc of the disc intercepted between these two circles of radius of the disc. represents the *āyana-valana* in terms of the radius of the disc.
70. This *valana* is equal to the difference between the sine of declination of the centre of the Sun and of the point of intersection of the disc and ecliptic; and it is thus found; multiply the radius of disc by the *bhogya-khaṇḍa* of the *bhuja* of the Sun's longitude and divide by 225.
71. Then multiply this result by sine of 24° and divide by the radius: the quotient is the difference of the two sine of declination. This again multiplied by the radius and divided by the radius of Sun's disc will give the value in terms of the radius (of a great circle).
72. Now in these proportions the radius of the Sun's disc and also radius are in one case multipliers (being in third places), and in the other divisors (being the first terms of the proportion) therefore cancel both. There will then remain rule, multiply the Sun's *bhogya khaṇḍa* by sine of 24° and divide by 225.
73. And this quantity is equal to the declination of a point of ecliptic 90° in advance of Sun's place. Thus you observe that the *valana* is found by the sine of declination as above alleged, (and not by the versed sine). Abandon therefore O foolish men, your erroneous rules on this subject.
74. The disc appears declined from the zenith like an umbrella; but the declination is direct to the equinoctial pole: the proportion of the *dynjyā* or complement of declination is therefore required to reduce the *valana* found to its proper value in terms of the radius.

End of Chapter VIII. In explanation of the cause or eclipses of the Sun and Moon.

---

# CHAPTER IX.

*Called DṚIKKARAMA-VĀSANĀ on the principles of the Rules for finding the times of the rising and setting of the heavenly bodies.*

*Object of the correction called the DṚIKKARMA which is requisite to be applied to the place of the planet, for finding the point of the ecliptic on the horizon when the planet reaches it.*

1. A planet is not found on the horizon at the time at which its corresponding point in the ecliptic (or that point of the ecliptic having the same longitude) reaches the horizon, inasmuch as it is elevated above or depressed below the horizon, by the operation of its latitude. A correction called *dṛikkarama* to find the exact time of rising and setting of a planet, is therefore necessary.

2. When the planet's corresponding point in the ecliptic reaches the horizon, the latitude then does not coincide with the horizon, but with the circle of latitude. The elevation of the latitude above and depression of it below the horizon, is of two sorts. Hence the *dṛikkarama* is two-fold, i. e. the *āyana* and the *akshaja* or *āksha*. The detail and mode of performing these two sorts of the correction are now clearly unfolded.

*DṚIKKARMA..*

3. When the two *valanas* are north and the planet's corresponding point in the ecliptic is the eastern horizon, the planet is thereby depressed below the horizon by south latitude, and elevated when the planet's latitude is north.
4. When the two kinds of *valana* are south, then the reverse of this takes place; the reverse of this also takes place when the planet's corresponding point is in the western horizon. If radius: *āyana-valana*: : what will celestial latitude give?
5. And if cosine of the latitude of the given place: *āksha-valana* : what will *spashṭa śara* give? Multiply the two results thus found by these two proportions, by the radius and divide the products by the *dyujyā* or cosine of declination.
6 and 7. Take the arcs of these two results (which are sines) and by the *asus* found from the sum of or the difference between these two arcs, the planet is depressed below or elevated above the horizon. The *lagna* or horoscope found by the direct process (as shown in the note on the verse 26, Chapter VII,) when the planet is depressed and by the indirect process (as shown in the same note) when it is elevated, by means of the *asus* above found, is its *udaya lagna* rising horoscope or the point of the ecliptic which comes to the eastern horizon at the same time with the planet. When the planet's corresponding point is in the western horizon, the *lagna* horoscope found then by the rule converse of that above given, by means of the place of the planet added with 6 signs, is its *asta lagna* setting horoscope or the point of the ecliptic which is on the eastern horizon when the plane comes to the western horizon.
8 and 9. For the fixed stars whose latitudes are very considerable the resulted time of the *dṛikkarma* is found in a different way. Find the ascensional difference from the mean declination of the star, i. e. from the declination of its corresponding point in the ecliptic, and also from that applied with the latitude, i. e. from the true declination. The *asus* found from the sum of or the difference between the ascensional differences just found, as the mean and true declinations are of the different or of the same denominations respectively, are the *asus* of depression or elevation depending on the *āksha dṛikkarma*. (Find also the time depending on the *āyana-dṛikkarma*): and from the sum of or the difference between them, as they may be of the same or different denominations, the *udaya lagna* or *asta lagna* may be ascertained as above found (in the 6th and 7th verses).

*To find the value of celestial latitude in terms of a circle of declination, to render it fit to be added to or subtracted from declination.*

10. The *Śara* or true latitude multiplied by the *Dyujyā* or cosine of declination of the point of the ecliptic, three signs in advance of the planet's corresponding point and divided by the radius becomes the *spashṭa* or rectified latitude. This rectified latitude is used when it is to be applied to the mean declination and also in the *āksha dṛikkarma*.

*Omission of the last mentioned correction or reduction of Celestial latitude to its value in declination, by Brahmagupta and others.*

11. The celestial latitude is not reduced by *Brahmagupta* and other early astronomers to its value in declination: and the reason of this omission, seems to have been its smallness of amount. And also it is the uncorrected latitude which is used in finding the half duration of the eclipses and in their projections etc.
12. As the constellations are fixed, their latitudes as given in the books of these early astronomers are the *spashṭa-śaras* i. e. the reduced values of the latitudes so as to render them fit to be added to or subtracted from the declination; and the *dhruvas* or longitude of these constellations are given, after being corrected by the *āyana dṛikkarma* so as to suit those corrected latitudes that is, the star will appear to rise at the equator at the same time with longitude found by the correction.

*Bhaskaracharya exposes the incorrect theory of certain of his predecessors, by quoting their own practice which is irreconcilable with their own theory.*

13. Those astronomers, who have mentioned that celestial latitude is an arc of a circle of declination, are stupid. Were the celestial latitude nothing more than an arc of a circle of declination, then why should they or others have ever had recourse to the *āyana dṛikkarma* at all? (The planets or stars would appear on the six o'clock line at the time that the corresponding degree of the ecliptic appeared there.)
14. How moreover have these same astronomers in delineating an eclipse marked off the Moon's latitude in the middle of the eclipse on *spashṭa-valana-sūtra* or on the line denoting the secondary circle to the ecliptic? and how also have they drawn perpendicularly on the *valana-sūtra* or the line representing the ecliptic, the latitudes of the Moon at the commencement and termination of the eclipse.
15. How moreover, have they made the latitude *koṭi*, i.e. perpendicular to the ecliptic and thus found the half duration of the eclipse? If the latitude were of this nature, it would never be ascertained by the proportion (which is used in finding it).

*Censure of the astronomers who erroneously used the verse sine in the* dṛikkarma *and* valana.

16. A certain astronomer has (first) erroneously stated the *dṛikkarma* and *valana* by the versed sign. This course has been followed by others who followed him like blind men following each other in succession:

*Praise of Brahmagupta.*

17. *Brahmagupta's* rule, however, is wholly unexceptionable, but it has been misinterpreted by his followers. My observations cannot be said to be presumptuous, but if they are alleged to be so, I have only to request able mathematicians to weigh them with candour.
18. The *dṛikkarma* and *valana* found by the former astronomers through the versed sine are erroneous: And I shall now give an instance in proof of their error.

*An instance in proof of the error.*

19 and 20. In any place having latitude less than *24°N.* multiply the sine of the latitude of the place by the radius and divide the product by the sine of 24° or the sine of the obliquity of the ecliptic and take the arc in degrees of the result found. And find the point of the ecliptic, the degrees just found in advance of the 1st Aries. Now, if from this point the planet's corresponding point on the ecliptic three signs backwards or forwards, be on the western or eastern horizon respectively, then the ecliptic will coincide with the vertical circle, and the horizon will consequently be secondary to the ecliptic. Hence the planet will not quit the horizon, though it be at a distance of extreme latitude from its corresponding point in the ecliptic, as the celestial latitude is perpendicular to the ecliptic.
21. In this case the resulted times of the *dṛikkarma* being of exactly the same amount but one being *plus* and the other *minus,* neutralize each other. Now this result would not be obtained by using the versed sine—hence let the right sine (as prescribed) be always used for the *dṛikkarma.*
22. Again here, in like manner, it is from the two *valanas* having different denominations, but equal values that they mutually destroy each other. By using the versed sine, they would not have equal amounts, hence the *valanas* must be found by the right sine.
23. When the Sun comes to the zenith, and consequently the ecliptic coincides with the vertical circle, the *spashṭa valana* then evidently appears to be equal to the sine of the amplitude of the ecliptic point 90° in advance of the Sun's place in the horizon. If you, my friend, expert in spherics, can make the *spashṭa valana* equal to the sine of amplitude by means of the versed sine, then I will hold the *valana* found in the *Dhīvṛiddhida tantra* by *Lalla* and in the other works to be correct.
24. In the place where the latitude is 66° N. when the Sun at the time of his rising is in 1st Aries, 1st Taurus, 1st Pisces, or in 1st Aquarius, he will then be eclipsed in his southern limb, because the ecliptic then coincides with the horizon.

Therefore, tell me how the *spashṭa valana* will be equal to the radius by means of the versed sine!

*Cause of error in Lalla and others stated.*

25. Even clever men are frequently led astray by conceit in their own quick intelligence, by their too hasty zeal and anxiety for distinction, by their confidence in others and by their own negligence or inadvertence, when it is thus with the wise; what need I say of fool? others, however, have said:—

26. Those given to the service of courtesans and bad poets, are both distinguished by their disregard of the criticisms and reflections of the world, by their breach of the rules of time and metres, and their destruction of their substance and of their subject, being beguiled by the vain delight they feel towards the object of their taste.

End of Chapter IX. called *DṚIKKARMA-VĀSANĀ*:.

---

# CHAPTER X.

*Called Śringonnati-vāsanā in explanation of the cause of the Phases of the Moon.*

1. This ball of nectar the Moon being in contact with rays of the Sun, is always illuminated by her shinings on that side turned towards the Sun. The side opposite to the Sun dark as the raven black locks of a young damsel, is obscured by being in its own shadow, just as that half of a water-pot which is turned from the Sun, is obscured by its own shadow.

2. At the conjunction, the Moon is between us and the Sun: and its lower half which is then visible to the inhabitants of the earth, being turned from the Sun is obscured in darkness. That half again of the Moon when it has moved to the distance of six signs from the Sun, appears to us at the period of full Moon brilliant with light.

3. Draw a line from the earth to the Sun's orbit at a distance of 90° from the Moon, and find also a point in the Sun's orbit (in the direction where the Moon is) at a distance equal to that of the Moon from the earth. When the Sun reaches the point just found, he comes in the line perpendicular at the Moon to that drawn from the earth to the Moon. Then the Sun illumines half of the visible side of the Moon. That is when the Moon is 85° . . 45' from the Sun east or west, it will appear half full to us.

4. The illuminated portion of the Moon gradually increases as it recedes from the Sun: and the dark portion increases as it approaches the Sun. As this sea-born globe of water (the Moon) is a sphere, its horns assume a pointed or cusped appearance (varying in acuteness according to its distance from the Sun).

*Diagram for illustrating the subject.*

5. (To illustrate the subject, a diagram should be drawl as follows). Let the distance north and south between the Sun and Moon represent the *bhuja*, the upright distance between them the *koṭi* and the line joining their centres the hypothenuse. The Sun is in the origin of the *bhuja* which stretches in the direction where the Moon is, the line perpendicular at the end a the *bhuja* is *koṭi* at the extremity of which is the Moon and the line stretching (from the Moon) in the direction of the Sun is the hypothenuse. The Sun gives light (to the Moon) through the direction of the hypothenuse.
6. I have thus only briefly treated of the principles of the subjects mentioned in the Chapters on *madhyagati* etc. fearing to lengthen my work; but the talented astronomer should understand the principles of all the subjects in completion, because this is the result to be obtained by a complete knowledge of the spheric.

End of Chapter X. called *ŚRINGONNATI-VĀSANĀ*.

---

# CHAPTER XI.

*Called Yantrādhyāya, on the use of astronomical instruments.*

*Object of the Chapter.*

1. As minute portions of time elapsed from sun-rise cannot be ascertained without instruments, I shall therefore briefly detail a few instruments which are of established use for this purpose.
2. The Armillary sphere, *nādī-valaya* (the equinoctial)) the *yashṭi* or staff, the gnomon, the *ghaṭi* or clepsydra, the circle, the semi-circle, the quadrant, and the *phalaka*: but of all instruments, it is "ingenuity" which is the best.

*Use of Armillary Sphere.*

3 and 4. (This instrument is to be made as before described, placing the *Bhagola* starry sphere, which consists of the ecliptic, diurnal circles, the Moon's path, and the circles of declination etc. within the *khagola* celestial sphere, which consists of the horizon, meridian, prime vertical, six o'clock line, and other circles which remain fixed in a given latitude). Bring the place of the Sun on the ecliptic to the eastern horizon: and mark the point of the equinoctial (in the *bhagola*) intersected by the horizon, viz. east point. Having made the horizon as level as water, turn the *bhagola* westward till the Sun throws its shadow on the centre of the Earth. The distance between the mark made on the equinoctial and the now eastern point of the horizon will represent the time from sun-rise.

*The Nādī-valaya.*

5 and 6. The *lagna* or horoscope will then be found in that point of the ecliptic which is cut by the horizon. Take a wooden circle and divide its outer rim into 60 *ghaṭikās*: Then place the twelve signs of the ecliptic on both sides, but instead of making each sign of equal extent, they must b e made each with such variable arcs as shall correspond with their periods of rising in the place of observation (the twelve periods are to be thus marked on either side, which are to be again each subdivided into two *horas* (or hours), three *dreshkāṇas*, into *navānśas* or ninths of 3°. . 20' each, twelfths of 2° . . 10' and into *trinsansas* or thirtieths. These are called the *shadvarga* or six classes). These signs, however, must be inscribed in the inverse order of the signs, that is 1st Aries, then Taurus to the west or right of Aries and so on. Then place this circle on the polar axis of the *khagola* at the centre of the Earth (the polar axis should be elevated to the height of the pole). Now find the Sun's longitude in signs, degrees, etc. for the sun-rise of the given day (by calculation) and find the same degree in the circle. Mark there the Sun's place, turn the circle round the axis, so that the shadow of the axis will fall on the mark of the Sun's place at sun-rise and then fix the circle. Now as the Sun rises, the shadow of the axis will advance from the mark made for the point of sun-rise to the nadir and will indicate the hour from sun-rise, and also the *lagna* (horoscope): the number of hours will be seen between the point of sun-rise and the shadow: and the *lagna* will be found on the shadow itself.

7. Or, if this circle marked as above, be placed on any axis elevated to the altitude of the pole, then the distance from the shadow of the axis to the lowest part of the circle will represent the time to or from midday.

*The ghaṭi or clepsydra.*

8. A *ghaṭi* made of copper like the lower half of a waterpot, should have a large hole bored in its bottom. See how often it is filled and falls to the bottom of the pail of water on which it placed. Divide 60 *ghaṭīs* of day and night by the quotient and it will give the measure of the clepsydra. (If it is filled 60 times, then the *ghaṭi* will be of one *ghaṭikā*; if 24 times it will be of one hour or 2 1/2 *ghaṭikās*.)

*Gnomon.*

9. For a gnomon take a cylindrical piece of ivory, and let it be turned on a lathe, taking care that the circumference be equal above and below. From its shadow may be ascertained the points of the compass, the place of observer, including latitude etc. and times (as has been elsewhere explained).

*The* chakra *circle.*

10. The circle should be marked with 360° on its outer circumference, and should be suspended by a string or chain moveable on the circumference. The horizon or Earth is supposed to be at the distance of three signs or 90° from the point at which it is suspended: the point opposite to that point being the zenith.
11. Through its centre put a thin axis: and placing the circle in a vertical plane, so as to catch the shadow of the Sun: the degrees passed over by the axis from the place denominated the Earth, will be altitude :
12. And the arc to the point denominated the zenith, will be that of the zenith distance. Some former astronomers have given the following rule for making a rough calculation of the time, viz. multiply the half length of day by the obtained altitude and divide the product by the meridian altitude, the quotient will be the time sought.

*To find the longitudes of planets by the circle.*

13. First let the circle be so held or fixed that any two of the following fixed stars appear to touch the circumference, viz. *Maghā* (Leonis, Regulus), *Pushya* (Cancri), *Revatī* (Piscium) and *Śatatārakā* (or Aquarii). Or, that any star (out of the *Chitrā* or Virginis Spica etc.) having very inconsiderable latitude, and the planet whose longitude is required and which is at a considerable distance from the star, appear to touch the circumference.

*Semi-circle and quadrant.*

14 and 15. Then look from the bottom of the circle along its plane, so that the planet appear opposite the axis; and still holding it on the plane of the ecliptic, observe also any of the above mentioned stars. The observed distance between the planet and the star, if added to the star's longitude, when the star is west, and subtracted when east of the planet, will give the planet's longitude. The half of a circle is called a *chāpa* or semicircle. The half of a semicircle is called *turīya* or a quadrant.

*PHALAKA-YANTRA.*

16. As others have not ascertained happily the apparent time by observations of altitudes in a vertical circle, I have therefore laboured myself in devising an instrument called *phalaka* yantra, the uses of which I now proceed to explain perspicuously. It contains in itself the essence of all our calculations which are founded on the true principles of the Doctrine of the Sphere.

*Addresses to the Sun.*

17. I *Bhaśkara* now proceed to describe this excellent instrument, which is calculated to remove always the darkness of ignorance, which is moreover the

delight of clever astronomers and is founded on the shadow of its axis: it is also eminently serviceable in ascertaining the time, and in illustrating truths of astronomy, and therefore valued by the professors of that science. It is distinguished by having a circle in its centre. I proceed to describe this instrument after invoking that bright God of day, the Sun, which is distinguished by the epithets I have above given to the instrument viz. he is eternal and removes obscurity and cold: he makes the lotus to flower and is ever shining: he easily points out the time of the day and season and year, and makes the planets and stars to shine. He is worthy of worship from the virtuous and resides in the centre of his orb.

18. Let a clever astronomer make a *phalaka* or board of a plane rectangular and quadrilateral form, the height being 90 digits, and the breadth 180 digits. Let him halve its breadth and at the point thus found, attach a moveable chain by which to hold it: from that point of suspension let him draw a perpendicular which is called the *lamba-rekhā.*

19. Let him divide this perpendicular into 90 equal parts which will be also digits, and through them draw lines parallel to the top and bottom to the edges: these are called sines.

20. At that point of the perpendicular intersected by the 30th sine at the 30th digit, a small hole is to be bored, and in it is to be placed a pin of any length which is to be considered as the axis.

21. From this hole as centre draw a circle (with a radius of 30 digits: the circle will then cut the 60th sine), 60 digits forming the diameter. Now mark the circumference of this circle with 60 *ghaṭis* and 360 degrees, each degree being subdivided into 10 *palas*.

22. Let a thin *paṭṭikā* or index arm with a hole at one end be made of the length of 60 digits and let it be so marked. The rough ascensional difference in *palas* determined by the *khanḍakas* or parts, being divided by 19, will here become the sine of the ascensional difference (adapted to this instrument.)

23. The numbers 4, 11, 17, 18, 13, 5 multiplied severally by the *āksha-karṇa* and divided by 12, will be the *khanḍakas* or portions at the given place; each of these being for each 15 degrees (of *bhuja* of the Sun's longitude) respectively.

24. Now find the Sun's true longitude by applying the precession of the equinoxes to the Sun's place, and adding together as many portions as correspond to the *bhuja* of the Sun's longitude above found, divide by 60 and add the quotient to *āksha-karṇa*. Now multiply the result by 10 and divide by 4 (or multiply by 2 1/2). The quotient is here called the *yashṭi* in digits and the number of digits thus found is to be marked off on the arm of the *paṭṭikā* counting from its hole penetrated by the axis.

25. Now hold the instrument so that the rays of the Sun shall illuminate both of its sides (to secure its being in a vertical circle): the place in the circumference marked out by the shadow of the axis is assumed to be the Sun's place.

26. Now place the index arm on the axis and putting it over the Sun's place, from the point at the end of the *yashṭi* set off carefully above or below (parallel to the

*lamba-rekhā*) on the instrument, the sine of the ascensional difference above found, setting it off above if the Sun be in the northern hemisphere, and below if it be in the southern hemisphere. The distance from the point where the sine which meeting the end of the sine of the ascensional difference thus set off, cuts the circle, to the lowest part of the circle will represent the *ghaṭis* to or after midday.

*To find the place of the shadow of axis from time.*

27. Set off the time from midday on the instrument counting from the *lamba-rekhā*; from the end of the sine of this time, set off the sine of ascensional difference in a line parallel to the *lamba-rekhā*, but below and above according as it was to be set off above or below in finding the time from the shadow, (this operation being the reverse of the former). The sine met, by the sine of ascensional difference, thus set off, is the new sine across which the *paṭṭikā* or index is now to be placed till the *yashṭi-chinha* or point of *yashṭi* falls on it. This position will assuredly exhibit the place of the shadow of the axis.

*The* YASHṬI *or Staff.*

28, 29, and 30. Having drawn a circle (as the horizon) with a radius equal to radius of a great circle, mark east and west points (and the line joining these points is called the *prāchyaparā* or east and west line) and mark off (from them) the amplitude at the east and west. Draw a circle from the same centre with a radius equal to cosine of declination i. e. with a radius of diurnal circle, and mark this circle with 60 *ghaṭis*. Now take the *yashṭi*, equal to the radius (of the great circle) and hold it with its point to the Sun, so that no shadow be reflected from it; the other point should rest in the centre. Now measure the distance from the end of the amplitude to the point of the *yashṭi* when thus held opposite to the Sun. This distance applied as a chord within the interior circle will cut off; if it be before midday, an arc of the number of *ghaṭikās* from sunrise, and if after midday an arc of the time to sunset.

*To find the palabhā with the Yashṭi.*

31. The perpendicular let fall from the point of the *yashṭi* is the *śanku* or sine of altitude: the place between the *śanku* and centre is equivalent to *dṛigya* or sine of zenith distance. The sine of amplitude is the line between the point, of horizon at which the Sun rises or sets, on which the point of the *yashṭi* will rest at sun-rise and sun-set, and the east and west line the *prāchyaparā*.

*To find PALABHĀ, declination, time, etc. from three observations by the Yashṭi, of three śankus.*

32 and 33. The distance between the *śanku* and the *udayāsta-sūtra*, multiplied by 12 and divided by the *śanku*, will be the *palabhā*. Take two altitudes of the Sun with the *yashṭi*: observe the *śanku*s of the two times and the *bhujas*. Add the two *bhujas*, if one be north and the other south, or subtract if they be both of the same denomination: multiply the above quantity (whether sum or difference) by 12 and divide by the difference of the two *śanku*s, the result will be the *palabhā*. The difference between the east and west line and the root of *śankus* is called *bhuja*. If the *śanku* be observed three different times by the *yashṭi*, then the time, declination etc. may be found (by simply observing the Sun).
34. First of all find three *śanku*s: draw a line from the top of the first to the top of the last; from the top of the second *śanku*, draw a line to the eastern point and a line to the western point of the horizon, so as to touch the first line drawn.
35. A line drawn so as to connect these two points in horizontal circumference will be the *udayāsta sūtra*. The distance between it and the centre will give the sine of amplitude. The line drawn through the centre parallel to the *udayāsta-sūtra* at the distance of the sine of amplitude is the east and west line.
36. Find the *palabhā* as before (and also the *āksha-karṇa*). Now the sine of amplitude multiplied by 12 and divided by *āksha-karṇa* will be the sine of declination. This again multiplied by the radius and divided by the sine of 24° or the sine of the Sun's greatest declination, will give the sine of the *bhuja* of the Sun's longitude.
37 and 38. Which converted into degrees is Sun's longitude, if the observation shall have been made in the 1st quarter of the year. If in the second quarter, the longitude will be found by subtracting the degrees found from 6 signs: if in the 3rd quarter, 6 signs must be added: if in the fourth quarter of the year, then the degrees found must be subtracted from 12 signs for the longitude. The quarters of the year will be known from the seasons the peculiarities of each of which I shall subsequently describe. It is declared (by some former astronomers) that the shadow of the gnomon revolves on the circle passing through the ends of the three shadows made by the same gnomon (placed in the centre of the horizon), but this is wrong, and consequently the east and west and north and south lines, the latitudes etc. found by the aid of the circle just mentioned are also wrong.
39. Whether the place of the Sun be found from the shadow or from the sine of the amplitude, it will be found corrected for precession. If the amount of precession be subtracted, the Sun's true place will be found. If the true place of the Sun be subtracted, the amount of precession will be ascertained.

*In praise of instrument called* dhīyantra *or genius instrument.*

40. But what does a man of genius want with instruments about which numerous works have treated.? Let him only take a staff in his hand, and look at any object along it, casting his eye from its end to the top, there is nothing of which he will not then tell its altitude, dimensions, etc. if it be visible, whether in the heavens, on the ground or in the water on the earth. Now I proceed to explain it.

41. He who can know merely with the staff in his hand the height and distance of a bamboo, of which he has observed the root and top, knows the use of that instrument of instruments—genius—(*dhīyantra*)and tell me what is there that he cannot find out.

*To find* palabhā.

42. Direct the staff lengthways to the north polar star; let drop-lines fall from both ends of staff, when thus directed to the star. Now the space between the two drops is the B*huja* or base of a right angled triangle, when the difference between the lines thus dropped is the *koṭi* or perpendicular.
43. The *koṭi* multiplied by 12 and divided by the *bhuja* gives the *palabhā*. Having in the same way observed the root of the bamboo; multiply the *bhuja* by the height of the man's eye.

*To find the distance and height of a bamboo.*

44 and 45. And divide the product by the *koṭi*, the result is, you know the distance to the root of the bamboo. Having thus observed the top of the bamboo (with the staff, and ascertained the *bhuja* and *koṭi*), multiply the distance to the root of the bamboo by the *koṭi*, and divide the product by the *bhuja*, the result is the height of the bamboo above the observer's eye: this height added with the eye's height will give the height of the whole bamboo.

*Example.*

For instance, suppose the staff 145 digits long, the height of observer's eye 68 digits; that in making the lower observation the *bhuja*=144 digits=6 cubits, and *koṭi*=17 digits; that in making the observation of the top of the bamboo, the *bhuja*= 116 digits and *koṭi*= 87 digits. Then tell me the height of bamboo and the distance of it. As,

68 X 144 / 17= 576 digits or 24 cubits distance to bamboo;
576 X 87 / 116 = 432 height of tree above observer's eye,
add 68 the eye's height

500 = height of tree.

Let a man, standing up, first of all observe the top of an object: then (with a staff, whether it be equal to the former or not in length), let him observe again the top of the same object whilst sitting.
46. Then divide the two *koṭis* by their respective *bhujas*: take the difference of these quotients, and by it divide the difference of the heights of observer's eye—this will give the distance to the bamboo: from this distance the height of the bamboo may be found as before.

*Question.*

47. There is a high famous bamboo, the lower part of which being concealed by houses etc. was invisible: the ground, however, was perfectly level: If you, my friend, remaining on this same spot by observing the top (first standing and then sitting), will tell me the distance and its height, I acknowledge you shall have the title of being the most skilful of observers and expert in the use of the best of instruments *dhīyantra*.

*Example.*

The observer, first standing, observes the top of the bamboo and finds the *bhuja*, with the first staff, to be 4 cubits or 96 digits: he then sits down and finds with another staff the *bhuja* to be 90 digits. In both cases the *koṭi* was one digit. Tell me, O you expert in observation, the distance of observer from the bamboo and the bamboo's height.

*Observation in water.*

48. So also the altitude may be observed in the surface of smooth water: but in this case the height of observer's eye is to be subtracted to find the true height of the object:—Or the staff may be altogether dispensed with: In which last case two heights of the observer's eye (viz. when he stands and sits) will be two *koṭis*: and the two distances from the observer to the places in the water where the top of the object is reflected the *bhujas*.

*Question.*

49. Having seen only the top of a bamboo reflected in water, whether the bamboo be near or at a distance, visible or invisible, if you remaining on this same spot, will tell me the distance and height of bamboo, I will hold you, though appearing on Earth as a plain mortal, to have attributes of superhuman knowledge.

*Example.*

An observer standing up first observes (with his staff) the reflected top of a bamboo in water. The *koṭi* = 3 digits and *bhuja* = 4 digits. Then sitting down he makes a second observation and finds the *bhuja* = 11 digits and *koṭi* = 8 digits. His eye's height standing = 3 cubits or 72 digits, and sitting = 1 cubit or 24 digits. Tell me height of bamboo and its distance.

*Example.*

A man standing up sees the shadow of a bamboo in the water—the point of the water at which the shadow appears is 96 digits off: then sitting down on the same spot he again observes the shadow and finds the distance in the water at which it appears to be 33 digits: tell me the height of the bamboo and his distance from the bamboo.

*A self-revolving instrument of* swayanvaha yantra.

50 and 51. Make a wheel of light wood and in its circumference put hollow spokes all having bores of the same diameter, and let them be placed at equal distances from each other; and let them also be all placed at an angle somewhat verging from the perpendicular: then half fill these hollow spokes with mercury: the wheel thus filled will, when placed on an axis supported by two posts, revolve of itself. Or scoop out a canal in the tire of the wheel and then plastering leaves of the *tala* tree over this canal with wax, fill one half of this canal with water and other half with mercury, till the water begins to come out, and then cork up the orifice left open for filling the wheel. The wheel will then revolve of itself, drawn round by the water.

*Description of a syphon.*

Make up a tube of copper or other metal, and bend it into the form of an *ankuśa* or elephant hook, fill it with water and stop up both ends.

54. And then putting one end into a reservoir of water, let the other end remain suspended outside. Now uncork both ends. The water of the reservoir will be wholly sucked up and fall outside.
55. Now attach to the rim of the before described self- revolving wheel a number of water-pots, and place the wheel and these pots like the water-wheel so that the water from the lower end of the tube flowing into them on one side shall set the wheel in motion, impelled by the additional weight of the pots thus filled. The water discharged from the pots as they reach the bottom of the revolving wheel, should be drawn off into the reservoir before alluded to by means of a water-course or pipe.
56. The self-revolving machine (mentioned by *Lalla* etc.) which has a tube with its lower end open is a vulgar machine on account of its being dependant, because that which manifests an ingenious and not a rustic contrivance is said to be a machine.
57. And moreover many self-revolving machines are to be met with, but their motion is procured by a trick. They are not connected with the subject under discussion. I have been induced to mention the construction of these, merely because they have been mentioned by former astronomers.

End of Chapter XI. called *Yantrādhyāya*.

# CHAPTER XII.

*Description of the seasons.*

*Spring.*

1. (This is the season in which) the *kokilas* (Indian black birds) amidst young climbing plants, thickly covered with gently swaying and brilliantly verdant sprouts of the mango (branches) raising their sweet but shrill voices say, "Oh travellers! how are you heart-whole (without your sweethearts, whilst all nature appears revelling) in the jubilee of spring *chaitra*, and the black bees wander intoxicated by the delicious fragrance of the blooming flowers of the sweet jasmine!"
2. The spring-born *mallikā* (Jasminum Zambac, swollen by the pride she feels in her own full blown beautiful flowers) derides (with disdain her poor) unadorned (sister) *mālatī* (Jasminum grandiflorum) which appears all black soiled and without leaf or flower (at this season), and appears to beckon her forlorn sister to leave the grove and garden with her tender budding arms, agitated by the sweet breezes from the fragrant groves of the hill of *Malaya*.

*The grīshma or mid-summer season.*

3. In the summer (which follows), the lovers of pleasures and their sweethearts quitting their stone built houses, betake themselves to the solitude of well wetted cottages of the *kuśakāśa* grass, salute each other with showers of rose-water and amuse themselves.
4. Now fatigued by their dalliance with the fair, they proceed to the grove, where *Kāma-deva* has erected the (flowering) mango as his standard, to rest (themselves) from the glare of the fierce heat, and to disport themselves in the (well shaded) waters of its *bowris* (or large wells with steps).

*Rainy season.*

5. (The rainy season has arrived, when the deserted fair one thus calls upon her absent lover: ) Why, my cruel dear one, why do you not shed the light of your beaming eye upon your love-sick admirer? The fragrance of the blooming *mālatī* and the turbid state of every passing torrent proclaims the season the rains and of all-powerful love to have arrived. Why, therefore, do you not have compassion on my miserable lot?
6. (Alas, cries the deserted wife, alas!) the peacocks (delighted by the thundering clouds) scream aloud, and the breeze laden with the honied fragrance of the

*kadamba* comes softly, still my sweet one comes not. Has he lost all delight for the sweet scented grove, has he lost his ears, has he no pity—has he no heart?
7. Such are the plaintive accusations of the wife in the season of the rains, when the jet black clouds overspread the sky:—angered by the prolonged absence of him who reigned over her heart, she charges him, but still smilingly and sweetly, with being cruelly heedless of her devoted love.

*The* śaratkāla *or season of early autumn.*

8. The mountain burning with remorse at the guilt of having received the forbidden embraces of his own *pushpavatī* daughter, forest appears in early autumn through its bubbling springs and streams sparkling at night with the rays of the Moon, to be shedding a flood of mournful tears of penitence.

Hemanta *or early winter.*

9. In the *hemanta* season, cultivators seeing the earth smiling with the wide spread harvest, and the grassy fields all bedecked with the pearl-like dew, and teeming with joyous herds of plump kine, rejoice (at the grateful sight).

Śiśira *or close of winter.*

10. When the *śiśira* season sets in what unspeakable beauty and what sweet and endless variety of red and purple does not the '*kachnār*' grove unceasingly present, when its leaf is in full bloom, and its bright glories are all expanded.
11. The rays of the Sun fall midday on the earth, hence in this *śiśira* season, they avail not utterly to drive away the cold:

*Sweets of poetry.*

12. Here, under the pretence of writing a descriptive account of the six seasons, I have taken the opportunity of indulging my vein for poetry, endeavouring to write something calculated to please the fancy of men of literary taste.
13. Where is the man, whose heart is not captivated by the ever sweet notes of accomplished poets, whilst they discourse on every subject with refinement and taste? or whose heart is not enchanted by the blooming budding beauties of the handsome willing fair one, whilst she prattles sweetly on every passing topic:—or whose substance will she not secure by her deceptive discourse?
14. What man has not lost his heart by listening to the pure, correct, nightingale-like notes of the genuine poets? or who, whilst he listens to the soft notes of the water-swans on the shores of large and overflowing lakes well filled with lotus flowers, is not thereby excited?

15. As holy pilgrims delight themselves, in the midst of the streams of the sacred Ganges, in applying the mud and the sparkling sands of its banks, and thus experience more than heaven's joys: so true poets lost in the flow of a fine poetic frenzy, sport themselves in well rounded periods abounding in displays of a playful taste.

End of Chapter XII.

## CHAPTER XIII.

*Containing useful question called praśnādhyaya.*

*Object of the Chapter and its praise.*

1. Inasmuch as a mathematician generally fails to acquire distinction in an assemblage of learned men, unless well practised in answering questions, I shall therefore propose a few for the entertainment of men of ingenuity, who delight in solving all descriptions of problems. At the bare proposition of the questions, he, who fancies in his idle conceit, that he has attained the pinnacle of perfection, is often utterly disconcerted and appalled, and finds his smiling cheeks deserted of their colour.
2. These questions have been already put and have been duly answered and explained either by arithmetical or algebra: processes, by the pulverizer and the affected square, i. e. methods for the solutions of indeterminate problems of the first and of the second degree, or by means of the armillary sphere, or other astronomical instruments. To impress and make them still more familiar and easy I shall have to repeat a few.

*Praise of ingenious persons.*

3. All arithmetic is nothing but the rule of proportion: and Algebra is but another name for ingenuity of invention. To the clever and ingenious then what is not known! I, however, write for
men and youths of slow comprehension.
4. With the exception of the involution and evolution of the square and cube roots, all branches of calculation may be wholly resolved into the rule of proportion. It indeed assumes many shapes, but it is universally prevalent. All this arithmetical calculation denominated *Pāṭī gaṇita*, which has been composed in many ways by the wisest of former mathematicians, is only for the enlightenment of simple men like myself.
5. Algebra does not consist in the letters (assumed to represent the unknown (quantities): neither are the different processes any part of its essential properties.

But Algebra is wholly and simply a talent and facility of invention, because the faculties of inventive genius are infinite.

*Question 1st.*

6. Why, O astronomer, in finding the *ahargaṇa*, do you add *saura* months to the lunar months *chaitra* etc. (which may have elapsed from the commencement of the current year): and tell me also why the (fractional) remainders of *adhimāsas* and *avamas* are rejected: for you know that to give a true result in using the rule of proportion, the remainders should be taken into account.

*Question 2nd.*

7. If you have a perfect acquaintance with the *miśra* or allegation calculations, then answer this question. Let the place of the Moon be multiplied by one, that of the Sun by 12 and that of Mars by 6, let the sum of these three products be subtracted from three times the Jupiter's place, then I ask what are the revolutions of the planet whose place when added to or subtracted from the remainder will give the place of Saturn ?

*RULE.*

8 and 9. In order to work this proposition in the first place proceed with the whole number of revolutions of the several planets in the *kalpa*, adding, subtracting and multiplying them in the manner mentioned in the question: then subtract the result from the revolutions of the planet given: or subtract the revolutions of the given planet from the result, according as the place of the unknown planet happen to be directed to be added or subtracted in the question. This remainder will represent the number of revolutions of the unknown planet in the *kalpa*. If the remainder is larger than the number from which it to be subtracted, then add the number of terrestrial days in a *kalpa*, or if the remainder exceed the number of terrestrial days in the *kalpa*, then reduce it into the remainder by dividing it by the number of days in the *kalpa*.

*Question 3rd.*

10.The algebraical learned, who knowing the sum of the additive months, subtractive days elapsed and their remainders, shall tell the number of days elapsed from the commencement of the *kalpa*, deserves to triumph over the student who is puffed up with a conceit of his knowledge of the exact pulverizer called *saṁslishta* united, as the lion triumphs over the poor trembling deer he tears to pieces in play.

*RULE.*

11. For the solution of this question, you must multiply the given number of additive months, subtractive days and their remainders, by 863374491684 and divide by one less than the number of lunar days in a *kalpa* i. e. by 1602998999999, the remainder will be the number of lunar days elapsed from the beginning of the *kalpa*. From these lunar days the terrestrial days may be readily found.

*Example.*

12. Given the sum of the elapsed additive months, subtractive days and their remainders, equal (according to *Brahmagupta*'s system) to 648426000171; to find the *ahargaṇa*. He who shall answer my question shall be dubbed a "*Brahma-siddhānta-vit*" i.e. shall be held to have a thorough knowledge of the *brahma-siddhānta*.

*Question 4th.*

13 and 14. Given the sum of the remainders of the revolutions, of the signs, degrees, minutes and seconds of the Moon, Sun, Mars, Jupiter, the *śīghrochchas* of Mercury and Venus and of Saturn according to the *dhīvṛiddhida*, including the remainder of subtractive days in finding the *ahargaṇa*, abraded (reduced into remainder by division) by the number of terrestrial days (in a *yuga*). He who, well-skilled in the management of *sphuṭa kuṭṭaka* (exact pulverizer), shall tell me the places of the planets and the *ahargaṇa* from the abraded sum just mentioned, shall be held to be like the lion which longs to make its seat on the heads of those elephant astronomers, who are filled with pride by their own superior skill in breaking down and unraveling the thick mazes and wildernesses which occur in mathematical calculations.

*RULE.*

15. If the given sum abraded by the number of terrestrial days in a *yuga*, on being divided by 4, leaves a remainder, then the question is not to be solved. It is then called a *khila* or an "impossible" question. If, on dividing by 4, no remainder remain, then multiply the quotient by 293627203, and divide the product by 394479375. The number remaining will give the *ahargaṇa*. If the day of the week does not correspond with that of the question, then add this *ahargaṇa* to the divisor (394479375) until the desired day of the week be found.

*Example.*

16. Tell me, my friend, what is the *ahargaṇa* when on a Thursday, Monday or Tuesday, the 35 remainders of the revolutions, signs, degrees, minutes and seconds of the places of the planets, (the Sun, the Moon, Mars, Jupiter and Saturn and the *śīghrochchas* of Mercury and Venus) together with the remainder of the

subtractive days according to the *dhīvṛiddhida*, give, when abraded by the number of terrestrial days in a *yuga*, a remainder of 1491227500.

*Question 5th.*

17. The place of the Moon is of such an amount, that

$$\frac{\text{The minutes}}{2} + 10 = \text{the seconds}$$

$$\text{the minutes} - \text{seconds} + 3 = \text{degrees}$$

$$\frac{\text{the degrees}}{2} = \text{signs.}$$

And the signs, degrees, minutes and seconds together equal to 130. On the supposition that the sum of these *four* quantities is of this amount on a Monday then tell me, if you are expert in rules of Arithmetic and Algebra, when it will be of the same amount on a Friday.

*RULE.*

18. Reduce the signs, degrees and minutes to seconds adding the seconds, then reducing the terrestrial days and the planet's revolutions in a *kalpa* to their lowest terms, multiply the seconds of the planet (such as the Moon) by the terrestrial days (reduced) and divide by the number of seconds in 12 signs: then omitting the remainder, take the quotient and add 1 to it, the sum will be the remainder of the *bhagaṇas* revolutions.
19. The remainder before omitted subtracted from the divisor will give the remainder of seconds: if that remainder of the seconds is greater than the terrestrial days in a *kalpa*, then the question is an "impossible one" (incapable of solution and the planet's place cannot be found at any sun-rise): but if less it may be solved. Then from the remainder of the seconds the *ahargaṇa* may be found (by the *kuṭṭaka* pulverizer as given in the *līlāvātī* and *bīja-gaṇita*) Or,
20. That number is the number of *ahargaṇa* by which the reduced number of revolutions multiplied, diminished by the remainder of the revolutions and divided by the reduced number of terrestrial days in the *kalpa*, will bear no remainder. The reduced number of terrestrial days in a *kalpa* should be added to the *ahargaṇa*. such a number of times as may make the day of the week correspond with the day required by the question.
21. If the Moon's *bhagaṇa-śesha* or the remainder after finding the complete revolutions admits of being divided by 1650000, without leaving any remainder, the question may then be solved: the reduced *bhagaṇa-śesha* on being multiplied by 886834 and divided by 951363, then the remainder will give the *ahargaṇa*, the

divisor should be added to this remainder till the day of the week found corresponds with that of the question.

*Impossible question.*

22. The mean place of the Moon will never be at any sun-rise, equal to 0 signs, 5 degrees, 36 minutes and 19 seconds.

*Question 6th.*

23. When will the square of the *adhimāsa-śesha* remainder of the additive months, multiplied by 10 and the product increased by one, be a square: or when will the square of the *adhimāsa-śesha* decreased by one and the remainder divided by 10 be a square? The man who shall tell me at what period of the *kalpa* this will take place—will be humbly saluted even by the wise, who generally speaking, gaze about in utter amazement and confusion at such questions, like the bee that wanders in the boundless expanse of heaven without place of rest.

*Remark on the preceding question.*

24. (In working questions of *kuṭṭaka* pulverizer, the augment must be reduced by the same number by which the *bhājya* dividend and *hara* divisor are reduced to their lowest terms, and when the augment is not reducible by the same number as the *bhājya* and *hara*, the question is always insoluble.) But here: in working questions of *kuṭṭaka*, those acquainted with the subject should know that the given augment is not to be reduced, i. e. it belongs to the reduced *bhājya* and *hara* otherwise in some places the desired answer will not be obtained, or in others the question will be impossible.

*Question.*

25. Tell me, O you competent in the spheric, considering it frequently in your mind for awhile, what is the latitude of the city (A) which is situated at a distance of 90° from *Ujjayinī*, and bears due east from that city (*Ujjayinī*)? What is the latitude of the place (B) distant also 90° from the city (A) and bearing due west from it? What also is the latitude of a place (C) also 90° from (B) and bearing N, E. from (B) : and of the place (D) which is situated at a distance of 90° from (C) and bears S. W. from (C)?

*Rule*

26 and 27, Convert the distance of *yojanas* (between the two cities, one is given and the other is that of which the latitude is to be found,) into degrees (of a large circle) and then multiply the sine and cosine of these degrees by the cosine of the

azimuth of the other city and *palabhā* at the given city, and divide the products by radius and 12 respectively, Take then the difference between these two quotients, if the other city be south of east of the given city; and if it be north of that, the sum of the quotients is to be taken, But the reverse of this takes place, if the distance between the cities be more than a quarter of the earth's circumference. The difference or sum of the quotients multiplied by 12 and divided by *akshakarṇa* will give the sine of the latitude sought.

*Question.*

28. Tell me quickly, O Astronomer, what is the latitude of a place (A) which is distant 1/6 of the earth's circumference from the city of *Dhārā* and bears 90° due east from it? What also is the latitude of a place distant 60° from *Dhārā*, but bearing 45° N, E, from it? What also is the latitude of a place distant 60° from *Dhārā* and bears S. E. from it? What also are the latitudes of three places 120° from *Dhārā* and bearing respectively due east, N. E., and S. E. from it?

*Question.*

29. Tell me, my friend, quickly, without being angry with me, if you have a thorough knowledge of the spheric, what will be the *palabhā* of the city where the Sun being in the middle of the *ardrā nakshatra* (i. e. having the longitude 2 signs 13° 20') rises in the north-east point.

*Question.*

30. Tell me the several latitudes in which the Sun remains above the horizon for one, two, three, four, five and six months before he sets again.
31. If you, O intelligent, are acquainted with the resolution of affected quadratic equations, then find the Sun's longitude, observing that the sum of the cosine of declination, the sine of declination, and the sine of the Sun's longitude: equal to 5000 is (the radius is assumed equal to 3438.)

*Rule.*

32. Multiply the sum of the cosine of declination, the sine of declination, and the sine of Sun's longitude by 4, and divide the product by 15, the quotient found will be what has been denominated the *ādya*. Next square the sum and double the square and divide by 337, the quotient is to be subtracted, from 910678, Take the square-root of the remainder. That root must then be subtracted from the *ādya* above found: the remainder will be the declination, when the radius is equal to 3438. From the declination the Sun's longitude may be found.

*Question.*

33. Given the sum of the sines of the declination and of the altitude of the Sun when in the prime
vertical; the *taddhṛiti*, the *kujyā* and sine of amplitude equal to 9500, at a place where the p*alabhā* or equinoctial shadow is 5 digits, tell me then, my clever friend, if quick in working questions of latitudinal triangles and capable of abstracting your attention, what are the separate amounts of each quantity?

*Rule.*

34. First assume the sine of declination to be equal to 12 times the shadow *palabhā*: and then find the amounts of the remaining quantities upon this supposition. Then these on the supposition made, multiplied severally by the given sum and divided by their sum on the supposition made, will respectively make manifest the actual amounts of those quantities the sum of which is given.

*Question.*

35. If you have a knowledge of mathematical questions involving the doctrine of the sphere, tell me what will be the several amounts of sines of amplitude, declination and the *kujyā* (where the *palabhā* is 5 digits) when their sum is 2000.
36. But dropping for a moment those questions of the *siddhāntas* involving a knowledge of the doctrine of the sphere, tell me, my learned friend, why in finding the point of the ecliptic rising above the horizon at any given time, (that is the *lagna* or horoscope of that time,) you first calculate the Sun's apparent or true place for that time. i.e. the Sun's instantaneous place: and further tell me, when the Sun's *sāvana* day, i.e. terrestrial day, consists of 60 sidereal *ghaṭikās* and 10 *palas*, the *lagna* calculated for a whole terrestrial day should be in advance of the Sun's instantaneous place, and the *lagna* calculated for the time equal to the terrestrial day *minus* 10 *palas* should be equal to the Sun's instantaneous place.
37. Are the *ghaṭikās* used in finding the *lagna*, *ghaṭikās* of sidereal or common *sāvana* time? If they are *sāvana ghaṭikās*, then tell me why are the hours taken by the several signs of the ecliptic in rising, i.e. the *rasyudaya* which are sidereal, subtracted from them, being of a different denomination? If on the other hand you say they are sidereal, then I ask why, in calculating the *lagna* for a period equal to a whole *sāvana* day i.e. 60 sidereal *ghaṭikās* and 10 *palas*, the *lagna* does not correspond with, but is somewhat in advance of, the Sun's instantaneous place and then why the Sun's instantaneous place is used in finding the *lagna* or horoscope.

*Question.*

38. Given the length of the shadow of gnomon at 10 *gaṭīs* after sun-rise equal to 9 digits at a place where the *palabhā* in 5 digits: tell me what is the longitude of the Sun, if you are *au fait* in solving question involving a knowledge of the sphere.
39. Tell me, O Astronomer, what is the *palabhā* at that place where the gnomon's shadow falling due west is equal to the gnomon's height when the Sun is in the middle of the sign Leo, i.e. when his longitude is 4 signs and 15 degrees.

*Question.*

40. When the Sun enters the prime vertical of a person at *Ujjayinī* either at 5 *ghaṭis* after sun-rise or 5 *ghaṭis* before or after midday, what are his declinations? If you will answer me this I will hold you to be the sharp *ankuśa* (goad) for the guidance of the intoxicated elephants, the proud astronomers.

*Question.*

41. In a place of which the latitude is unknown and on a day which is unknown, the Sun was observed, on entering the prime vertical, to give a shadow of 16 digits from a gnomon (12 digits long) at 8 *ghaṭikās* after sun-rise. If you will tell me the declination of the Sun, and the *palabhā* I will hold you to be expert without an equal in the great expanse of the questions on directions space and time.

*Question.*

42. O Astronomer, tell me, if you have a thorough knowledge of the latitudinal figures, the *palabhā* and the longitude of the Sun at that place, where (at a certain time) the *kujyā* is equal to 245 and the *taddhṛiti* is equal to 3125.

*Question.*

43. Given the sum of the 3 following quantities, viz. of the sines of declination, and of the altitude of the Sun (when in the prime vertical) and of the *taddhṛiti* decreased by the amount of the *kujyā* equal to 6720, and given also the sum of the *kujyā*, the sines of amplitude and declination (at the same time) equal to 1960. I will hold him, who can tell me the longitude of the Sun and also *palabhā* from the given sums, to be a bright instructor of astronomers, enlightening them as the Sun makes the buds of the lotus to expand by his genial heat.

*Question.*

44. Given the sum of the sine of declination, sine of the Sun's altitude in prime vertical and the *taddhṛiti minus kujyā* equal to 1440', and given also the sum of the sine of amplitude, the sine of the Sun's altitude in prime vertical and the *taddhṛiti* equal to 1800'. I will hold him, who having observed the given sums.

*Question.*

45. Given the equinoctial shadow equal to 9. What longitude must the Sun have in that latitude to give an ascensional difference of three *ghaṭis*? I will hold you to be the best of astronomers if you will answer me this question.

*Question.*

46. Hitherto it has been usual to find the length of the Sun's midday shadow, of the shadow of the Sun when in the prime vertical, and when in an intermediate circle (i. e. when he has an azimuth of 45°) by three different modes of calculation: now he who will by a single calculation tell me the length of these three shadows and of the shadows at any intermediate point at the wish of the querist, shall be held to be a very Sun on the Earth to expand the lotus-intellects of learned astronomers.

*Question.*

47. He who, knowing both the azimuth and the longitude of the Sun, observes one shadow of the gnomon at any time, or he who knowing the azimuth observes two shadows and can find the *palabhā*, I shall conceive him to be a very *Garuḍa* in destroying conceited snakes of astronomers.
48. First of all multiply on *bhuja* of the shadow by the hypothenuse of the other, and the second *bhuja* by the hypothenuse of the first: then take the difference of these two *bhujas* thus multiplied, if they are both north or if both south, and their sum if of different denominations, and divide the difference or the sum by the difference of the two hypothenuses; it will be the *palabhā*.

*Question.*

49. How should he who, like a man just drawn up the bottom of a well, is utterly ignorant of the *palabhā*, the place of the Sun, the points of the compass, the number of the years elapsed from the commencement of the *yuga*, the month, the *tithi* or lunar day and the day of week, being asked by others to tell quickly the points of the compass, the place of the Sun, etc., give a correct answer? He, however, who can do so, has my humble reverence, and what astronomers will not acknowledge him worthy of admiration?

*Question.*

50. He, who can know merely with the staff in his hand the height and distance of a bamboo which he has observed the root and top, knows the use of that

instrument of instruments—Genius (the *dhīyantra*): and tell me what is there that he cannot find out!

*Question.*

51. There is a high famous bamboo, the lower part of which, being concealed by houses, etc. was invisible: the ground, however, was perfectly level. If you, my friend remaining on this same spot, by observing the top, will tell me the distance and its height, I acknowledge you shall have the title of being the most skilful of observers, and expert ill the use of the best of instruments, *dhīyantra.*

*Question.*

52. Having seen only the top of a bamboo reflected in water, whether the bamboo be near or at a distance, visible or invisible, if you, remaining on this same spot, will tell me the distance and height of the bamboo, I will hold you, though appearing on the Earth as a plain mortal, to have attributes of superhuman knowledge.

53. Given the places of the Sun and the Moon increased by the amount of the precession of the equinox, i. e. their longitudes, equal to four and two signs (respectively) and the place of the Moon increased by the place of the ascending node equal to 8 signs, tell me whether the Sun and the Moon have the same declination (either both south or one north and one south), if you have a perfect acquaintance with the *Dhīvṛiddhida Tantra.*

54. If the place of the Moon with the amount of the precession of the equinox be equal to 100 degrees, and the place of the of the Moon diminished by that of the ascending node equal to 200 degrees, tell me whether the Sun and the Moon have the same declination, if you have a perfect acquaintance with the *Dhīvṛiddhida Tantra*.

55. If you understand the subject of the *pāta* i. e. the equality of the declinations (of the Sun and the Moon), tell me the reason why there is in reality an impossibility of the *pāta* when there is its possibility (in the opinion of *Lalla*), and why there is a possibility when there is an impossibility of it (according to the same author).

56. If the places of the Sun and the Moon with the amount of the precession of the equinox be equal to 3 signs *plus* and minus 1 degree (i. e. 2s. 29° and 3s.1° respectively) and the place of the Moon decreased by that of the ascending node equal to 11s. 28°, tell me whether the Sun and the Moon have the same declination, if you perfectly know the subject.

57. (In the *Dhīvṛiddhida Tantra*), it is stated that the *pāta is to come* in some places when it has already taken place (in reality), and also it *has happened* where it *is to come.* It is a strange thing in this work when the possibility and impossibility of the *pāta* are also reversely mentioned. Tell me, O you best of astronomers, all this after considering it well.

*Date of the Author's birth and his work.*

58. I (*Bhaśkara*), born in the year of 1036 of the *Śālivāhana* era, have composed this *Siddhānta-śiromaṇi*, when I was 36 years old.

*Author's apology.*

59. He who has a penetrating genius like the sharp point of a large *darbha* straw, is qualified to compose a good work in mathematics: excuse, therefore, my impudence, O learned astronomers, (in composing this work for which I am not qualified).
60. I, having lifted my folded hands to my forehead, beg the old and young astronomers (who live at this time) to excuse me for having refuted the (erroneous) rules prescribed by my predecessors; because, those who fix their belief in the rules of the predecessors will not know what is the truth, unless I refute the rules when I am going to state astronomical truths.

*Author's birth-place, etc.*

61. The learned *Maheśwara*, the head of all astronomers, the most good humoured man, the store of all sciences, skilful in the discussion of acts connected with law and religion, and a *brāhmaṇa* descended from *Śāndilya* (a *muni*) flourished in a city, thickly inhabited by learned and dull persons, virtuous men of all sorts, and men competent in the three *Vedas*, and situated near the mountain *Sahya*.
62. His son, the poet and intelligent *Bhaśkara*, made this clear composition of the *Siddhānta* by the favour of the lotuslike feet of his father; this *Siddhānta* is the guidance for ignorant persons, propagator of delight to the learned astronomers, full of easy and elegant style and good proofs, easily comprehensible by the learned, and remover of mistaken ideas.
63. I have repeated here some questions, which I have stated before, for persons who wish to study only this *Praśnādhyāya*.
64. The genius of the person who studies these questions becomes unentangled, and flourishes like a creeping plant watered at its root by the consideration of the questions and answers, by getting hundreds of leaves of clear proofs, shooting from the Spheric as from a bulbous root.

End of the 13th and last Chapter of the *Golādhyāya* of the *Siddhānta-śiromani.*

# APPENDIX.

ON THE CONSTRUCTION OF THE CANON OF SINES.

1. As the Astronomer can acquire the rank of an *Ācharya* in the science only by a thorough knowledge of the mode of conducting the canon of sines, *Bhaskara* therefore now proceeds to treat upon this (interesting and manifold) subject in the hope of giving pleasure to accomplished astronomers.
2 and 3. Draw a circle with a radius equal to any number of digits: mark on it the four points of the compass and 360°. Now by dividing 90° by the number of sines (you wish to draw in a quadrant), you will get the arc of the first sine. This arc, when multiplied by 2, 3 etc., will successively be the arcs of other sines. Now set off the first arc on the circumference on both sides of one of the points of the compass and join the extremities of these arcs by a transverse straight line, the half of which should be known the sine of the first arc: All the other sines are thus to be known.
4. Or, now, I proceed to state those very sines by mathematical precision with exactness. The square-root of the difference between the squares of the radius and the sine is cosine.
5. Deduct the sine of an arc from the radius the remainder will be the versed sine of the complement of that arc, and the cosine of an arc deducted from the radius will give the versed sine of that arc. The versed sine has been compared to the arrow between the bow and the bow-string: but here it has received the name of versed-sine.
6. The half of the radius is the sine of 30°: the cosine of 30° will then be the sine of 60°. The square-root of half square of radius will be the sine of 45°.
7. Deduct the square-root of five times the fourth power of radius from five times the square of radius and divide the remainder by 8: the square-root of the quotient will be the sine of 36°.
8. Or the radius multiplied by 5878 and divided by 10000 will give the sine of 36°, (where the radius = 3438.) The cosine of this is the sine of 54°.
9. Deduct the radius from the square-root of the product of the square of radius and five and divide the remainder by 4: the quotient thus found will give the exact sine of 18°.
10. Half the root of the sum of the squares of the sine and versed sine of any arc, is the sine of half that arc. Or, the sine of half that arc is the square-root of half the product of the radius and the versed sine.
11. From the sine of any arc thus found, the sine of half the arc may he found (and so on with the half of this last). In like manner from the complement of any arc may be ascertained the sine of half the complement (and from that again the sine of half of the last arc).

Thus the former Astronomers prescribed a mode for determining the other sines (from a given one), but I proceed now to give a mode different from that stated by them.

*12. Deduct and add the product of radius and sine of* bhuja *from and to the square of radius and extract the squareroots of the halves of the results (thus found), these roots will respectively give the sines of the half of 90° decreased and increased by the*

bhuja. *In like manner, the sines of half of 90° decreased and increased by the* koṭi *can be found from assuming the cosine for the sine of* bhuja.
*13. Take the sines of* bhujas *of two arcs and find their difference, then find also the difference of their cosines, square these differences, add these squares, extract their square-root and halve it. This half will be the sine of half the difference of the sines. Thus sines can be determined by several ways.*
14. The square-root of half the square of the difference of the sine and the cosine of the *bhuja* of an arc is equal to the sine of half the difference of the *bhuja* and its complement. I will now give some rules for constructing sines without having recourse to the extraction of roots.
15. Divide the square of the sine of the *bhuja* by the half radius, The difference between the quotient thus found and the radius is equal to the sine of the difference between thedegrees of *bhuja* and its complement. In this way several sines may be found here.

*Rules for finding the sine of every degree from 1° to 90°.*

16. Deduct from the sine of *bhuja* its 1/ 6569 part and divide the ten-fold sine of *koṭi* by573.
17. The sum of these two results will give the following sine (i.e., the sine of *bhuja* one degree more than original *bhuja* and the difference between the same results will give the preceding sine, i.e., the sine of *bhuja* one degree less than original *bhuja*). Here the first sine, i.e., the sine of 1°, will be 60 and the sines of the remaining arcs may be successively found.
18. The rule, however, supposes that the radius = 3438. Thus the sines of 90° of the quadrant may be found. Multiply the cosine by 100 and divide the product by 1529.

*Rules for finding the 24 sines viz., 3°¾, 7°½, 11°¼, 15°, etc.*

19. And subtract the 1/467 part of the sine from it. The sum of these will be the following sine (i.e., the sine of arc of 3°¾ degrees more than original arc): and the difference of them will be preceding sine (i.e., the sine of arc 3°¾ degrees less than original arc).
20. But the first sine (or the sine of 3°¾ ) is here equal to 224 6/7 (and not to 225 as it is usually stated to be). By this rule 24 sines may be successively found.

*Rules for finding the sines of sum and difference of any two arcs.*

21 and 22. If the sines of any two arcs of a quadrant be multiplied by their cosines reciprocally (that is the sine of the first arc by the cosine of the 2nd and the sine of the 2nd by cosine of the first arc) and the two products divided by the radius, then the quotients will, when added together, be the sine of the sum of the two arcs,

and the difference of these quotients will be the sine of their difference. This excellent rule called *jya-bhāvanā* has been prescribed for ascertaining the other sines.

23. This rule is of two sorts, the first of which is called *samāsa-bhāvanā* (i.e., the rule for finding the sine of sum of two arcs) and the second *antara-bhāvanā* (i.e., the rule to find the sine of difference of arcs). Find the first line by the aid of the above-mentioned rule *pratibhāgajyakā-vidhi.*

24 and 25. And then reduce it to the value of any new radius by applying the proportion. After that apply the *jya-bhāvanā*, rule through the aid of the first sine and the cosine thus found, for as many sines as are required. The sines will thus be successively eliminated to the value of any new radius. The rule given in my *Patī* or *Līlāvatī* is not sufficiently accurate (for nice calculations) I have not therefore repeated here that rough rule. (The end)

* (Translation of Siddhānta-śiromaṇi by Pundit Bapu Deva Sastri and Lancelot Wilkinson. Please see Chapter 8 for more information.)

# 14

# The Aryabhatiya of Aryabhata

## *An Ancient Indian Work on Mathematics and Astronomy*

## TABLE OF CONTENTS

8-9. Dimensions of Epicycles of Apsides and Conjunctions of Planets; Circumference of Earth Wind

10. Table of Sine-Differences

C. Colophon

II. *Gaṇitapāda* or Mathematics.

1. Invocation.
2. Names and Values of Classes of Numbers Increasing by Powers of Ten
3. Definitions of Square *(varga)* and Cube *(ghana)*
4. Square Root.
5. Cube Root
6. Area of Triangle; Volume of Pyramid
7. Area of Circle; Volume of Sphere.
8. Area of Trapezium; Length of Perpendiculars from Intersection of Diagonals to Parallel Sides.
9. Area of Any Plane Figure; Chord of One-sixth Circumference Equal to Radius.
10. Relation of Circumference of Circle to Diameter
11. Method of Constructing Sines by Forming Triangles and Quadrilaterals in Quadrant of Circle
12. Calculation of Table of Sine-Differences from First One
13. Construction of Circles, Triangles, and Quadrilaterals; Determination of Horizontal and Perpendicular
14. Radius of *khavṛtta* (or *svavṛtta);* Hypotenuse of Right-Angle Triangle Formed by Gnomon and Shadow

15-16. Shadow Problems

17. Hypotenuse of Right-Angle Triangle; Relation of Half-Chord to Segments of Diameter Which Bisects Chord.
18. Calculation of *sampātaśaras* When Two Circles Intersect

19-20. Arithmetical Progression

21. Sum of Series Formed by Taking Sums of Terms of an Arithmetical Progression
22. Sums of Series Formed by Taking Squares and Cubes of Terms of an Arithmetical Progression
23. Product of Two Factors Half the Difference between Square of Their Sum and Sum of Their Squares
24. To Find Two Factors When Product and Difference Are Known
25. Interest
26. Rule of Three (Proportion)
27. Fractions
28. Inverse Method
29. To Find Sum of Several Numbers When Results Obtained by Subtracting Each Number from Their Sum Are Known

30. To Find Value of Unknown When Two Equal Quantities Consist of Knowns and Similar Unknowns
31. To Calculate Their Past and Future Conjunctions from Distance between Two Planets
32-33. Indeterminate Equations of First Degree *(kuṭṭaka)*

III. *Kālakriyā* or the Reckoning of Time
1-2. Divisions of Time; Divisions of Circle Correspond
3. Conjunctions and *vyatīpātas* of Two Planets in a *yuga*
4. Number of Revolutions of Epicycles of Planets; Years of Jupiter
5. Definition of Solar Year, Lunar Month, Civil Day, and Sidereal Day
6. Intercalary Months and Omitted Lunar Days
7-8. Year of Men, Fathers, and gods; *yuga* of All the Planets; Day of Brahman
9. *Utsarpiṇī, avasarpiṇī, suṣamā,* and *duṣṣamā* as Divisions of *yuga*
10. Date of Writing of *Aryabhaṭīya;* Age of Author at Time
11. *Yuga,* Year, Month, and Day Began at First of *Caitra;* Endless Time Measured by Movements of Planets and Asterisms
*12. Planets Move with Equal Speed; Time in Which They Traverse Distances Equal to Orbit of Asterisms and Circumference of Sky*
*13. Periods of Revolution Differ because Orbits Differ in Size*
*14. For Same Reason Signs, Degrees, and Minutes Differ in Length*
*15. Order in Which Orbits of Planets Are Arranged (beneath the Asterisms) around Earth as Center*
16. Planets as "Lords of Days" (of Week)
17. Planets Move with Their Mean Motion on Orbits and Eccentric Circles Eastward from Apsis and Westward from Conjunction
18-19. Eccentric Circle Equal in Size to Orbit; Its Center Distant from Center of Earth by Radius of Epicycle
20. Movement of Planet on Epicycle; When ahead of and When behind Its Mean Position
21. Movement of Epicycles; Mean Planet (on Its Orbit) at Center of Epicycle
22-24. Calculation of True Places of Planets from Mean Places
25. Calculation of True Distance between Planet and Earth

IV. Gola or the Sphere
1. Zodiacal Signs in Northern and Southern Halves of Ecliptic; Even Deviation of Ecliptic from Equator
2. Sun, Nodes of Moon and Planets, and Earth's Shadow Move along Ecliptic
3. Moon, Jupiter, Mars, and Saturn Cross Ecliptic at Their Nodes; Venus and Mercury at Their Conjunctions.
4. Distance from Sun at Which Moon and Planets Become Visible
5. Sun Illumines One Half of Earth, Planets, Asterisms; Other Half Dark

6-7. Spherical Earth, Surrounded by Orbits of Planets and by Asterisms, Situated in Center of Space; Consists of Earth, Water, Fire, and Air
8. Radius of Earth Increases and Decreases by a *yojana* during Day and Night of Brahman
9. At Equator Stationary Asterisms Seem To Move Straight Westward; Simile of Moving Boat and Objects on Shore
10. Asterisms and Planets, Driven by Provector Wind, Move Straight Westward at Equator—Hence Rising and Setting
11-12. Mount Meru and Vaḍavāmukha (North and South Poles); Gods and Demons Think the Others beneath Them
13. Four Cities on Equator a Quadrant Apart; Sunrise at First Is Midday, Sunset-Midnight at Others
14. Laṅkā (on Equator) 90° from Poles; Ujjain 22 ½° North of Laṅkā
15. From Level Place Half of Stellar Sphere *minus* Radius of Earth Is Visible; Other Half *plus* Radius of Earth Is Cut Off by Earth
16. At Meru and Vaḍavāmukha Northern and Southern Halves of Stellar Sphere Visible Moving from Left to Right or Vice Versa
17. At Poles the Sun, after It Rises, Visible for Half Year; on Moon the Sun Visible for Half a Lunar Month
18. Definition of Prime Vertical, Meridian, and Horizon
19. East and West Hour-Circle Passing through Poles *(unmaṇḍala)*
20. Prime Vertical, Meridian, and Perpendicular from Zenith to Nadir Intersect at Place Where Observer Is
21. Vertical Circle Passing through Planet and Place Where Observer Is *(dṛṅmaṇḍala);* Vertical Circle Passing through Nonagesimal Point *(dṛkkṣepamaṇḍala)*
22. Construction of Wooden Globe Caused To Revolve So as To Keep Pace with Revolutions of Heavenly Bodies
23. Heavenly Bodies Depicted on This; Equinoctial Sine (Sine of Latitude) Is Base; Sine of Colatitude *(śaṅku* at Midday of Equinoctial Day) Is *koṭi* (Perpendicular to Base)
24. Radius of Day-Circle
25. Right Ascension of Signs of Zodiac
26. Earth-Sine Which Measures Increase and Decrease of Day and Night
27. Oblique Ascension of Signs of Zodiac
28. *Śaṅku* of Sun (Sine of Altitude on Vertical Circle Pas8ing through Sun) at Any Given Time
29. Base of *śaṅku* (Distance from Rising and Setting Line)
30. Amplitude of Sun *(agra)*
31. Sine of Altitude of Sun When Crossing Prime Vertical
32. Midday *śaṅku* and Shadow
33. Sine of Ecliptic Zenith-Distance *(dṛkkṣepajyā)*
34. Sine of Ecliptic Altitude *(dṛggatijyā);* Parallax
35-36. *Dṛkkarman (ākṣa* and *āyana)*

# CHAPTER I

*Daśagītika* or *the Ten Gīti Stanzas*

A. Having paid reverence to Brahman, who is one (in causality, as the creator of the universe, but) many (in his manifestations), the true deity, the Supreme Spirit, Āryabhaṭa sets forth three things: mathematics [*gaṇita*], the reckoning of time [*kālakriyā*], and the sphere [*gola*].

B. Beginning with *ka* the *varga* letters (are to be used) in the *varga* places, and the *avarga* letters (are to be used) in the *avarga* places. *Ya* is equal to the sum of *ṇa* and *ma*. The nine vowels (are to be used) in two nines of places *varga* and *avarga*. *Navāntya-varge vā.*

1. In a *yuga* the revolutions of the Sun are 4,320,000, of the Moon 57,753,336, of the Earth eastward 1,582,237,500, of Saturn 146,564, of Jupiter 364,224, of Mars 2,296,824, of Mercury and Venus the same as those of the Sun.

2. of the apsis of the Moon 488,219, of (the conjunction of) Mercury 17,937,020, of (the conjunction of) Venus 7,022,388, of (the conjunctions of) the others the same as those of the Sun, of the node of the Moon westward 232,226 starting at the beginning of Meṣa at sunrise on Wednesday at Laṅkā.

3. There are 14 Manus in a day of Brahman, and 72 *yugas* constitute the period of a Manu. Since the beginning of this *kalpa* up to the Thursday of the Bhārata battle 6 Manus, 27 *yugas,* and 3 *yugapādas* have elapsed.

4. The revolutions of the Moon (in a *yuga)* multiplied by 12 are signs *[rāśi].* The signs multiplied by 30 are degrees. The degrees multiplied by 60 are minutes. The minutes multiplied by 10 are *yojanas* (of the circumference of the sky). The Earth moves one minute in a *prāṇa.* The circumference of the sky (in *yojanas)* divided by the revolutions of a planet in a *yuga* gives the *yojanas* of the planet's orbit. The orbit of the Sun is a sixtieth part of the circle of the asterisms.

5. A *yojana* consits of 8,000 times a *nṛ* [the height of a man]. The diameter of the Earth is 1,050 *yojanas.* The diameter of the Sun is 4,410 *yojanas.* The diameter of the Moon is 315 *yojanas.* Meru is one *yojana.* The diameters of Venus, Jupiter, Mercury, Saturn, and Mars are one-fifth, one-tenth, one-fifteenth, one twentieth, and one-twenty-fifth of the diameter of the Moon. The years of a *yuga* are equal to the number of revolutions of the Sun in a *yuga.*
6. The greatest declination of the ecliptic is 24 degrees. The greatest deviation of the Moon from the ecliptic is 4½ degrees, of Saturn 2 degrees, of Jupiter 1 degree, of Mars 1½ degrees, of Mercury and Venus 2 degrees. Ninety-six *aṅgulas* or 4 *hastas* make 1 *nṛ.*
7. The ascending nodes of Mercury, Venus, Mars, Jupiter, and Saturn having moved (are situated) at 20, 60, 40, 80, and 100 degrees from the beginning of Meṣa. The apsides of the Sun and of the above-mentioned planets (in the same order) (are situated) at 78, 210, 90, 118, 180, and 236 degrees from the beginning of Meṣa.
8. Divided by 4½ the epicycles of the apsides of the Moon, the Sun, Mercury, Venus, Mars, Jupiter, and Saturn (in the first and third quadrants) are 7, 3, 7, 4, 14, 7, 9; the epicycles of the conjunctions of Saturn, Jupiter, Mars, Venus, and Mercury (in the first and third quadrants) are 9, 16, 53, 59, 31;
9. the epicycles of the apsides of the planets Mercury, Venus, Mars, Jupiter, and Saturn in the second and fourth quadrants are 5, 2, 18, 8, 13; the epicycles of the conjunctions of the planets Saturn, Jupiter, Mars, Venus, and Mercury in the second and fourth quadrants are 8, 15, 51, 57, 29. The circumference within which the Earth-wind blows is 3,375 *yojanas.*
10. The (twenty-four) sines reckoned in minutes of are 225, 224, 222, 219, 215, 210, 205, 199, 191, 183, 174, 164, 154, 143, 131, 119, 106, 93, 79, 65, 51, 37, 22, 7.
C. Whoever knows this *Daśagītika Sūtra* which describes the movements of the Earth and the planets in the sphere of the asterisms passes through the paths of the planets and asterisms and goes to the higher Brahman.

## CHAPTER II

*Gaṇitapāda* or Mathematics.

1. Having paid reverence to Brahman, the Earth, the Moon, Mercury, Venus, the Sun, Mars, Jupiter, Saturn, and the asterisms, Āryabhaṭa sets forth here the science which is honored at Kusumapura.
2. The numbers *eka* [one), *daśa* [ten), *śata* [hundred), *sahasra* [thousand), *ayuta* [ten thousand], *niyuta* [hundred thousand], *prayuta* [million), *koṭi* [ten million), *arbuda* [hundred million], and *vṛnda* [thousand million] are from place to place each ten times the preceding.
3. A square, the area of a square, and the product of two equal quantities are called *varga.* The product of three equal quantities, and a solid which has twelve edges are called *ghana.*

4. One should always divide the *avarga* by twice the (square) root of the (preceding) *varga.* After subtracting the square (of the quotient) from the *varga* the quotient will be the square root to the next place.
5. One should divide the second *aghana* by three times the square of the (cube) root of the (preceding) *ghana.* The square (of the quotient) multiplied by three times the *pūrva* (that part of the cube root already found) is to be subtracted from the first *aghana,* and the cube (of the quotient of the above division) is to be subtracted from the *ghana.*
6. The area of a triangle is the product of the perpendicular and half the base. Half the product of this area and the height is the volume of a solid which has six edges (pyramid).
7. Half of the circumference multiplied by half the diameter is the area of a circle. This area multiplied by it's own square root is the exact volume of a sphere.
8. The two sides (separately) multiplied by the perpendicular and divided by their sum will give the perpendiculars (from the point where the two diagonals intersect) to the parallel sides. The area is to be known by multiplying half the sum of the two sides by the perpendicular.
9. The area of any plane figure is found by determining two sides and then multiplying them together. The chord of the sixth part of the circumference is equal to the radius.
10. Add 4 to 100, multiply by 8, and add 62,000. The result is approximately the circumference of a circle of which the diameter is 20,000.
11. One should divide a quarter of the circumference of a circle (into as many equal parts as are desired). From the triangles and quadrilaterals (which are formed) one will have on the radius as many sines of equal arcs as are desired.
12. By what number the second sine is less than the first sine, and by the quotient obtained by dividing the sum of the preceding sines by the first sine, by the sum of these two quantities the following sines are less than the first sine.
13. The circle is made by turning, and the triangle and the quadrilateral by means of a *karṇa*; the horizontal is determined by water, and the perpendicular by the plumb-line.
14. Add the square of the height of the gnomon to the square of its shadow. The square root of this sum is the sum is the radius of the *khavṛtta.*
15. Multiply the length of the gnomon by the distance between the gnomon and the *bhujā* and divide by the difference between the length of the gnomon and the length of the *bhujā.* The quotient will be the length of the shadow measured from the base of the gnomon.
16. The distance between the ends of the two shadows multiplied by the length of the shadow and divided by the difference in length of the two shadows gives the *koṭi.* The *koṭi* multiplied by the length of the gnomon and divided by the length of the shadow gives the length of the *bhujā.*
17. The square of the *bhujā* plus the square of the *koṭi* is the square of the karṇa. In a circle the product of two *śaras* is the square of the halfchord of the two arcs.

18. (The diameters of) two circles (separately) minus the *grāsa,* multiplied by the *grāsa,* and divided separately by the sum of (the diameters of) the two circles after the *grāsa* has been subtracted from each, will give respectively the *sampātaśaras* of the two circles.
19. The desired number of terms minus one, halved, plus the number of terms which precedes, multiplied by the common difference between the terms, plus the first term, is the middle term. This multiplied by the number of terms desired is the sum of the desired number of terms. Or the sum of the first and last terms is multiplied by half the number of terms.
20. Multiply the sum of the progression by eight times the common difference, add the square of the difference between twice the first term and the common difference, take the square root of this, subtract twice the first term, divide by the common difference, add one, divide by two. The result will be the number of terms.
21. In the case of an *upaciti* which has one for the first term and one for the common difference between the terms the product of three terms having the number of terms for the first term and one as the common difference, divided by six, is the *citighana.* Or the cube of the number of terms plus one, minus the cube root of this cube, divided by six.
22. The sixth part of the product of three quantities consisting of the number of terms, the number of terms plus one, and twice the number of terms plus one is the sum of the squares. The square of the sum of the (original) series is the sum of the cubes.
23. One should subtract the sum of the squares of two factors from the square of their sum. Half the result is the product of the two factors.
24. Multiply the product (of two factors) by the square of two (4), add the square of the difference between the two factors, take the square root, add and subtract the difference between the two factors, and divide the result by two. The results will be the two factors.
25. Multiply the sum of the interest on the principal and the interest on this interest by the time and by the principal. Add to this result the square of half the principal. Take the square root of this. Subtract half the principal and divide the remainder by the time. The result will be the interest on the principal.
26. In the rule of three multiply the fruit by the desire and divide by the measure. The result will be the fruit of the desire.
27. The denominators of multipliers and divisors are multiplied together. Multiply numerators and denominators by the other denominators in order to reduce fractions to a common denominator.
28. Multipliers become divisors and divisors become multipliers, addition becomes subtraction and subtraction becomes addition in the inverse method.
29. If you know the results obtained by subtracting successively from a sum of quantities each one of these quantities set these results down separately. Add them all together and divide by the number of terms less one. The result will be the sum of all the quantities.

30. One should divide the difference between the pieces of money possessed by two men by the difference between the objects possessed by them. The quotient will be the value of one of the objects if the wealth of the two men is equal.
31. The two distances between two planets moving in opposite directions is divided by the sum of their daily motions. The two distances between two planets moving in the same direction is divided by the difference of their daily motions. The two results (in each case) will give the time of meeting of the two in the past and in the future.
32-33. Divide the divisor which gives the greater *agra* by the divisor which gives the smaller *agra*. The remainder is reciprocally divided (that is to say, the remainder becomes the divisor of the original divisor, and the remainder of this second division becomes the divisor of the second divisor, etc.). (The quotients are placed below each other in the so-called chain.) (The last remainder) is multiplied by an assumed number and added to the difference between the *agras*. Multiply the penultimate number by the number above it and add the number which is blow it. (Continue this process to the top of the chain.) Divide (the top number) by the divisor which gives the smaller *agra*. Multiply the remainder by the divisor which gives the greater *agra*. Add this product to the greater *agra*. The result is the number which will satisfy both divisors and both *agras*.
(According to Ganguly's interpretation, the translation would be:

32-33. Divide the divisor corresponding to the greater remainder by the divisor corresponding to the smaller remainder. The remainder (and the divisor) are reciprocally divided. (This process is continued until the last remainder is 0.) (The quotients are placed below each other in the so-called chain.) Multiply any assumed number by the last quotient of the reciprocal division and add it to the difference between the two remainders. (Interpreted as meaning that this product and this difference are placed in the chain beneath the quotients.) Multiply the penultimate number by the number above it and add the number which is below it. (Continue this process to the top of the chain.) Divide (the lower of the two top numbers) by the divisor corresponding to the smaller remainder. Multiply the remainder by the divisor corresponding to the greater remainder. Add the product to the greater remainder. The result is the (least number) which will satisfy the two divisors and the two remainders.

## CHAPTER III

*Kālakriyā* or the Reckoning of Time

1. A year consists of twelve months. A month consists of thirty days. A day consists of sixty *nāḍīs*. A *nāḍī consists of sixty vināḍikās.*
2. Sixty long letters or six *prāṇas* make a sidereal *vināḍikā*. This is the division of time. In like manner the division of space beginning with a revolution.
3. The difference between the number of revolutions of two planets in a *yuga* is the number of their conjunctions.Twice the sum of the revolutions of the Sun and Moon is the number of *vyatīpātas*.

4. The difference between the number of revolutions of a planet and the number of revolutions of its *ucca* is the number of revolutions of its epicycle. The number of revolutions of Jupiter multiplied by 12 are the years of Jupiter beginning with Aśvayuja.
5. The revolutions of the Sun are solar years. The conjunctions of the Sun and Moon are lunar months. The conjunctions of the Sun and the Earth are [civil] days. The revolutions of the asterisms are sidereal days.
6. Subtract the solar months in a *yuga* from the lunar months in a *yuga.* The result will be the number of intercalary months in a *yuga.* Subtract the natural [civil] days in a *yuga* from the lunar days in a *yuga.* The result will be the number of omitted lunar days in a *yuga.*
7. A solar year is a year of men. Thirty of these make a year of the Fathers. Twelve years of the Fathers make a year of the gods.
8. Twelve thousand years of the gods make a *yuga* of all the planets. A thousand and eight *yugas* of the planets make a day of Brahman.
9. The first half of a *yuga* is called *utsarpiṇī* [ascending]. The latter half is called *avasarpiṇī* [descending]. The middle part of a *yuga* is called *suṣamā.* The beginning and the end are called *duṣṣamā.* Because of the apsis of the Moon.
10. When three *yugapādas* and sixty times sixty years had elapsed (from the beginning of the *yuga)* then twenty-three years of my life had passed.
11. The *yuga,* the year, the month, and the day began all together at the beginning of the bright fortnight of Caitra. Time, which has no beginning and no end, is measured by (the movements of) the planets and the asterisms on the sphere.
12. The planets moving equally (traversing the same distance in *yojanas* each day) in their orbits complete the circle of the asterisms in sixty solar years, and the circle of the sky in a divine age [*caturyuga*].

13. The Moon, being below, completes its small orbit in a short time. Saturn, being above all the others, completes its large orbit in a long time.
14. The zodiacal signs (a twelfth of the circle) are to be known as small in a small circle and large in a large circle. Likewise the degrees and minutes are the same in number in the various orbits.
15. Beneath the asterisms are Saturn, Jupiter, Mars, the Sun, Venus, Mercury, and the Moon, and beneath these is the Earth situated in the center of space like a hitching-post.
16. These seven lords of the hours, Saturn and the others, are in order swifter than the preceding one, and counting successively the fourth in the order of their swiftness they become the Lords of the days from sunrise.
17. All the planets move by their (mean) motion on their orbits and their eccentric circles from the apsis eastward and from the conjunction westward.
18. The eccentric circle of each planet is equal to its *kakṣyāmaṇḍala* [the orbit on which the mean planet moves]. The center of the eccentric circle is outside the center of the solid Earth.

19. The distance between the center of the Earth and the center of the eccentric circle is equal to the radius of the epicycle. The planets move with their mean motions on their epicycles.
20. The planet in its swift motion from its *ucca* has a *pratiloma* motion on its epicycle. In its slow motion from its *ucca* it has an *anuloma* motion on its epicycle.
21. The epicycles move eastward from the apsis and westward from the conjunctions. The mean planet, situated on its orbit, appears at the center of its epicycle.
22-23. (The corrections) from the apsis are minus, plus, plus, minus (in the four quadrants). (The corrections) from the conjunctions are just the reverse. In the case of Saturn, Jupiter, and Mars in the first process half of the *mandaphala* obtained from the apsis is minus and plus to the mean planet. Half (the correction) from the conjunction is minus and plus to the *manda* planets. (By applying the correction) from the apsis they become *sphuṭamadhya.* (By applying the correction) from the conjunction they become *sphuṭa.*
24. Half (the correction) from the conjunction is to be applied minus and plus to the apsis. (By applying the correction) from the *manda* [apsis] thus obtained Venus and Mercury become *sphuṭamadhya.* They become *sphuṭa* (by applying the correction from the conjunction),
25. The product of its hypotenuses divided by the radius will give the distance between the planet and the Earth. The planet has the same speed on its epicycle that it has on its orbit.

## CHAPTER IV

*Gola or the Sphere*

1. From the beginning of Meṣa to the end of Kanyā is the northern half of the ecliptic. The other half from the beginning of Taulya to the end of Mīna is the southern half of the ecliptic. Both deviate equally from the Equator.
2. The Sun, the nodes of the planets, and the node of the Moon move constantly along the ecliptic. The shadow of the Earth moves along the ecliptic at a distance of 180 degrees from the Sun.
3. The Moon, from its nodes, moves northward and southward of the ecliptic. Likewise Jupiter, Mars, and Saturn. Venus and Mercury do the same from their conjunctions.
4. When the Moon has no declination it is visible when 12 degrees from the Sun. Venus when 9 degrees. The other planets in succession according to their decreasing sizes when at 9 degrees increased by two's.
5. Half of the spheres of the Earth, the planets, and the asterisms is darkened by their shadows, and half, being turned toward the Sun, is light (being small or large) according to their size.
6. The sphere of the Earth, being quite round, situated in the center of space, in the middle of the circle of asterisms, surrounded by the orbits of the planets, consists of water, earth, fire, and air.

7. Just as a ball formed by a Kadamba flower is surrounded on all sides by blossoms just so the Earth is surrounded on all sides by all creatures terrestrial and aquatic.
8. During a day of Brahman the sphere of the Earth increases a *yojana* in size all around. During a night of Brahman, which is equal in length to a day of Brahman, there is a decrease by the same amount of the Earth which has been increased by Earth.
9. As a man in a boat going forward sees a stationary object moving backward just so at Laṅkā a man sees the stationary asterisms moving backward (westward) in a straight line.
10. The cause of their rising and setting is due to the fact that the circle of the asterisms, together with the planets, driven by the provector wind, constantly moves straight westward at Laṅkā.
11. In the center of the Nandana forest is Mount Meru, a *yojana,* in measure (diameter and height), shining, surrounded by the Himavat Mountains, made of jewels, quite round.
12. Heaven and Meru are at the center of the land, hell and Vaḍavāmukha are at the center of the water. The gods and the dwellers in Hell both think constantly that the others are beneath them.
13. Sunrise at Laṅkā is sunset at Siddhapura, midday at Yavakoṭī, and midnight at Romaka.
14. Laṅkā is 90 degrees from the centers of the land and water [north and south poles]. Ujjain is straight north of Laṅkā by 22 ½ degrees.
15. From a level place half of the sphere of the asterisms decreased by the radius of the Earth is visible. The other half, plus the radius of the Earth, is cut off by the Earth.
16. The gods, who dwell in the north on Meru, see the northern half of the sphere of the asterisms moving from left to right. The *Pretas,* who dwell in the south at Vaḍavāmukha, see the southern half of the sphere of the asterisms moving from right to left.
17. The gods and the *Pretas* see the Sun after it has risen for half a solar year. The Fathers who dwell in the Moon see it for half a lunar month. Here men see it for half a natural [civil] day.
18. There is a circle east and west (the prime vertical) and another north and south (the meridian) both passing through zenith and nadir. There is a horizontal circle, the horizon, on which the heavenly bodies rise and set.
19. The circle which intersects the east and west points and two points on the meridian which are above and below the horizon by the amount of the observer's latitude is called the *unmaṇḍala*. On it the increase and decrease of day and night are measured.
20. The east and west line and the north and south line and the perpendicular from zenith to nadir intersect in the place where the observer is.

21. The vertical circle which passes through the place where the observer is and the planet is the *dṛṅmaṇḍala.* There is also the *dṛkkṣepamaṇḍala* which passes through the nonagesimal point.
22. A light wooden sphere should be made, round, and of equal weight in every part. By ingenuity one should cause it to revolve so as to keep pace with the progress of time by means of quicksilver, oil, or water.
23. On the visible half of the sphere one should depict half of the sphere of the asterisms by means of sines. The equinoctial sine is the sine of latitude. The sine of co- latitude is its *koṭi.*
24. Subtract the square of the sine of the given declination from the square of the radius. The square root of the remainder will be the radius of the day-circle north or south of the Equator.
25. Multiply the day-radius of the circle of greatest declination (24 degrees) by the sine of the desired sign of the zodiac and divide by the radius of the day-circle of the desired sign of the zodiac. The result will be the equivalent in right ascension of the desired sign beginning with Meṣa.
26. The sine of latitude multiplied by the sine of the given declination and divided by the sine of co-latitude is the earth-sine, which, being situated in the plane of one's day-circle, is the sine of the increase of day and night.
27. The first and fourth quadrants of the ecliptic rise in a quarter of a day (15 *ghaṭikās)* minus the ascensional difference. The second and third quadrants rise in a quarter of a day plus the ascensional difference, with regular increase and decrease.
28. The sine of the Sun at any given point from the horizon on its day-circle multiplied by the sine of co-latitude and divided by the radius is the *śaṅku* when any given part of the day has elapsed or remains.
29. Multiply the given sine of altitude of the Sun by the sine of latitude (the equinoctial sine) and divide by the sine of co-latitude. The result will be the base of the *śaṅku* of the Sun south of the rising and setting line.
30. The sine of the greatest declination multiplied by the given base-sine of the Sun and divided by the sine of co-latitude is the Sun's *agra* on the east and west horizons.
31. The measure of the Sun's amplitude north of the Equator [i.e., when. the Sun is in the Northern hemisphere], if less than the sine of latitude, multiplied by the sine of co-latitude and divided by the sine of latitude gives the sine of the altitude of the Sun on the prime vertical.
32. The sine of the degrees by which the Sun at midday has risen above the horizon will be the sine of altitude of the Sun at midday. The sine of the degrees by which the Sun is below the zenith at midday will be the midday shadow.
33. Multiply the meridian-sine by the orient-sine and divide by the radius. The square root of the difference between the squares of this result and of the meridian-sine will be the sine of the ecliptic zenith-distance.
34. The square root of the difference of the squares of the sines of the ecliptic zenith-distance and of the zenith-distance is the sine of the ecliptic-altitude.

*kuvaśāt kṣitije svā dṛk chāyā bhūvyāsārdhaṁ nabhomadhyāt*
35. The sine of latitude multiplied by the sine of celestial latitude and divided by the sine of co-latitude is minus and plus to the Moon when it is north of the ecliptic depending on whether it is in the Eastern or Western hemisphere, plus and minus when it is south of the ecliptic under the same circumstances.
36. Multiply the versed nine (of the Moon) by the celestial latitude and by the (greatest) declination, and divide by the square of the radius. The result is minus or plus to the Moon when it is in the northern *āyana* depending on whether its celestial latitude is north or south, and plus or minus when it is ill the southern *āyana* under the same conditions.
37. The Moon consists of water, the Sun of fire, the Earth of earth, and the Earth's shadow of darkness. The Moon obscures the Sun and the great shadow of the Earth obscures the Moon.
38. When at the end of the true lunar month the Moon, being near the node, enters the Sun, or when at the end of the halfmonth the Moon enters the shadow of the Earth that is the middle of the eclipse which occurs sometimes before and sometimes after the exact end of the lunar month or half-month.
39. Multiply the distance between the Earth and the Sun by the diameter of the Earth and divide by the difference between the diameters of the Earth and the Sun. The result will be the length of the shadow of the Earth (measured) from the diameter of the Earth.
40. The difference between the length of the Earth's shadow and the distance of the Moon from the Earth multiplied by the diameter of the Earth and divided by the length of the Earth's shadow is the diameter of the Earth's shadow (in the orbit of the Moon).
41. Subtract the square of the celestial latitude of the Moon from the square of half the sum (of the diameters of the Sun and Moon and the shadow). The square root of the remainder is known as the *sthityardha.* From this the time is calculated by means of the daily motions of the Sun and Moon.
42. Subtract the radius of the Moon from the radius of the Earth's shadow. Subtract from the square of the remainder the square of the celestial latitude. The square root of this remainder will be the *vimardardha.*
43. Subtract the radius of the Moon from the radius of the Earth's shadow. Subtract this remainder from the celestial latitude. The remainder is the part of the Moon which is not eclipsed.
44. Subtract the given time from half of the duration of the obscuration. Add this to the square of the celestial latitude. Take the square root. Subtract this from half the sum of the diameters. The remainder will be the obscuration at the given time.
45. The sine of the latitude multiplied by the sine of the hour-angle and divided by the radius is the deflection due to latitude. It is south.

*sthityardhāc cārkendos trirāśisahitāyanāt sparśe.*

46. At the beginning of an eclipse the Moon is *dhūmra,* when half obscured it is *kṛṣṇa,* when completely obscured it is *kapila*, at the middle of an eclipse it is *kṛṣṇatāmra*.
47. When the Moon eclipses the Sun even though an eighth part of the Sun is covered this is not perceptible because of the brightness of the Sun and the transparency of the Moon.
48. The Sun has been calculated from the conjunction of the Earth and the Sun, the Moon from the conjunction of the Sun and Moon, and all the other planets from the conjunctions of the planets and the Moon.
49. By the grace of God the precious sunken jewel of true knowledge has been rescued by me, by means of the boat of my own knowledge, from the ocean which consists of true and false knowledge.
50. He who disparages this universally true science of astronomy, which formerly was revealed by Svayambhū, and is now described by me in this *Āryabhaṭīya,* loses his good deeds and his long life. (The end)

*( Translation of The Aryabhaṭīya of Āryabhaṭa; An Ancient Indian Work On Mathematics and Astronomy by Walter Eugene Clark. Please see Chapter 8 for further information)



# Glossary

**Antarikṣa**—the narrow space between the upper and lower hemispheres of the Brahmānda, literally "inner space."

**Aryabhata**—famous Indian astronomer who was born in A.D. 476 and who wrote *Aryabhatiya.*

**Bha-gaṇāḥ**—the luminaries, like the sun, the moon, Venus, Mercury, Mars and Jupiter;

**Bhaskaracarya**—medieval Indian astronomer who wrote the *Jyotisa Sastra* known as *Siddhānta-śiromaṇi.*

**Bhāgavata Purāṇa**—one of the most famous and popular of the *Purāṇas,* especially among the Vaiṣṇavas, and one which contains an extensive section on cosmology. It is also known as *Śrīmad Bhāgavatam,* or simply *Bhāgavatam.*

**Bhāgavatam**—same as *Bhāgavata Purāṇa.*

**Bhārata-varṣa**—southernmost of the nine subdivisions *(varṣas)* of Jambū-dvīpa, said to refer to the earth as a whole.

**Bhū-maṇḍala**—the earthly lotus-shaped planetary system bisecting the sphere of the Brahmāṇda, the Purānic universe. Its central region consists of seven alternating ring-shaped oceans and islands *(dvīpas),* known as Sapta-dvīpa.

**Bhū-pātāla**—underneath the land;

**Brahmā, Lord**—the secondary creator of the universe as described in Vedic literature and in the Hindu trinity of Brahmā, Viṣṇu, and Śiva. Also traditionally seen as the original disseminator of the *Vedas* in human society, after his enlightenment by Viṣṇu from within.

**Brahmāṇḍa**—literally Brahmā egg, one of the innumerable universes, described in the *Purāṇas.* It consists of a spherical shell bisected by an earthly lotus-shaped planetary system, Bhū-maṇḍala, of the nearly the same radius, with a region of space above and a watery region below.

**Cakraṁ**—encirclement;

**Cakra-sthaḥ**—in the orbit;

**Candra**—the moon;

**Caranti**—traverse;

**Celestial sphere**—an imaginary sphere much larger in diameter than the earth and centered on the earth's center. The stars and planets can be thought of as points projected onto the surface of this sphere. Positions on the celestial sphere are measured by celestial longitude, called right ascension, and celestial latitude, called declination.

**Copernicus, Nicolaus (1473-1543)**—the Polish astronomer famous for introducing the heliocentric (sun-centered) theory of planetary motion.

**Dakṣiṇam**—to the right.

**Dhruvaloka**—the spiritual abode of King Dhruva which is said to situated at the north celestial pole.

**Diva-ākara**—sun;

**Dvīpa**—island.

**Ecliptic**—the geocentric orbit of the sun projected onto the celestial sphere as a great circle. This great circle is tilted at an angle of about 23.5° to the celestial equator. It intersects the celestial equator at two points, known as the vernal and autumnal equinoxes, as shown on right.

**Equinox**—one of the two points on the celestial sphere where the ecliptic intersects the equator, known as vernal and autumnal equinoxes. In one year, the sun travels around the ecliptic, traveling counterclockwise as seen looking down from the north. At the vernal equinox the sun cuts the celestial equator going from south to north, and at the autumnal equinox it cuts the celestial equator going from north to south. The length of the day equals the length of the night at the equinoxes.

**Galileo Galilei (1564-1642)**—Italian astronomer who accepted the heliocentric model and first used the telescope for astronomical observations.

**Geocentric orbit of a planet**—The theory that from the view point of a person standing on the earth, the sun moves around the earth and a planet orbits the sun. The combination of the sun's motion around the earth and the planet's motion around the sun forms the apparent orbit of the planet around the earth. This is called the planet's geocentric orbit. It lies between two curves: the inner boundary curve (A) and the outer boundary curve (B).

**Graha**—the planets; planetary system; influential planets like the moon; the nine planets (Mercury, Venus, Earth, Mars, Jupiter, Saturn, Uranus, Neptune and Pluto);

**Graha-grastaḥ**—being caught by an eclipse;

**Graha-ulka**—a falling star

**Great circle**—any circle on the surface of a sphere that is centered on the sphere's center. All other circles on a sphere are smaller than a great circle.

**Hasta**—Indian cubit.

**Indra**—the ruler of the heavenly region as described in Vedic literature.

**Jagat**—the entire universe;

**Jambūdvīpa**—central island *(dvīpa)* of Bhū-maṇḍala having a central mountain (Meru) and nine subdivisions *(varṣas)*.

**Jyotiṣām**—by luminaries;

**Jyotiṣā Śastras**—medieval Indian astronomy texts which contain instructions for calculating planetary motions and solar and lunar eclipses. *Sūrya-Siddhānta* is a text of this type.

**Kakup**—the four sides of the heavens;

**Kali-yuga**—the age of quarrel and degradation in the Purāṇic system of cyclic time. There are four cyclically repeating *yugas,* Satya (Kṛta), Tretā, Dvāpara, and Kali, with lengths proportional to 4:3:2:1. Kali-yuga is said to last 1,200 divine years, which comes to 432,000 solar years. The major *Jyotiṣā Śastras* all assume that Kali-yuga began on February 18, 3102 B.C., as measured by the Julian calendar. They assume a mean alignment of the planets on this date.

**Kṛṣṇa, Lord**—the original transcendental Supreme Lord, as described in the *Bhāgavata Purāṇa* and *Bhagavad-gītā.*

**Lokāloka Mountain**—circular mountain dividing the inner, illuminated region of Bhū-maṇḍala from the dark, uninhabited, outer region, called Aloka-varṣa, which extends from Lokaloka to the shell of the Brahmānda.

**Mahābhārata**—an epic of some 100,000 verses, said to be the history book of planet earth.

**Mānasottara Mountain**—circular mountain in the center of Puṣkaradvīpa which is traversed by the wheel of the sun's chariot.

**Maṇḍala** or **Maṇḍalāni**—orbits; chain, central, encirclement, system, or planetary system, globe, groups, entire scale, orb, arch, realm, atmosphere, society

**Meridian**—any great circle of the earth passing through the poles.

**Meru, Mount**—the central mountain of Jambūdvīpa. It is shaped like an inverted pyramid, with residences of demigods, and Brahmapurī on top. Also called Sumeru.

**Nabhasi**—in the sky;

**Nakṣatras**—the stars; 27 or 28 star constellations, situated roughly along the ecliptic, which play a role in Indian astronomy similar to that of the 12 constellations of the zodiac in the west. They are particularly connected with the motion of the moon, which completes one sidereal orbit in about 27.3 days.

The *nakṣatra* constellations are irregular in shape, but the *nakṣatras* are also defined to be angular subdivisions of the ecliptic.

**Paricaṅ kramanti**—revolve all around;

**Pariyanti**—circumambulate;

**Paryeti**—completes its orbit;

**Planisphere**—a plane map of features on a sphere, produced by stereo-graphic projection.

**Planetary influence**; ghost

**Polestar**—a star located near the celestial north pole—the center of daily rotation of the heavens. At present the polestar is called Polaris or Alpha Ursa Minoris by Western astronomy.

**Precession of the equinoxes**—the slow shift of the equinoxes relative to the fixed stars. Astronomers attribute this motion to the gradual rotation of the earth's spin axis like a spinning top. This slow rotation takes a "great year" of about 25,700 years to complete one revolution, or about 71.4 years per degree.

**Ptolemy, Claudius**—second-century Alexandrian astronomer known for his geocentric (earth-centered) epicyclic models for the motion of the planets. These were widely accepted until the heliocentric (sun-centered) theory was introduced by Copernicus in the sixteenth-century.

**Purāṇas**—Sanskrit texts from India dealing with ten topics, including the creation and annihilation of the universe, the history of humans and demigods, cosmology, and the relation between the soul and God. The word *"purāṇa"* designates ancient knowledge.

**Ṛkṣa**—luminaries like Aśvinī;

**Saṁvatsara-avasānena**—by the end of one year;

**Sapta-dvīpa**—the central region of Bhū-maṇḍala consisting of seven *(sapta)* alternating ring-shaped oceans and islands *(dvīpas).*

**Sapta-ṛṣi**—the Big Dipper constellation, literally the "seven sages."

**Siddhānta-śiromaṇi**—a *Jyotiṣā Śastra,* written by the medieval Indian astronomer Bhaskaracarya, which maps the seven oceans and *dvīpas* around the Earth globe.

**Sidereal**— Measured or determined by means of the apparent daily motion of the stars.

**Śiva, Lord**—the destroyer of the universe as described in Vedic literature and in the Hindu trinity of Brahmā, Viṣṇu, and Śiva, often pictured in meditation on Lord Viṣṇu.

**Solstice**—As the sun travels around the ecliptic in its yearly course, it reaches its northernmost point 90° after the vernal equinox. This is called the summer solstice, and it is the time of the longest day in the northern hemisphere. The sun reaches its southernmost point 90° after the autumnal equinox. This is the winter solstice, the time of the shortest day in the Northern Hemisphere.

**Śrīmad-Bhāgavatam**—same as *Bhāgavata Purāṇa.*

**Stereographic projection**—a method of mapping the surface of a globe onto a plane. One form of Stereographic projection is defined as follows: Assume that we have a sphere and a plane tangent to the North Pole of the sphere. Given a point on the surface of a sphere, extend a straight line from the South Pole of the sphere through this point, until it intersects the plane. The point on the sphere is mapped to this point.

**Sūrya-Siddhānta**—a Sanskrit astronomical text *(Jyotiṣā Śastra)* presenting planetary motion. It also discusses the Indian system of world chronology known as the *yuga* system, the periodic motions of the sun, moon, and planets, distances from the earth to the sun, the moon, and the planets, rules for calculating the times of solar and lunar eclipses, and cosmology and geography.

**Tārāṇām**—luminaries;

**Udīryamāṇāḥ**—being propelled;

**Varṣān**—continents; any one of the nine subdivisions of Jambūdvīpa, separated from one another by mountain ranges.

**Vāyunā**—by the wind;

**Vedas**—ancient Sanskrit texts of revealed knowledge.

**Viṣṇu, Lord**—the transcendental Supreme Lord, described in the *Bhāgavata Purāṇa* as an expansion of Lord Kṛṣṇa. He is also known as the maintainer in the Hindu trinity of Brahmā, Viṣṇu, and Śiva.

**Viṣṇu-padam**—planet known as Vaikuṇṭhaloka or Viṣṇuloka;
**Viṣṇu Purāṇa**—one of the eighteen major *Purāṇas.* It contains an extensive section on cosmology which complements the cosmological section of the *Bhāgavatam.*

**Vyoma**—the sky;

**Yojana**—an standard unit of length, which is used to express distances throughout the universe and is used in all Vedic writings including the Jyoti Śastras. A standard *yojana* length is about eight miles.

# Bibliography

"A Brief Account of the Solar System" in English; with the Translation into Hindustani; Arranged as Reading Lessons for the Use of Schools (Calcutta, 1836: Baptist Mission Press)

Aryabhaṭīya of Āryabhaṭa; Walter Eugene Clark, 1930.

Astronomy in India, 1800-1850; "Political" Science: Richard F. Young

Bhaktisiddhānta, Bhakti Vikāṣa Swami (Manuscript)

Bhaktivedānta VedaBase, a collection of the teachings of A.C. Bhaktivedānta Swami Prabhupāda, 1998. Bhaktivedānta Book Trust International. (for Śrīla Prabhupāda's letters, room conversations, morning walks, etc.)

*Bibliotheca Indica. Trans. of the Golādhyāya of the Siddhānta-śiromaṇi*

Brahmāṇḍa Purāṇa

Brahma-saṁhitā

Bṛhad-bhāgavatāmṛta: Gopīparāṇadhana Dāsa, 2002. Śri Bṛhad-bhāgavatāmṛta of Śrīla Sanātana Gosvāmī. Los Angeles: Bhaktivedānta Book Trust.

Caitanya-caritāmṛta: Bhaktivedānta Swami Prabhupāda, A.C., 1974. *Śri Caitanya-caritāmṛta*, Los Angeles: Bhaktivedānta Book Trust.

Caitanya-caritāmṛta by Śrīla Bhaktisiddhānta Sārasvatī Goswāmī Mahārāja.

Easy Journey to Other Planets: Bhaktivedānta Swami Prabhupāda, A.C., 1997. Revised Edition. Los Angeles: Bhaktivedānta Book Trust.

Hari-bhakti-vilāsa: Śrīla Sanātana Gosvāmī

Http://www.math.sfu.ca/histmath/India/12thCenturyAD/Siddhanta_Siroman.html

Indian Conceptions of Immortality; Walter Eugene Clark, Wales Professor of Sanskrit at Harvard and Master of Kirkland House. Cambridge: Harvard University Press, 1934. 49 p.

Introduction to Sūrya-siddhānta; Prabodh Chandra Sengupta; Burgess's Sūrya-siddhānta. 1935

Kūrma Purāṇa

Mahābhārata: Krishna Dharma, 1999. Badger: Torchlight Publishing.

Pandit as Public Intellectual, The: The Controversy over virodha or Inconsistency in the Astronomical Sciences, Christopher Z. Minkowski: Cornell University Heidelberg: South Asia Institute; New Delhi: Manohar Publications (South Asian Studies, No. ?), 2000 or 2001.]

Perfect Questions Perfect Answers: Bhaktivedānta Swami Prabhupāda, A.C., 1983. Mumbai: Bhaktivedānta Book Trust.

Quest, Arthur Ross, freelance science writer describing Yale Scientists' Project

Roget's Thesaurus

Sandarbha, Śrīla Jīva Gosvāmī

Satvata Tantra

Science of Self Realization: Bhaktivedānta Swami Prabhupāda, A.C., 2000. Los Angeles: Bhaktivedānta Book Trust.

Śrī Caitanya Bhāgavata (with Gauḍīya Bhāṣya):

Śrīmad-Bhāgavatam (also known as Bhāgavata Purāṇa): Bhaktivedānta Swami Prabhupāda, A.C., 1977. Los Angeles: Bhaktivedānta Book Trust.

"Stars as Suns," (Internet article) Florida State University, Physics Department.

Sūrya-siddhānta by A.D. by Reverend Ebeneezer Burgess (1860), Journal of the American Oriental Society and later edited by Phanindralal Gangooly

Sūrya-siddhānta, The or An Ancient System of Hindu Astronomy Followed by the Siddhānta Śiromaṇi: Pundit Bapu Deva Sastri & Lancelot Wilkinson. Bibliotheca Indica 32 Calcutta: Asiatic Society; 1861.

Teachings of Queen Kuntī: Bhaktivedānta Swami Prabhupāda, A.C., 1995. Bombay: Bhaktivedānta Book Trust.

TKG's Diary, Tamal Krishna Goswāmī

Vāyu Purāṇa

Vedic Cosmography and Astronomy, Richard L. Thompson

Viṣṇu Purāṇa

His Divine Grace

# A.C. Bhaktivedanta Swami Prabhupada

His Divine Grace A.C. Bhaktivedanta Prabhupada appeared in this world in 1896 in Calcutta, India. He first met his spiritual master, Srila Bhaktisiddhanta Sarasvati Goswami, in Calcutta in 1922. Bhaktisiddhanta Sarasvati, a prominent religious scholar and the founder of sixty-four Gaudiya Mathas (Vedic institutes), liked this educated young man and convinced him to dedicate his life to teaching Vedic knowledge. Srila Prabhupada became his student and, eleven years later (1933) at Allahabad he became his formally initiated disciple.

At their first meeting, in 1922, Srila Bhaktisiddhanta Sarasvati Thakura requested Srila Prabhupada to broadcast Vedic knowledge through the English language. In the years that followed, Srila Prabhupada wrote a commentary on the *Bhagavad-gita,* assisted the Gaudiya Matha in its work and, in 1944, started *Back to Godhead,* an English fortnightly magazine. Maintaining the publication was a struggle. Single-handedly, Srila Prabhupada edited it, typed the manuscripts, checked the galley proofs and even distributed the individual copies. Once begun, the magazine never stopped; it is now being continued by his disciples in the West and is published in over thirty languages.

Recognizing Srila Prabhupada's philosophical learning and devotion, the Gaudiya Vaisnava Society honored him in 1947 with the title "Bhaktivedanta." In 1950, at the age of fifty-four, Srila Prabhupada retired from married life, adopting the *vanaprastha* (retired) order to devote more time to his studies and writing. Srila Prabhupada traveled to the holy city of Vrndavana, where he lived in very humble circumstances in the historic medieval temple of Radha-Damodara. There he engaged, for several years, in deep study and writing. He accepted the renounced order of life *(sannyasa)* in 1959. At Radha-Damodara, Srila Prabhupada began work on his life's masterpiece: a multivolume commentated translation of the eighteen-thousand verse *Srimad-Bhagavatam (Bhagavata Purana).* He also wrote *Easy Journey to Other Planets.*

After publishing three volumes of the *Bhagavatam,* Srila Prabhupada came to the United States, in September 1965, to fulfill the mission of his spiritual master. Subsequently, His Divine Grace wrote more than sixty volumes of authoritative commentated translations and summary studies of the philosophical and religious classics of India.

When he first arrived by freighter in New York City, Srila Prabhupada was practically penniless. Only after almost a year of great difficulty did he establish the International Society for Krishna Consciousness, in July of 1966. Before his passing away on November 14, 1977, he guided the Society and saw it grow to a worldwide confederation of more than one hundred *asramas,* schools, temples, institutes and farm communities.

In 1968 Srila Prabhupada created New Vrindaban, an experimental Vedic community in the hills of West Virginia. Inspired by the success of New Vrindaban, now a thriving farm community of more than two thousand acres, his students have since founded several similar communities in the United States and abroad.

In 1972 His Divine Grace introduced the Vedic system of primary and secondary education in the West by founding the Gurukula school in Dallas, Texas. Since then, under his supervision, his disciples have established children's schools throughout the United States and the rest of the world, with the principal educational center now located in Vrndavana, India.

Srila Prabhupada also inspired the construction of several large international cultural centers in India. The center at Sridhama Mayapura in West Bengal is the site for a planned spiritual city, an ambitious project for which construction will extend over many years to come. In Vrndavana, India, are the magnificent Krsna-Balarama Temple and International Guesthouse and the Srila Prabhupada Memorial and Museum. There is also a major cultural and educational center in Bombay. Other centers are planned in a dozen important locations on the Indian subcontinent.

Srila Prabhupada's most significant contribution, however, is his books. Highly respected by the academic community for their authority, depth and clarity, they are used as standard textbooks in numerous college courses. His writings have been translated into over fifty languages. The Bhaktivedanta Book Trust, established in 1972 to publish the works of His Divine Grace, has thus become the world's largest publisher of books in the field of Indian religion and philosophy.

In just twelve years, in spite of his advanced age, Srila Prabhupada circled the globe fourteen times on lecture tours that took him to six continents. In spite of such a vigorous schedule, Srila Prabhupada continued to write prolifically. His writings constitute a veritable library of Vedic philosophy, religion, literature and culture.

# About the Author

Danavir Goswami (Dr. Dane Holtzman) was born in Los Angeles in 1949. He studied sociology and economics at the University of California, Los Angeles (UCLA). While there he excelled in athletics and was selected the Most Valuable Player of 1970 (indoor volleyball) by the National Collegiate Athletics Association (NCAA) and won the international 2-man beach volleyball tournament at Santa Cruz.

On June 11, 1970, he met His Divine Grace A.C. Bhaktivedanta Swami Prabhupada and received initiation from him in February of 1971. Later that year, he was appointed temple president of the Portland, Oregon branch of ISKCON. Then he became the national sales manager for ISKCON's Spiritual Sky Scented Products Company, until he was asked to institute the world's first Bhakta Program in October of 1974 at the Los Angeles temple. Under the direction of Srila Prabhupada, Danavir Goswami proceeded to inaugurate Bhakta Programs elsewhere in America, Europe, Africa, India, etc. Presently this comprehensive educational system for new devotees is standard throughout ISKCON.

In 1980, Danavir Goswami inaugurated Friends of Lord Krishna (FOLK), a lay congregational network in the UK, likely the Society's first. Thereafter he became director of ISKCON Scotland, Israel, Greece, Cyprus, Holland and Belgium. In 1987 he was elected to ISKCON's thirty-man world Governing Body Commission and authorized as an initiating spiritual master.

In 1989 he returned to America to co-direct KrishnaFest, a national traveling Vedic cultural festival which included music, lectures and multimedia theater. In the 1990's he commissioned the establishment of a permanent center in Osafia, Israel to teach Israeli Druze about Krishna consciousness.

He has earned the Bhakti-sastri and Bhakti-vaibhava degrees from ISKCON and his Masters and Doctoral degrees in Vaisnava Philosophy and Vaisnava Administration in 1996 from the Florida Vedic College. Danavir

Goswami has taught courses in five major universities, lectured at over 100 colleges and many diverse religious institutions, and appeared on numerous TV and radio talk shows.

He is the author of *Fortunate Souls, His Divine Grace, Change of Heart, Vedic Cosmology, RVC Prospectus* and co-author/editor of *Vedic Paradigm, Diksa Diksa, Poison Antidote, Uttama Sloka, If Attacked... I'll Chant Hare Krishna, Pocket Temple Song Book,* and six volumes of *Vaisnava Society Journal.*

He currently serves as ISKCON's Global Director of Monastic Recruitment and Education as well as inaugural president of Rupanuga Vedic College (RVC) in Kansas City, Missouri and RVC-Europe in Athens, Greece.

# OTHER PUBLICATIONS FROM RVC

**FORTUNATE SOULS – THE BHAKTA PROGRAM MANUAL:** by HH Danavir Goswami. The popular, standard guide to recruiting and training new devotees in ISKCON. Congregation devotees also use it. (760 pages, hardcover, stitch binding, 64 pages of illustrations including dozens of original drawings)($40)

"It is a masterpiece." – Indradyumna Swami

"*Fortunate Souls* provides an important spiritual road map helping devotees to assist others on the journey back to Godhead."
– Professor E. Burke Rochford; Middlebury College

**THIS IS GOD?** (Vaisnava Society Journal #1): By HH Danavir Goswami and other ISKCON writers. 70 pages, 8.5 x 11 inches, color covers, excellent graphics, includes an amazing article of Vedic evidence predicting Lord Caitanya; the size of the universe; the author of the Vedas, etc. ($4)

**DIKSA DIKSA:** By leading ISKCON devotees. Thoroughly analyzes and discredits the modern "Ritvic Theory" using scripture, logic, evidence and history. 120 pages, coilbound. ($5)

**THE LEGACY GOES ON** (Vaisnava Society Journal #2): By HH Danavir Goswami and other ISKCON writers. Includes an interesting article debunking anti-dairy beliefs; plus a special look at Visnujana Swami, an all-ladies temple, etc. 60 pages, 8.5. x 11 inches. ($4)

**HIS DIVINE GRACE:** Danavir Goswami tells of his unique experiences with Srila Prabhupada from 1970 to 1977. Includes episodes from Los Angeles, San Francisco, Berkeley, Portland, Chicago, London, Mayapur and Vrndavana. Hardbound, 240 pages includes 24 color pages of photographs. ($15)

"*His Divine Grace* is a treasure."
–Kumsi Krishna (retired engineer)

"Fascinating account of life within the movement."
–Dru Sefton (Feature writer for Kansas City Star)

**VEDIC PARADIGM:** A marvelous panoramic survey of Vedic knowledge and culture systematically discussing topics from Vedic morality to mystic powers. Designed for college-level students by Danavir Goswami using writings from Srila Prabhupada and his disciples. Hardbound, 384 pages. ($20)

"*Vedic Paradigm* should be distributed widely and given to every college student in the land."
– Wallace Dorian (freelance writer)

**POISON ANTIDOTE:** More than a dozen astute ISKCON writers exhaustively repudiate a modern conjecture that Srila Prabhupada was poisoned by his disciples. 114 pages, coilbound. ($5)

**THE KRISHNA ERA 5227** (Vaisnava Society Journal #3): Primarily written by HH Danavir Goswami. A compilation of essays discussing reinitiation, sports, kirtana, the exact date of Lord Krisna's appearance, etc. 128 pages, coilbound. ($5)

**RUPANUGA VEDIC COLLEGE PROSPECTUS:** A complete and fascinating description of ISKCON's first Vedic Seminary College. Includes course outlines, degree programs, bhakti sastri and shakti vaibhava programs, application forms, student handbook, etc. 160 pages, softbound. ($2)

**CHANGE OF HEART/ THE AGE OF KALI** (125 pages; softbound) By His Holiness Danavir Goswami and His Grace Yamunacarya dasa. Contains two plays written for and performed by KrishnaFest Theatre Company across the US. Change of Heart tells of a young man's attempts to become a devotee of Krishna. The Age of Kali depicts the evil influence of the age we live in and how it can be counteracted. ($5)

**POCKET TEMPLE SONG BOOK** (58 pages; soft) For new devotees in the temple and in the congregation. Contains all the songs sung in ISKCON temple programs. Ideally fits in the pocket. Designed and used by the Bhakta Program Institute and Rupanuga Vedic College. ($2)

**UTTAMA SLOKA** (310 pages; hardcover and coilbound; ) An extensive array of verses quoted by Srila Prabhupada. Includes the entire Vaisnava Song Book, Brahma Samhita, etc. Compiled and edited by His Holiness Danavir Goswami. ($20)

**RAH TAH TAH HARE KRISHNA** (Vaisnava Society Journal Vol. 4) (125 pages; softbound) Includes articles by ISKCON devotees on chanting the holy names, Vedic etiquette, asramas, academic studies in Vaisnavism, personal closeups, contemporary issues, and book distribution. ($5)

**ISKCON IN THE MIDDLE EAST** (Vaisnava Society Journal Vol. 4) 125 pages softbound; Includes articles by ISKCON devotees on remembering Tribhuvannatha Prabhu, the Druze leaders visit ISKCON India, Guru in the Morning, brahmacarya, infallible Srila Prabhupada, pramanam—Vedic or experimental evidence, the Karmic Trust Fund, etc. ($5)